D0812610

Food Composition and Analysis

Food Composition and Analysis

Leonard W. Aurand
Department of Food Science
North Carolina State University
Raleigh, North Carolina

A. Edwin Woods
Department of Chemistry
Middle Tennessee State University
Murfreesboro, TN

Marion R. Wells
Department of Biology
Middle Tennessee State University
Murfreesboro, TN

An avi Book
Published by Van Nostrand Reinhold Company,
New York

An AVI Book
(AVI is an imprint of Van Nostrand Reinhold Company Inc.)

Library of Congress Catalog Card Number 86-26463

ISBN 0-442-20816-2

Printed in the United States of America

Van Nostrand Reinhold Company Inc.
115 Fifth Avenue
New York, New York 10003

Van Nostrand Reinhold Company Limited
Molly Millars Lane
Wokingham, Berkshire RG11 2PY, England

Van Nostrand Reinhold
480 La Trobe Street
Melbourne, Victoria 3000, Australia

Macmillan of Canada
Division of Canada Publishing Corporation
164 Commander Boulevard
Agincourt, Ontario M1S 3C7, Canada

16 15 14 13 12 11 10 9 8 7 6 5 4 3 2 1

Library of Congress Cataloging-in-Publication Data

Aurand, Leonard W. (Leonard William)
Food composition and analysis.

Bibliography: p.
Includes index.
1. Food—Composition. 2. Food—Analysis. I. Woods, A. Edwin. II. Wells, Marion R. III. Title.
TX531.A93 1987 664′.07 86-26463
ISBN 0-442-20816-2

Contents

Preface

There is an increasing demand for food technologists who are not only familiar with the practical aspects of food processing and merchandising but who are also well grounded in chemistry as it relates to the food industry. Thus, in the training of food technologists there is a need for a textbook that combines both lecture material and laboratory experiments involving the major classes of foodstuffs and food additives. To meet this need this book was written. In addition, the book is a reference text for those engaged in research and technical work in the various segments of the food industry.

The chemistry of representative classes of foodstuffs is considered with respect to food composition, effects of processing on composition, food deterioration, food preservation, and food additives. Standards of identity for a number of the food products as prescribed by law are given. The food products selected from each class of foodstuffs for laboratory experimentation are not necessarily the most important economically or the most widely used. However, the experimental methods and techniques utilized are applicable to the other products of that class of foodstuff.

Typical food adjuncts and additives are discussed in relation to their use in food products, together with the laws regulating their usage. Laboratory experiments are given for the qualitative identification and quantitative estimation of many of these substances.

The quantitative determination of the common constituents of foods and food products is thoroughly discussed. The generally accepted methods of analysis involved are given in detail. This permits a comparison of the available methods and the selection of one for a particular sample or condition. Instrumental methods of analysis are widely used in the food industry and have to a large extent replaced the older methods involving gravimetric and volumetric analysis. The methods commonly used in food analysis are discussed with respect to the principles upon which they operate and their use in the food industry. Whenever feasible, specific directions for the use of these instruments are given.

Some equipment and chemicals used in this book are hazardous. Standard texts on laboratory safety should be reviewed. In addition, AOAC 14th Edition (1984) contains a brief discussion on laboratory safety.

Acknowledgments

The authors are particularly indebted to Perkin-Elmer Corp., The Foxboro Company (Analabs), Pierce Chemical Co., Alltech Associates, Inc. (Applied Science Labs), Hoffman-LaRoche, Inc., and AVI Publishing Company for their permission to use certain tables and methods found in their instruction manuals, books, and catalogs.

A special acknowledgment is appropriate to the Association of Official Analytical Chemists for their permission to use certain methods from their Official Methods of Analysis.

The authors are also indebted to their associates who have given them advice and assistance in the preparation of the text and to Misses Liz Coder, Sue Coder, Beth Wells, Mary Mason, and Pam Martin for their help in preparation of the manuscript.

Finally, the authors wish to express their deep appreciation to their wives, Eleanor, Saundra, and Tommie for words of encouragement and for support during the writing of the book.

1

Food Laws and Regulations

☐ INTRODUCTION

A paramount objective of the food industry is to provide a continuing supply of safe, wholesome foods for the public. To achieve this goal, food producers must be concerned about possible deterioration of foods, their contamination with microorganisms or potentially hazardous chemicals (e.g., residues of pesticides used in agriculture), and the safety of additives or other materials used in food processing.

Ultimately, a complete and accurate chemical, biological, and physical analysis of a food product is the only way to determine its quality and safety for human consumption. The Association of Official Analytical Chemists (AOAC) has been instrumental in efforts to develop convenient and reliable methods for analyzing agricultural products. To help assure the safety of our food supply, new and better analytical methods are constantly being introduced.

An important aim of this book is to aquaint the reader with the wide variety of procedures available for analyzing foods and to present specific examples from the AOAC and other sources that illustrate these varied methods. In this first chapter, the laws and regulations concerning the safety and quality of food products are discussed; in some cases these dictate the type of analytical procedure that must be used.

DEVELOPMENT OF FEDERAL FOOD LAWS

The efforts of the food industry to protect the food supply has been reinforced by federal laws and regulations. In addition, most states and municipalities have laws and regulations designed to maintain a wholesome food supply. The Federal Food, Drug, and Cosmetic (FD&C) Act of 1938 protects the consumer's health by prohibiting the use of certain substances in foods and by requiring that the presence of all ingredients be stated on the label. The FD&C Act also serves as an economic safeguard by specifying certain standards of identity and quality and labeling requirements. The Bureau of Animal Industry, through the Federal Inspection Service, monitors the wholesomeness of meat products, and the Bureau of Dairy Industry, along with the FDA, protects the milk supply.

There were no federal food laws governing the entire food supply in the United States before 1906. The adulteration of food products was common and in some instances harmful to health. Professor E. F. Ladd, in 1904, reported that more than 90% of the local meat markets in North Dakota were using chemical preservatives to slow spoilage (Ladd 1904). He found that 70% of the cocoas and chocolates examined were adulterated, and 90% of the French peas examined contained copper salts. Only one brand of pure ketchup was found on the market.

As a result of investigations by various state chemists, particularly Harvey W. Riley, Chief of the Bureau of North Dakota Agriculture Experiment Station (Ladd 1904), consumers became aware of the unscrupulous practices of many food producers. In 1906, Upton Sinclair wrote a book, *The Jungle*, that dealt in part with the unsanitary conditions of the Chicago stockyards and meat processing plants. This capped a 27-year drive for federal legislation, and on June 30, 1906, Congress passed the Food and Drugs Act, which went into effect June 4, 1907.

Food and Drugs Act of 1906

The act of 1906 was initially passed for only 1 year, but the measure was reenacted in 1907 with the inclusion of the word "hereafter." This extended the provisions of the law for an indefinite period.

The Food and Drugs Act made illegal interstate commerce in adulterated or misbranded, manufactured or natural foods, beverages, stock feeds, drugs, and medicines. It stated that substances could be added to food only if they were not likely to render the food injurious to health. Foods and drugs produced locally for local consumption remained under the jurisdiction of the individual states.

Responsibility for administering the 1906 act was given to the U.S. Department of Agriculture (USDA), but no fines or penalties were provided for. The government had the right to inspect food plants and publish the results of investigations of food products. However, because analytical methods were quite crude and limited at that time, the USDA often could not prove the presence of deleterious substances in foods. In addition, limited personnel and the lack of enforcement powers lessened the effectiveness of the act.

In 1931, the Food and Drug Administration (FDA) was created, within the USDA, to administer the food and drug law. Creation of the FDA did not significantly enhance the effectiveness of the law because of its inherent weakness—the lack of fines or other penalties for violators. Recognition of the law's limitations led to the introduction of a new bill in 1933. After many legislative hearings and revisions, Congress passed the Food, Drug, and Cosmetic Act in 1938.

Food, Drug, and Cosmetic Act of 1938

The scope of the original act of 1906 was broadened in the 1938 Act. The 1938 Act provided positive requirements on informative labeling in the interest of consumers in addition to the negative prohibitions against mislabeling contained in the 1906 Act. The 1938 Act amplified and strengthened provisions designed to safeguard the public health and prevent deception, and it also extended the law to include cosmetics and therapeutic devices that formerly escaped regulation. Since we are concerned here with foods; cosmetics, devices, and drugs are not considered further. The new act empowered the FDA to seize and enjoin illegal products and to fine and imprison violators of these laws. The Act applied to exports and imports and to commerce between the states and within the District of Columbia and the territories.

As noted above, responsibility for the administration of the food and drug laws rested in FDA within the USDA. However, because there was bickering about the authority of the FDA within the USDA, the FDA was transferred to the Federal Security Agency in 1940. This agency was reorganized into the Department of Health, Education, and Welfare in 1953. A later reorganization resulted in creation of the Department of Health and Human Services in which the FDA currently operates.

The Food, Drug, and Cosmetic Act of 1938 is a long and highly technical law. Several aspects of the law that relate to foods and are pertinent to consumers are discussed in the remainder of this section.

Adulterated Foods

According to the 1938 Act, food must not contain any added poisonous or deleterious substance that may render it injurious to health unless the added substance is required or cannot be avoided by good manufacturing practices and a safe level has been established; any food that does not meet these conditions is considered adulterated. Food is also considered adulterated if it is putrid or decomposed; if it is packed under unsanitary conditions; if it is, in whole or in part, the product of a diseased animal or of an animal that has died otherwise than by slaughter; or if its container consists of or contains anything deleterious. Adulteration is also defined to include all kinds of economic cheating; for example, the removal of a constituent that would ordinarily be expected to be present; the addition of a substance to increase the bulk or weight of a product, to make it appear better than it is or to have a greater value, or to reduce its strength or quality; the partial or complete substitution of one product for another; and the concealment of an inferior or damaged product.

Misbranded Food

A food is deemed to be misbranded if its label is false or misleading in any particular way; if it is offered for sale under the name of another food; if it is an imitation of another food, unless it is labeled as an imitation; and if its container is misleading. The law specifies that packaged food must bear a label containing the name and place of business of the manufacturer, packer, or distrubutor and an accurate statement of the quantity of the contents in terms of weight, measure, or numerical count. Any information that is required on the label must be conspicuous and in such terms as to render it likely to be read and understood by the ordinary individual under customary conditions of purchase and use. In addition, if a standard of identity has been established for a food and the label bears the name of the food, then the food in the package must conform to the standard, and all ingredients except the mandatory ingredients under the standard must be named on the label. If there is no standard set for a food, the ingredients must be named on the label.

Definitions and Standards of Identity

Food definitions and standards of identity were authorized for the first time in the 1938 Act. Food definitions and standards of identity are established by the FDA and have the full effect of law.

Such administrative rules and regulations are commonly promulgated by the executive departments and agencies of the federal government to carry out the general provisions of laws passed by Congress. Proposed rules and regulations are published in the *Federal Register*, and the public is allowed a period of time, usually 60 to 90 days, to comment on published rules or regulations. Sometimes changes suggested by the public comments are incorporated into the final rule. The Code of Federal Regulations (CFR) is a codification of the general and permanent rules published in the *Federal Register*. The FDA operates under Title 21 of this code (21 CFR). The FDA regulations concerning standards of identity, quality, and fill are described briefly in the following paragraphs. Copies of the standards may be obtained from the FDA.

Standards of identity define what a particular food product shall consist of, often setting minimum levels for valuable ingredients and in some instances maximum levels for less valuable ingredients that are also common to the product. For example, the standard of identity for fruit jelly (21 CFR 150.140) requires not less than 45 parts by weight of fruit juice ingredients to each 55 parts by weight of sugar. If less than that proportion of fruit juice is used, the product must be labeled *imitation*. The standard of identity for any food may also contain a list of optional ingredients. Standards have been established for chocolate and cocoa products, cereal flours and related products, macaroni and noodle products, bakery products, milk and cream, cheese and cheese products, fruit butters, jellies and preserves, canned fruit and canned fruit juices, food flavors, and many other foods. The standards may include enriched varieties of foods, insuring that the amount of enrichment is a substantial one.

Standards of quality have been established for a number of canned fruits and vegetables. Minimum standards and specifications are established for such factors as tenderness, color, freedom from defects, and weight of units in the container. If the food does not meet the standards, the label must include the general statement of substandard quality, "Below Standard in Quality Good Food—Not High Grade." In lieu of such a general statement of substandard quality, the label may bear the alternative statement "Below Standard in Quality______", the blank to be filled with a description of how the product fails to meet the standard.

Standards of fill specify how full a container must be to avoid deception of the consumer. Standards of fill of container have been established for seafoods, tomato products, and some canned fruits and vegetables. Substandard fill of container must be indicated on the label by the phrase "Below Standard of Fill."

Amendments to the 1938 Act

Miller Pesticide Amendment

Enacted in 1954, the Miller Pesticide Amendment authorized the FDA (now EPA) to specify the amount of pesticides that may remain on fresh fruits, vegetables, and other raw agricultural products once they enter the marketplace. Under the Federal Insecticide, Fungicide, and Rodenticide Act (FIFRA), the EPA certifies that a pesticide to be marketed for use in destroying insects, weeds, or other destructive agricultural pests is safe and effective for the purpose claimed. Under the Miller Amendment, EPA then sets the maximum amount of pesticide residue that may lawfully remain on or in a raw agricultural commodity. These residue tolerances are based on an evaluation of scientific evidence submitted by the pesticide manufacturer.

The Food Additives Amendment

The chief purposes of the Food Additives Amendment enacted in 1958 are to protect the health of consumers by requiring proof of safety before a "food additive" may be added to a food and to advance food technology by permitting the use of food additives that are safe at the levels of intended use. The law covers intentional additives, incidental additives, and sources of radiation. The Delaney clause, a rider attached to this amendment, prohibits the addition of carcinogenic additives to foods; in other words, the FDA is prohibited from setting any tolerance as a food additive for substances known to be carcinogenic. The Delaney clause has caused considerable controversy among scientists because they believe it precludes the exercise of scientific judgment and prevents the use of some substances for which safe tolerances of use could be established. The FDA has used it only twice, however, and has often interpreted it not to apply to important food items found to be carcinogenic in test animals.

The term *food additive* is legally defined as follows:

> any substance the intended use of which results or may reasonably be expected to result directly or indirectly in its becoming a component or otherwise affecting the characteristic of any food (including any substance intended for use in producing, manufacturing, packing, processing, preparing, treating, packaging, transporting, or holding food; and including any source of radiation intended for any such use) if such substance is not generally recognized, among experts qualified by scientific training and experience to evaluate its safety, as having been adequately shown through scientific procedures (or, in the case of a substance used in food prior to January 1, 1958, through either scientific procedures or experience based on common use in food) to be safe under the conditions of its intended use; except that such a term does not include: (1) a pesticide chemical in or on a raw agricultural commodity; or (2) a pesticide chemical to the extent that it is intended for use or is used in the production, storage, or transportation of any raw agricultural commodity; or (3) a color additive.

This highly complex and technical definition has been the subject of numerous interpretations, regulations, and court decisions.

Color Additives Amendment

Passed in 1960, the Color Additives Amendment was primarily designed to correct the inflexibility of the old law, which prohibited use of coal tar dyes unless the FDA listed them as "harmless and suitable for use in foods." The old law also contained no provisions for limiting the amounts of added dyes, a shortcoming that was dramatized in the mid-1950s when a number of cases of diarrhea in children occurred as result of overuse of FD&C Orange No. 1 and FD&C Red No. 32 in candy and on popcorn. These incidents and chronic toxicity animal-feeding studies led to the removal of several colors from the approved list. Recommendations also were made to establish quantity limitations for colors, but at that time the FDA took the position that it did not have the authority to set quantity limitations. A Supreme Court decision supported the Secretary's lack of authority to regulate the amounts of added colors; to correct this deficiency and generally improve the law, Congress passed the Color Additives Amendment.

The Amendment sets up uniform rules for both certified and uncertified colors. The term *color additive* is defined to include "any dye, pigment or other substance capable of coloring a food, drug or cosmetic on any part of the body." The Amendment provides for the approval (called "listing") of color additives that must be certified or are exempt from certification. The FDA is empowered to list color additives for specific uses, and to set quantity limitations (tolerances).

The Color Additives Amendment, like the Food Additives Amendment, includes a Delaney cancer clause, which states that no color additive may be approved for food use in any amount whatsoever if it is found to induce cancer when ingested by man or animal. The Delaney clause also provides that whenever a question arises as to whether a color additive may induce cancer, an advisory committee will be established to serve as a fact-finding body. The committee is to be composed of experts selected by the National Academy of Sciences, and reports its findings to the FDA which makes the final decision.

☐ OTHER FDA REGULATIONS, GUIDELINES, AND ACTION LEVELS

In addition to the regulations discussed already, the FDA has promulgated many other administrative rules, guidelines, and action levels for implementing the basic principles of the Food, Drug, and Cosmetic Act of 1938. A summary of those established as of April 11, 1981, appears in the Code of Federal Regulations, Title 21, Part 100 to 169. The FDA guidelines in several important areas are described briefly in this section.

Good Manufacturing Practice Guidelines

The Good Manufacturing Practice (GMP) Regulations (21 CFR 110) establish the criteria that apply in determining whether the facilities, methods, practices, and controls used in the manufacture, processing, packing, or holding of foods are in conformance with or are operated or administered in conformity with good manufacturing practices to assure that food for human consumption is safe and has been prepared, packed, and held under sanitary conditions. In addition to the general GMP regulations that apply to all food, specific GMP regulations have been developed for low-acid foods (21 CFR 113) and acidified foods (21 CFR 114).

Food Labeling

These regulations (21 CFR 101) define the specific information that must be on a food package to insure that the consumer is not misled. Included in these regulations are provisions for identity of food in packaged form; designation of ingredients; name and place of business manufacturer, packer, or distributor; labeling of food with number of servings; nutrition labeling of food; and food labeling warning statements.

Freedom of Information Act

Passed in 1971, the Freedom of Information Act specifies that all actions of the Federal government must be public unless they are in the interest of national security.

This legislation created much concern when the public learned that the FDA had established levels for natural and unavoidable defects in food for human use that present no health hazard (FDA 1978). For example, the defect action levels for corn meal are as follows: average of 1 whole insect (or equivalent) per 50 g or average of 25 insect fragments per 25 g; or average of 1 rodent hair per 25 g or average of 1 rodent excreta per 50 g. The FDA set the action levels under 21 CFR 110.99 levels because it is not now possible, and never has been possible, to grow in open fields, harvest, and process crops that are totally free of natural defects. The FDA emphasizes that compliance with defect levels will not prevent it from acting against a manufacturer who does not observe good manufacturing practices.

Nutritional Quality Guidelines

The nutritional quality guidelines (21 CFR 104) prescribe the minimum level or range of nutrient composition appropriate for a given class of foods.

Recall Guidelines

Specific procedures have been developed by the FDA (21 CFR 7.40) to facilitate the removal, or recall, from the market of products that are not in compliance with the Food, Drug and Cosmetic Act. A numerical designation indicates the relative degree of the health hazard associated with the product being recalled.

A Class I Recall is "a situation in which there is a reasonable probability that the use of, or exposure to, a violative product which will cause serious, adverse health consequences or death." The depth of recall extends to the consumer level, including individual consumers, hospitals, restaurants, etc. A Class I Recall is a priority procedure in which full press coverage is given because consumption of the food is an imminent health hazard. For example, this procedure would be used in the event that a case of canned food was found to be contaminated with botulin.

A Class II Recall is "a situation in which the use of, or exposure to, a violative product may cause temporary or medically reversible adverse health consequences or where the probability of serious adverse health consequences is remote." The depth of recall is to the level immediately preceding the consumer or user level, such as retail establishments. This procedure would be followed to recall canned foods in which pesticides or filth levels were above the guidelines.

A Class III Recall is "a situation in which the use of, or exposure to, a violative product is not likely to cause adverse health consequences." This type of recall extends only to the wholesale level. An example of a Class III Recall would be mold in a carbonated beverage or insect fragments in flour.

Other Laws

Meat Inspection Act of 1906

The Federal Meat Inspection Act was enacted as a result of public clamor stimulated by Upton Sinclair's description of a Chicago packinghouse in his novel *The Jungle*. This legislation, which was modernized in 1967, makes inspection mandatory for meat and meat food products entering interstate commerce. It also contained provisions for sanitary conditions in the plants to assure cleanliness in their operation. The wording "U.S. INSP'D and P'S'D" on fresh meat and "U.S. Inspected and Passed by Department of Agriculture" certifies that the meat and meat food products came from healthy animals and were prepared under sanitary conditions, and that no harmful preservative, chemical, dye or contaminant is allowed in the preparation. The 1906 Act carried a special exemption for meat and meat products prepared by farmers, but, if such meat and meat products are to enter interstate commerce, farmers are required to furnish a statement in duplicate to the shipper indicating the animal was healthy and wholesome at the time of slaughter.

Animal parts as well as whole carcasses that fail to pass inspection must be destroyed in the presence of an inspector, and a report must be filed on the destruction. If products are found to be unsuitable for human food, they are marked "Inspected and Condemned" by the inspector. The inspectors who check for diseased animals and plant sanitation are part of the Food Safety Inspection Service of the USDA.

Imported Meat Act

According to the Imported Meat Act, a section of the Tariff Act of 1930, all imported meat must be inspected under the same standards as domestic meats. Import inspectors are stationed by the USDA at ports of entry to inspect each shipment of meat (and poultry) products to assure that a certificate accompanies the shipment attesting that the product was inspected before and after slaughter, and is wholesome and free from adulteration. They also examine the product's label to assure it has been approved. Once an imported meat enters the United States, it is treated as a federally inspected product.

The Poultry Products Inspection Act

Similar to the Meat Inspection Act, the Poultry Products Inspection Act, which became law in 1957, is administered by the USDA Food Safety Inspection Service. The law requires that all slaughtered poultry moving in interstate or foreign commerce must be inspected. The law also is applicable in major consuming areas (designated by the USDA) as to whether or not the slaughtered poultry moves across state lines.

The Poultry Products Inspection Act, similar to the Federal Meat Inspection Act, provides for the following essential functions: sanitary inspection of the processing plant, inspection of birds before slaughter, inspection of the carcass and internal organs at the time of slaughter, processing inspection, stamping and labeling, and inspection of imported poultry products.

Fair Packaging and Labeling Act of 1966

This legislation authorizes the federal government to regulate, in detail, certain aspects of the labeling on all packaged products, including foods, package sizes, "cents off" promotions, and slack fill. The FDA has jurisdiction over the kind of information that can be put on the package label itself, while the Federal Trade Commission (FTC) has jurisdiction over advertising of foods in newspapers, magazines, television, and radio.

☐ OTHER REGULATORY AGENCIES

In addition to the FDA, several other federal agencies have responsibility for monitoring various activities involved in the food production system. Furthermore, as noted earlier, the states have food-related laws that are quite similar to many of the federal laws. Consequently, a close working relationship exists between state and federal agencies responsible for assuring a safe, wholesome food supply.

The National Marine Fisheries Service of the U.S. Department of Commerce is responsible for quality assurance in seafood processing.

Inspection is voluntary although the FDA can enforce regulations under the GMP guidelines.

The Occupational Safety and Health Administration (OSHA) of the U.S. Department of Labor is authorized to regulate the safety and health aspects of working conditions in the food industry (29 CFR 1910). Difficulties sometimes arise when safety standards create an unsanitary condition that food regulations might not allow. For example, all mechanical transmission shifting, gearing, and sprocket drives must be completely guarded, but these safety devices may serve as incubators for cockroaches and other pests. Similarly, difficulties in cleaning "nonslip" walkway surfaces, which are required by OSHA regulations, may allow buildup of undesirable microorganisms. Guidelines have been established by OSHA for bakery equipment (29 CFR 1910.263).

The Environmental Protection Agency (EPA), established in 1970, regulates air pollution, water pollution, pesticide residues on food, and chemical safety.

The Bureau of Alcohol, Tobacco, and Firearms (BATF) of the U.S. Department of Treasury is responsible for regulations concerning the production and sales of wine, liquor and beer.

☐ OPERATING UNDER THE FOOD ADDITIVES AMENDMENT

The primary objective of the Food Additives Amendment to the 1938 Act is to insure that any substance, intended for use in or in connection with food shall not be used until its safety has been first established and a regulation issued by the FDA defining the safe use of the substance. Thus, the first consideration a food processor must give to the proposed use of a new substance is whether or not the substance would be classed as a food additive. The following series of questions should be asked to determine whether a substance meets the legal definition of a food additive:

1. Is the substance intended for use in producing, manufacturing, packing, processing, preparing, treating, transporting, packaging, or holding a food? If the answer to any one of these criteria is "yes," the substance would fall within the definition of a food additive and additional questions must be asked.
2. Is the substance to be added intended to perform a specific function? Will the substance, which may have no specific function in the finished food product, indirectly become a part of the food product

through some phase of food production, processing, storage, or packaging? If the answer to either of these questions is "yes," the substance may qualify as a food additive, since the law covers both intentional additives and incidental additives.

3. Has the substance been found to induce cancer when ingested by man or test animals? If the answer is "yes," the substance cannot legally be classed as a food additive and cannot be used. The reason for this, as discussed earlier, is that the Delaney clause of the Food Additives Amendment prohibits the FDA from setting any tolerance for any amount of a substance that is carcinogenic. If the answer to this question is "no," additional questions need to be asked.

4. Is the substance one which was expressly permitted by the USDA or FDA for use in processing of food during 1938–1958? If the answer is "yes," the substance is "prior sanctioned" and not subject to the Food Additives Amendment; if the answer is "no," proceed to the next question.

5. Is the substance recognized, among experts qualified by scientific training and experience to evaluate its safety, as having been adequately shown through scientific procedures (or, in the case of a substance used in food prior to January 1, 1958, through either scientific procedures or experience based on common use in food) to be safe under the conditions of its intended use? If the answer is "yes," the substance is classified in the GRAS (generally recognized as safe) category and is not subject to the Food Additives Amendment. If the answer is "no," and the answers to the preceding questions are as indicated, then the substance meets the legal definition of a food additive.

In summary, a substance or mixture of substances is subject to the provisions of the Food Additives Amendment and regulation by the FDA if the answers to the preceding questions are as follows: No. 1, yes; No. 2, yes; No. 3, no; No. 4, no; No. 5, no.

Once it is established that a substance for a proposed use is a food additive as defined by law, a petition must be filed with the FDA and the intended use approved before the substance can be used. Procedural Regulations and Interpretations (21 CFR 121.51) prescribe the form in which the petition is to be submitted and its required contents. Submission of the petition by certified mail is recommended in order to establish the date of its receipt by the FDA.

The petition must be accompanied by specified pertinent information, concerning the composition of the food additive, its safety, its effect on the food, the quantity to be used, the conditions of use, manufacturing methods involved, and methods of analyzing the food for the additive. If the petition is complete and in the prescribed form, the petitioner will be notified by mail within 15 days that it is accepted. The date of the letter of notification serves as filing date of record.

Within 30 days after determining that the petition is complete and ready to be filed, the FDA is required to publish a notice describing in general terms the regulation proposed. This notice, which often appears months or years after the petition is actually submitted, will include the name of the petitioner and a general description of the petition. Within 90 days of the filing of the petition, the FDA is required by statute to (1) establish a regulation permitting the use of the additive, or (2) deny the petition. A food additive regulation may specify the food or classes of food in or on which the additive may be used, the maximum amount that may be used or be permitted to remain in or on the food, the manner of addition to or use in the food, and labeling or packaging requirements necessary to assure safety.

APPROVED FOOD ADDITIVES

Over the years, many substances have been approved as food additives by the FDA or classified generally regarded as safe. Table 1.1 contains a representative list of approved food additives, or those generally regarded as safe, and descriptions of their functions. In addition

Table 1.1. Representative Food Additives Approved for Use in the United States

Type/function	Common example	Products
Acidulants: Control pH, flavoring agents, preservatives to prevent growth of microorganisms	Vinegar, citric acid, fumaric, malic and sorbic acid	Salad dressings, desserts, natural cheese, margarine, and baking powder
Anti-caking agents: Keep many salts and powders free flowing	Aluminum calcium silicate, calcium silicate, and tricalcium phosphate	Table salt, baking powder, and coffee whiteners
Antimicrobial additives: Control growth of bacteria, mold, and yeasts	Sodium benzoate, esters of *p*-hydroxybenzoic acid, proprionic acid	Syrups, jams, jellies, fruit juices, baked goods, pickle products
Antioxidants: Prevent changes in color or flavor due to oxidation	Butylated hydroxyanisole, butylated hydroxytoluene, tocopherols, ethoxyquin, and nordihydroguaiaretic acid	Vegetable oils and shortenings, essential oils, potato chips, and dessert mixes

Table 1.1. *(continued)*

Type/function	Common example	Products
Color additives: Impart color acceptable to consumers	Annatto, caramel color, carotene, approved color additives (FD&C), and titanium dioxide	Carbonated beverages, fruit flavored gelatin, candy, cake mixes, and dairy products
Firming agents: Improve the texture of processed fruits and vegetables	Calcium lactate, mono- and di-calcium phosphate, calcium citrate and calcium chloride	Pickles, canned peas, tomatoes, potatoes, and apples
Flavoring agents (natural and synthetic): Added to a substance to enhance flavor	Allspice, parsley, garlic, peppermint, sage, citrus oils, vanillin, hydrolyzed vegetable proprotein, cinnamic aldehyde, and benzaldehyde	Meat flavors, beverage flavors, baked goods, salad dressings, soups, and hot sauces
Flavor enhancers: Stimulate perception of a flavor already in the food	Monosodium glutamate, maltol, and 5′-inosine monophosphate	Canned meats, gravies, fruits, beverages, stews, and vegetables
Foam inhibitors: Keep liquids from foaming during bottling and canning	Sodium alginate and propylene glycol esters	Beer and fruit juices
Humectants: Keep moisture in foods	Glycerol, propylene glycol, sorbitol, polyols, and invert sugar	Coconut, marshmallows, candy, dietary foods, and flavor solvents
Nonnutritive sweeteners: Replace sugar in special dietetic foods and beverages	Saccharin, Nutrasweet®, or aspartame	Low-calorie desserts mixes, dietetic jams, jellies and preserves, soft drinks and canned fruits
Nutritive additives: Improve the nutritive value of foods where such supplementation is approved	Thiamin, riboflavin, niacin, ascorbic acid, vitamin A and D, potassium iodide, and ferrous sulfate	Enriched cereals, flour and bread, milk, margarine, macaroni and noodle products
Sequestrants: Prevent fats and oils from undergoing oxidation and to keep soft drinks clear by sequestring metal ions	Sodium citrate, calcium acetate, EDTA, calcium gluconate, sodium hexametaphosphate, calcium phytate and tartaric acid	Salad oils, soft drinks, bread, and baked goods
Surfactants: Provide and maintain consistency; act as emulsifiers, wetting agents, solubilizers, and detergents	Lecithin, mono- and di-glycerides, glycerol monostearate, gelatin, sucrose esters, and gum arabic	Cake mixes, ice cream, chocolate milk, cream cheese, coffee whiteners, and bread

to intentional additives, such as those listed in Table 1.1, incidental additives, which have no specific function but find their way into foods as a result of some phase of production, processing, packaging, or storage, also are regulated by the FDA.

Every intentional food additive serves a specific function in accordance with legal requirements. Such additives include preservatives, antioxidants, sequestrants, surfactants, stabilizers, bleaching and maturing agents, buffers, acids, alkalies, colors, special sweeteners, nutrient supplements, flavoring agents (synthetic and natural), and miscellaneous materials.

For consumer appeal and industrial processing, physical state modifiers are used. Since a specific consistency is desired in many food products, hydrophilic gums may be used in thickening and gelling processes. Carbohydrates and starches are used to change the consistency of foods and as food binders. Emulsion stabilizers and emulsifiers help to maintain the desired consistency of mayonnaise, candies, and salad dressings. Table salt and other products whose flow properties are important may contain limited amounts of anticaking agents (e.g., magnesium or calcium silicate or calcium aluminum silicate).

The color, aroma, and flavor of foods are critical to their acceptance by consumers. Specific compounds are added to a variety of food products to enhance, maintain, or modify these organoleptic properties. In the future, aroma will become an important consideration in food development, since the compounds contributing to aroma can now be measured by gas chromatography.

Nutritive additives are being widely used to improve the biological value of certain foods. Vitamins D and A, thiamin, riboflavin, niacin, and ascorbic acid are approved additives, which have effectively eliminated rickets and pellagra from our population. Mineral additives include iron salts and potassium iodide. These are needed where food is grown on mineral-poor soils. Other useful minerals in animal nutrition are arsenic, chromium and cobalt, copper, manganese, selenium, and zinc. Iron supplementation prevents incipient anemia among many females. Iodized salt reduces simple goiter and many communities fluoridate water as a means to control dental caries.

☐ SUMMARY

Congress has enacted laws to protect the safety of foods and government agencies to enforce those laws. Important aspects of the regulation of foods in the United States can be summarized as follows:

1. The law requires food to be clean, wholesome, and prepared under sanitary conditions. Government inspectors must periodically visit food processing and food storage plants to see that violations do not occur.
2. The law requires food to be safe to eat.
3. The safety of food additives and color additives must be demonstrated to the FDA in advance of their use in foods. This requirement also applies to additives in animal feeds and veterinary drugs to avoid the carry-over of harmful additives to milk, meat, eggs, or other products consumed by humans.
4. A food label must contain specified information. It must carry the common or usual name of the food. If the product is made from two or more ingredients, the ingredients must be listed (except for standardized foods). The label must also state the amount of food in the package. It must give the name and address of the manufacturer, packer, or distributor. If the food is intended for special dietary use, the label must give the special information the consumer needs for proper use. The label must not have any information that is false and misleading.
5. The package must not deceive the consumer regarding the quantity of food in the package. Any picture or brief description on the label must be an accurate representation of the food in the package.
6. Many basic foods have been standardized by government regulations. All brands of a standardized food must contain the ingredients named in the food and in the amounts expected. All standardized foods, therefore, will have basically the same nutritional value, although different brands may vary as to flavor, texture, or some other factor that is permitted under optional ingredients.
7. Minimum standards of quality have been set for most canned fruits and vegetables. If the quality of the product falls below the minimum required by the standard, special labeling must indicate this fact.
8. Standards of fill of container have also been assigned for certain foods to insure reasonable uniformity in the quantity of food in a container. Such standards are particularly useful if a food is to be packed in a liquid.

One can readily see from this brief review of food laws and regulatory procedures that accurate methods of analyzing foods are vital in assuring the safety and quality of foods. Food analysis is relevant to several aspects of food regulation: (1) qualitative and quantitative characterization of the chemical components in foods; (2) the standardization of production and manufacture of food products via quality control; and (3) laboratory techniques to help in law enforcement. Although food analysis techniques in all these areas have gradually been developed and improved, further improvements are continually sought.

SELECTED READINGS

Environmental Protection Agency, Code of Federal Regulations, Title 5. 1985 ed.

FDA. 1976. Regulatory Procedure Manual, Part 5. Food and Drug Administration, Washington, DC.

The Federal Food, Drug, and Cosmetic Act of 1938, 52 Stat. 1040. 1938. 21 U.S.C. 321 *et seq.* 1976.

Food and Drugs, Code of Federal Regulations, Title 21. 1985 ed.

FURIA, T. E. (editor). 1968. Handbook of Food Additives. Chemical Rubber Co. Cleveland, OH.

HUTT, P. B. and HUTT, P.B. II. 1984. A History of Government Regulation of Adulteration and Misbranding of Food. Food, Drug, and Cosmetic Law Journal *39,* 2.

LADD, E. F. 1904. Some Adulterations and Frauds in the Food Markets. North Dakota Agric. Exp. Stn. Bull. *63.*

Occupational Safety and Health Administration. Code of Federal Regulations, Title 29. 1985 ed.

2

Sampling and Proximate Analysis

☐ INTRODUCTION

Considerable information about a food sample can be gained through a general analysis of its main components—moisture, crude fat, crude protein, ash, and crude fiber. The determination of the percentages of these components is termed a *proximate analysis*. In some cases, a proximate analysis may be all that is required and the more sophisticated instrumental methods discussed in Chapter 3 may be unnecessary. For example, a proximate analysis is usually sufficient to establish the general category of foodstuff to which a particular sample belongs and the similarity of a particular food sample to materials previously reported in the literature.

Whether a proximate analysis or more sophisticated analyses are to be performed, careful sampling is required to obtain accurate and reproducible values. The chemical and physical properties of foodstuffs exhibit a certain inherent variability among different samples; variability within a given sample, however, can be minimized by proper sampling. Analyses usually are performed on small, discrete samples rather than on the entire amount of a foodstuff (a so-called perfect sample). Various techniques (grinding, mixing, etc.) are used to insure that such small samples are representative of the entire material and provide a true measure of its overall content.

The analyst should consult standard texts for the statistics of sampling. Usually a particular food product has peculiarities that must be recognized by the sampler. These include heterogeneity of the sample *per se* and heterogeneity of the dissolving or suspending media, (e.g., inconsistent moisture content). Other factors may be relevant in a particular food.

☐ SAMPLING

Two basic types of sampling procedures are used: manual and continuous. Manual sampling is accomplished with instruments such as triers, probes, or sampling tubes, which are available in different designs and sizes. Usually a sampling tube varies from 12 to 60 in. in length. The sample can enter through the end of the tube or through slots or openings in the side of the tube; then the end or slots are closed, thus trapping the sample. There are slots or openings at intervals in the tube to allow simultaneous sampling at various depths of the product being sampled. Auger- or drill-type samplers can be used to remove a sample from solid materials such as cheese, butter, or frozen food products. To obtain liquid samples, the food product is thoroughly mixed and then the sample may be removed with a syringe-type sampler or simply by submerging a vessel under the surface (a so-called "grab" sample.) Commercially available liquid samplers may also be used.

In continuous sampling procedures, now used by many industries, samplers or sample boxes mechanically divert a fraction of the material being sampled. For solid materials, a riffle cutter often is used. A liquid sample may be obtained by "bleeding off" a fraction of the mainstream line through a smaller diversion line. The rate of the sampling can be controlled by adjusting flow rates via valves.

An excellent description of sampling devices and their use is given by Johnson (1963), who discusses the problems of sampling and factors that must be considered. Generally, the analyst should follow the AOAC method (or other published method) for sampling if one is available. Specific directions for sampling various types of foods are given in later chapters.

Types of Sampling Devices

A *riffle cutter* is a box-like device with equally spaced dividers which divide the stream of sample equally. The sample can be proportionally reduced by passing through successive riffles or recycled through the same riffle. This "cutting and quartering" is used in laboratories to reduce larger samples to convenient laboratory size.

Vezin samplers can be used for intermittent or continuous sampling of either wet or dry materials. A Vezin sampler is a truncated wedge of a circle that passes through the stream of material once each revolution. If the cutter wedge is 5% of the volume of the circle, then 5% of the sample stream is removed. The peripheral speed of the cutter is usually approximately 30 in./sec. The formula for determining the pounds of sample per cut is

$$S = \frac{0.0925FA}{N}$$

where S = pounds of sample per revolution of cutter; N = revolutions per minute; A = total angle of the cutter(s) expressed in degrees of the entire circle; and F = feed rate of sample in tons per hour.

Straight-line samplers move in a straight line and at a uniform speed across and completely through the stream of sample. A detailed discussion of the geometry of variations of this type is given by Johnson (1963).

Preparing Samples

The composite sample usually must be reduced to laboratory size. Mechanical grinding, mixing, rolling, agitation, stirring, or any logical means of making the sample more homogeneous is desired.

Grinding, which reduces particle size, helps reduce variability in the weight and size of particles. Usually, a particle diameter of 0.5–1.0 mm is recommended. Good results are obtained if the material is ground to pass through a 35-mesh sieve. None of the sample should be discarded because this could remove components that concentrate in the discarded particles, and lead to erroneous analyses. The Wiley mill, ball mill, mortar and pestle, mechanical high-speed beaters or blenders, and meat grinders are commonly used for sample disintegration.

Liquid samples can be mixed by magnetic stirrers or sonic oscillators, similar to the devices used to disintegrate cells.

☐ MOISTURE

Moisture is the measure of the water content of a material. Compounds that volatilize under the same physical conditions as water also would be included; however, these are usually negligible. Determination of moisture content is necessary for the analyst in order to calculate the nutritive value (e.g., vitamin content) of a food product and to express the results of analytical determinations on a uniform basis.

Moisture is an important factor in food quality, preservation, and resistance to deterioration. Grain that contains too much water tends to deteriorate rapidly due to mold growth, heating, insect damage, and sprouting. The browning rate of dehydrated fruits and oxygen absorption by powdered eggs also increase with an increase in moisture content.

Water exists in three major forms: (1) solvent and dispersing media; (2) adsorbed on the internal or external surfaces or as fine capillaries by capillary condensation; and (3) water of hydration. Solvent water is the solvent for soluble compounds such as sugars, amino acids, and carboxylic acids. Compounds that do not dissolve are dispersed in water; these include proteins, gums, and polysaccharides. Adsorbed water is a very thin film or a fine capillary and is commonly not removed in normal moisture determinations. Hydrate water is a chemical component of sugars such as glucose, maltose, and lactose, of salts such as potassium tartrate, and of proteins and polysaccharides which form gels with water firmly bound.

Solvent water (free water) is the most easily removed, but special precautions are necessary for all samples. Drying temperature, particle size, vacuum, crust formations on the surface, and surface area of the sample all affect the rate at which moisture is removed from foods. Vacuum ovens significantly reduce the deterioration of samples during heating. At elevated temperatures, chemical reactions such as hydrolysis can occur and cause significant errors in the moisture determination, since the water of hydrolysis is not released from the sample. Sugar solutions (e.g., honey and fruit syrups and fructose solutions) decompose at elevated temperatures; thus, the use of an air oven is not recommended for such materials. Glucose solutions are relatively stable at 98°C. Residual (bound) water of most foods (1%) is quite difficult to remove without vacuum drying. Pressures of less than 25 mm Hg are most desirable. Most foods require long drying times (up to 16 hr).

Various methods for determining moisture content are described in detail in the following sections. In some cases, the analysis of a specific food product is used to illustrate a method.

Vacuum Oven Method

Determination of moisture in wheat flour (AOAC 14.002 and 14.003) is a good example of the vacuum oven method.

Apparatus

1. Metal dish approximately 15 mm high by 55 mm in diameter with cover (according to AACC 44-15A: Sargent-Welch S-25705).
2. Desiccator with drying agent in bottom (dry CaO).
3. Vacuum oven with a pump capable of maintaining 25 mm Hg. A temperature measuring device should be in place near the sample. Only dry air should be admitted to the oven after drying. An H_2SO_4 gas-drying bottle will accomplish this.

Procedure

Dry the metal dish and tare. Weigh the flour to nearest 0.1 mg with cover loose. Dry at 100°C to constant weight and 25 mm Hg vacuum. Break vacuum with dry air and place lid on dish. Transfer immediately to desiccator and weigh after the sample and dish reach room temperature. Calculate the percentage moisture and percentage solids.

Lyophilization Method

Commercial lyophilizers (freeze dryers) are quite convenient for moisture measurements. A vacuum of approximately 5 μm of Hg and a cold trap (finger) temperature of −50°C are possible. National Bureau of Standard Reference Material 1577-Bovine Liver is prepared by removing fat, major blood vessels, and skin, then grinding. The ground liver is then transferred to polyethylene trays and lyophilized, at a pressure not greater than 30 Pa (0.2 mm Hg) with a cold trap temperature of −50°C, for 24 hr. If analyses are contemplated beyond moisture measurement, the lyophilized liver is ground in a Toronado mill then relyophilized for at least 24 hr using a cold trap at or below −50°C and at a pressure not greater than 30 Pa (0.2 mm Hg). This method is particularly good for preparing tissue for metals analysis or for nonvolatiles. A sample of at least 250 mg should be used for metals analysis.

Distillation Method

Foods that contain only small amounts of water or may contain significant amounts of substances other than water volatile at 100°C require special methods for determining moisture. One of these is the distillation method in which a solvent immiscible with water is codistilled from a weighed sample. The solvent usually is toluene, xylene, or a mixture of these with other solvents. The solvent itself or the solvent mixture has a boiling point slightly higher than that of water. Upon boiling, the solvent and water distill over, are condensed, and drop into a graduated collection tube (the Bidwell-Sterling tube is most commonly used).

The determination of moisture in blue and similar cheeses illustrates the distillation method.

Apparatus and Solvent

1. Receiver—Bidwell & Sterling, or modified Bidwell & Sterling, with 5-ml volumetric tube with $ or upper 24/40, lower 24/40 (Sargent-Welch No. S-28317-A).
2. Condenser—Cold-finger type with $ 24/40 joint.
3. Boiling flask—250 ml, round bottom, short neck; $ 24/40 joint with receivers with lower joint $ 24/40.
4. Heating mantle connected to voltage controller.
5. Distillation solvent—Toluene (boiling point = 111°C).

Standardization of Apparatus

Support the apparatus. Lubricate lower joints with silicone stopcock grease. Clean and dry the interior of the apparatus. Rinse with toluene, and fill volumetric tube of receiver with solvent. Also rinse the interior of the condenser, dry, and immediately insert condenser into apparatus. Remove all moisture from exterior of apparatus. Add glass beads to boiling flask to prevent bumping, and heat flask to redry.

Add 5 ±0.0001 g of water and 75 ml of toluene to the sample flask and connect to the receiver. Heat until refluxing starts and adjust heat to distill at a rate of 0.25–0.5 ml of water per min. Increase the rate of refluxing gradually to maximum. When no more water distills, add a few milliliters of toluene to the sample flask and bring to boil. Repeat this procedure until no more water is distilled over.

Cool the Bidwell-Sterling tube and read the volume of water collected (repeat whole procedure at least four times). From these data, a distillation recovery factor is calculated: $f = \text{g } H_2O \text{ added/g } H_2O$ recovered. The reproducibility of this factor is determined by the operator and the acceptable variability is determined by the accuracy required in the analysis.

Procedure

Set up apparatus as outlined under *Standardization*. Weigh into the 250-ml sample flask a sample that will give 2–5 ml of water on distillation. Add boiling beads. Add sufficient toluene to cover the sample completely (approximately 75 ml). Connect the flask to the side arm of the Bidwell-Sterling tube. Pour toluene through the condenser until the collecting tube is filled. Heat the flask to boiling and distill slowly (2 drops/sec) until most of the water is in the Bidwell-Sterling tube.

Then increase the distillation rate to 4 drops/sec until no more water comes over. Wash down the condenser with toluene. If water droplets are evident in the condenser, pour more toluene through the condenser. Continue the distillation to determine if additional water is present in the sample or apparatus. Remove the heating mantle and cool the Bidwell-Sterling collector to approximately 25°C. Be sure that all water droplets are in the tube. Read the volume of water collected and calculate the percentage of water in the sample:

$$\% \ H_2O = f \times \frac{\text{g } H_2O \text{ distilled from sample} \times 100}{\text{g of sample}}$$

Fischer Method

The Fischer method is particularly applicable to foods that give erratic results when heated or under a vacuum. Low-moisture foods such as dried fruits and vegetables, candies, chocolate, roasted coffee, oils, and fats are commonly analyzed for moisture by the Fischer titration method; high-moisture foods are not analyzed by this method.

The method is based on the reduction of I_2 by SO_2 in water according to the equation

$$2H_2O + SO_2 + I_2 \rightarrow H_2SO_4 + 2HI$$

Karl Fischer modified and quantitized the procedure to include I_2, SO_2, pyridine (C_5H_5N), and methanol in a four-component system. Ethylene glycol monomethyl ether has been substituted for anhydrous methanol. The overall equation for the reaction has been given as

$$C_5H_5N{\cdot}I_2 + C_5H_5N{\cdot}SO_2 + C_5H_5N + H_2O$$

$$\rightarrow 2\ C_5H_5N{\cdot}HI + C_5H_5N{\cdot}SO_3\ C_5H_5{\cdot}SO_3 + CH_3OH$$

$$\rightarrow C_5H_5N{\cdot}SO_4CH_3$$

The titration is carried out using a commercial titrimeter equipped with platinum electrodes. The water extraction is achieved with solvents such as methanol, formamide, pyridine, dioxane, and dimethylformamide.

Apparatus and Reagents

1. Any commercial titrimeter equipped with two platinum electrodes.
2. Karl Fischer reagent (Sargent-Welch SC 12960) with water equivalent of approximately 5 mg H_2O/ml (for preparation see AOAC method 32.048).

3. Water in methanol standard (Sargent-Welch SC 12966) with 1 ml = 1 ± 0.01 mg H_2O at 25°C. (This may be used to recheck the water concentration of the Karl Fischer reagent.)

Procedure

Weigh a 3.0-g sample of dried beans ground to powder into a 50- to 100-ml glass-stoppered flask. Add 20 ml N,N-dimethylformamide. Close top and seal. Place in 90°C oven for 1 hr, then mechanically shake flask for 10 min. Cool to 25°C. Place supernatant in a glass-stoppered centrifuge tube and centrifuge to remove debris.

Place 100 ml of formamide into a 250-ml flask and titrate to Karl Fischer reagent end point. Add 15 ml of the supernatant from the sample into the same 250-ml flask and titrate to the same end point. Determine a dimethylformamide blank (15 ml). Calculate percentage water in the sample as follows:

$$\%\ H_2O = \{[\text{mg } H_2O \text{ in 15-ml sample supernatant} - \text{mg } H_2O \text{ in 15-ml blank } (20/15)] \div \text{mg sample}\} \times 100$$

CRUDE FAT

Crude fat is the term used to refer to the crude mixture of fat-soluble materials present in a sample. The lipid materials may include triglycerides, diglycerides, monoglycerides, phospholipids, steroids, free fatty acids, fat-soluble vitamins, carotene pigments, chlorophyll, etc.

The two methods most commonly used to determine crude fat are wet extraction and dry extraction. Wet extraction is performed with the water remaining in the sample. The Babcock method and the Mojonnier method both are wet extraction methods used for crude fat determinations in milk and milk products. These methods have also been applied to other food products such as raw, canned, and frozen fish (AOAC method 18.045). A common dry extraction method is described in AOAC 7.055. Soxhlet extraction is performed with anhydrous ether. This technique extracts the crude fat into the ether which is finally evaporated. Details of the procedure are given in Woods and Aurand (1977). Dry extraction is preferred when it is inconvenient to remove most of the water from a food.

Babcock Method

The Babcock method is a rapid method for fat determination in various food products. The test depends upon the fact that when milk or other foods are treated with concentrated sulfuric acid, the proteins are first precipitated and then dissolved, permitting the fat to rise in a layer at the top. The fat layer is measured in a calibrated tube; from this value, the percentage of fat present may be calculated.

Reagents and Apparatus

1. Concentrated sulfuric acid with a specific gravity of 1.82 at 20°C.
2. Centrifuge capable of being electrically or otherwise heated to 55°–60°C. The proper speed of the centrifuge depends upon the size of the head. The following rpms should be used for the corresponding head diameters: 10 in., 1074; 12 in., 980; 14 in., 909; 16 in., 848; 18 in., 800; 20 in., 759; 22 in., 724; and 24 in., 693.
3. Divider or calipers for measuring the height of the fat column.
4. Graduated cylinder or pipette graduated to deliver 17.5 ml of sulfuric acid.
5. Standard Babcock test milk bottle approximately 6 in. in height, with a neck not less than 63.5 mm long and graduated from 0 to 8 in percent units with intermediate graduations between the units representing tenths of a percent.
6. Standard milk pipette graduated to contain 17.6 ml of water at 20°C, with a delivery time of 5–8 sec and a maximum error in graduation not to exceed 0.05 ml. Check the pipette by measuring from a burette the volume of 20°C water that it holds up to the graduation mark.
7. Water bath held at 55°–60°C.

Procedure

Secure with the aid of the milk pipette, a 17.6-ml sample of well-mixed milk. Transfer the sample to the standard test milk bottle, blowing out the milk in the tip after it has ceased flowing. Hold the test bottle at an angle and pour the sulfuric acid (15°–20°C) into it slowly, in small portions, while rotating the bottle to wash all milk into the bulb of the bottle. Thoroughly mix the acid and milk with a rotary motion (so that no liquid gets into the neck of the bottle) until all traces of curd have disappeared. When the acid and milk are properly mixed, the mixture becomes hot and turns uniformly dark colored.

Transfer the bottle to a centrifuge and counterbalance with another sample bottle or bottle containing water. When the centrifuge has attained the proper speed (depending upon the diameter of the head), centrifuge the sample for 5 min. Remove the bottle and add hot distilled water (60°C or higher) until the bulb of the bottle is filled. Centrifuge for 2 min at the proper speed. Again add hot water until the fat column is forced into the neck of the bottle, and the liquid approaches the top graduation on the bottle. Centrifuge 1 min at 55°–60°C.

Place the bottle into the water bath, immersing it to the level of the top of the fat column. When equilibrium is attained, as evidenced by no change in the lower fat surface, remove the bottle from the bath, dry, and measure the height of the fat column with dividers or calipers. The fat column is measured from its lower surface to the highest point of the upper meniscus and is read directly as percentage by weight of fat in the milk.

Remarks

When measured, the fat column should be translucent golden yellow in color and free of visible suspended particles. Light-colored fat with white particles beneath it indicates that the acid was too weak, the milk was too cold when the acid was added, or insufficient acid was used. Dark-colored fat containing dark specks indicates that too much acid or too strong an acid was used.

Dry Extraction of Peanuts

Determine the moisture content of the sample by an appropriate method. Weigh a 2- to 3-g sample into an extraction thimble and place into a Soxhlet extraction apparatus. Attach the apparatus to a weighed flask. Half-fill the flask with anhydrous ether. Using a heating mantle, bring the ether to boil and extract for 16 hr. Rearrange the apparatus for distillation and evaporate the ether to near dryness. Evaporate the remaining ether under a hood and with gentle heating in the mantle. (Care should be taken to remove peroxides from the ether prior to use.) Dry the flask containing the crude fat extract for 2 hr at 90°C. Weigh and calculate percentage of crude fat.

$$\%\ \text{Crude fat} = \frac{\text{Wt. of crude fat}}{\text{Wt. of dry sample}} \times 100$$

☐ CRUDE PROTEIN

For routine analysis, the determination of protein *per se* is not performed due to the difficulty of extracting protein from a sample. Generally, uncomplicated samples that do not contain unusually high concentrations of nonprotein nitrogen-containing compounds (e.g., NH_4^+, free amino acids, urea, and other more complex nitrogenous compounds) may be analyzed by simply determining the percentage of nitrogen (as NH_3) and making the assumption that this nitrogen was released from protein during digestion. The standard method for determining nitrogen, the Kjeldahl procedure, can be adapted for application to a wide variety of foodstuffs.

Kjeldahl Nitrogen Determination

In the Kjeldahl procedure, the nitrogen in an organic compound is converted to ammonium salts, from which the ammonia is liberated by adding a nonvolatile alkali. After distillation, the ammonia is determined by titration. The initial decomposition of the organic compound is usually accomplished by acid digestion, often in the presence of catalysts.

Apparatus and Reagents

1. 500-ml Kjeldahl flasks.
2. Digestion apparatus with fume exhaust and distillation apparatus. A combination unit specifically designed for Kjeldahl determinations is convenient (Sargent-Welch S-63215).
3. Catalyst mixture (96% Na_2SO_4, 3.5% $CuSO_4$, and 0.5% SeO_2).
4. 0.10 *N* Sulfuric acid.
5. 2% Boric acid solution.
6. Methyl red–bromcresol green indicator (0.016% and 0.083%, respectively, in ethanol).

Procedure

Weigh a sample that is known to contain 0.03–0.04 g N into a 500-ml Kjeldahl digestion flask. Add 8–10 g of catalyst mixture and 20 ml of concentrated H_2SO_4. Heat gently in the digestion apparatus, then vigorously until boiling begins. Continue heating at least 1 hr after the mixture has cleared.

Add approximately 400 ml of deionized distilled water to the digestion flask and a large piece of metallic zinc. The receiving flask should contain 50 ml of 2% boric acid solution and a few drops of methyl red–bromcresol green indicator. The delivery tube of the distillation apparatus should be below the surface of the boric acid solution.

Add approximately 75 ml of 50% NaOH to make the mixture basic and distill the ammonia into the boric acid solution; collect at least 300 ml of distillate. Wash the walls of the receiver and the condenser. Titrate the distillate with 0.10 *N* sulfuric acid.

Calculations

The milliequivalents (meq) of NH_3 released from the protein sample equals the meq of acid required in the titration step; thus, meq NH_3 = ml acid × normality acid. Since the meq of N in the protein sample equals the meq of NH_3, the grams of N can be obtained as follows: g N = meq N × (0.014077 g N/meq N). The percentage nitrogen in the sample is then calculated from the expression: % N = (g N/g sample) × 100. These formulas can be combined to give

$$\% \text{ N} = \frac{(\text{ml acid} \times \text{normality acid}) \times 1.4}{\text{g sample}}$$

The weight of nitrogen in a sample can be converted to protein by using the appropriate factor based on the percentage of nitrogen in the protein. Protein in most foods ranges from approximately 15% to 20% N; the average is near 16%. Thus, to convert g of N to g of protein, the common factor is 6.25 (100 ÷ 16). However, the nitrogen-to-protein conversion factor does vary among different food products, as shown in Table 2.1. It is advisable to check appropriate sources and to use the most accurate factor known for converting g of N to g of protein.

Remarks

Potassium or sodium sulfate is added to the digestion mixture to increase the boiling point; this permits a shorter digestion time. Me-

Table 2.1. Factors for Converting Nitrogen to Protein in Various Food Products

Food product	Conversion factor
Milk	6.38
Rye	6.25
Oats	6.25
Corn	6.25
Buckwheat	6.25
Rice	6.25
Barley	6.25
Wheat	5.70
Wine	6.25
Peanuts	5.46
Brazil nuts	5.46
Almonds	5.18
Other tree nuts	5.30
Coconut	5.30

Source: AOAC 1980.

tallic zinc granules are added to prevent bumping. The zinc slowly reacts with sodium hydroxide to produce hydrogen bubbles, which stir and prevent superheating of the digestion mixture.

AOAC method 2.057 uses sulfuric acid as an oxidizing-digestion agent and mercury or HgO as the metal catalyst. This alternative method, in which ammonia is distilled into a known volume of standard acid and the excess acid back-titrated with standard alkali, is described in Chapter 6. Other metals (e.g., copper and selenium) have been used as catalysts; mercury appears to be superior but suffers from its slowness to catalyze the digestion. Selenium promotes a faster digestion, but losses of nitrogen may occur.

CRUDE FIBER

Crude fiber is a measure of the quantity of indigestible cellulose, pentosans, lignins, and other components of this type present in foods. These components have little food value but provide the bulk necessary for proper peristaltic action in the intestinal tract.

Recent research into the roles of dietary fiber components have caused reevaluation of some of the traditional concepts concerning fiber. Actually, *dietary fiber* may be an unfortunate misnomer; a term such as *nondigestible* portion may be more correct since this dietary component may or may not have a fibrous structure.

The nondigestible or dietary fiber fraction is a complex mixture of different substances. The major ones are cellulose, the glucose polymer that is the predominant material of plant cells; hemicellulose, a shorter version of cellulose; pectin, the glue that binds plant cells together; and lignins, amorphous, aromatic polymers that together with cellulose form the woody cell walls of plants.

Dietary fiber acts to lower the concentration of low-density lipoprotein cholesterol in the blood, possibly by binding with bile acids. The lignin fraction has been identified as the possible binding agent.

The most common technique for measuring a food's fiber content is the crude fiber method, which dates from the 1800s and does not differentiate one polysaccharide from another. AOAC method 7.073 describes the details of this frequently used method for determining crude fiber in grains and similar products.

Crude Fiber Determination in Grains

The sample is digested with boiling dilute acid to hydrolyze the carbohydrate and protein materials contained in it. Further digestion with boiling dilute alkali causes the saponification of the fatty materials not extracted by ether. Both treatments contribute to the solution of most of the mineral matter. The residue after digestion—consisting mainly of fiber and a little mineral matter—is filtered off, dried, and weighed. It is then ignited to constant weight and again weighed. The difference in the two weights represents the weight of crude fiber present in the sample.

Reagents and Equipment

1. Sulfuric acid solution containing exactly 1.25 g of H_2SO_4/100 ml of solution.
2. Sodium hydroxide solution containing exactly 1.25 g of carbonate-free NaOH/100 ml of solution.
3. Filtering asbestos—prepare by digesting on a steam bath for 8 hr or longer with a 5% NaOH solution and then thoroughly wash with hot water. Again digest on a steam bath for 8 hr or longer with a dilute HCl solution (1 part acid + 3 parts H_2O) and thoroughly wash with hot water. Dry and ignite at a bright red heat.
4. Erlenmeyer flasks—700 to 750 ml in capacity.

Procedure

Carefully transfer the dried residue remaining in the extraction thimbles after dry extraction of the crude fat to a 750-ml Erlenmeyer flask; add about 0.5 g of the ignited asbestos along with the sample. Add 200 ml of boiling sulfuric acid solution, connect the flask with a reflux condenser, and digest the sample at boiling temperature for 30 min. (The contents of the flask should come to boiling within 1 min and boil for exactly 30 min.) Rotate the flask at 5-min intervals to keep the contents of the flask thoroughly mixed. The solid matter has a tendency to stick to the sides of the flask, out of contact with the solution, and this must be avoided as much as possible. A blast of air directed into the flask will help to reduce frothing, which may carry the sample up the walls of the flask and even into the condenser at times.

At the completion of the digestion period, filter the contents of the flask through a linen filtering cloth, using an appropriate filtering device, and wash free from acid with boiling water. Bring a quantity of the sodium hydroxide solution to boiling and keep it at this temperature under a reflux condenser until used. Transfer as much as possible of the washed residue on the filter back to the Erlenmeyer flask by means of a spatula; the remainder of the residue is washed into the flask with 200 ml of the boiling sodium hydroxide solution. A wash bottle marked to deliver 200 ml of solution is helpful in effecting this transfer.

Immediately connect the flask to a reflux condenser and boil for 30 min, rotating the flask at 5-min intervals. At the end of this period, remove the flask and filter the contents through the same linen filtering cloth used at the completion of the acid hydrolysis. Wash thoroughly with boiling water and then transfer the contents of the filter, with the aid of a spatula and wash water, to a Gooch crucible prepared with a thin asbestos mat. Wash the crucible and contents thoroughly with water and then with 15 ml of ethanol.

Dry the crucible and contents at 100°–110°C until a constant weight is attained, cooling the crucible in a desiccator before weighing. Then ignite the crucible and contents in a muffle furnace at a dull red heat (approximately 600°C) until all organic matter has been destroyed (approximately 20 min). The loss in weight during incineration represents the weight of crude fiber in the sample from which the percentage present in the sample may be calculated.

Calculations

The % crude fiber is calculated by the equation

$$\%\ \text{Crude fiber} = \frac{(\text{loss in wt of sample} - \text{loss in wt of asbestos blank}) \times 100}{\text{wt of sample}}$$

The asbestos mat blank is necessary to eliminate error due to ignitable materials in the asbestos. The % crude fiber also may be expressed on a desired moisture basis, as follows:

$$\%\ \text{Crude fiber on desired moisture basis} = \%\ \text{Crude fiber} \times \frac{(100 - \%\ \text{moisture desired})}{(100 - \%\ \text{moisture in ground sample})}$$

☐ ASH

When either organic or inorganic compounds are decomposed or released at high temperatures (500°–600°C), the remaining residue is the ash. This residue consists of oxides and salts containing anions such as phosphates, chlorides, sulfates, and other halides and cations such as sodium, potassium, calcium, magnesium, iron, and manganese.

During the ashing process organic salts decompose, losing the carbon-containing moiety. The metal from such salts forms an oxide or reacts with other anions of the matrix. Some metals (e.g., cadmium and lead) may be volatilized during ashing; therefore, if the ash is to be examined for trace elements, care should be exercised to prevent losses during ashing. For most foods, ashing at 485°C or less for 12 hr will give acceptable results for trace element analysis. Losses of minerals due to carbon (soot) release can be appreciable from carbon-containing samples. This mechanical loss of ash can be avoided by starting the incineration in the muffle furnace at a low temperature (room temperature) and allowing the temperature to rise slowly. Air currents may be a problem if the door of the furnace is opened suddenly. Generally, samples should not be placed in areas of the furnace closer than 1 in. from the rear wall and 1.5 in. from the front.

Ashing Sugar and Sugar Products

Ashing of sugar and sugar products is described in AOAC methods 31.012 and 31.013. AOAC method 31.014 gives the procedure for sulfated ash, and 31.015 describes the determination of soluble and insoluble ash.

Procedure

Heat a 5- to 10-g sample in a previously tared 100-ml platinum or Vycor crucible to 110°C until the residual water is lost. Add 5 drops of olive oil and heat for 30 min under a heat lamp. Place the dish or crucible in a cold muffle furnace and bring slowly to 525°C for 12 hr. Cool and weigh the crucible and ash residue.

☐ SELECTED REFERENCES

AACC. 1976. Approved Methods of the American Association of Cereal Chemists. Vols. I and II. American Assoc. of Cereal Chemists, St. Paul, MN.

AOCS. 1979. Official and Tentative Methods of the American Oil Chemists' Society. Vols. I and II. American Oil Chemists' Society, Champaign, IL.

AOAC. 1980. Official Methods of Analysis. 13th ed. W. Horowitz (editor). Assoc. of Official Analytical Chemists, Arlington, VA.

AOAC. 1984. Official Methods of Analysis. 14th ed. S. Williams (editor). Assoc. of Official Analytical Chemists, Arlington, VA.

JOHNSON, N. L. 1963. Sampling devices. Food Technool. *17,* 1516–1520.

POMERANZ, Y. and MELOAN, C. E. 1978. Food Analysis: Theory and Practice. Rev. ed. AVI Publishing Co., Westport, CT.

WOODS, A. E. and AURAND, L. W. 1977. Laboratory Manual in Food Chemistry. AVI Publishing Co., Westport, CT.

3

Instrumental Methods Used in Food Analysis

☐ INTRODUCTION

The development of solid state electronics, grating optics, microprocessors, and minicomputers represents but a few of the outstanding changes that have occurred in analytical instruments over the past decade. Instruments are, as a result of these developments, more stable and reliable, and afford the food chemist more convenient techniques of data acquisition.

Particularly outstanding advancements have been made in atomic absorption, visible, ultraviolet, and infrared spectrophotometry; nuclear magnetic resonance and electron spin resonance; mass and x-ray fluorescent spectrometry; and gas–liquid partition, high-pressure, and conventional chromatography. Other instrumental techniques that have undergone significant improvements include polarography, polarimetry, and weighing devices.

WEIGHING DEVICES

Analytical balances are usually classified on the basis of the range of weight (or mass) that the instrument is designed to measure. Balances of low sensitivity are commonly top loading; low-capacity balances of this type have a weighing range of 0–2000 ± 0.01 g and large-capacity ones have a range of 0–60,000 ± 1 g.

Micro and semimicro analytical balances now have many convenient features such as automatic taring and digital readout. Mettler has balances with a weighing range of 0–160 g with a digital reading device that allows the operator to read directly ±0.01 mg. Mettler also makes an ultra microbalance with a weighing range of 0–3050 mg and readability of 1.0 μg. Cahn automatic electrobalances give even more sensitivity. They are capable of variable weighing ranges from 0–1000 mg with a sensitivity to 0.0001 mg.

Balances should be chosen to meet the needs of a particular analysis. Obviously, microanalytical procedures require microbalances. Less precise techniques require balances of lesser sensitivity.

Proper care and maintenance are absolutely essential. Periodic calibration checks with proper weight standards are also mandatory. The accuracy of an analytical procedure can be no better than that of the poorest quality measurement. Unfortunately, the poorest quality measurement often is the first step, i.e., the weighing of the sample or one of the reagents.

VISIBLE AND ULTRAVIOLET SPECTROPHOTOMETRY

The most often used instrumental technique in analytical chemistry is spectrophotometry. Almost all classes of organic compounds absorb energy either in the visible (400–700 nm), ultraviolet (<350 nm), or infrared (>700 nm) regions of the electromagnetic spectrum.

Instruments used for colorimetry measurements are limited generally to the visible region of the spectrum. Spectrophotometric measurements in the ultraviolet or infrared region require more sophisticated instruments, such as the Cary Model 17D, which is capable of measurements from <200 nm up to 2500 nm. Solid state electronics and dual monochrometers contribute greatly to the resolution and stability of such an instrument. Other equally sophisticated instruments are manufactured by various companies.

The stability of a spectrophotometer can be affected by external factors including extreme temperature variations, contaminants (acid fumes, corrosive organics, etc.), and extreme variations in line voltage. Such instability has been virtually eliminated, even in single-beam spectrophotometers, with the introduction of solid state electronics.

An older spectrophotometer with Beckman DU or DU-2 optics can be upgraded by the addition of solid state electronics and improved lamps. Gilford Instrument Laboratories, Inc., furnish a conversion unit for either of these optical units that is stable, easily installed, and very reliable. These units can convert a high-grade optical system to a versatile spectrophotometer at a minimum cost. These converted units are sufficiently stable in the single-beam mode to be used for monitoring liquid chromatography or enzymatic reactions.

Spectrophotometers especially designed to monitor liquid chromatography have undergone radical improvements in recent years. These spectrophotometers use special constant pathlength and constant volume cells (usually approximately 8–10 μl). Exceptional stability can be observed for absorbance ranges of 0–0.01 A or less.

Another recently developed ultraviolet-visible spectrophotometric technique is diode array spectrophotometry. At least two companies (Hewlett-Packard and Perkin-Elmer) manufacture excellent research-grade instruments. These instruments have the capability of producing a spectrum over the full wavelength range (approximately 200–800 nm) in less than a second. They are equipped with computer systems and CRTs that allow the user to collect and process rapid kinetic data.

The Hewlett-Packard system uses a deuterium lamp which emits radiation over a range of 190 to 820 nm. The emitted light is focused onto the sample cell by an ellipsoidal mirror, then reflected onto a monozone holographic grating by a second ellipsoidal mirror. The grating disperses the light onto a linear photodiode array (see Fig. 3-1).

Diode array spectrophotometers are also used to monitor liquid chromatographs. The advantage of this application is the continuous monitoring of full wavelength range instead of one or two wavelengths monitored by conventional detectors. This offers the user the capability of detecting components of a mixture which may not be known to be present. A typical diode array kinetic experiment is illustrated in Fig. 3.2. This figure is a copy of computer-processed spectra of lactoperoxidase-catalyzed iodination of tyrosine. Triiodide (I_3^-) absorbs strongly at 292 and 325 nm and is formed from I^- and H_2O_2 and catalyzed by lactoperoxidase. As I_3^- reaches a maximum concentration, it begins to decompose. This in turn leads to the formation of I_2 (which absorbs at 458 nm) and monoiodotyrosine. (Tyrosine is one of the original substrates.) As monoiodotyrosine continues to form, I_2 goes to a maximum

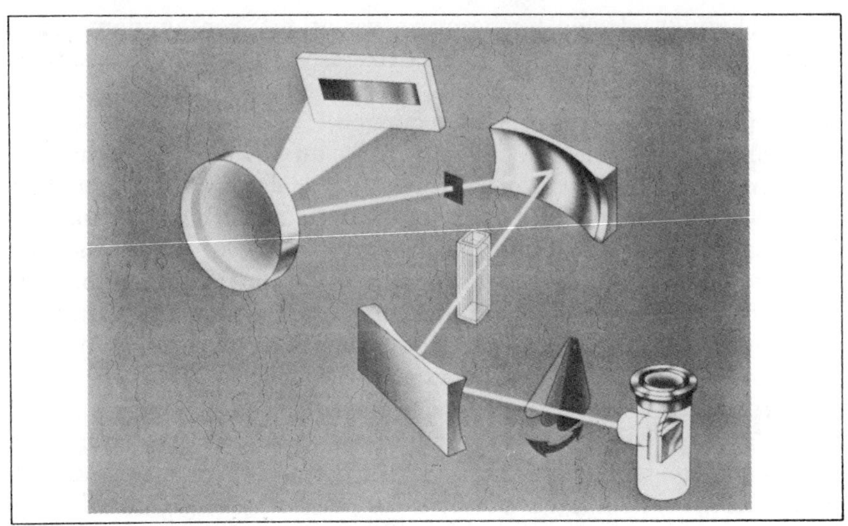

Figure 3.1. Hewlett-Packard 8451 Optical System

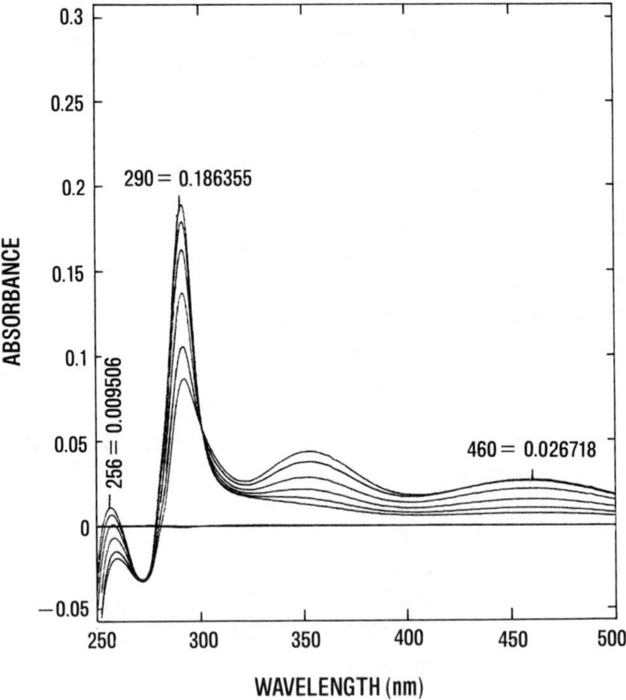

Figure 3.2. Computer-processed spectra of lactoperoxidase-catalyzed iodination of tyrosine.

concentration and then goes to equilibrium with a maximum absorbing peak at 292 nm. Ultimately I_3^- and I_2 drop below detection levels.

Infrared spectrophotometers (see next section) are primarily designed to determine the characteristic absorption spectra of organic (and a few inorganic) molecules. Although quantitative determinations can be made on these instruments, it is usually more convenient to use visible or ultraviolet measurements for determining the concentration of a compound.

Sample Cells

In visible and ultraviolet spectrophotometry, the cells (cuvettes) that hold the sample are constructed of plastic, glass, fused silica, or quartz. Plastic and glass absorb strongly in the ultraviolet region of the spectrum. Fused silica or quartz are usable down to 200 nm, but below 200 nm fused silica is superior. Before using any brand of cell, it is advisable to scan the wavelength range of interest to determine if the cell is transparent in that region.

Special pathlength cells and special volume cells are available. It is often more convenient and in some cases necessary to use a short pathlength cell; this is especially true in enzyme kinetics where the concentration of the substrate and/or coenzyme is critical to the reaction. Reduced pathlength can also be accomplished by the use of precision quartz or silica blocks that can be placed in a 1-cm cell to reduce the pathlength of the solution (Fig. 3.3D). If the volume of sample is limited, then it is often preferable to reduce the sample size without reducing the pathlength because when the pathlength is reduced, the new pathlength must be considered in the calculations. Cells are avail-

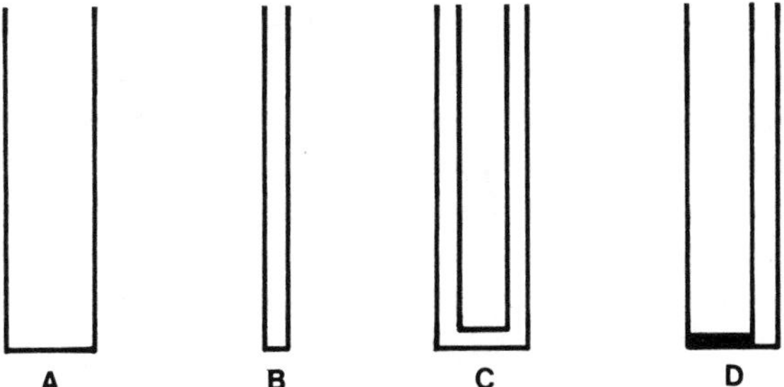

Figure 3.3. Various configurations of spectrophotometer cells: (A) side view, 1-cm pathlength; (B) side view, 0.1-cm pathlength; (C) path view, 1-ml cavity cell with 1-cm pathlength; (D) side view, 0.1-cm pathlength with 0.9-cm quartz block.

able that have a 1-cm pathlength and sample cavity of 1 ml or less (Fig. 3.3C). Often it is necessary to adjust an aperture to focus the light beam on the transparent area of a cell with a smaller sample cavity.

Recorders

Most spectrophotometers are constructed so that external or built-in strip chart recorders can be attached for recording the change in absorbance at a fixed or varying wavelength. Recorders usually have a variable range of response to the output voltage of the spectrophotometer (e.g., 0–1 mV, 0–10 mV, etc.). The spectrophotometer electronics are constructed so that the changing absorbance is converted to a millivolt output signal, which is then displayed on the recorder chart.

Some instruments have an integral recorder synchronized with the wavelength drive so that the wavelength is displayed both on the recorder and on the wavelength meter on the spectrophotometer. Add-on-type recorders frequently do not show the actual wavelength and no provisions are provided to have a wavelength marker. In this case the operator must be certain that the wavelength displayed by the recorder is calibrated to correspond to the monochrometer of the spectrophotometer. The operator also must check the display of the recorder to see that the absorbance display on the recorder corresponds to the absorbance displayed by the spectrophotometer.

Quantitative Analysis

The amount of light absorbed (or transmitted) by a solution depends upon the concentration of the solution and its thickness. This relationship is expressed by the Lambert–Beer law, which is the mathematical basis for the spectrophotometric determination of concentration. The amount of light absorbed is directly (linearly) proportional to the solution concentration but is exponentially related to the thickness through which the light passes. This exponential relationship occurs because each layer of equal thickness absorbs an equal fraction of the light that traverses it. For example, suppose a sample of solution A in a 1-cm pathlength cell is found to absorb 87.4% of the incident light (I_0), then I_1 (the proportion of light remaining after passage through 1 cm of solution) would be 12.6%. If the pathlength of the cell is increased to 2 cm, then the absorption of the incident light (I_0) will not be twice the absorption in a 1-cm cell. Since 87.4% of the original light (I_0) is absorbed when passing through 1 cm, then 87.4% of the remaining 12.6% (I_1) will be absorbed when passing through the second centimeter. Thus, I_2 would be 0.126 × 12.6% = 1.58%, i.e., 1.58% of the incident light (I_0). This is illustrated as follows:

$$I_1 = 12.6\% \text{ of } I_0$$
$$= 100\% - 87.4\%$$

$I_0 = 100\%$ ——————→ $I_2 = 1.58\%$ of I_0

←---1 cm -----→|←----- 1cm---→ $= 0.126 \times 12.6\%$

According to the Lambert-Beer law

$$\frac{dI}{I_0} = -kcdl$$

where dI/I_0 = the fraction of incident light absorbed. Thus, a change either in the pathlength (dl) or in the concentration (dc) will cause a change in the amount of light transmitted through a solution. Each compound has a proportionality constant (k), specific for the compound at a given wavelength. Since l is usually held constant in a particular measurement, the differential form of the Lambert-Beer law may be integrated and rearranged to give

$$2.303 \log \frac{I_0}{I} = klc$$

$$\log \frac{I_0}{I} = \frac{klc}{2.303}$$

$$\log \frac{I_0}{I} = a_m lc$$

where a_m is the molar absorptivity, i.e., the absorption of a 1 M solution at a given wavelength. (Note that a_m is also given the symbol ϵ.) AOAC (1980) and other sources list values of a_m.

The most convenient form of the Lambert–Beer relationship is

$$A = \log \frac{I_0}{I} = a_m lc$$

where A = absorbance or optical density (O.D.). However, most older instruments displayed the absorption of light in percent transmittance: $\%T = (I/I_0) \times 100$. This required the operator to convert $\%T$ to A or to record spectral data on logarithmic paper. Modern spectrophotometers display almost exclusively in A, usually as a digital or LED direct readout. With such instruments it is quite simple to determine sample concentration. If a_m and l are held constant, then $A = c$. The linear relationship between A and c is illustrated in Fig. 3.4.

The slope of the line in a plot of absorbance vs concentration equals $a_m l$ and is characteristic for the compound being measured and the conditions under which it is measured. The linear relationship does not hold for all concentrations, especially higher ones. Measurements should generally be made so that A values of 1.0 or less are observed, although better quality instruments have extended the linear range to $A = 2$ or greater. Before attempting to extend the linear range of a measurement, the analyst should construct a graph of A vs c to establish the extent of linearity for the specific compound and conditions.

Typical spectrophotometric measurements used in food analysis are presented in AOAC methods 26.003–26.095. The analyst should first carefully determine the molar absorptivity (ϵ, a_m) of known concentrations of $K_2Cr_2O_7$ at 350 nm. This is to insure proper technique and/or instrument calibration. Sources of error include the following:

1. Poor technique in weighing, diluting, etc.
2. Cells that are not exactly 1 cm in pathlength.
3. Dirty cells, i.e., absorbing substances on either the exterior or interior of the cells.

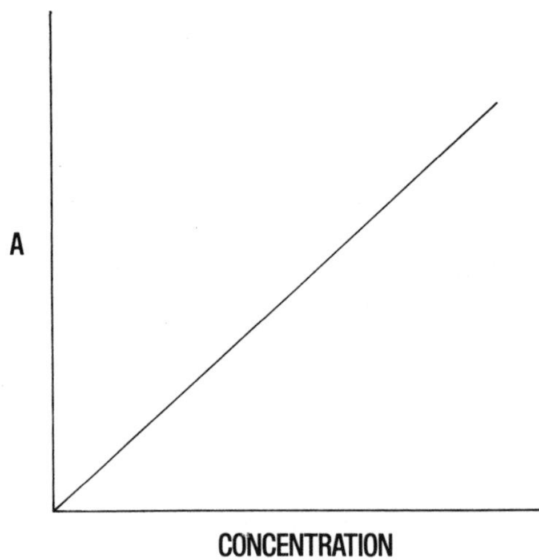

Figure. 3.4 Lambert-Beer plot of absorbance (A) versus concentration.

4. Improperly functioning microbalance or spectrophotometer.
5. Impure $K_2Cr_2O_7$.

The importance of the cell pathlength and construction is illustrated by the AOAC warning: "Use same set of cells in calibration and detn. of purity."

Determining Purity of Aflatoxins

The care needed when performing microspectrophotometric analyses is illustrated by AOAC method 26.007 in which samples of known aflatoxins are prepared and their purity determined by comparing their molar absorptivity to that of standard aflatoxins with known molar absorptivity.

Since aflatoxins are extremely potent carcinogens to animals, and their effects on humans is essentially unknown, extreme precautionary measures are recommended.[1] Manipulation under a hood and in a glove box is recommended. When weighing small samples, the experimenter should use a dry glove box, particularly when the sample is known to absorb moisture readily.

Procedure

Weigh approximately 1 mg of aflatoxin to the nearest 0.001 mg and transfer quantitatively to a 100-ml volumetric flask. Dissolve in and dilute to volume with methanol. Calculate the concentration of the solution in μg/ml (mM). Measure the A of the solution at the wavelength of maximum absorption (Table 3.1). Calculate the molar absorptivity of the sample:

$$a_m = (A \times \text{MW} \times 1000)/(\mu\text{g aflatoxin/ml})$$

where MW = molecular weight. The aflatoxins have the following molecular weights: B_1, 312; B_2, 314; G_1, 328; and G_2, 330. The AOAC uses the symbol ϵ, rather than a_m, for molar absorptivity. The factor 1000 is used to convert mM absorptivity to molar absorptivity since the original calculation of concentration, c, is in mM rather than M

[1] If a less toxic compound is desired to illustrate spectrophotometric procedures, the authors suggest using alkyl parabens (alkyl p-hydroxybenzoates). These are available from Poly Science corporation (6366 Gross Point Road, Niles, IL 60648) as a kit containing 15 different food preservatives and additives. This kit may also be used for liquid chromatography experiments. The parabens have an a_m near 10^4, thus a 10^{-4} M solution would give an A of near 1.0.

Table 3.1. Wavelengths (λ) of Maximum Absorption and Molar Absorptivities of Aflatoxins

Aflatoxin	Max. λ (nm)	Molar absorptivity (in methyl alcohol) ($\times 10^{-4}$)
B_1	223	2.21
	265	1.24
	360	2.18
B_2	222	1.86
	265	1.21
	362	2.40
G_1	216	2.74
	242	0.96
	265	0.96
	362	1.77
G_2	214	2.53
	244	1.05
	265	0.90
	362	1.93

Source: AOAC 1980.

units. The purity of each aflatoxin sample is calculated as

$$\text{Purity} = \frac{a_m \text{ of sample}}{a_m \text{ of standard}}$$

☐ INFRARED SPECTROPHOTOMETRY

Infrared spectrophotometers became commercially available in the 1950s and since then have become widely used for both qualitative and quantitative determination of food components and contaminants. AOAC (1980) lists 38 entries under infrared methods, some of which are repetitions. Nevertheless, these serve to illustrate the usefulness of the infrared region in analysis.

Molecules absorb energy in the infrared region of the electromagnetic spectrum from approximately 0.8 to 40 μm (800 to 40,000 nm). Although the region between 800 nm and 2000 nm is considered infrared, it is less useful for analytical purposes. Most ultraviolet and visible instruments limit the wavelength range to 800 nm or less. Most infrared spectrophotometers have a wavelength range of 2 to 40 μm.

Infrared spectrophotometers are presently constructed using grating optics. Originally, the early instruments had NaCl or KBr prisms. The source of radiation is a hot filament. In contrast to ultraviolet and visible spectrophotometers, most infrared analyses are carried out by scanning the wavelength region of interest, then selecting the specific peak for qualitative and quantitative interpretation. A spectrum such as that shown in Fig. 3.5 is produced.

The origin of the infrared spectrum of a molecule is the absorption of radiation at a specific wavelength by vibrating bonds within the molecule. This vibration can be compared to a spring holding two masses together. The frequency of the vibration is similar to a harmonic oscillator and is dependent on the masses of the atoms and the force constant of the bond holding the atoms together. There are only a few specific frequencies allowed, and the molecule vibrates at one or more of these frequencies. Various tables have been compiled showing the region in the infrared spectra where absorption by various functional groups of molecules occur. The scales on some instruments are constructed so that wavelength is a linear scale, while other instruments are constructed so that wave number (frequency) is linear. (The wave number is the reciprocal of the wavelength.) Wavelength is usually expressed as μ (micron) or μm (micrometer) and frequency as cm^{-1}. This is particularly confusing for the neophyte since both expressions are commonly used.

The infrared absorption frequencies of some common functional groups and structural features are listed in Table 3.2. Spectroscopists often can deduce the molecular structure of an organic compound by correlating the absorption peaks of its infrared spectrum with the known absorption frequencies of various functional groups and structural features. The absorption peaks of various types of organic compounds are described in Table 3.3.

Table 3.2. Infrared Absorption Frequencies of Common Functional Groups and Structural Features

Functional group/structural feature	Frequency (cm^{-1})
O—H, N—H and C—H stretching	3600–2700
Hydrogen bonded O—H---X stretching and ammonium ion N^+—H stretching	3300–2500
Triple bond and cumulated double bond stretching (C≡C, C≡N, N=C=O, C=C=O, N=C=N)	2400–2000
Double bond stretching (C=O, C=N, C=C)	1850–1550
Single bond bending (NH_2, CH_3, C—C—C) and single bond stretching (C—C, C—O, C—N)	1600–650

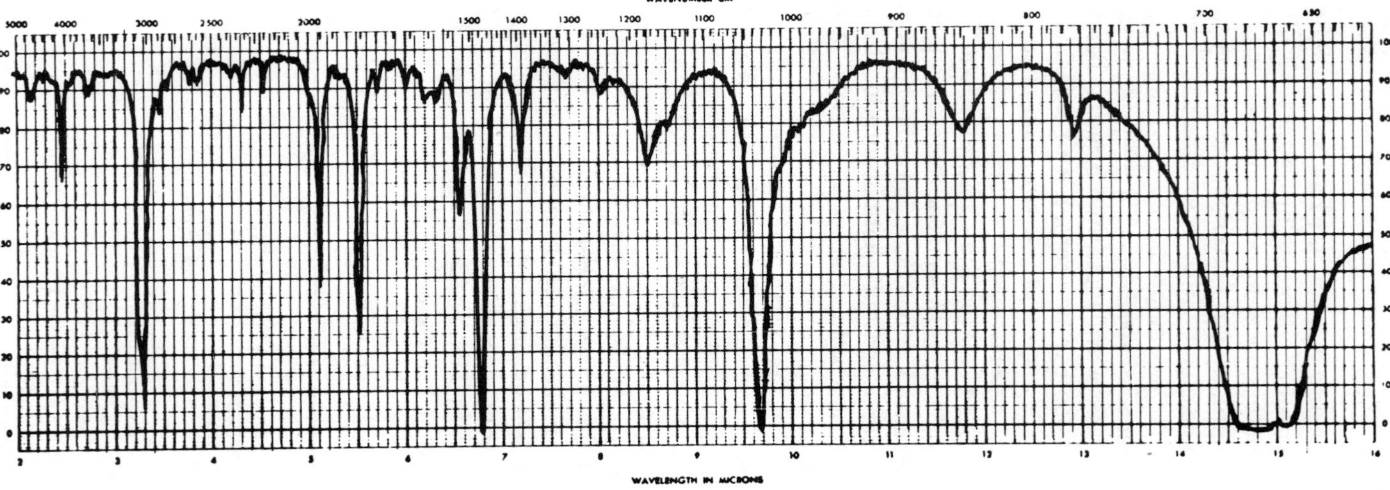

Figure 3.5. Infrared spectrum of benzene. Note that chart paper is calibrated in both wavelength (microns) and wave number (cm^{-1}). The sharp peak between 1500 and 1600 cm^{-1} is typical of benzenoid compounds.

Table 3.3. Infrared Absorption Peak Positions in Various Types of Organic Compounds

Compound type	Absorbing feature	Frequency (cm^{-1})
Alkanes	C—H stretching and bending	3000–2900 and 2900–2800
	—CH_2 and —CH_3 bending	1470–1420 and 1380–1340
	—CH_2 rocking	750
Alkenes	Olefinic C—H stretching	3100–3000
	Olefinic C—H out-of-plane bending	
	a. R—CH=CH_2	1000–900 and 940–900
	b. R_2C=CH_2	915–870
	c. *trans* R—CH=CHR	990–940
	d. *cis* R—CH=CHR	790–650
	e. R_2C=CHR	850–790
Alkynes	Acetylenic C—H	3300
	R—C≡CH	2150–2100
	R—C≡C—R	2270–2150
Aromatics	Aromatic C—H stretching	3100–3000
	Aromatic C—H out-of-plane bending	900–690
	a. monosubstitution	775–730 and 710–690
	b. 1,2-disubstitution	765–730
	c. 1,3-disubstitution	800–750
	d. 1,4-disubstitution	840–800
	e. 1,2,3-trisubstitution	800–760 and 740–700
	f. 1,2,4-trisubstitution	880–860 and 820–800
	Benzenoid	1600–1500
	Conjugated ring	1580
Alcohols, phenols, and enols[a]	O—H stretching	
	a. Solutions	3650–3600
	b. Neat liquids or solids	3500–3200
	O—H bending	1500–1300
	C—O stretching	1220–1000
Ethers	C—O asymmetric stretching	1280–1050
	Phenol and enol ethers	1275–1200
	Dialkyl ethers	1150–1050
	Epoxides	1270–1240, 950–810, and 850–750
Amines	Primary and secondary N—H stretching[b]	3500–3300

Table 3.3. (*continued*)

Compound type	Absorbing feature	Frequency (cm^{-1})
Aldehydes and ketones	C=O stretching for aldehydes	1735–1710
	Acylic ketones[c]	1720–1700
	C—H vibration (gives two peaks)	2850–2700
Carboxylic acids and lactones	C=O stretch	1740–1720
	C—O—C stretch for esters (gives two bands)	1280–1050 regions
Anhydrides	Two C=O stretching frequencies	1830–1800 and 1775–1740
Amides and lactams	C=O stretching	1670–1640
	Primary and secondary amides N—H stretching	3500–3100
	N—H bending	1640–1550

[a] Frequency of phenols>tertiary alcohols>secondary>primary.
[b] Primary amines generally have two bonds approximately 70 cm^{-1} apart due to asymmetric and symmetric stretching modes. Secondary amines have only one band. Hydrogen bonding broadens the absorptions and lowers the frequency. NH_2 groups give an additional broad band at 900–700 cm^{-1} due to out-of-plane bending.
[c] Unsaturation lowers the frequency by 25–50 cm^{-1}

Determination of Isolated Trans Isomers

In most naturally occurring vegetable fats and oils, there are found nonconjugated double bonds in the cis configuration. Isomerization of the cis bonds to the trans isomers occurs during extraction and processing. Also, oxidation and partial hydrogenation promote isomerization from cis to trans. Animal and marine fats can contain measurable amounts of naturally occurring trans isomers.

Isolated trans bonds in long-chain fatty acids, esters, and triglycerides have an infrared absorption band with a maximum absorption near 10.3 μm (970.8 cm^{-1}). This absorption, due to C–H deformation about the trans double bond, does not appear in cis bonds or saturated compounds. By comparing the absorption of this peak with that of standards containing a known percentage of trans isomer, the percentage of trans isomer in a sample may be determined. This method is usable for long-chain fatty acids under the following conditions: (1) the sample contains only isolated trans bonds and has less than 5% conjugation; (2) a correction for carbonyl vibration at 10.6 μm is made; and (3) the percentage of trans isomer is less than 15% in methyl esters.

Apparatus

1. Infrared spectrophotometer covering the spectral region from 9 to 11 μm, with a wavelength scale readable to 0.01 μm, and equipped with a cell compartment for holding 0.2- to 2.00-mm cells. Any of several commercially availabe infrared spectrophotometers are satisfactory.[2]
2. Absorption cells with a fixed thickness and NaCl or KBr windows from 0.2 to 2.0 mm. For use in null-type instruments pairs of cells matched to within 0.01 absorbance units are required. In split-beam instruments electronic balance of the two beams with both cells filled with the CS_2 solvent to within these limits, should be attained.
3. Volumetric flasks of 5-ml and 10-ml capacity, accurately calibrated.
4. Hypodermic syringes with blunted needles for filling absorption cells.
5. Chart paper that is linear in either wave number or wavelength (depending upon the instrument used) and calibrated in either transmittance or absorbance.
6. Analytical balance capable of weighing 0.2 g to an accuracy of ± 0.0002 g.

Reagents

1. Carbon disulfide (dry ACS grade).
2. Primary standards of elaidic acid, methyl elaidate, and trielaidin. As the entire quantitative accuracy of the procedure depends upon the absorptivity values obtained from the primary standards, these materials must be of highest possible purity (99 + %).
3. Secondary standards containing a known proportion of the trans isomer calibrated against a primary standard. As samples of sufficiently pure elaidic acid, methyl elaidate, and trielaidin are not readily available, secondary acid, ester, and triglyceride standards may be used.

[2] Use of a modern infrared spectrophotometer and data-collecting system, such as the Perkin-Elmer 983 with a 3600 data station, makes this procedure quite easy. A quantitative software program is available to simplify spectra collection and calculations. This software eliminates the need for manually determining the peak absorption. Other companies offer similar instruments and software. The details of each computer-controlled infrared operation are given in the manufacturer's instructions; these should be consulted and followed.

Preparation of Sample and Standard Solutions

Melt solid fats on a steam bath and mix before sampling; filter samples that appear cloudy. If the diluted sample appears cloudy due to the presence of water, add a small quantity of anhydrous sodium sulfate, mix, and allow to settle before removing the sample for analysis. Weigh approximately 0.2000 ± 0.0002 g of standard or sample into a glass-stoppered 10-ml volumetric flask. Dilute to volume with carbon disulfide and mix thoroughly. The transmittance at the trans absorption maximum should be between 20 and 70%. If it is outside this range, use a different sample weight or cell thickness (pathlength).

Procedure

Fill a clean absorption cell with carbon disulfide and fill a matching cell with the prepared sample or standard solution, using a hypodermic syringe. With the cell in an upright position, inject the sample from the bottom allowing any trapped air bubbles to pass up through the cell. Place the cells in the reference and sample beam sample holders of the spectrophotometer. Measure the transmittance or absorbance from 9 to 11 μm. The exact programming of the instrument to obtain this curve will depend upon the particular instrument selected.

Once the transmittance (or absorbance) curve is obtained for the primary standard, all samples subsequently analyzed must be measured with the same instrument with all programming controls set at identical positions. It is not practical to specify exact operating conditions for the numerous infrared spectrophotometers commercially available and suitable for these analyses. However, the analyst is cautioned that established techniques must be employed in any method for quantitative analysis by means of infrared absorption spectra. Thus, the slit widths must not be too wide to permit sufficient resolution, the scanning speeds must not be so fast as to prevent adequate pen response, etc.

The type of standard used (acid, ester, or triglyceride) should correspond to the type of sample being analyzed. Although a standard curve, once obtained, need not be repeated as long as the same instrument is used with the same programming controls, it is recommended that the standard be rechecked from time to time, especially if any adjustment in the instrument has been made (e.g., replacement of glower or detector). If, for any reason, the exact programming cannot be duplicated when measuring a specific sample, the calibration curve for the standard must be redetermined, and the new one used for analyses.

Calculation

From the recorded spectra, the percent transmittance at the 10.36 μm maximum is determined for the primary elaidic acid, methyl elaidate, or trielaidin standard (or the appropriate secondary acid, ester, or triglyceride standard) and for each sample and converted to absorbance, as follows. On the charts draw a base line from 10.10 μm to 10.65 μm for acids, from 10.02 μm to 10.59 μm for methyl esters, or from 10.05 μm to 10.67 μm for triglycerides, as shown in Figs. 3.6 and 3.7.

Measure the distance from the zero line of the recorder chart to the absorption peak (distance *ab*). Calculate the fractional *T* (distance *bc*) as the distance to the absorption peak (distance *ab*) divided by the distance to the base line (distance *ac*), i.e., fractional $T = ab/ac$. convert this value to *A* by the equation

$$A = 2 - \log T.$$

For instruments whose charts are calibrated in *A*, subtract *A* at the base line from *A* at the peak to obtain *A* of the sample which has been

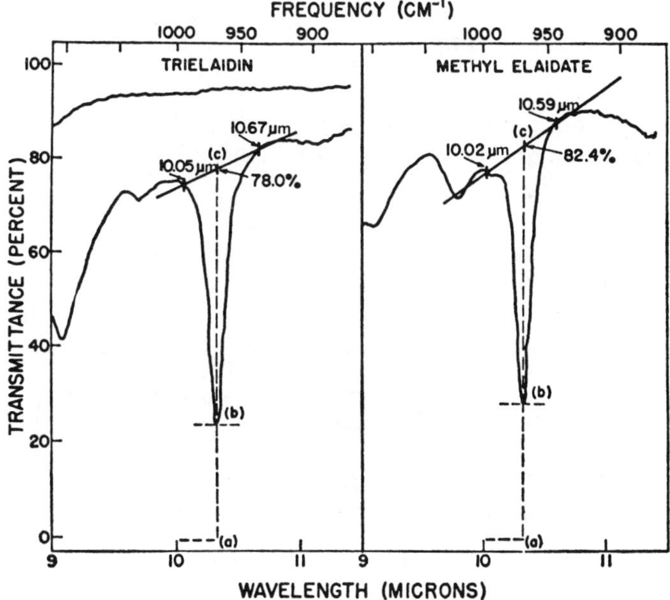

Figure 3.6. Infrared absorption peak for isolated trans double bond in trielaidin and methyl elaidate.

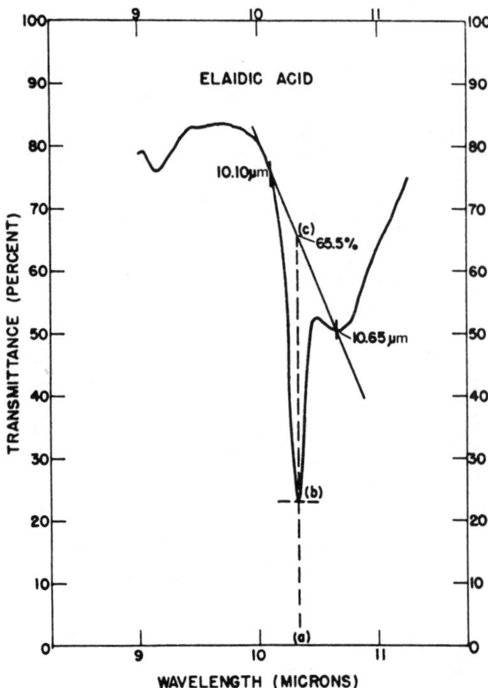

Figure 3.7. Infrared absorption peak for isolated trans double bond in ealaidic acid.

corrected for background. Using the value of A, calculate the background-corrected *absorptivity* (a_1) as follows:

$$\text{Absorptivity } (a_1) = A/lc$$

where A = corrected absorbance, l = cell path length (cm), and c = concentration of the solution (g/liter).

From the corrected absorptivity (a_1), the % trans isomer (as elaidic acid, methyl elaidate, or trielaidin) may be calculated by the equation:

$$\%\text{ trans isomer} = \frac{\text{Absorptivity } (a_1) \text{ of sample}}{\text{Absorptivity } (a_1) \text{ of standard}} \times 100$$

Corrections must be made for triglycerides containing long-chain fatty acids and also long-chain methyl esters derived from triglycerides:

$$\text{Triglyceride, \% trans (corrected)} = (\%\text{ trans} - 2.5)/0.975$$

$$\text{Methyl esters, \% trans (corrected)} = (\%\text{ trans} + 1.5)/1.015$$

For triglycerides containing large proportions of lower and medium-chain fatty acids and also methyl esters derived from these triglycerides the correction is

$$\text{Triglycerides, \% trans (corrected)} = (\text{\% trans} - 3.0)/0.970$$

$$\text{Methyl esters, \% trans (corrected)} = (\text{\% trans} + 3.0)/1.030$$

Absence of a peak at 10.3 μm, regardless of the baseline A, indicates no trans isomers in the sample.

☐ FLUOROMETRY

When some compounds absorb light, they immediately reemit energy as light of a longer wavelength. This phenomenon, called fluorescence, is the basis of fluorometry, which can be more than 1000 times more sensitive than ultraviolet spectrophotometry.

In a fluorometer, the sample is excited by an incident source at the excitation wavelength (I_0). The sample absorbs this energy and emits light at the emission wavelength; this emission is 90° to the transmitted light, I (Fig. 3.8). A fluorometer must have the proper filters, gratings, or prisms to allow both the excitation and emission wavelengths to be selected. Some simple fluorometers use filters to isolate both the excited and emitted wavelengths; these must be chosen to avoid high blanks. Chemical treatment of the sample can eliminate some of the background interference. Monochromators may be used to select both the excitation and emission wavelengths using diffraction gratings. A spectrofluorometer can record both excitation and emission spectra. This convenience allows the analyst to easily locate the optimum excitation and emission wavelengths. Modern spectrofluoro-

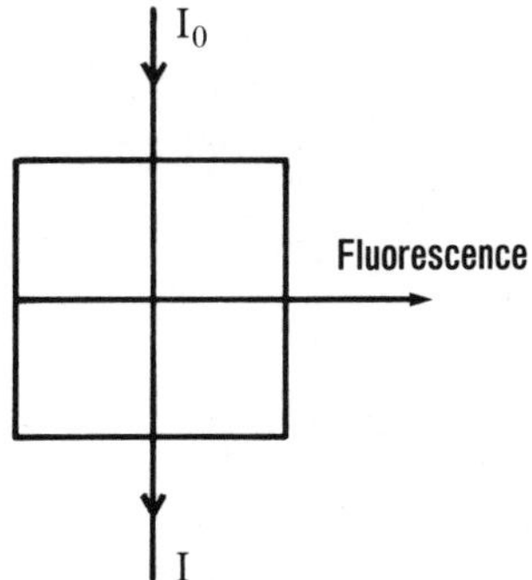

Figure 3.8. Schematic diagram of excitation and fluorescence of a sample.

meters, such as the Perkin-Elmer 650-40 with microprocessor-controlled operation, aids the operator in optimizing all the critical parameters (e.g., slit width and wavelengths).

As in absorption spectrophotometry, the concentration of a fluorescing compound is directly related to the intensity of fluorescence. One problem that arises in fluorescence that is not encountered in ordinary absorption spectrophotometry is quenching. This is caused by the absorption of the fluorescence emitted by the sample by another compound in the solution. For example, if the fluorescence intensity, in arbitrary fluorescence units, of a fluorescing compound in a nonquenching medium is 100, then in a more complex and quenching medium the intensity might be only 75. Because of this quenching, blanks are necessary for all analyses and should contain the same chemicals used in the samples and standards.

The quantities and relationships involved in a fluorometric analysis can be summarized as follows:

I_b = Fluorescence of blank

I_u = Fluorescence of unknown plus blank

I_s = Fluorescence of internal standard plus unknown plus blank

$I_u - I_b$ = Fluorescence of unknown under assay conditions

$I_s - I_u$ = Fluorescence of standard under assay conditions

$$\frac{\text{Fluorescence of unknown}}{\text{Fluorescence of standard}} = \frac{\text{Amount of unknown}}{\text{Amount of standard}}$$

$$\frac{I_u - I_b}{I_s - I_u} = \frac{\text{Amount of unknown}}{\text{Amount of standard}}$$

Sources of high blank values include cuvettes with fluorescing materials incorporated, solvents and reagents, detergents used to wash the glassware, and stopcock grease, to list just a few.

☐ GAS CHROMATOGRAPHY

Gas chromatography (GC) is an accurate and convenient instrumental technique for the separation and quantitation of numerous classes of organic compounds found in foods. Fatty acid esters and volatile carboxylic acid esters (both mono- and dicarboxylic) were among the first compounds analyzed by this technique. Volatile aldehydes, ketones, and amines were also examined by gas chromatography. Gas chromatography is now a routine procedure for analyzing fatty acids and pesticides (AOAC 1980; AOCS 1979). Today, compounds with extremely low vapor pressures (e.g., steroids such as β-sitosterol, stigmasterol, and campesterol found in plants) may be separated by gas chromatography.

In gas chromatography, which is based on the principles of both distillation and extraction, compounds partition between a stationary phase (solid or liquid) and a moving phase (liquid or gas). The components of a gas chromatograph are shown in Fig. 3.9. The sample (usually less than 5 μl) is injected into the injection block assembly (usually through a neoprene septum). The sample is swept by the helium or nitrogen carrier gas into the column, which is packed with a support coated with the stationary phase. The support is a diatomaceous earth that has been specially treated (e.g., acid washed). The particle size of the support and the percentage stationary phase used to coat the support are quite critical to the separation.

Stationary phases vary from very polar to nonpolar. By careful selection of the stationary phase, most compounds can be separated, even if they have identical boiling points. There are literally hundreds of stationary phases available. Most analysts do not have the time to test all of the logical phases to determine the best one for a particular separation. Table 3.4 lists the results of a study of seven compounds

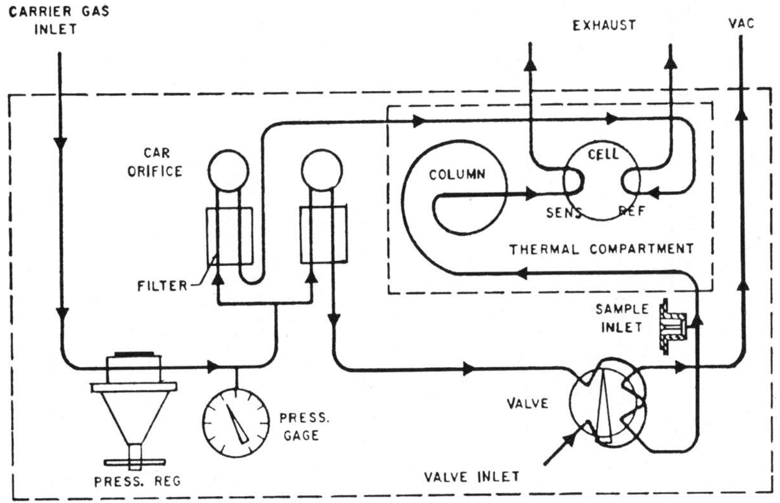

Figure 3.9. Diagram of the basic components of a gas chromatograph.

Table 3.4. Selectivity Values for Stationary Phases in Gas Chromatography[a]

Phase #	Phase name	Values with reference compounds[b]							
		1	2	3	4	5	6	7	$\sum^{c}$
1	Squalane	0	0	0	0	0	0	0	0
2	Paraffin oil	9	5	2	6	11	2	2	33
2A	APOLANE-87	21	10	3	12	25	0	0	71
3	Apiezon M	31	22	15	30	40	12	10	138
4	Apiezon L	32	22	15	32	42	13	11	143
5	SF 96	12	53	42	61	37	31	21	205
6	Apiezon J	38	36	27	49	57	23	13	207
7	Apiezon N	38	40	28	52	58	25	15	216
8	SE 30	15	53	44	64	41	31	22	217
9	OV-1	16	55	44	65	42	32	23	222
9A	OV-73	16	55	44	65	42	32	23	222
10	M and B silicone oil	14	57	46	67	43	33	22	227
11	DC 200 (12,500 cstks.)	16	57	45	66	43	33	23	227
12	OV-101	17	57	45	67	43	33	23	229
13	DC-410	18	57	47	68	44	34	24	234
14	Versilube F-50	19	57	48	69	47	36	23	240
15	DC 11	17	86	48	69	56	36	23	276
16	SE 52	32	72	65	98	67	44	36	334
17	SE 54	33	72	66	99	67	46	36	337
19	OV-3	44	86	81	124	88	55	35	423
20	Dexsil 300	47	80	103	148	96	55	46	474
21	Fluorolube HG 1200	51	68	114	144	118	68	53	495
22	Kel F wax	55	67	114	143	116	73	57	495
23	Apiezon H	59	86	81	151	129	46	23	506
24	OV-7	69	113	111	171	128	77	66	592
25	DC 550	74	116	117	178	135	81	72	620
26	Di(2-ethylhexyl) sebacate	72	168	108	180	125	132	49	653

27	Diisodecyl adipate	71	171	113	185	128	134	52	668
28	Octyl decyl adipate	79	179	119	193	134	141	57	704
29	Dilauryl phthalate	79	158	120	192	158	120	52	707
30	Bis(2-Ethylhexyl)tetrachlorophthalate	112	150	123	168	181	110		734
31	Diisodecyl phthalate	84	173	137	218	155	133	59	767
32	Dinonyl phthalate	83	183	147	231	159	141	65	803
33	DC 710	107	149	153	228	190	107	98	827
34	Dioctyl phthalate	92	186	150	236	167	143	66	831
35	POLY-I 110	115	194	122	204	202	152	55	837
36	Hallcomid M-18	79	268	130	222	146	202	48	845
38	OV-17	119	153	162	243	202	112	105	884
39	UCCN LB-550-X	118	271	158	243	206	177	91	996
40	Span 80	97	266	170	216	268	107	66	1017
41	Castorwax	108	265	175	229	246	202	73	1023
42	POLY-A 103	115	331	149	263	214	221	62	1072
43	OV-22	160	188	191	283	253	133	132	1075
44	Polypropylene glycol	128	294	173	264	226	196	98	1085
45	Trimer acid	94	271	163	182	378	234	57	1088
46	POLY-A 101A	115	357	151	262	214	233	64	1099
48	UCON LB-1715	132	297	180	275	235	201	100	1119
49	Acetyl tributyl citrate	135	268	202	314	233	214	102	1152
50	Didecyl phthalate	136	255	213	320	235	201	101	1159
51	OV-25	178	204	208	305	280	144	147	1175
52	Polyphenyl ether OS-124 (5 rings)	176	227	224	306	283	117	135	1216
53	Tributyl citrate	135	286	213	324	262	226	102	1220
54	Polyphenyl ether OS-138 (6 rings)	182	233	228	313	293	181	136	1249
55	POLY-A 135	163	389	168	340	269	282		1329
56	Neopentyl glycol sebacate (HI-EFF-3CP)	172	327	225	344	326	257	109	1394
57	Squalene	152	341	238	329	344	248	101	1404
58	UCON 50-HB-280X	177	362	227	351	302	252	130	1419
59	Tricresyl phosphate	176	321	250	374	299	242	131	1420
60	Sucrose acetate isobutyrate	172	330	251	378	295	264	128	1426

(continued)

Table 3.4. *(continued)*

		Values with reference compounds[b]							
Phase #	Phase name	1	2	3	4	5	6	7	$\sum^c$
61	QF-1	144	233	355	463	305	203	53	1500
62	OV-210	146	238	358	468	310	206	56	1520
63	OV-215	149	240	363	478	315	208	56	1545
64	Igepal CO-630	192	381	253	382	344	277	136	1552
65	DC LSX-3-0295	152	241	366	479	319	208	55	1557
67	UCON 50-HB-2000	202	394	253	392	341	277	147	1582
68	Emulphor ON-870	202	395	251	395	344	282	140	1587
69	Triton X-100	203	399	268	402	362	290	145	1634
70	UCON 50-HB-5100	214	418	278	421	375	301	155	1706
71	Siponate DS-10	99	569	320	344	388	466	61	1720
72	Tween 80	227	430	283	438	396	310		1747
73	XE 60	204	381	340	493	367	289	120	1785
74	OV-225	228	369	338	492	386	282	150	1813
75	Neopentyl glycol adipate (HI-EFF-3AP)	232	421	311	461	424	335	156	1849
76	UCON 75-H-90000	255	452	299	470	406	321	180	1882
77	Igepal CO 880	259	461	311	482	426	334	180	1939
78	Triton X-305	262	467	314	488	430	336	183	1961
79	HI-EFF-8BP	271	444	330	498	463	346	175	2006
80	Quadrol	214	571	357	472	489	431	142	2103
81	Neopentyl glycol succinate (HI-EFF-3BP)	272	469	366	539	474	371	184	2120
82	Igepal CO 990	298	508	345	540	475	366	205	2166
83	EGSP-Z	308	474	399	548	549	373	220	2278
84	Carbowax 20M	322	536	368	572	510	387	221	2308
85	Carbowax 20M (TPA)	321	537	367	573	520	387	220	2318
86	Epon 1001	284	489	406	539	601	378	207	2319

87	Carbowax 6000	322	540	369	577	512	390	222	2320
88	Carbowax 4000	325	551	375	582	520	399	224	2353
89	SILAR-5CP	319	495	446	637	531	379	216	2428
90	Ethylene glycol isophthalate (HI-EFF-2EP)	326	508	425	607	561	400	213	2427
91	XF-1150	308	520	470	669	528	401	174	2495
92	Carbowax 1000	347	607	418	626	589	449	240	2587
93	Ethylene glycol adipate (HI-EFF-2AP)	372	576	453	655	617	462	250	2673
94	Butane-1,4-diol succinate (HI-EFF-4BP)	369	591	457	661	629	476	243	2707
95	Phenyldiethanolamine succinate (HI-EFF-	386	555	472	674	654	437	242	2741
	10BP)	364	619	449	647	671	482	245	2750
96	Reoplex 400	377	601	458	663	655	477	253	2754
97	LAC-1-R-296	378	603	460	665	658	479	254	2764
98	Diethylene glycol adipate (HI-EFF-1AP)	371	639	453	666	641	479	255	2770
99	Carbowax 1540								
100	LAC-2-R-446	387	616	471	679	667	489	257	2820
101	EGSS-Y	391	597	493	693	661	469	261	2835
102	Hyprose SP-80	386	742	492	639		565	227	2936
103	EGSP-A	397	629	519	727	700	496	278	2972
104	SILAR-7CP	440	638	605	844	673	492	268	3200
105	ECNSS-M	421	690	581	803	732	548	259	3227
106	ECNSS-S	438	659	566	820	722	530	286	3205
108	EGSS-X	484	710	585	831	778	566	316	3388
109	Ethylene glycol phthalate (HI-EFF-2GP)	453	697	602	816	872	560	306	4310
110	SILAR-9CP	489	725	631	910	778	566	292	3536
111	Diethylene glycol succinate (HI-EFF-1BP)	499	751	593	840		595	323	3543
112	LAC-3-R-728	502	755	597	849	852	599	329	3555
113	SILAR-10C	523	757	659	942	801	584	298	3682
114	THEED	463	942	626	801	893	746	269	3725
115	Tetracyanoethylated pentaerythritol	526	782	677	920	837	621	333	3742
116	Ethylene glycol succinate (HI-EFF-2BP)	537	787	643	903	889	633	348	3759
117	1,2,3,4,5,6-Hexakis (CYCLO-N)	567	825	713	978	901	620		3984
118	1,2,3-Tris(2-cyanoethoxy)propane	593	857	752	1028	915	672	375	4145

(continued)

Table 3.4. *(continued)*

Phase #	Phase name	Values with reference compounds[b]							
		1	2	3	4	5	6	7	$\sum^c$
119	1,2,3,4-Tetrakis (CYANO-B)	617	860	773	1048	941	685		4239
120	Cyanoethylsucrose	647	919	797	1043	976	713	388	4382
121	N,N-Bis(2-cyanoethyl)formamide	690	991	853	1110	1000	773	371	4644
122	OV-275	781	1006	885	1177	1089			4938
NA	Absolute *I* values for squalane	653	590	627	652	699	690	841	NA

Source: Applied Science Catalog No. 22 (1979). Used with permission.

[a] The reference compounds 1 through 7 were analyzed by GC on the phases listed. The values shown are differences in retention index units (*I*) between the reference compound run on squalane and on the other phases listed. The last entry in the table shows the absolute retention indices (*I*) for the reference compounds on squalane (e.g., benzene elutes at a *I* of 1343 from a column packing coated with phase 121.

[b] Identity of reference compounds: (1) benzene; (2) 1-butanol; (3) 2-pentanone; (4) nitropropane; (5) pyridine; (6) 2-methyl-2-pentanol; and (7) 2-octyne.

[c] The value of $\sum$ should be treated as a general selectivity guide. The higher the value, the more selective overall is the phase. A 30 unit difference in the value of $\sum$ is almost insignificant; similarly, a difference of 20 units in i is also not very significant. As an example, if phase #8 ($\sum$ is 217) does not effect a separation, then do not expect to be able to do it on phases #1–15. You would do best to look at phases numbered between 20 and 60 with $\sum$ values of 474–1426. Likewise, if phase #50 does not produce separation, then try phase #104 or phase #12.

of varying polarities that were analyzed by gas chromatography on many phases. By examination of Table 3.4, or other similar data, an analyst can narrow the selection of possible stationary phases to those most likely to afford satisfactory separations.

Column materials, *per se*, also are important in separations. Glass is generally considered to be most inert; however, since glass is quite polar due to the hydroxyl groups on the surface, it is often necessary to "silanize" glass columns using dichlorodimethylsilane. This will reduce the attraction of certain compounds to the interior surface of the column.

As the compounds to be separated pass through a column, they partition between the stationary liquid phase and the moving gas phase. Since different compounds have differing distribution coefficients, they separate. The distribution coefficient is related to the vapor pressure of an individual compound and its solubility in the liquid phase. Because of the vapor pressure effect, it is usually necessary to heat the column in a carefully thermostated oven; therefore, the stationary phase must have a low vapor pressure, otherwise it would be lost when heated. Columns often "bleed" out the stationary phase at temperatures much below the boiling point of the stationary phase. Even though two compounds may have identical boiling points, they may separate in the column if their distribution coefficients differ sufficiently.

Recent developments in capillary chromatography systems has revolutionized gas chromatography. Microbore glass capillary and fused silica columns have recently been introduced. Table 3.5 summarizes the properties and characteristics of typical gas chromatographic columns. Bonded FSOT columns have some advantages over conventional FSOT columns, primarily in their resistance to phase stripping from on-column or splitless injection of large samples. There are several commonly available bonded FSOT columns: (1) polydimethyl siloxane (e.g., SE-30, OV-1); (2) polydiphenyldimethyl siloxane (e.g., SE-54, SE-52); (3) polyphenylmethyl siloxane (e.g., OV-17, OV-1701); (4) polytrifluoropropyl siloxane (e.g., OV-202, QF-1); and (5) polyphenylcyanopropylmethyl siloxane (e.g., OV-225, DB-225). These columns are in order of increasing polarity.

As the separated compounds emerge from a column, they are detected by one of a variety of detectors. The earliest detectors, of the thermal conductivity type, contained a thermistor, which changes electrical resistance with varying temperatures. With this type of detector, the thermal conductivity of the carrier gas (He) measured at constant temperature serves as the base line measurement. When an emerging

Table 3.5. Properties and Characteristics of Typical Gas Chromatographic Columns

	Column type[a]			
Property	FSOT	WCOT	SCOT	Packed
Typical inside diameter (mm)	0.25 0.32	0.25 0.50 0.75	0.50	2 4
Typical length (m)	10–100	10–100	10–100	1–6
Typical efficiency (plates/meter)	2000–4000	1000–4000	600–1200	500–1000
Sample size	10–75 ng	10 ng–1 μg	10 ng–1 μg	10 ng–1 mg
Pressures required	Low	Low	Low	High
Detector make-up gas	Usually required	Usually required	Usually required	Not required
Speed of analysis	Fast	Fast	Fast	Slow
Chemical inertness	Best ⟶			Poorest
Permeability	High	High	High	Low
Flexible	Yes	NO	NO	NO

Source: Alltech Associates, Inc. Deerfield, IL.
[a] FSOT = Fused silica open tubular; WCOT = Wall coated open tubular; SCOT = Support coated open tubular.

organic compound is mixed with the carrier gas, the thermal conductivity of the carrier gas is lowered and this is recorded via amplification. Common strip chart recorders (usually 0–1 mV full scale) or microcomputer data systems are used to record the changing thermal conductivity. The curves are usually symmetrical but may be unsymmetrical due to improper operating conditions or stationary phase. Typical asymmetry is due to "tailing" where the emerging compound is released slowly from the stationary phase.

Other commonly used detectors are flame ionization and electron capture devices, which are much more sensitive than thermal conductivity detectors. An electron capture detector only detects certain classes of compounds (e.g., polyhalogenated or polynitrated compounds) that have an affinity for electrons. The detector furnishes a constant source of electrons and measures their loss as the sample flows through the electron field. The sensitivity is thus determined by the relative electronegativity of the compounds being detected. An electron capture detector is up to 1000 times more sensitive than a flame ionization detector but unfortunately only detects molecules containing electronegative groups. A flame ionization detector measures the ion population produced when the emerging compounds burns in a hydrogen–air flame. As the sensitivity of such a detector increases, concomitant problems of contamination, electronic stability, etc., must be considered.

Routine measurement of picogram quantities of pesticides and other compounds are possible with modern gas chromatography instruments. These sophisticated instruments often include temperature programming, data collection systems, and auto-injection systems. By combining gas chromatography with mass spectrometry, identification and quantitation of literally hundreds of compounds in foods and food products are possible. Among these are the volatiles in coffee and the components of several essential oils. AOAC methods (19.047, 19.048, and 19.049) describe the separation and quantitative determination of the quality of vanilla extracts. This method uses gas chromatography and derivative formation to more thoroughly characterize the components of vanilla extracts. Temperature programming is used to promote more complete separation of the components. Other useful applications are the determination of sugars by gas chromatographic analysis of their trimethylsilyl (TMS) ether derivatives and the separation of the cis and trans isomers of fatty acids.

The techniques of gas chromatography (GC) and gas chromatography/mass spectrometry (GC/MS) are frequently utilized in the analysis of foods. Applications include the analysis of certain food additives contaminants, adulterants, and natural constituents. In addition, the volatile components of certain foods have been analyzed to yield information on the various flavors and fragrances (Yeransian *et al.* 1985). Both packed column and capillary column GC have been used. Capillary GC often yields a chromatogram of approximately 100 peaks. It is not always necessary to identify every peak in the chromatogram but only necessary to compare the "fingerprint patterns" of two chromatograms. This method is called profile analysis or pattern recognition. Many times only the peaks responsible for the differences in the pattern are identified. Pattern recognition has been utilized in the classification of wines according to their geographic origin (Kwoan and Kowalski 1980) or variety (Noble *et al.* 1980). Flath *et al.* (1969) have applied this method to the differentiation of six varieties of apple. Engel and Tressl (1983) used this method to aid in distinguishing between two varieties of mango. In addition, they identified 114 components of the mango aroma.

Orange essence is an aqueous distillate collected during the production of juice concentrate. It may be used to add natural flavor to synthetic products of to restore natural flavor to processed food. Moshonas and Shaw (1984) collected orange essence from oranges that were harvested before and after the January 1982 freeze. Gas chromatographic and organoleptic data were compared. The gas chromatographic data clearly showed the detrimental effect that the freeze had on the flavor of the orange essences. These authors have shown that instrumental data can be used to objectively determine the strength and quality of aqueous fruit essences. These techniques may be utilized in flavor and fragrance analysis to identify the major components so that artificial flavors can have a more natural taste and fragrance. They may also serve to aid in the identification of the cause of "off-flavors." In addition, they may serve to supplement or replace the more subjective organoleptic methods, particularly in the area of blending. In the preparation of a particular brand of blended whiskey or coffee, for example, several varieties are blended to achieve a specific flavor. Gas chromatography could be very useful in giving more objective data on the blend rather than relying wholly on professional tasters. In addition, it would be very useful in determining the changes caused by various storage techniques, in the determination of the presence of any adulterants such as artificial flavoring in the natural fruit essence or in the determination of contaminants such as pesticides or plasticizers from the package.

Some other applications include the study of the effects of aging on various alcoholic beverages, the determination of food additives in vegetable oil (Yu *et al.* 1984), the presence of toxic aflatoxins in peanuts (Rosen *et al.* 1984), and the detection of organic sulfur compounds in shellfish as an indication of oil pollution (Ogata and Fujisawa 1983).

Analysis of Sugars

Trimethylsilyl ether derivatives of sugars are sufficiently volatile to allow gas chromatographic analysis. These derivatives are formed by treating sugars with hydroxylamine hydrochloride and then converting the resulting oximes to the trimethylsilyl (TMS) ethers by the addition of a silylating agent. The use of TMS-sugar-oxime derivatives for the analysis of sugars in food products that contain mixtures of fructose, glucose, sucrose, maltose, and lactose offers the following advantages:

1. Fast, single-vial reaction of 1.5- to 2-hr analysis time. Oximes are silylated directly; there is no isolation of oximes before silylation.
2. Single peaks simplify chromatograms and calculations. Multiple peaks, due to the tautomeric forms of reducing sugars, are largely eliminated, and single peaks are obtained with fructose, glucose, and maltose.

3. Good separations, especially useful in separating fructose and glucose.
4. Direct analysis of aqueous solutions. Syrups containing 20–30% or more water need not be concentrated before analysis as long as no more than 20 mg of water are present in sample being derivatized. Samples containing 70–75% water should be concentrated.
5. Quantitative and reproducible. Multiple chromatograms of known solutions should have an accuracy of ±3%.

Reagents and Supplies

1. Sugars or syrup for analysis.
2. STOX oxime reagent with internal standard (Pierce Chemical Co., Rockford, IL).
3. Trimethylsilylimidazole.
4. Chromatography columns—6 ft × $\frac{1}{8}$ O.D., stainless steel packed with 2% or 3% OV-17 on Chromosorb W(HP) 800/100 mesh.

Derivative Formation

Accurately weigh 10–15 mg (dry basis) of sugars or syrup mixture into a 3.5-ml screw cap septum vial. Add 1.0 ml of STOX oxime–internal standard reagent. Heat for 30 min at 70°–75°C. Cool to room temperature and form ether derivatives as follows: Add 1.0 ml trimethylsilylimidazole. Cap and shake for 30 sec. Allow to stand at room temperature for 30 min (no precipitate forms).

GC Analysis

Inject sample and program chromatograph immediately from 150° to 325°C at a rate of 10°C/min. When analyzing for fructose, glucose, and sucrose, using 3% OV-17, a program from 140° to 250°C with an initial 2- to 3-min hold before programming is sufficient to separate all sugars. The carrier gas (helium) should flow at 40 ml/min through a flame ionization detector. Typical chromatograms illustrating the separation of sugars by this method are shown in Fig. 3.10.

Separation of Cis/Trans Isomers of Fatty Acids

Recent developments in capillary column technology allows the separation of cis and trans isomers of fatty acids. In this procedure, the fatty acid triglycerides or free fatty acids, are extracted, hydrolyzed (if necessary), and then esterified. The fatty acid methyl esters are appropriately diluted with solvent and injected into a suitable gas chro-

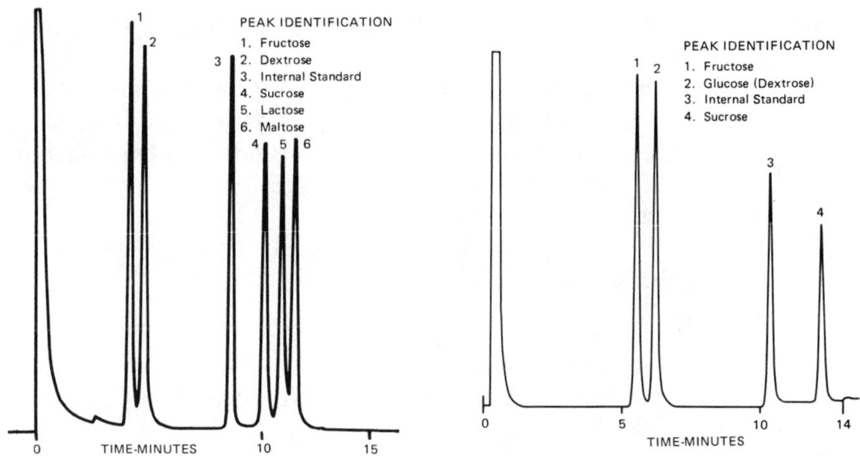

Figure 3.10. Typical separation of sugars achieved by gas chromatography of trimethylsilyl ether derivatives on columns packed with 2% (*left*) or 3% (*right*) OV-17 on Chromosorb W(HP) 80/100 mesh.

matograph (e.g., Varian 6000) equipped with a flame ionization detector (FID). The following conditions produce good separations:

Column: 37 m × 0.25 mm WCOT glass silar SCP
Column Temperature: 195°C
Injection temperature: 270°C
Detector temperature: 210°C
Carrier gas: Helium (23 psig)
Detector: (FID) 4×10^{-11} AFS

Under these conditions, methyl esters elute in the following approximate orders:

1. Methyl palmitate (C16)—11 min
2. Methyl palmitelaidate (*trans* $C_{16}{}^{1}$)—12 min
3. Methyl palmitoleate (*cis* $C_{16}{}^{1}$)—12.5 min
4. Methyl stearate (C_{18})—17 min
5. Methyl elaidate (*trans* $C_{18}{}^{1}$)—18 min
6. Methyl oleate (*cis* $C_{18}{}^{1}$)—18.5 min
7. Methyl vaccenate (*cis* 11-$C_{18}{}^{1}$)—19 min
8. Methyl linoleate ($C_{18}{}^{2}$)—22 min
9. Methyl linolenate ($C_{18}{}^{3}$)—27 min
10. Methyl arachidate (C_{20})—30 min

☐ ATOMIC ABSORPTION SPECTROPHOTOMETRY

Atomic absorption (AA) spectrophotometry has probably made the greatest impact on routine food analysis of any technique that has appeared within the last twenty years. Although not new in concept, the instrumentation for AA analysis was not readily available until the early 1960s. Before that time, the determination of trace quantities of metals was laborious, expensive, and often inaccurate. Flame photometry (emission spectroscopy) was the most widely used routine method for measuring elements such as sodium, potassium, and calcium. But relatively few elements could be determined at low levels in a complex matrix (the media in which the metal is distributed) with flame photometry, and interferences by other elements and the matrix itself often complicated the analyses. Quantitative emission spectrographic analysis was, and is, widely used, but it also is laborious and the equipment is quite expensive. The various colorimetric techniques for metal ion analysis are exceptionally laborious, frequently inaccurate (for a variety of reasons), and lack sensitivity to exceptionally low concentrations. In contrast with these methods, atomic absorption spectrophotometry is a simple, accurate, and relatively inexpensive method for analyzing microquantities of metals.

The most common AA instruments contain a burner with an aspiring device that, when burning, aspirates a quantity of solution (usually aqueous) into the burner gas-mixing chamber. In the burner chamber a large portion of the metal ion-containing sample is condensed and drains from the burner. A portion passes through the burner and is mixed with a combustible gas mixture (acetylene–air, acetylene–nitrous oxide, and hydrogen–argon are the most used mixtures).

Ground-state atoms in the flame are detected by the absorption of energy of a specific wavelength furnished by an appropriate radiation source. Because ground-state atoms of a particular element absorb only discrete wavelengths, a high degree of specificity can be achieved in atomic absorption. The radiation source is usually a hollow cathode lamp or an electrodeless discharge lamp that radiates specific wavelengths. The other wavelengths are isolated by a monochromater and the specific energy of the element being analyzed is detected by a photomultiplier (Fig. 3.11). The difference between the energy absorbed by the solvent without sample and the solvent with sample is displayed electronically as absorbance (A) on the meter or digital readout. A plot of absorbance vs concentration gives a linear Beer's relationship as in ordinary spectrophotometry (Fig. 3.12).

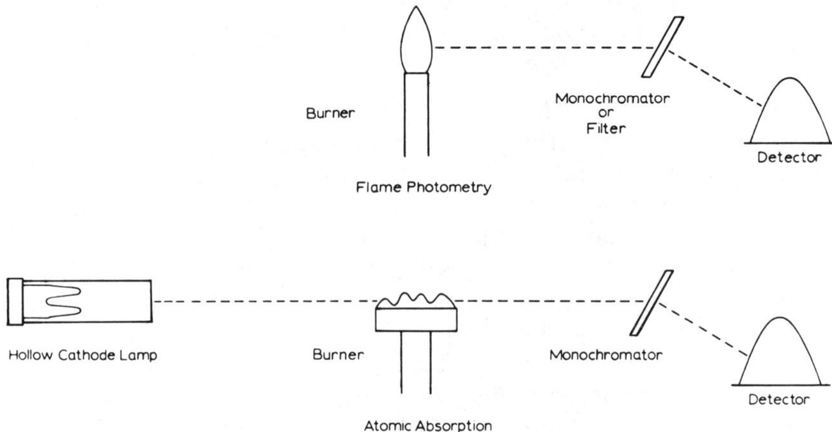

Figure 3.11. Schematic diagram comparing flame photometry and atomic absorption.

Since the metals must be in the ground state to absorb energy, every effort is made to inhibit ionization. Ground-state atoms of an element exist in equilibrium with excited atoms; however, it has been estimated that the ratio of ground-state to excited atoms for zinc at 3000°K is 10^9:1, while at 3500°K the ratio is 10^9:30. This makes flame temperature relatively noncritical.

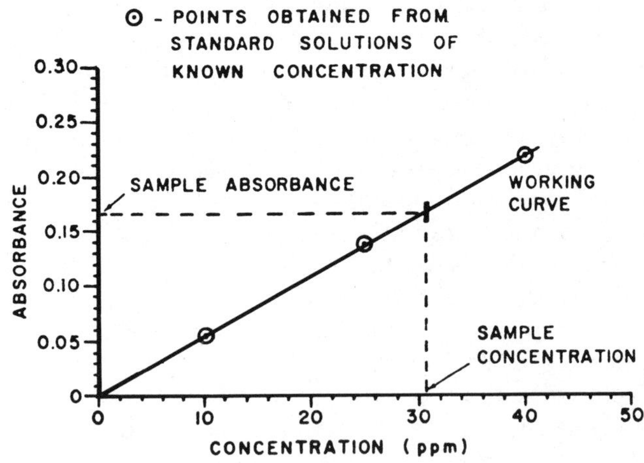

Figure 3.12. Plot of absorbance vs concentration in atomic absorption analysis.

Flame Method

The sample should be in a solution that can be easily aspirated through the nebulizer. The concentration of the metal to be determined should give 0.1–0.5 absorbance units. The concentration required can be estimated from the listed sensitivity (the concentration necessary to give an absorbance of 0.0044) for the element being analyzed. Approximately 2 ml of solution is the minimum volume for analysis. Graphite furnace methods require 100 μl or less.

Standard solutions containing known concentrations of the metal to be determined should be prepared using the same solvent as the sample being analyzed. The standard concentrations should bracket the expected concentration of the sample. If very dilute standards (less than 0.05 μg/ml) are needed for a period longer than 1 day, stock solutions in concentrations greater than 500 μg/ml should be prepared. The stock solutions can be diluted as required when dilute standards are needed. This will avoid changes in concentration (2–3% in 3 days or more) that occur in very dilute solutions upon aging.

Determine the absorbance of the standards and the sample. Standards should be determined repeatedly during the analysis of samples. Instrument variation or malfunction can easily be detected in this manner. Solvent absorbance is determined after each sample to check baseline stability or zero. If a standard should start decreasing in absorbance, this can be an indication of nebulizer or aspirator tube obstruction.

Sampling

Food samples containing high concentrations of cellulose, protein, fat, or other organics and inorganics are usually digested or diluted to reduce the problems of matrix interference. Ashing or a combination of wet digestion and ashing may be used, but losses due to volatilization of metals such as cadmium can be significant. Certain metals can be chelated with chelating agents such as ammonium pyrrolidine dithiocarbamate (also called ammonium 1-pyrrolidinecarbodithionate or APDC) and then extracted into an organic solvent such as methyl isobutyl ketone (MIBK) or butyl acetate (BuOAc). An example of such a procedure is AOAC method 25.068. In this method, milk is dry-ashed at 500°C for 16 hr to remove carbon-containing compounds. The ash is then chelated with APDC and the chelate is extracted into BuOAc. The extracted lead is then analyzed by conventional flame AA. This method is somewhat archaic in light of the more convenient and more accurate graphite furnace method.

Graphite Furnace (Flameless) Method

Graphite furnace or carbon furnace techniques present some unique problems. The furnace is an electrical resistance-heated graphite tube in which a small amount (usually less than 50 μl) of sample is placed. The furnace then is programmed to dry the sample to remove the solvent, char the sample to remove carbon-containing and other interfering substances, and then atomize the ash to vaporize the element in the light path of the hollow cathode or electrodeless discharge source. The atomization temperature and time is selected so that all of the element is volatilized in the shortest convenient time. The optimum time and temperature must be determined experimentally.

The absorbance signal is a momentary one since the sample is small. This momentary signal necessitates a fast-responding recorder and/or a peak read mode on the spectrophotometer. The peak read mode records the highest absorbance value that was detected by the photomultipher during the analysis sequence. If matrix components contribute nonspecific absorbance this is recorded as absorbance of the sample. Early measurements of copper in milk failed to take this into consideration and the values were higher than actually present. This extraneous absorbance can be essentially eliminated by the use of a deuterium lamp background corrector, which will compensate for nonspecific absorbance if the background is not too high. Unfortunately the emission spectrum of the deuterium lamp is centered near 240 nm and is weak above 300 nm. This offers little energy for wavelengths above 320 nm.

Careful selection of charring temperature and time will normally remove most of the matrix components that can volatilize during the atomization step. The optimum charring temperature and time may be determined if two conflicting goals are observed: (1) charring temperature and time are sufficient to volatilize as completely as possible any interfering or particulate-producing sample matrix, and (2) charring time is sufficiently short and the temperature low enough to prevent the loss of the element being analyzed.

Another way of preventing premature atomization and also allowing higher charring temperatures is the formation of a less volatile compound of the element to be analyzed. Cadmium normally has a low atomization temperature (1500°C for $CdCl_2$). If ammonium phosphate is added to the sample, the resulting cadmium phosphate, which is less volatile, will permit a higher charring temperature, thereby removing matrix interferences. The melting point of cadmium chloride is 560°C, and that of cadmium phosphate is 1500°C.

Since the graphite furnace method is 10–100 times more sensitive than flame analysis, more attention must be paid to contamination from solvent, acids, etc. It is recommended that especially purified acids such as Ultrex (J. T. Baker Co.) be used for all sample and standard preparation. Analysis of ACS reagent grade acids may show that they could be used for certain metal determinations, but it is recommended that each lot be analyzed to confirm the acid's purity.

Glassware should be cleaned normally and then soaked in 5% nitric acid to remove metals adhering to the surface. This should be followed by a deionized water wash.

Water purity can be a problem for the analyst also. Water purifiers, such as those manufactured by Millipore Corporation and Barnsted, with carbon filters and ion exchange cartridges prepare water of sufficient purity for most applications. However, constant monitoring of the water quality is recommended. Simple conductivity meters used to monitor water purifier performance are not specific for trace impurities.

Determination of Lead in Milk

The determination of lead (Pb) in evaporated milk is an excellent example of the care needed in selecting the optimum charring temperature and time in graphite furnace AA analyses. For evaporated milk, the following conditions have been established using a 20-μl sample at 283.31 nm with deuterium background correction:

Drying—110°C for 30 sec
Charring (ashing)—400°C for 60 sec
Atomizing—2300°C for 10 sec

These conditions, however, may vary with different instruments and need to be determined for each analysis, as described below. The following procedure is for a Perkin-Elmer 503 or equivalent instrument equipped with an EDL power source, graphite furnace, and deuterium background corrector. If another instrument is being used, the specific manufacturer's instructions should be consulted and the procedure modified as necessary.

Standard Solutions

Prepare standard Pb solutions (25, 50, and 75 μg/l) in class A volumetric flasks washed with 8 *M* nitric acid. Millipore system (or equivalent) water should be used, and the final standard made 1% (v/v) with Ultrex HNO_3.

Drying Temperature and Time

In general, select a drying temperature slightly above the boiling point of the solvent. The drying time generally is related to the volume of sample as follows: 10 μl, 15 sec; 20 μl, 20 sec; 50 μl, 40 sec; and 100 μl, 60 sec. The drying temperature and time is usually 105°–125°C for 10 sec for a 20-μl sample.

Atomization Temperature and Time

Choose a standard solution having a concentration sufficient to give an absorbance of 0.2–0.5. Dry at the temperature and time determined in the previous section; then char 10 sec at 300°C. Atomize 15 sec at 2700°C (8 sec for volatile elements) and determine the absorbance. Repeat the determinations at successively lower atomization temperature, keeping the drying and charring conditions constant. Plot a graph of atomization temperature versus absorbance. Choose the lowest atomization temperature giving maximum absorbance. Choose an atomization time sufficient to allow the atomization signal to return to baseline.

Charring Temperature and Time

Choose an aliquot of sample expected to give an absorbance of 0.2–0.5. Select the drying and atomization parameters determined in the previous steps. Change the wavelength to that of a nearby nonabsorbable; do not have the background corrector in operation. Char 60 sec or longer at 200°C and atomize the sample; the absorbance signals represent broad band absorption from the matrix. Repeat this step at successively higher charring temperatures, keeping the drying and atomizing conditions constant. Plot a graph of the broad band absorption atomization signal versus charring temperature. The minimum usable charring temperature is the lowest value giving reasonably low (less than about 0.5 absorbance) broad band absorption signals. Using the minimum charring temperature, observe the broad band atomization signals using successively shorter charring times. The minimum charring time is the lowest value giving reasonably low broad band absorption atomization signals (less than about 0.5 absorbance).

Reset the wavelength to the analytical line for the element and turn on the background corrector. Reanalyze the sample, charring at the minimum charring temperature and time previously determined; the signals now represent atomic absorption by the element of interest. Repeat this step at successively higher charring temperatures. Plot a graph of absorbance versus charring temperature. The maximum charring temperature is the highest value on the plateau of the curve.

Select an operating charring temperature between the minimum and maximum values.

Procedure

Prepare a standard curve to determine the linear range of lead concentration versus absorbance (see Fig. 3.10). Usually the relationship is linear up to 20 μl of 75 μg/l Pb for aqueous solutions. Using conditions established previously, check the Pb concentration of an undiluted sample. If the concentration of Pb is outside the linear range of the standard curve, dilute as necessary. Determine the concentration of Pb in concentrated evaporated milk or other types of milk from the standard curve. Report the results as μg/liter or μg/ml of original milk.

Determination of Mercury in Fish

With the introduction of flameless AA techniques it became possible to analyze very small amounts of contaminants in foods. Of particular interest is mercury (Hg) contamination of fish and other marine foods. Instances of mercury poisoning caused by contaminated seafoods, such as that observed in Minamata, Japan in the early 1950s, stimulated public concern about this problem and have led to closer monitoring of contamination levels by public authorities.

The method described here is generally useful for detecting 0.001–1.0 ppm Hg in a 0.5–1.0-g sample. This procedure is generally applicable to the Perkin-Elmer (303-0830) Mercury Analysis System (Fig. 3.13). Other systems may vary; their instruction manuals should be consulted and the procedure modified as necessary.

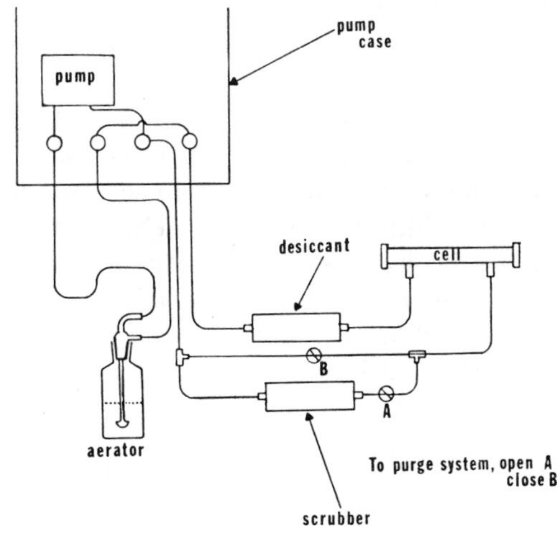

Figure 3.13. Diagram of apparatus for determination of mercury by flameless atomic absorption analysis.

Reagents

1. Potassium permanganate crystals and 5% (w/v) solution (Hg free).
2. 5.6 *N* Nitric acid.
3. 18 *N* Sulfuric acid.
4. 1.5% Hydroxylamine hydrochloride solution.
5. 10% Stannous chloride solution.
6. Concentrated sulfuric acid.
7. Mercury standard.

Sample Pretreatment

Weigh a 1.0-g sample of homogenized fish tissue into a 125-ml Erlenmeyer flask. Slowly add 30 ml of concentrated sulfuric acid. Stopper loosely with a polyethylene stopper and allow to stand at room temperature for approximately 15 min. Swirl the contents to disperse them, and then place flask in a 50°–60°C water bath for at least 2 hr. If the colored solution contains undissolved matter after the 2-hr digestion, add an additional 5 ml of concentrated sulfuric acid and heat for an additional hour.

Cool to room temperature and carefully transfer to a 300-ml Biochemical Oxygen Demand (BOD) bottle containing 50 ml of Hg-free distilled water. Rinse the flask with 20 ml of Hg-free distilled water (two 10-ml rinses) and add the rinses to the BOD bottle. Slowly add potassium permanganate crystals to the bottle. Heat in a 50°–60°C water bath. The sample will turn brown and froth. When frothing subsides, add more potassium permanganate until the purple color persists. Swirl the sample throughout the addition.

Procedure

To the BOD bottle containing 100 ml of the prepared sample add 5 ml of 5.6 *N* nitric acid and mix well. Wait approximately 15 sec, then add 5 ml of 18 *N* sulfuric acid and again mix well. Wait 45 sec. Add 5 ml of hydroxylamine hydrochloride solution and swirl. The sample should turn clear in approximately 15 sec. If not, add hydroxylamine hydrochloride crystals until a clear, colorless solution is obtained. Add 5 ml of stannous chloride solution and turn the air flow on and immediately insert the aerator into the BOD bottle (Fig. 3.13). No appreciable undissolved matter should be present. If any matter is noted, a new sample should be prepared.

Record the absorbance reading as the mercury is aspirated in the cell.

Preparation of Standards and Blanks

Prepare a reagent blank that contains all reagents but omits the fish sample.

Prepare standards by adding 2 drops of potassium permanganate solution to each of six 300-ml BOD bottles. Prepare duplicate standards containing 0.0, 0.5, and 1.0 μg Hg in a total volume of 1.00 ml. These serve as a blank and two standards. They should be treated exactly as those in the section on *Procedure.*

Results

Prepare a standard curve of μg of mercury versus maximum absorbance for each mercury standard. Read μg of mercury in the unknown directly from the standard curve and subtract the value obtained for the reagent blank. Report the results in μg mercury/g of tissue. This would give the concentration of mercury in μg/g (wet weight).

☐ COLUMN LIQUID CHROMATOGRAPHY

Column liquid chromatography (LC) technology can be divided into two major classes: (1) low-pressure or gravity flow liquid chromatography (LPLC) using relatively large-diameter columns (5 mm or greater) and (2) high-pressure liquid chromatography (HPLC) using small diameter columns with special packing techniques. Commercial liquid chromatographs are quite sophisticated and have features such as gradient solvent programming, microprocessors for controlling all modes of function, and programmable detectors. Basically, however, a liquid chromatography system is composed of (1) one or more solvent reservoirs, (2) pump (not needed for gravity flow), (3) column (with or without injector), and (4) detector.

Figure 3.14 shows a typical basic chromatograph in schematic form. Pump B and Reservoir B may be isolated when gradient elutions are unnecessary. The pump(s) transports the solvent(s) to the injector. The sample is injected and moves by solvent flow through the column, where it is separated by one or more of a variety of processes into the individual components in the sample. As the sample emerges, the detector senses a specific chemical or physical property of the sample and converts this into a signal via electronics and a strip chart recorder. Except in relatively simple separations, a solvent gradient may be necessary to achieve efficient separation. The gradient is usually accomplished by adding a second pump (Pump B) and reservoir. Gradient

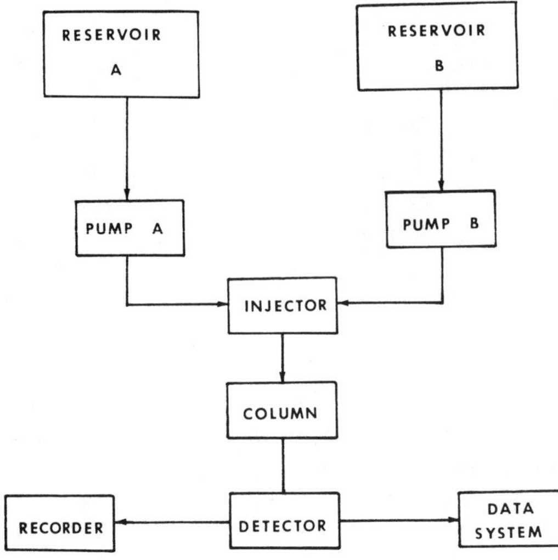

Figure 3.14. Schematic diagram of the components of a liquid chromatography system. If gradient elution is unnecessary, Reservoir B and Pump B are isolated.

elution can be varied to reduce analysis time and increase resolution. The pumps should furnish a constant pulse-free flow of solvent. Constancy and repeatability of flow rate are essential in order to use elution volumes as an index for identification. Solvent flow constancy is also necessary for optimum detector performance.

The detector most commonly employed is a photometric type with either fixed or variable wavelength. A wavelength of 254 nm is frequently used since this is the wavelength of an intense emission line of a mercury source lamp. Although a sample being detected may not have an absorbance peak centered at 254 nm, the emission is of sufficient intensity that there is sufficient sensitivity to the compound. Variable-wavelength detectors are more versatile, and detectors with programmable wavelength and stop-flow scanning are now available. Detectors using the refractive index principle are available; these can be used with certain classes of compounds that are not readily detected by visible, ultraviolet, or fluorescence photometric detectors.

Injectors and column end fittings must be designed to avoid "dead" space between the injector and the column and between the column effluent and the detector. Such spaces cause peak broadening.

Columns are usually constructed of 316 stainless steel and require a smooth precision bore. Packing materials must have a uniform particle size. Columns for modern small-bore, high-pressure liquid chromatographs are difficult to pack uniformly and require special equipment. Most chromatographers now use commercially prepared columns rather than pack their own. This is heartily recommended.

Larger-diameter, low-pressure columns used for preparative work can be packed by the chromatographer. These are constructed of glass, stainless steel, or plastic (depending on the application). One problem with this type of column is buildup of pressure by the restricted flow through the column when a constant-flow pump is used to pump the solvent. This type of column is not designed to withstand higher pressures.

Means of Separation

Several types of LC have been developed differing in the means by which separation is achieved and in the nature of the phases.

Adsorption Chromatography

Adsorption chromatography is accomplished by competition between the sample and the solvent molecule for active sites on a small particle (400 mesh down to 10 μm) of materials such as alumina or silica gel. This technique is applicable to molecules of low to high polarity with a molecular weight up to 1000. Typical columns are 15–25 cm long and have an inside diameter of 2–3 mm. Larger columns using lower pressures may use particle sizes up to 80–100 mesh. These larger particles offer much less resistance to flow, thus giving lower back pressures. Mixed solvent capability is essential and gradient elution capability is desired. Gradient elution allows a differential adsorption of the compounds to the adsorbing material in the column.

By blending a nonpolar solvent with a more polar one, the adsorption of the sample of the column material can be controlled. Table 3.6 lists commonly used liquid chromatographic solvents and their polarity (as adsorptive energy), along with their boiling points, viscosity, refractive index, and ultraviolet wavelength of minimum use. Manufacturer's instructions and texts on liquid chromatography give specific instructions for selecting a suitable solvent for a particular separation. Generally, the following procedure is used:

1. Match the polarity of the sample and the solvent.
2. Chromatograph the sample. If the sample appears at the solvent front the solvent is too polar to allow adsorption. Lower the solvent polarity and rechromatograph.
3. If the sample chromatographed with the original solvent does

Table 3.6. Properties of Common Liquid Chromatographic Solvents

Solvent	Adsorptive energy	Boiling point (°C)	Viscosity	Refractive index	UV cutoff (nm)
Hexane	0.00	69	0.33	1.375	210
Isoctane	0.01	99	0.3	1.404	210
Petroleum ether	0.01	175–240	0.3		210
Skellysolve B, etc.				1.412	
	0.04	174	0.92	1.427	210
n-Decane	0.04	81	1.00	1.406	210
Cyclohexane	0.05	49	0.47	1.466	265
Cyclopentane	0.18	77	0.97	1.413	225
Carbon tetrachloride	0.26	108	0.43	1.436	220
Amyl chloride	0.26	77	0.47		
Butyl chloride					
Xylene	0.26	138–144	0.62–0.81	~1.50	290
i-Propyl ether	0.28	69	0.37	1.368	220
Toluene	0.29	111	0.59	1.496	285
Chlorobenzene	0.30	132	0.80	1.525	330
Benzene	0.32	80	0.65	1.501	280
Chloroform	0.40	61	0.57	1.443	245
Methyl-*i*-	0.43	117		1.394	330
butylketone	0.45	65	0.35	1.408	220
Tetrahydrofuran					
Ethylene dichloride	0.49	84	0.79	1.445	230
Methylethylketone	0.51	80	0.3	1.381	330
Acetone	0.56	57	0.32	1.359	330
Dioxane	0.56	101	1.54	1.422	220
Ethyl acetate	0.58	77	0.45	1.370	260
Amyl alcohol	0.61	138	4.1	1.410	210
Dimethyl sulfoxide	0.62	189	2.24		
Nitromethane	0.64	101	0.67	1.394	380
Acetonitrile	0.65	82	0.37	1.344	210
Pyridine	0.71	115	0.94	1.510	305
i-Propanol, *n*-propanol	0.82	83	2.3	1.38	210
Ethanol	0.88	79	1.20	1.361	210
Methanol	0.95	65	0.60	1.329	
Ethylene glycol	1.11	198	19.9	1.427	210
Acetic acid	Large	118	1.26	1.372	
Water	Larger	100	1.0	1.333	190
Salts and buffer	Very large	100+			

not emerge in a reasonable time, then increase the polarity of the solvent by substituting another solvent of higher polarity or by blending.

Refinement of these general procedures can ultimately lead to a suitable separation.

Figure 3.15 shows a particle of silica gel with Si-O-H polarities. If a solvent such as acetone is used as a desorption solvent for an adsorbed compound such as an amine, then the electronegative end of the amine (the nitrogen has an unshared pair of electrons) would be attracted to the δ^+ hydrogen of the silica gel. Acetone with its δ^- oxygen could compete with the amine to cause desorption, thus achieving separation.

Liquid Partition Chromatography

Liquid-liquid partition separates compounds on the basis of their differential solubility in the eluting solvent and the liquid phase on the column packing *per se*. This partitioning of the sample between the phases delays some compounds more than others. There are basically two types of liquid partition chromatography:

1. Forward liquid phase. A stationary liquid is coated on an inert support. The immiscible mobile liquid can then flow over the coated support. Compounds of the sample can partition between the solvent and the stationary phase-coated bed. Typical columns would be 1 m × 2.4 mm packed with 1% β,β′-oxydipropionitrile on 30–40 μm Perisorb A (Fig. 3.16) or 1% Carbowa × 400 on Perisorb A. This technique is rarely used presently.
2. Reverse liquid phase. A nonpolar stationary liquid and a polar mobile liquid make up this pair of phases. The stationary phase is chemically bonded to the packing; C-18 ODS (18-carbon octadecylsilane) is widely used. C-18 columns are quite versatile with solvents such as acetonitrile and water gradients; C-8 columns are recommended for samples of higher polarity (Fig. 3.17).

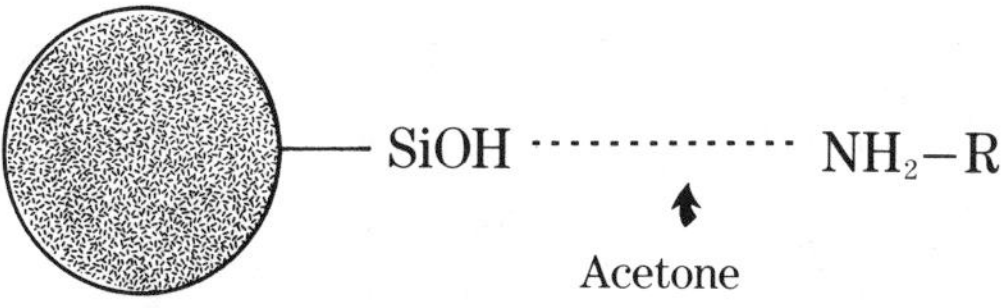

Figure 3.15. Particle of silica gel with bound NH_2-R sample (adsorption chromatography).

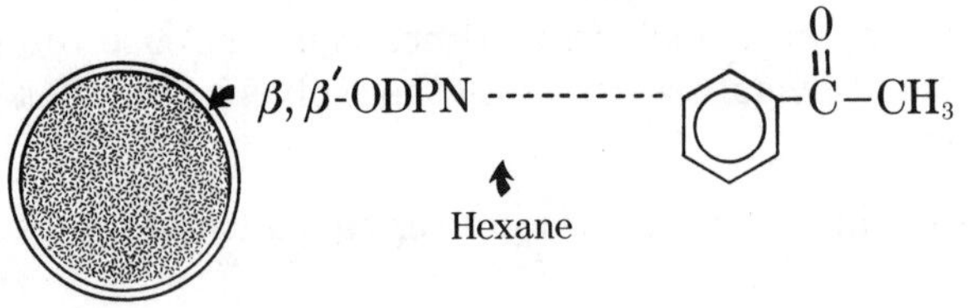

β,β'-ODPN

Hexane

$-C(=O)-CH_3$

Figure 3.16. Particle with liquid phase adsorbed (forward liquid-phase chromatography).

Exclusion Chromatography

Exclusion chromatography relies on the physical size of the molecules for separation. Column packings are porous gels having pores similar in size to the size of the sample molecules (Fig. 3.18). Small molecules permeate more pores and are retained longer than larger molecules. Compounds with molecular weights from 200 to several million can be analyzed by exclusion chromatography. This technique has been used primarily for separation of biomolecules such as proteins and enzymes using conventional low-pressure or gravity chromatography. Sephadex (Pharmacia Fine Chemicals) and Bio-Gel (Bio-Rad Corp) contain chemically crosslinked dextran (polysaccharide from *L. mesinteroides*) and a polyacrylamide, respectively, as the gel network. Recent advances have produced packings that are usable at high pressures; these are applicable to separation of biopolymers and synthetic polymers.

Ion-Exchange Chromatography

Ion-exchange chromatography most frequently uses crosslinked polystyrene resins as a support for the acid or base functional groups involved in the ion-exchange reaction. Sephadex or Bio-Gel are available with acid groups (e.g., carboxymethyl) or base groups (e.g., diethylaminoethyl) attached to the gel structure. Only the synthetic organic polymer supports are usable for high-pressure liquid chromatography.

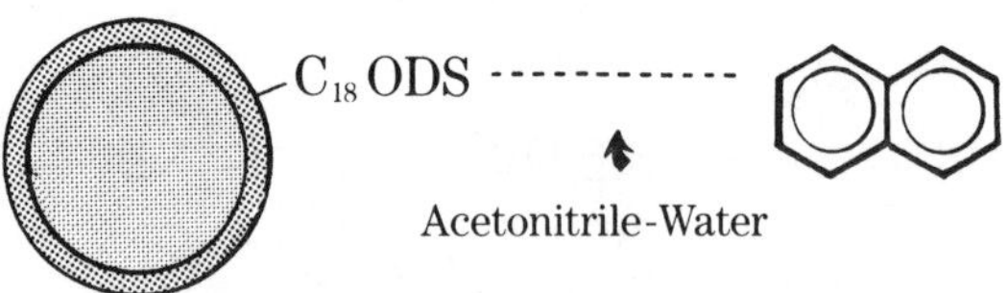

Figure 3.17. Particle with liquid phase chemically attached (reverse liquid-phase chromatography).

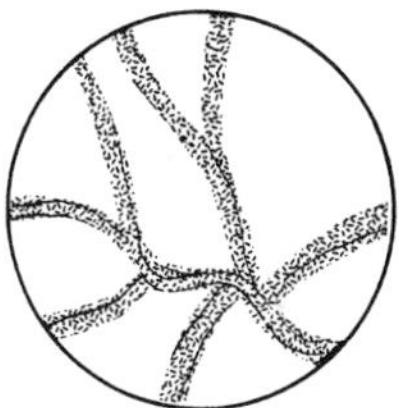

Figure 3.18. Gel particles showing pores (exclusion chromatography).

The basic principle behind this technique is the exchange of a cationic compound for a counter cation on the cation ion-exchange material or the exchange of an anionic compound for a counter ion on an anion ion-exchange material. In Fig. 3.19, for example, the cationic form of the amino acid exchanges for Na^+ on the gel. The amino acid cation then is displaced from the column by decreasing the pH or by increasing the ionic strength of the buffer by adding NaCl. The increased $[Na^+]$ displaces the amino acid cation. AOAC method 20.189 for determining monosodium glutamate uses conventional ion-exchange chromatography, and AOAC method 31.228 incorporates HPLC with ion-exchange.

Determination of Vitamin A in Milk

A number of methods (e.g., rat bioassay) are available for analyzing vitamin A in foods. Because of the presence of interfering substances such as vitamin A decomposition products, carotenoids, vitamin A isomers, and vitamin esters, preliminary separation of vitamin A from such interfering substances usually is necessary.

Liquid chromatography is a fast and reliable method for separating vitamin A from interfering substances. The fat-soluble vitamins, including vitamin A, are best chromatographed by the reverse-phase technique in which a polar water-alcohol solvent is used with a non-

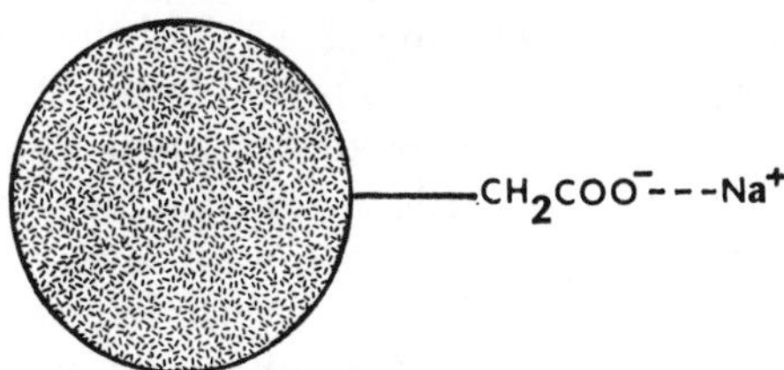

Figure 3.19. Particle with cationic exchange group and an amino acid sample (ion-exchange chromatography).

polar stationary phase (see Fig. 3.17). In this type of chromatography, compounds usually elute in order of their decreasing solubility in water. Once chromatographic separation is effected, the concentration of vitamin A can be determined from its ultraviolet absorption; at 325 nm the absorbance of vitamin A is directly proportional to its concentration.

Chromatographic Apparatus and Conditions

1. Instrument	Perkin-Elmer 3-B liquid chromatograph.
2. Column	Reverse-phase C-18 ODS.
3. Mobile phase	Start program at 90% methanol and 10% water. (Other solvents such as acetonitrile may be substituted.)
4. Column temperature	25°C.
5. Column pressure	Depends on flow rate.
6. Flow rate	1.0 ml/min.
7. Detector	LC75 at 325 nm.
8. Sample size	50 microliters (or less).
9. Retention time	Depends on solvent composition and is determined by chromatographing a solution of standard vitamin A.

Sample Preparation

Weigh by difference, 50 g of vitamin D homogenized milk and place into a saponification flask equipped with a standard taper joint. Add 50 ml of alcoholic KOH solution, prepared by dissolving 12 g of KOH in 100 ml methanol, and attach to a reflux condenser. Heat for 15 min under a nitrogen atmosphere or until saponification is complete. Then allow the solution to cool to room temperature without external cooling. Wash the condenser with 10 ml of water. Transfer contents of the saponification flask to a separatory funnel. Wash the saponification flask with 100 ml acetone–hexane (1:1). Add the solvent mixture to the separatory funnel and allow the layers to separate; then drain the aqueous layer into another separatory funnel. Repeat extraction of aqueous layer twice with 75–100 ml of hexane. Shake moderately. Combine all solvent extracts and wash with several portions of water until free of alkali, testing with phenolphthalein test paper. The hexane extracts are then dried over 20 g anhydrous sodium sulfate. Filter the hexane extracts through anhydrous sodium sulfate, placed on a plug of Pyrex glass wool in a funnel, into a 500-ml round bottom flask.

Rinse the separatory funnel with two 10-ml portions of hexane and add the rinses to the 500-ml flask, through the funnel.

Evaporate the hexane extracts on a water bath (30°C) to a low volume (about 25 ml), under a nitrogen atmosphere, then transfer to a 100-ml round bottom flask and continue evaporation to dryness. Take up the residue immediately in 10 ml of chloroform.

The vitamin A standards for liquid chromatographic analysis are prepared by adding one capsule of U.S.P. vitamin reference standard to a 250-ml round bottom flask. (Each capsule contains approximately 250 mg of cottonseed oil which is standardized to contain in each gram 34.4 mg of trans retinyl acetate equivalent to 30.0 mg of retinol.) Add 25 ml of H_2O and 25 ml of methanol saturated with KOH and a few mg of EDTA. Attach a condenser and reflux for 15 min and cool in a nitrogen atmosphere.

Extract once with acetone: hexane (1:1); reextract the aqueous layer twice with hexane. Combine all extracts and wash free of alkali with water until neutral. Dry with anhydrous sodium sulfate, then evaporate and make to a volume of 100 ml (stock solution). Dilute 1 ml of stock solution to 50 ml of hexane and read absorbance at 325 nm with a spectrophotometer. The following equation can be used to determine the concentration of vitamin A alcohol.

$$\frac{\text{A at 325 nm}}{0.182} = \mu\text{g/ml or mg/L} \quad \text{vitamin A alcohol}$$

Dilute 4 ml of diluted vitamin A solution to 10 ml with chloroform. Inject 50 μl of this solution into the LC column and measure the absorbance at 325 nm. Peak heights may be compared manually and the concentration of vitamin A determined; or, an internal standard such as vitamin E succinate can be added to all standards and the unknown. The ratio of the peak height of the vitamin E succinate peak to the standard peaks and the unknown peaks are then determined. This internal standard method eliminates the error in sample size injection. A typical protocol for the internal standard method is given in the following.

Standard concentration (μg/ml)	Internal standard concentration (μg/ml)
2	4
4	4
6	4
8	4
Unknown	4

A standard curve then can be plotted where Y = concentration of vitamin A and

$$X = \frac{\text{Peak height (absorbance) standard}}{\text{Peak height (absorbance) internal standard}}$$

If the liquid chromatograph is equipped with a computer and quantitative software, a calibration run can be made with standards and the above calibration plot can be avoided.

The preliminary specifics of the recorder sensitivity, flow rate, etc. are determined by the analyst who optimizes these parameters.

Determination of Carbohydrate

Carbohydrates, particularly trisaccharides and higher, are well suited for high-pressure liquid chromatography (HPLC) due to their low volatility. Methods such as AOAC 31.228 involving columns packed with cross-linked resins materials such as Dowex 50W-X4 give reasonable acceptable results but require considerable sample preparation and strict control of operating conditions. Column life is generally fairly short with wide variations in retention times and resolution as the column deteriorates.

One company (Alltech Associates) offers a carbohydrate analysis column that is based on amino bonded-phase silica. This column offers a rapid, highly selective separation of both lower and higher saccharides, without the drawbacks of cross-linked resins. By varying the ratio of the acetonitrile–water solvent, a high degree of selectivity can be achieved. Since carbohydrates are fairly hydrophilic compounds, increasing the water content of the mobile phase tends to decrease the retention times of the solutes. Separation of several carbohydrates has been achieved using the amino bonded-phase silica column (Fig. 3.20).

All HPLC instruments should be fitted with a high-performance guard column installed and used according to the manufacturers recommendation.

Apparatus

A Perkin-Elmer Model 3-B liquid chromatograph with refractive index (RI) detector is suitable. The signal from the detector can be interfaced with the 3600 with computer chromatography software. The CRT of the 3600 may be used to monitor the signal. Any suitable computer with chromatography software may be used for the analysis. An integrating recorder may also be used.

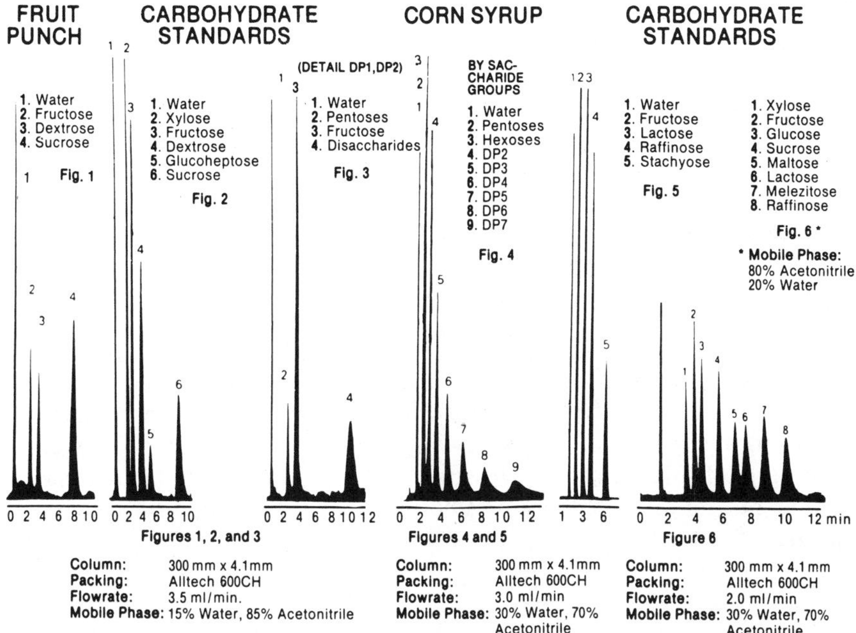

Figure 3.20. Sample chromatograms illustrating separation of sugars possible with high-pressure liquid chromatography. In Figs. 1, 3, and 4, samples were prepared by diluting the sample with water and removing particulate matter with a filter syringe.

Quantitative Measurements

Three methods for quantitatively determining the amount of different carbohydrates in a sample are possible:

1. Normalization. Each peak area is calculated and expressed as a percentage of the total area of all peaks.
2. Internal standard. A compound that is similar to the compounds being chromatographed is added and used as an internal standard.
3. External standard. An external standard is chromatographed separately. This method is particularly susceptible to error due to sample injection because the standards are introduced and analyzed independently of the sample. Error due to changes in the operating conditions can also be serious.

The normalization method has the disadvantage that the reported concentration of any one component depends on the concentration of every other peak analyzed in the mixture. Since the concentration of one peak is expressed as the ratio of the peak area to the total area of all peaks in the file, any change affecting total peak area, due to noise, baseline shifts, or any other factor can affect the reported concentration for a particular peak.

The internal standard method is the preferred method whenever it is practical. The component being analyzed and the internal standard component are the only two that affect the concentration reported for the component of interest.

The simplest form of the internal standard calculation is

$$C_i = \frac{f_i A_i}{f_s A_s} \times \frac{W_s}{W}$$

where C_i = amount of component i; f_i = relative response factor; A_i = area of peak identified as component i; f_s = relative response factor of internal standard; A_s = area of internal standard peak; W_s = weight (volume) of internal standard; and W = weight (volume) of sample. A sugar not found in the product being analyzed can be used as the internal standard.

☐ THIN-LAYER CHROMATOGRAPHY

In thin-layer chromatography, the solid support (silica gel, paper, or alumina) is spread in a thin layer over a large surface layer. The substances to be separated partition between the solid support and a moving liquid phase. Exchange of molecules between the solid and liquid phases occurs by adsorption–desorption as the solvent front moves over the surface; exchange between phases also may involve partition between solvent and an immiscible adsorbed liquid (e.g., water on cellulose).

Commercially prepared thin-layer plates consist of a glass or plastic plate coated with a thin coat of silica gel. The gel may contain fluorescing substances, which aid in locating the sample spots after separation. Applicators are available for preparing TLC plates. These applicators are adjustable to allow the preparation of thin layers for analytical separations or thicker layers for preparative work. Most manufacturer's of TLC apparatus furnish product application guides. We suggest using these products and guides rather than trying to prepare your own plates and reagents. Alltech Associates, Inc. (Deerfield, IL) offers a wide variety of special accessories and TLC plates. Different manufacturers furnish similar equipment.

Analysis of Food Additives

Many food additives can be separated and identified by TLC. The procedure described is based on AOAC method 19.041.

Materials and Reagents

1. Alltech Adsorbosil R– Plus 1 Hard Layer Prekote plates, 20 × 20 cm.
2. Migrating solvents—A 4:1 mixture (by volume) of hexane and ethyl acetate or a 97 + 3 v/v benzene and methanol.
3. Visualizing sprays—(a) 9:1 mixture of 0.1 *M* $KMnO_4$ and 0.1*N* NaOH; (b) hydrazine sulfate solution prepared by saturating 1 *M* HCl with $H_2NNH_2 \cdot H_2SO_4$ (*warning*: severe poison and suspected carcinogen); (c) 50 g/liter KOH in methanol; and (d) 10% phosphomolybdic acid in methanol.
4. Developing tanks and sprayers (for visualization reagents) (AOAC 29.006).

Procedure

Prepare mixtures of known food additives such as *p*-hydroxybenzaldehyde, vanillin, ethyl vanillin, veratraldehyde, piperonal coumarin, and vanitrope containing about 1.0 μg/μl of each additive. Spot samples (1–10 μl) of the known mixtures on TLC plates. The spots should be at least 3 cm from the edge of the plate, no larger than 0.25 cm in diameter, and at least 2 cm apart. Place spotted plates in developing tank containing solvent about 2 cm deep. Cover the tank and let stand until the solvent front has moved to within 2 cm of the side opposite the point of sample application. Carefully remove the plate, mark the solvent front, and air dry.

After drying, spray the plate with the $KMnO_4$ and NaOH mixture; this spray causes all of the additives except coumarin and vanitrope to appear tan on pink, which later turns to brown. When sprayed with a hydrazine sulfate solution, these additives are visualized as yellow spots. When dried and viewed under UV light (ca. 360 nm), *p*-hydroxybenzaldehyde may appear yellow; vanillin and ethyl vanillin, orange-brown; veratraldehyde, orange; and piperonal, blue-yellow. Shades of these colors are variable. Coumarin can be visualized by spraying with a methanolic KOH solution to give a bluish color when viewed under long wavelength UV light. Vanitrope reacts with a methanolic phosphomolybdic acid spray 10% to give blue color after drying at 100°C.

Results

The increasing order of migration in hexane–ethylacetate is vanillin, ethyl vanillin, veratraldehyde, coumarin, piperonal, vanitrope. The increasing order of migration in benzene–methanol is *p*-hydroxybenzaldehyde, ethyl vanillin, coumarin, veratraldehyde, piperonal-vanitrope.

Separation of Cis/Trans Isomers of Fatty Acid Esters

In this method, pretreatment of the silica gel layer with silver nitrate provides a complexing medium that affects, to varying degrees, the migration of similar compounds having different configurations. It is for this reason that the cis and trans methyl esters of oleic acid, which migrate together on untreated layers, are resolved completely when this method is applied.

Materials

1. Eastman Chromagram without indicator.
2. Eastman Chromagram developing apparatus.
3. Vacuum oven (or suitable substitute).
4. One large sheet of filter paper.
5. Ultraviolet light source (366 nm).
6. Sprayer (for visualization reagent).

Reagents

1. Silver nitrate solution—Dissolve 40 g of silver nitrate in 100 ml of water and adding 300 ml of ethyl alcohol.
2. Migrating solvent—Hexane/ethyl ether (9:1 v/v)
3. Visualization reagent—0.2% solution of 2′,7′-dichlorofluorescein in ethyl alcohol.

Procedure

Dip the Chromagram sheet into the silver nitrate solution for 1 min., then carefully blot dry with filter paper. Activate the sheet in a vacuum oven at room temperature for 1 hr prior to use. Spot the sheet about 2 cm from the lower edge with 5 to 10 μl of a solution that contains approximately 1 mg/ml of the isomers to be separated. Develop the sheet in the developing apparatus using a migrating solvent of hexane/diethyl ether (9:1 v/v). Visualization of the separated zones is accomplished by spraying with dichlorofluorescein followed by examination under UV light (366 nm).

☐ pH

The pH is defined as the negative logarithm of the hydrogen ion activity:

$$pH = -\log a_{H^+}$$

Since individual ionic activities cannot be evaluated easily, the operational National Bureau of Standards (NBS) scale of acidity has been developed. Using a glass electrode and a saturated calomel reference electrode, the operational definition of pH in a aqueous solution is

$$pH = pH_s + \frac{(E - E_s)F}{RT \ln 10}$$

where E = EMF of a cell containing the unknown solution; E_s = EMF of a cell containing a standard reference buffer solution of known or defined pH_s; and F = Faraday's constant. The NBS pH_s values were assigned from measurements so as to make pH_s as near as possible to $-\log a_{H^+}$. The pH is temperature dependent because K_a of the buffer system varies with temperature.

NBS primary standards cover the pH range from near 3.5 to 10.5 and were chosen for their reproducibility, stability, buffer capacity, and ease or preparation. Table 3.7 lists the pH_s values of NBS primary standards.

Table 3.7. NBS pH_s of Primary Standards

Temperature (°C)	pH_s					
	KH tartrate[a]	KH phthalate[b]	Phosphate (equimolal)[c]	Phosphate (3.5:1)[d]	Borax[e]	Carbonate[f]
0	—	4.003	6.982	7.534	9.460	10.321
10	—	3.996	6.921	7.472	9.331	10.181
20	—	3.999	6.878	7.430	9.227	10.064
25	3.557	4.004	6.863	7.415	9.183	10.014
30	3.552	4.011	6.851	7.403	9.143	9.968
40	3.547	4.030	6.836	7.388	9.074	9.891
50	3.549	4.055	6.831	7.384	9.017	9.831

[a] Sat. at 25°C (NBS 188).
[b] 0.05 *m* KH tartrate (NBS 185e).
[c] 0.025 *m* $KH_2 PO_4$; 0.025 *m* Na_2HPO (NBS 186-I-c; 186-II-c).
[d] 0.008695 *m* KH_2PO_4; 0.03043 *m* Na_2HPO_4; (NBS 186-I-c; 186-II-c).
[e] 0.01 *m* $Na_2B_4O_7 \cdot 10H_2O$ (NBS 187b).
[f] 0.025 *m* Na HCO_3; 0.025 *m* Na_2Co_3 (NBS 191; 192).

At best, pH is an estimate of $-\log a_{H^+}$, subject to error of instruments and conditions. It has been estimated that pH may correspond to the true a_{H^+} to within about $\pm 5\%$. As shown in Fig. 3.21 the glass pH electrode is constructed of a thin H^+-ion responsive glass membrane tip on nonresponsive glass and an internal reference electrode with a constant internal $[H^+]$. The internal electrode may be either Ag/AgCl in HCl or Hg/Hg_2Cl_2 in HCl. The cell requires an external reference electrode for operation. Combination electrodes are quite convenient because they have the external reference electrode built in with the glass electrode, thus simplifying maintenance and operation.

Glass is a noncrystalline three-dimensional arrangement of silicate tetrahedra in which each oxygen atom is shared by two silicate groups. Glass pH electrodes usually contain Li^+ and Ba^{2+} in place of Na^+ and Ca^{2+}. This makes the electrode more selective to H^+. When placed in an aqueous solution, cations from the glass membrane are leached and replaced by H^+ to form a hydrated silica-rich layer about 500Å thick. This layer can function as a cation-exchange membrane which gives a highly selective sensitivity to H^+. An electrical potential difference develops across the glass membrane that is dependent on the activities and particular characteristics of the ions present in the two solutions, the composition of the glass, etc.

Specific ion electrodes are currently being used for a wide variety of applications. Among the ions being determined are mono- and divalent cations such as Na^+, K^+, Ca^{2+}, and Mg^{2+} and anions such as NO_3^-, NO_2^-, F^-, Cl^-. Figure 3.20 also shows a diagram of a typical divalent cation specific ion electrode (liquid membrane electrode). The mode of

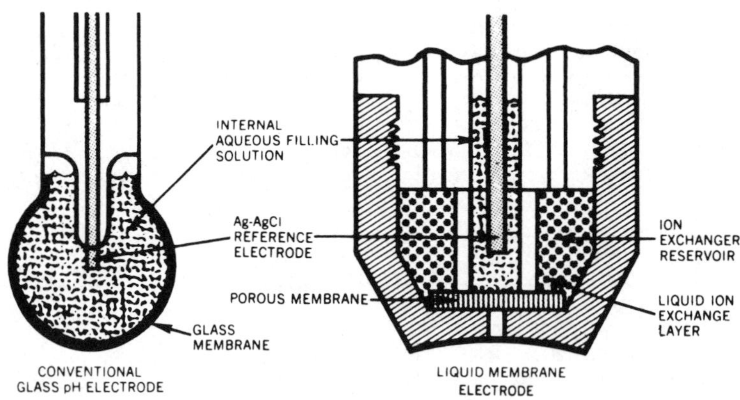

Figure 3.21. Conventional glass pH electrode (*left*) and specific ion, or liquid membrane, electrode (*right*).

detection is much like a conventional glass pH electrode. Instead of developing a potential across a glass membrane, this electrode develops a potential across a thin layer of water immiscible liquid cation exchanger. The ion exchanger, which is selective for divalent cations, is held in place by an inert porous membrane disk. An aqueous filling solution containing fixed levels of calcium and chloride ion contacts the inside surface of the membrane disk, which is saturated with the ion exchanger. The calcium ions in the filling solution provide a stable potential between the inside surface of the membrane and the internal filling solution, and the chloride ions provide a constant potential between the filling solution and the internal Ag/AgCl reference element. The electrode therefore develops potentials only in response to sample divalent cation activity.

Determinations of pH are now most frequently performed with solid state electronic pH meters using combination electrodes. Various sizes of electrodes are available.

The microelectrodes are quite reliable and require less sample. AOAC method 11.036 describes measuring the pH of wine; the same general procedure can be used for all types of aqueous solutions. Unfortunately, AOAC method 11.036 does not give a true measurement of pH ($-\log[H^+]$ or $-\log a_{H^+}$) since wine contains approximately 9–12% ethanol. Measurements in a partially aqueous environment really are only valid for comparison. Another source of error in any pH measurement is the response of the electrode to Na^+ or other ions similar to H^+; a solution is high in these ions, corrections must be made to obtain a true pH value.

The general procedure for any pH measurement includes these steps:

1. Turn on pH meter and allow to equilibrate.
2. Check electrode(s) for proper electrolyte levels.
3. Calibrate pH meter using pH 4, 7, and 10 standards.
4. Note temperature of standards since pH is temperature dependent. NBS standard buffers are available and will list the accepted pH at various temperatures.
5. Use deionized distilled water to wash the electrodes before measuring pH of sample.
6. Determine the pH of the sample to the nearest 0.01 pH unit.

☐ POLAROGRAPHY

Interest in polarographic methods for the quantitative analysis of metals and certain organics has increased recently due to the availability of low-cost and reliable commercial polarographs. Newer instruments permit analyses that are quite selective down to the μg/liter levels. AOAC 25.089 is a procedure for determining lead in fish. Other applications include AOAC 29.050 for the determination of diazinon, malathion, methyl parathion, and parathion.

Figure 3.22 presents a schematic of a simple polarograph. The dropping mercury electrode is a glass capillary attached to a mercury reservoir (Fig. 3.23). Drops of mercury fall from the top of the capillary at a constant rate (5–30 drops/min). Each drop of mercury acts as the electrode as long as it is attached to the column of mercury.

A typical polarogram is shown in Fig. 3.24. The limiting current is proportional to the concentration of the sample being analyzed. By comparing i_1 of different samples vs standard additions, the concentration of an unknown may be determined (Fig. 3.25). The limiting current (i_1) is the sample with a known amount of added standard (C_{std}), while i_2 is the sample *per se* (C_{unk}). A volume (V) of the sample solution (C_{unk}) is transferred into the polarographic cell and the curve recorded to give i_2. Then a volume (v) of the standard solution (C_{std}) is added and a second curve is recorded to give i_1. The concentration of the sample solution can be calculated according to

$$C_{\text{unk}} = C_{\text{std}} \frac{i_2}{i_1 + [(i_1 - i_2)(V/v)]}$$

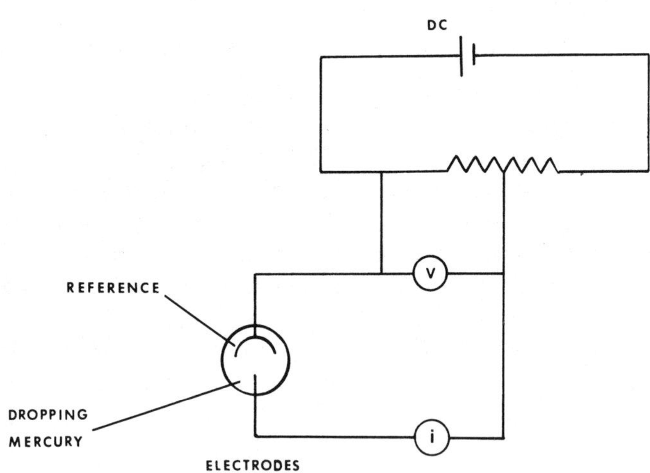

Figure 3.22. Schematic diagram of simple polarograph.

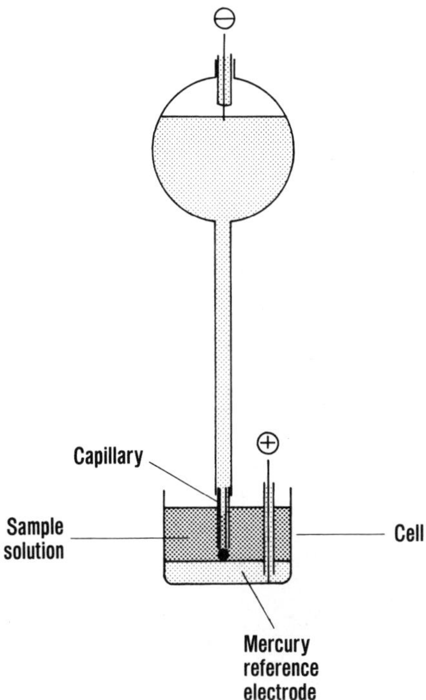

Figure 3.23. Dropping mercury electrode used in polarography.

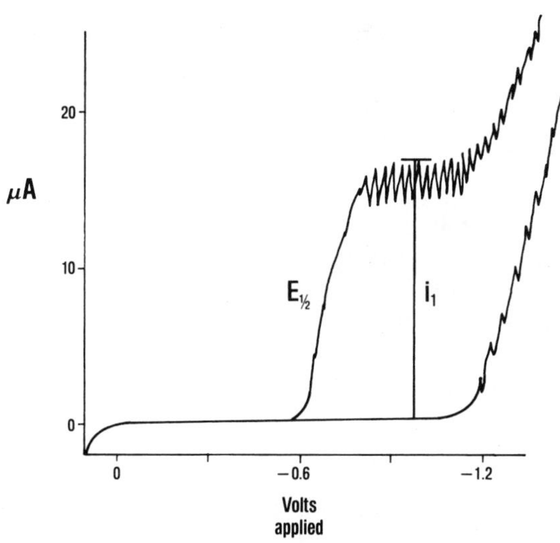

Figure 3.24. A typical polarogram.

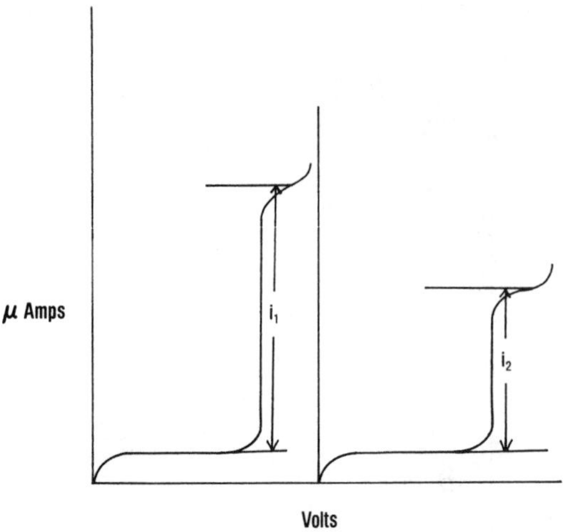

Figure 3.25. Polarograms illustrating determination of concentration by method of standard additions. See text for explanation.

If v is 1% or less of V then the equation simplifies to

$$C_{\text{unk}} = C_{\text{std}} \frac{i_2 v}{(i_1 - i_2)V}$$

Determination of Lead in Milk

Apparatus

Any commercial fast-sweep polarographed properly equipped with cells, electrodes, capillaries, connectors, and related accessories may be used. Pulsed DC polarographs may also be used. Follow the manufacturer's instructions and recommendations.

Standard Curve

Prepare standard solutions containing 0, 0.1, 0.2, 0.4, 0.8, 1.0, 2.0, 5.0, and 10.0 μg Pb/ml in 1 *N* HCl. Place 5 ml of the standard solution in the polarographic cell (temperature should be maintained at 25°C). Bubble N_2 through the solution for 5 min, then polarograph between −0.1 and −0.8 volts using a H_g reference electrode. The peak voltage is near −0.45. Prepare a standard curve by plotting μg Pb/ml cell solution vs maximum wave height (at appropriate sensitivity setting).

Sample Determination

Dilute a sample of whole skim milk and add 5 ml to the polarographic cell. After adjusting to 25°C, bubble N_2 through the solution for 5 min (or until all oxygen is removed). Polarograph as for the standard curve determination. Determine the maximum wave height at the potential of Pb (approximately −0.45 volts). Check for the proper potential by adding a known quantity of Pb to the unknown and determining the total wave height.

Some samples will have interfering elements. Tin polarographs at a peak potential that is essentially identical to Pb. To avoid this interference problem, add 1 ml of NH_4OH (conc.) and 0.4 g of tartaric acid to the cell solution and treat as in the standard curve measurement. A standard curve should be prepared with each standard containing 1 ml of NH_4OH and 0.4 g of tartaric acid. The peak potential for Pb is shifted to −0.54 volts at which no response is observed for Sn.

Calculate the Pb concentration in the unknown as

$$C_{\mathrm{unk}} = \frac{C_{\mathrm{std}} \times ht_{\mathrm{u}} \times \text{scale factor}_{\mathrm{u}}}{ht_{\mathrm{s}} \times \text{scale factor}_{\mathrm{s}}}$$

where C_{unk} = concentration of unknown in μg Pb/ml cell solution; C_{std} = concentration of standard in μg Pb/ml cell solution; ht_{u} = wave height of unknown; ht_{s} = wave height of standard; and scale factor = sensitivity factor setting for respective sample standard. Concentration of the original sample must take into account the dilution factor.

If a sample is a solid or insoluble, ashing, sonication, or any technique of solubilization may be used to solubilize the Pb. Care should be taken to keep ashing temperatures below the volatilization temperature for Pb. Recovery studies indicate that 400°–475°C for 12 hr will ash most tissues without loss of Pb.

☐ X-RAY FLUORESCENCE

X-ray fluorescence is a useful and convenient method for determining both metals and nonmetals either qualitatively or quantitatively. Basically, this method involves the observation of characteristic x-rays emitted from a sample subsequent to fluorescence by photons from a suitable excitation radioactive source. For fluorescence via low-energy photon bombardment, the photoelectric effect is predominantly responsible for the removal of K electrons from atoms in the sample. Radiation produced by an L electron falling into the K vacancy is referred to as a K_α x-ray. The x-rays produced when K-shell vacancies are filled by electrons from the M and N shells are called K_β x-rays.

A typical source-excited x-ray fluorescence system has (1) a 25-millicurie Cd-109 source or similar source, (2) an energy-dispersive spectrometer utilizing a high-resolution Si–Li detector, and (3) a multichannel pulse height analyzer.

Determination of Selected Elements in Tissues

Bromine in fish is an example of an element that is not analyzable by atomic absorption or other conventional methods. However, bromine produces K_α x-rays of 11.923 keV and K_β of 13.290 keV when irradiated with a Cd-109 fluorescing source.

After identification of the peak in the observed spectrum that corresponds to the bromine K_α energy (also K_β), the sum of the counts per unit time, corrected for background, can be related to the bromine concentration in the sample. For absolute quantitative analysis, matrix effects, sample density, and sample geometry must be considered. By keeping conditions constant, a relative method of comparison is easily established that permits quantitative evaluation of samples.

Counting intervals of 1000 seconds are generally adequate for samples that contain bromine in a concentration range of 25–250 μg/g of dry tissue. Samples that contain less than 25 μg/g require 2000-sec intervals.

Procedure

Fish or other tissue samples are washed in deionized water and ground with an equal volume of deionized water in a glass homogenizer (Ten-Broeck). The resultant slurries are lyophilized at −50°C and 50 μm Hg.

Plastic planchets (25 mm diam × 20 mm) are fitted with an x-ray transparent window (4-μm Mylar film). Approximately 0.25–0.5 g of dry tissue is placed in the sample cup.

Sample irradiation is accomplished by a 25-millicurie Cd-109 excitation source. Excitation time is usually >1000 sec. A typical spectrum is shown in Fig. 3.26. Using the Cd-109 source, elements 19 through 38 are easily determined.

The concentration of bromine (all forms, organic and inorganic) in a particular sample is best determined by the method of standard additions. This can be determined by adding the proper amount of a 1-mg/ml solution of bromide (KBr) to separate samples, grinding to insure homogeneity, and relyophilizing at −50°C and 50 μm Hg.

When this procedure was conducted on a freshwater fish sample collected in the Middle Tennessee area, extrapolation of the graph from the data obtained by counting each standard addition sample gave 29 ± 1.5 μg Br/g of dry tissue. Neutron activation analysis of the same

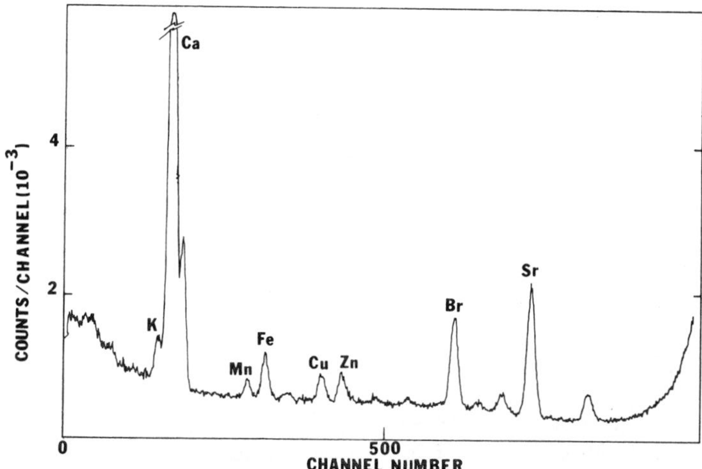

Figure 3.26. X-ray fluorescence spectrum of crayfish tissue.

sample was carried out for comparison. A Cf-252 source with a neutron flux of 1.42×10^8 neutrons/cm^2/sec was used. Sample irradiation time was 111 hr with a waiting time of 32 hr before counting. The sample was then counted for approximately 17 hr to ensure sufficient statistical accuracy. The neutron activation analysis of the sample gave a bromine content of 21 ± 1 μg Br/g of dry tissue.

X-ray fluorescence analysis can be used to determine elements other than bromine, including elements atomic number 19–38.

☐ SCANNING ELECTRON MICROSCOPY

The scanning electron microscope (SEM) has greatly facilitated the examination of food surface characteristics, particle size, and shape (see Chapter 4) and insect contamination (see Chapter 13). Large specimens can be accommodated in the specimen chamber and orientation is possible from many angles. The instrument provides images characterized by a great depth of field, sharp focus, and unique contrast. Specimens can be magnified from 10 to 100,000 times, with a routinely obtainable resolution of 100 Å. The SEM images of surfaces are normally portrayed photographically to allow convenient interpretation of structure and determination of morphological relationships. By making photographic pairs separated by a small angle, the microscopist can obtain stereoscopic photographs that allow for three-dimensional observation and analysis.

The SEM may also be used in conjunction with an energy dispersive x-ray system for quantitative elemental analysis. Using a computer-based system such as the TN 5500 (Tracor-Northern, Middletown, WI), elements may be analyzed simultaneously and displayed as true intensity x-ray maps. Energy dispersive x-ray analysis allows one to analyze particulate in foods as well as an overall elemental mapping of the total SEM field.

SELECTED REFERENCES

ALLTECH ASSOCIATES, INC. 1983. Chromatography Catalog No. 50. Deerfield, IL.

AACC. 1976. Approved Methods of the American Association of Cereal Chemists. Vols. I and II. American Assoc. of Cereal Chemists, St. Paul, MN.

AOCS. 1979. Official and Tentative Methods of the American Oil Chemists' Society, Vols. I and II. American Oil Chemists' Society, Champaign, IL.

AOAC. 1980. Official Methods of Analysis, 13th ed. W. Horwitz (editor). Assoc. of Official Analytical Chemists, Arlington, VA.

AOAC. 1984. Official Methods of Analysis. 14th ed. S. Williams, (editor). Assoc. of Official Analytical Chemists, Arlington, VA.

AURAND, L. W. and WOODS, A. E. 1973. Food Chemistry. AVI Publishing Co., Westport, CT.

BARBI, N. C. 1980. Electron Probe Microanalysis Using Energy Dispersive X-Ray Spectroscopy, Princeton Gamma-Tech, Inc., Princeton, NJ.

BAUER, H. H., CHRISTIAN, G. D., and O'REILLY, J. E. 1978. Instrumental Analysis. Allyn and Bacon, Inc., Boston, MA.

EASTMAN KODAK CO. Analytical Procedure Bulletin-TLC. Rochester, NY.

ENGEL, KARL-HEING and TRESSL, ROLAND. 1983. Studies on the volatile components of two mango varieties. J. Agric. Food Chem. *31,* 796–801.

FLATH, R. A., FORREY, R. R., and TERANISHI, R. 1969. High resolution vapor analysis for fruit variety and fruit product comparisons. J. Food Sci. *34,* 382–386.

FOXBORO/ANALABS. 1983. Chromatography Catalog No. 23. North Haven, CT.

KWAN, WING-ON and KOWALSKI, BRUCE R. 1980. Pattern recognition of gas chromatograph data. Geographic classification of wines of *Vitis vinifera* cv. Pinot Noir from France and the United States. J. Agric. Food Chem. *28,* 356–359.

MOSHONAS, MANUEL G. and SHAW, PHILLIP E. 1984. Direct gas chromatographic analysis of aqueous citrus and other fruit essences. J. Agric. Food Chem. *32,* 526–530.

NOBLE, ANN C., FLATH, ROBERT A., and FORREY, RALPH. 1980. Wine headspace analysis. Reproducibility and application to varietal classification. J. Agric. Food Chem. *28,* 346–353.

OGATA, MASANA and FUJISAWA, KUNIYASU. 1983. Capillary GC/MS determination of organic sulfur compounds detected in oyster and mussel caught in the sea as an oil pollution index. J. Chromat. Sci. *21,* 420–424.

PIERCE CHEMICAL CO. 1982. Chromatography Catalog. Rockford, IL.

POMERANZ, Y. and MELOAN, C. E. 1971. Food Analysis: Theory and Practice AVI Publishing Co., Westport, CT.

ROSEN, ROBERT T., ROSEN, JOSEPH D., and DIFROSIMO, VINCENT P. 1984. Confirmation of aflatoxins B_1 and B_2 in peanuts by gas chromatography/mass spectrophotometry/selected ion monitoring. J. Agric. Food Chem. *32,* 276–278.

SEGEL, I. H. 1976. Biochemical Calculations. 2nd ed. John Wiley and Sons, New York.

WILLARD, H. H., MERRITT, L. L., JR., and DEAN, J. A. 1980. Instrumental Methods of Analysis. 6th ed. D. Van Nostrand Co., New York.

WOODS, A. E., and AURAND, L. W. 1977. Laboratory Manual in Food Chemistry. AVI Publishing Co., Westport, CT.

WOODS, A. E., CARLTON, R. F., CASTO, M. E. and GLEASON, G. I. 1979. Environmental bromine in freshwater and freshwater organisms: Factors affecting bioaccumulation. Bull. Environm. Contam. toxicol. *23,* 179–185.

YERANSIAN, JAMES A., SLOMAN, KATHERINE G., and FOLTZ, ARTHUR K. 1985. Food. Analt. Chem. Appl. Rev. *57,* 278R–315R.

YU, LIANG ZU, INOKO, MASANORI, and MATSUNO, TAKEO. 1984. A method for rapid determination of food additives in vegetable oils. J. Agric. Food Chem. *32,* 683–685.

ZAPSALIS, C. and BECK, R. A. 1985. Food Chemistry and Nutritional Biochemistry. John Wiley and Sons, New York.

4

Carbohydrates

☐ INTRODUCTION

The word *carbohydrate* was originally derived from the fact that the greater part of the compounds in this class had the empirical formula $C_n(H_2O)_n$. The values of n ranged from three to many thousands. This formula is now considered too restrictive, and a more useful definition might be "polyhydroxy aldehydes or ketones and their derivatives." This latter definition encompasses deoxy sugars, sugar alcohols, sugar acids, and amino sugars.

Carbohydrates occur in fruits and vegetables as storage reserves in seeds, roots, and tubers; in the sap; and as constituents of the structural tissues. They are also found in the milk, blood, and tissues of animals.

Carbohydrates are the most abundant food in the world and the most economical as an energy food source. In the American diet, carbohydrate foods constitute approximately 50% of the daily caloric intake. However, carbohydrates make up an even greater proportion of the diet of peoples of other countries where cereals are a staple. In addition to their value as an inexpensive source of energy, carbohydrates are important for several other reasons. The body needs carbohydrates in order to use fat efficiently. Many foreign substances are removed from the body through the intermediate formation of glycosides of glucuronic acid. Diseases such as diabetes develop when the body is unable to utilize sugar properly. Some carbohydrates have an effect on the type of bacteria that will grow in the intestine. In some of the lower animals (crab and lobster), a major constituent of the exoskeleton is a polymer of glucosamine.

Certain industries (such as milling, baking, brewing, syrup, and sugars) are based on carbohydrates. Thus, it is readily seen that carbohydrates are at the very foundation of the economic structure of our society.

The carbohydrates are divided into two large groups—simple sugars and compound sugars—based upon the number of sugars that are obtained when the various carbohydrates undergo hydrolysis. Carbohydrates such as glucose and fructose, which cannot be hydrolyzed into simpler compounds, are called *monosaccharides*. The compound sugars are made up of two or more molecules of monosaccharides and are further divided into two broad categories—*oligosaccharides* and *polysaccharides*. Oligosaccharides, which are composed of two to nine monosaccharides, include sucrose (a disaccharide) and raffinose (a trisaccharide). Polysaccharides consist of ten or more monosaccharide units. They may be separated into three broad groups: homopolysaccharides (one kind of monosaccharide unit); heteropolysaccharides (two or more kinds of monosaccharides); and nitrogen-containing polysaccharides. The monosaccharides and disaccharides are sweet tasting and soluble in water. In contrast, the polysaccharides are colloidal, dispersible under certain conditions in water, tasteless, and vary greatly in digestibility.

The carbohydrates of interest to the food chemist include the following:

- Simple Sugars or Monosaccharides
    - Pentoses—arabinose, xylose, ribose
    - Hexoses
        - Aldohexoses—galactose, glucose
        - Ketohexose—fructose
- Oligosaccharides
    - Disaccharides
        - Reducing—maltose, lactose
        - Nonreducing—sucrose
    - Trisaccharides
        - Nonreducing—raffinose, gentianose

- Polysaccharides
  - Homo—(one kind of monosaccharide unit)
    - Pentosans—xylan, araban
    - Hexosans
      - Glucosans—starch, dextrin, glycogen, cellulose
      - Fructosan—inulin
      - Mannan
      - Galactan
  - Hetero—(two or more kinds of monosaccharide units) pectins, gums, mucilages
  - Nitrogen-containing—chitin

These various types of carbohydrates are discussed in detail later in this chapter. But first, several important general properties of carbohydrates are considered.

☐ GENERAL PROPERTIES

Optical Activity

One very important characteristic of sugars is their ability to rotate rays of polarized light. This property is referred to as optical rotation, and such compounds are said to be optically active.

The instrument used for measuring optical activity is called a polarimeter. The polarimeter employs monochromatic light from a sodium light and, in its simplest form, has two Nicol prisms. The prism nearest the light source is in a fixed position and is known as the polarizing prism; the prism nearer the eye of the observer is movable and is called the analyzer. If the two prisms are arranged so that their optical axes are in the same plane, essentially all of the radiation will pass through. If the optical axis of the analyzer is at right angles to that of the polarizing prism, the radiation will be totally absorbed. This is termed total extinction. The instrument is also equipped with a scale to indicate the number of degrees through which the analyzer is rotated. The zero point may be set at the point where the two Nicol prisms are crossed without the sample in the polarized beam. If an optically active compound is placed between the prisms, the plane of polarized light is rotated either to the right or to the left. The operator rotates the analyzer until the prisms are again crossed, and the angle through which the analyzer is turned is equal to the angular rotation of the optically active compound. If it is necessary to rotate the analyzer to the right to accomplish total extinction, the optically active compound is dextrorotatory (+). If, on the other hand, the analyzer is rotated to the left, the compound is said to be levorotatory (−).

The angular rotation of an optically active compound in solution is directly proportional to the concentration of the compound, the length of the column of solution through which the light passes, and the rotating power of the substance. The term *specific rotation*, [α], was introduced to define this relationship and may be defined as the angular rotation in degrees of a solution containing 100 g of solute in 100 ml of solution when read in a tube 1 decimeter long:

$$[\alpha] = \frac{100\,A}{lc}$$

where A is the rotation (plus or minus), l is the length of the tube in decimeters, and c is the concentration in grams per 100 ml of solution. Specific rotation is a function of temperature as well as the wavelength of light. Therefore, in order to standardize data, specific rotation is indicated as $[\alpha]_D^{20}$, which refers to the specific rotation at 20°C with light from the D line of the spectrum.

The optical activity of an organic compound is due to asymmetry of the molecule. An asymmetric carbon atom may be defined as an atom that has four different atoms or atomic groups attached to it. For example, glyceraldehyde, contains an asymmetric carbon atom to which are attached four different radicals. Molecules possessing an asymmetric carbon atom exist in two mirror images, one of which is dextrorotatory and the other levorotatory. Formulas **4.1** and **4.2** represent the two forms of glyceraldehyde. Since these sugars are isomeric with each other, they are said to be *stereoisomers*. This term refers to the geometrical arrangement of the atoms and atomic groups in space. The stereochemistry is designated D and L and will be discussed later.

```
        O                          O
        ||                         ||
        C—H                        C—H
        |                          |
H ▬▬▬ C ▬▬▬ OH        HO ▬▬▬ C ▬▬▬ H
        |                          |
      CH₂OH                      CH₂OH
```

D(+)-Glyceraldehyde

4.1

L(−)-Glyceraldehyde

4.2

A mixture of equal parts of D and L isomers is optically inactive because the isomers rotate polarized light the same number of degrees in opposite directions, the net effect being zero. The mixture is referred to as a *racemic mixture* and is an example of external compensation.

When a molecule contains more than one asymmetric carbon atom, it is optically active if the rotatory effects of the asymmetric carbon atoms do not neutralize one another. If the molecule is symmetrical in structure, that is, if a plane of symmetry can be passed through it, the molecule is optically inactive. To illustrate this, consider formulas **4.3**, **4.4**, and **4.5** for the different forms of tartaric acid. If one constructs models of **4.3**, **4.4**, and **4.5**, it may be noted that **4.5** possesses a plane of symmetry; i.e., the molecule can be divided into two mirror image halves and is optically inactive. Conversely, **4.3** and **4.4** are asymmetrical and thus are optically active.

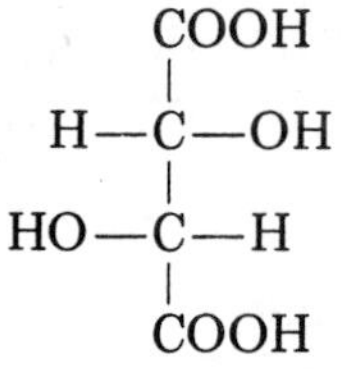

$$\begin{array}{c} COOH \\ | \\ H-C-OH \\ | \\ HO-C-H \\ | \\ COOH \end{array} \qquad \begin{array}{c} COOH \\ | \\ HO-C-H \\ | \\ H-C-OH \\ | \\ COOH \end{array} \qquad \begin{array}{c} COOH \\ | \\ H-C-OH \\ | \\ H-C-OH \\ | \\ COOH \end{array}$$

L(+)-Tartaric acid **4.3** D(−)-Tartaric acid **4.4** *meso*-Tartaric acid **4.5**

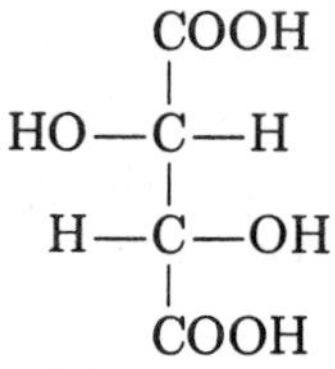

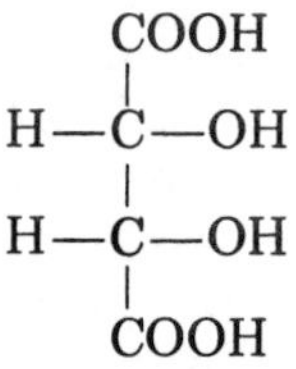

Levo, dextro, and *meso*-tartaric acids are stereoisomers; that is, they have exactly the same functional groups but these groups are arranged differently in space. *Dextro*- and *levo*-tartaric acids are optical isomers (*enantiomorphs*) in which the groups about the asymmetric carbon atoms are exactly reversed in space as mirror images and rotate polarized light equally and oppositely. Racemic tartaric acid is composed of an equal mixture of the dextro and levo acids. It is optically inactive because of the external compensation. Whenever organic compounds that contain asymmetric carbon atoms are synthesized, a racemic mixture of optical isomers is produced.

We have illustrated this discussion of optical activity by glyceraldehyde and tartaric acids, both of which are rather simple compounds. Most organic compounds of biological importance are much more complicated, usually having several asymmetric carbon atoms. Thus, the possibilities for optical isomerism are greatly increased. In a simple compound like a hexose there are four asymmetric carbon atoms and sixteen possible optical isomers.

Structure

A discussion of the determination of carbohydrate structure would be involved and beyond the scope of this book. For such information, the reader is referred to texts on carbohydrates (e.g., Pigman and Horton 1972). The food chemist is concerned with the identification of carbohydrates primarily because their properties and reactions are dependent on the structure of the molecule. For example, the properties and reactions of monosaccharides depend partly on the position of the hydroxyl group and the number of carbon atoms in the molecule. Similarly, the properties and reactions of complex carbohydrates depend on the number of units in the chain, the kind of units in the chain, and the position of the linkages that join the individual units together. This involves some knowledge of carbohydrate structure; consequently, structural formulas will be given in the discussion of individual carbohydrates.

When the structures of the monosaccharides were first determined, they were pictured as straight-chain aldehyde or ketone polyols. For example, D-glucose and L-glucose were assigned formulas **4.6** and **4.7**.

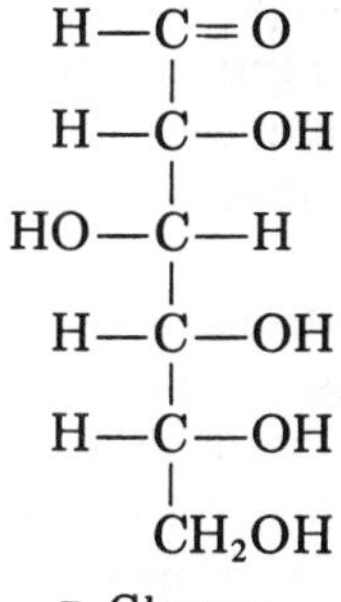

$$\begin{array}{rcl} H- & C & =O \\ & | & \\ H- & C & -OH \\ & | & \\ HO- & C & -H \\ & | & \\ H- & C & -OH \\ & | & \\ H- & C & -OH \\ & | & \\ & CH_2OH & \end{array} \qquad \begin{array}{rcl} O= & C & -H \\ & | & \\ HO- & C & -H \\ & | & \\ H- & C & -OH \\ & | & \\ HO- & C & -H \\ & | & \\ HO- & C & -H \\ & | & \\ & CH_2OH & \end{array}$$

D-Glucose **4.6**   L-Glucose **4.7**

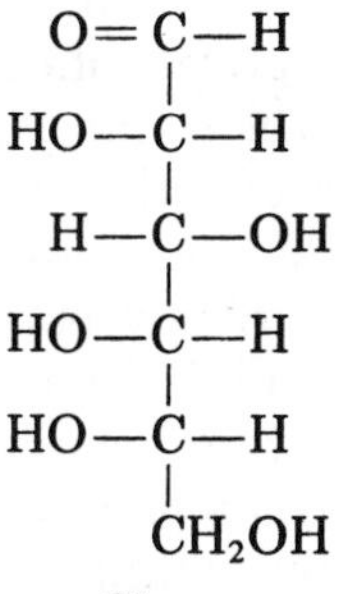

When the structure of the glucose molecule is arranged and projected according to the Fischer system (aldehyde on top and primary alcohol at bottom; hydrogens and hydroxyl groups on the other carbon atoms projecting up out of the plane of the paper), we find that the assignment to the D or L series of compounds depends on whether the hydroxyl group of the asymmetric carbon atom farthest from the aldehyde group is projected to the right (D) or the left (L).[1] Stereoisomers that are structurally related to D-glyceraldehyde are designated D-sugars (**4.8**)

$$\begin{array}{c} O \\ \| \\ C{-}H \\ | \\ H \blacktriangleright C \blacktriangleleft OH \\ | \\ CH_2OH \end{array} \qquad \begin{array}{c} R \\ | \\ (CHOH)_n \\ | \\ H \blacktriangleright C \blacktriangleleft OH \\ | \\ CH_2OH \end{array}$$

D-Glyceraldehyde **D-Sugar**

4.8

and the corollary L-sugars are related to L-glyceraldehyde. These D and L designations do *not* indicate the direction in which the sugar rotates the plane of polarized light. If it is desired to indicate the direction of optical rotation, (+) indicates to the right or dextrorotatory, (−) to the left or levorotatory.

The straight-chain structure for the sugars is not consistent with several of their properties. For example, sugars fail to respond to Schiff's reagent (test for aldehydes) under the usual conditions of the test, indicating that no appreciable amount of free or potential aldehyde is present. In addition, freshly prepared solutions of sugars exhibit a change in optical rotation on standing, indicating changes in the asymmetry of the sugar. Further evidence is the isolation of two crystalline isomers of both D and L forms of the sugars. When the two isomers of D-glucose (designated α and β) are dissolved in water, they are found to have different optical rotations: $[\alpha]_D^{20} = +112.2°$ for α-D-glucose; $[\alpha]_D^{20} = +18.7°$ for β-D-glucose. When aqueous solutions of either the α or β forms are allowed to stand, the specific rotations of each changes to a common rotation of $[\alpha]_D^{20} = +52.7°$. Thus the β form increases in positive rotation, while the α form decreases in positive rotation. It should be noted that $[\alpha]_D^{20} = +52.7°$ represents approximately 63% of β and 37% of α with probably minute amounts of the free aldehyde form.

[1] According to the Cahn-Ingold-Prelog conventions, R and S are used rather than D and L, respectively. Thus, D-glyceraldehyde is designated as R-glyceraldehyde and L-glyceraldehyde is designated S-glyceraldehyde. The convention is particularly useful when a compound contains several asymmetric carbon atoms. The configuration of each group around a specific asymmetric carbon atom can be specified. For example, D-erythrose becomes 2(R),3(R)-erythrose. Thus, the configuration of both hydroxyl groups are designated.

The interconversion of the α form of any sugar to its β form, or vice versa, is termed *mutarotation*. This phenomenon is catalyzed by the addition of dilute acid or alkali. The enzyme mutarotase, present in extracts from the mold *Penicillium notatum*, also catalyzes the mutarotation of glucose. Discovery of mutarotation led to the representation of the sugars in Fischer-Tollens cyclic formulas (**4.9** and **4.10**), which permitted an additional asymmetric carbon atom in the chain at the carbonyl functional group. It should be noted, however, that this functional group still exists as a potential aldehyde or ketone group.

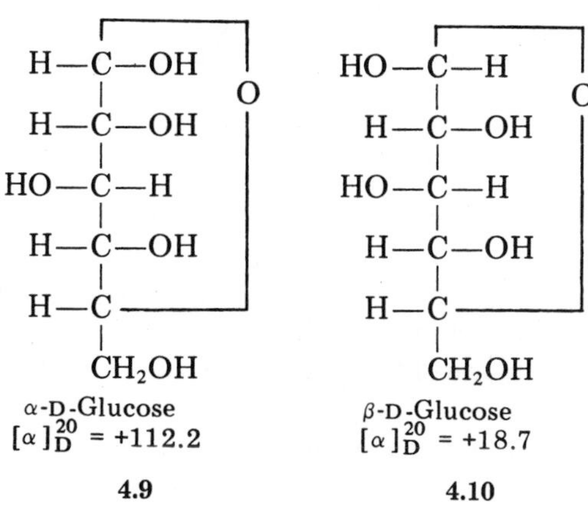

α-D-Glucose $[\alpha]_D^{20} = +112.2$

4.9

β-D-Glucose $[\alpha]_D^{20} = +18.7$

4.10

In depicting the structure of an α isomer, we place the hydroxyl group of the hemiacetal carbon on the same side of the carbon chain as the hydroxyl group that determines whether the sugar is D or L. In practical terms, for sugars for the D series, the hemiacetal hydroxyl group of α forms would be shown with the hydroxyl group projecting to the right. The potential aldehyde or ketone carbon is called the *anomeric carbon*, and α and β isomers thus are referred to as *anomers*.

The use of the cyclic formula introduced the question of ring size into the chemistry of the sugars. It is known that a cyclic hemiacetal involving either carbon 4 or carbon 5 can be formed. In the first case, a ring is created consisting of four carbon atoms and one oxygen atom; this is called a *furanose* ring. In the second case, the resulting ring consists of five carbon atoms and one oxygen atom; this is a *pyranose* ring. These names were derived from the five- and six-membered cyclic ethers furan and pyran. The pyranose form is the more stable of the two and is the one most frequently found in nature.

The cyclic Fischer formula, while useful to indicate differences in structure at the different asymmetric centers, does not properly represent the actual molecular configuration (bond angles and distances) of the sugars. For example, in D-glucose, carbons 1 and 5, which are involved in the oxygen bridge, must be close together for the existence of this bridge. In an attempt to illustrate better the structure and configuration of the sugar molecule, Haworth proposed a system of symmetric rings (**4.11** and **4.12**) that avoids the impossible stretching of the electron pairs of the oxygen in the ring.

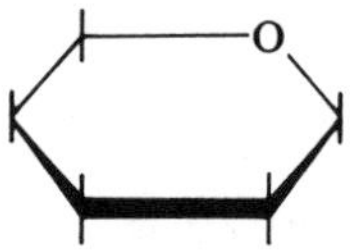

Pyranose

4.11

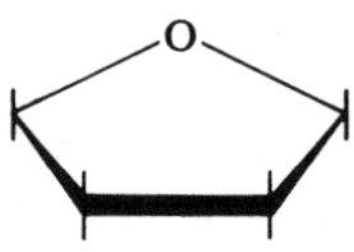

Furanose

4.12

In this system, the plane of the ring is conceived to be perpendicular to the plane of the page, and the attached groups lie either above or below the plane of the ring. The thin lines of the ring are conceived as being behind the plane of the page while the thick lines are in front of the plane of the page.

We may apply certain rules to translate the Fischer projection to the Haworth form:

1. Any group to the right of the carbon chain is written below the plane of the ring; those to the left are written above the plane of the ring. For example:

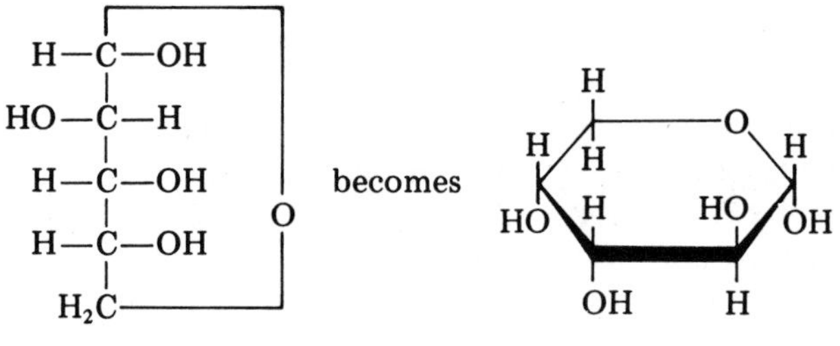

H—C—OH
HO—C—H
H—C—OH
H—C—OH
H_2C—
O

α-D-Arabinose

becomes

H, H, H, HO, H, O, H, HO, OH, OH, H

2. When there are more carbon atoms in the sugar than are involved in ring formation (e.g., hexose), it is necessary to determine whether these atoms should be written above or below the plane of the ring. If the ring is to the right, the extra carbon will be up: conversely,

if the ring is to the left, the extra carbon will be down (i.e., below the plane of the ring). For example:

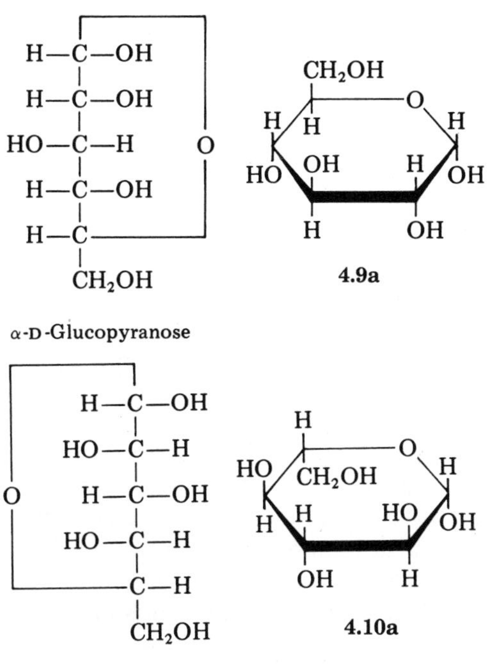

α-D-Glucopyranose

β-L-Glucopyranose

3. In the case of a ketohexose in the furanose form, the position of carbon 6 is determined as in (2). The position of carbon 1 is up for the α form and down for the β form. For example:

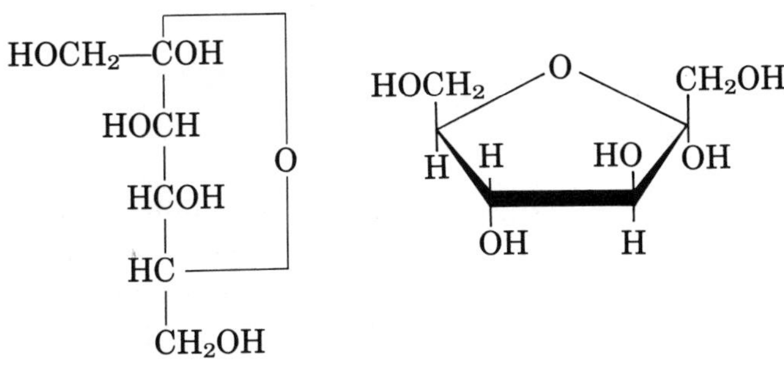

α-D-Fructofuranose

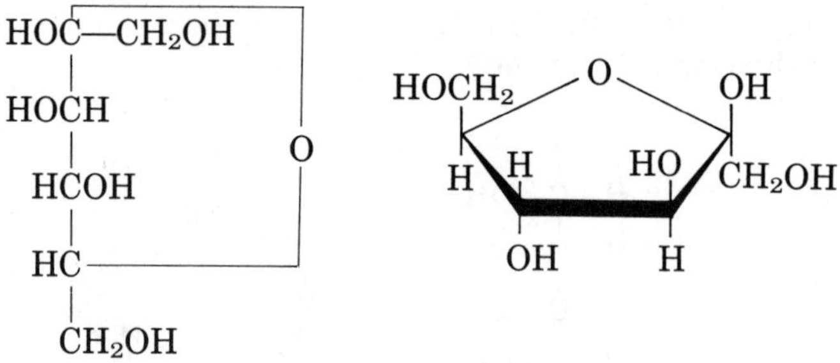

HOC—CH_2OH
HOCH
HCOH
HC
CH_2OH
O

$HOCH_2$ O OH
H H HO CH_2OH
OH H

β-D-Fructofuranose

Although the Haworth type of formula represents the structure of a sugar molecule better than the Fischer-type formula, it does not reveal the true geometry of the pyranose rings or furanose rings. A Haworth formula is oversimplified in that it places all atoms in a single plane; however, the normal valence angle of the carbon atom prevents a stable planar arrangement for the atoms. Studies of molecular models have shown that the five-membered furanose ring is strained and non-planar while the six-membered pyranose ring is strainless and exists in two arrangements in space, the *chair* and *boat* forms. The chair form is the structure of lower energy and thus more stable. In contrast, the boat form is flexible and has an unlimited number of conformations. The pyranose ring can be compared to cyclohexane, which also can exist in both chair and boat forms.

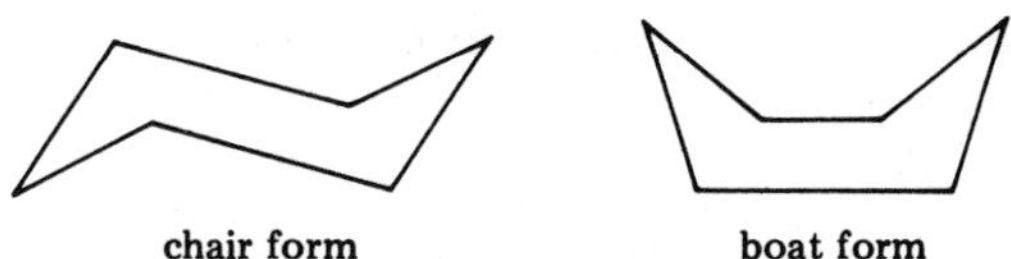

chair form **boat form**

Examination of structural models of cyclohexane reveals that hydrogen atoms are arranged on the ring in two different ways: parallel to an imaginary plane of the ring (equatorial) and perpendicular to this plane (axial). These relationships are shown in **4.13**, the chair form of cyclohexane.

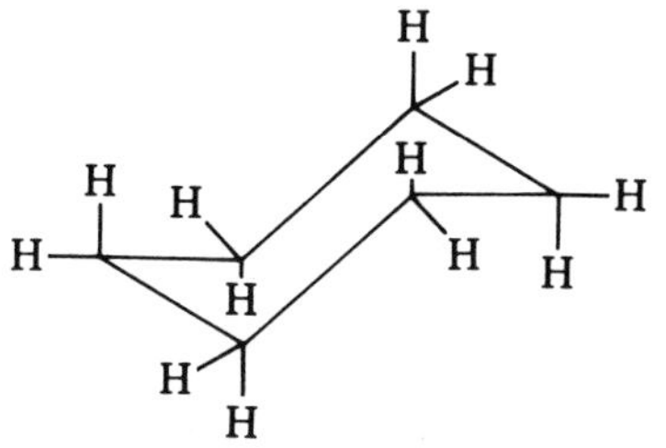

H H H H H H H H H H H H

Cyclohexane

4.13

The conformational representation of α-D-glucopyranose in the chair form is given by **4.9b**.

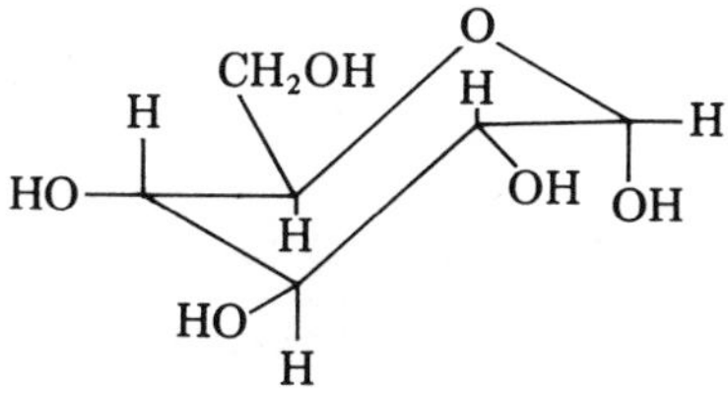

CH_2OH O H H H HO OH OH H HO H

α-D-Glucopyranose

4.9b

The structures of oligo- and polysaccharides are more complex than those of the simple sugars because of the additional bonds between their component monosaccharide units and the multiplicity of configurations they can assume. Because of these additional structural features, compound sugars exhibit some physical and chemical properties that simple sugars do not have.

Reactions

The most important, and common, reactions of sugars involve the carbonyl and hydroxyl groups. The discussion in this section focuses on the reactions of monosaccharides; however, once a compound sugar is degraded, the resulting simple sugars can undergo their usual reactions.

Carbonyl Reactions

Oxidation. The aldehyde and ketone groups of sugars can be oxidized to yield different products depending on the conditions. Complete oxidation, which also involves oxidation of the hydroxyl groups, yields carbon dioxide and water.

The aldehyde group of an aldose (e.g., glucose) may be oxidized to a carboxyl group with bromine water to form an aldonic acid (e.g., gluconic acid).

```
        CHO                               COOH
         |                                 |
     H—C—OH                            H—C—OH
         |                                 |
    HO—C—H                            HO—C—H
         |          Br2 + H2O              |
     H—C—OH       ——————————→          H—C—OH   + 2HBr
         |                                 |
     H—C—OH                            H—C—OH
         |                                 |
       CH2OH                             CH2OH
    D-Glucose                       D-Gluconic acid
```

4.14

The aldonic acids, when heated, readily lose water to form an equilibrium mixture of gamma (γ) and delta (δ) lactones:

```
      ┌────────┐
      C=O      |                COOH                    ┌──────────┐
      |        O                 |                      C=O        |
  H—C—OH      |     Δ       H—C—OH        Δ            |          O
      |        |  + H2O ⇌        |           ⇌      H—C—OH        |
 HO—C—H       |            HO—C—H                      |          |
      |        |                 |                 HO—C—H         |
  H—C ────────┘             H—C—OH                     |          |
      |                          |                  H—C—OH        |
  H—C—OH                    H—C—OH                     |          |
      |                          |                  H—C ──────────┘
    CH2OH                      CH2OH                   |
                                                     CH2OH
γ-Gluconolactone          D-Gluconic acid        δ-Gluconolactone
```

4.15 4.16

As cation-sequestering agents, the aldonic acids can be used to introduce metal ions into the body in a neutral and easily assimilable form. Calcium gluconate is valuable as a dietary supplement during the oral administration of calcium.

The oxidation of aldoses with an oxidant such as concentrated nitric acid produces a dicarboxylic acid with the same number of carbon atoms. These acids are referred to as the saccharic or aldaric acids. This reaction can be used to distinguish galactose (**4.17**) from its isomers because the galactaric acid (mucic acid) produced is insoluble in acidic solutions, whereas the acids produced from the other hexoses are soluble.

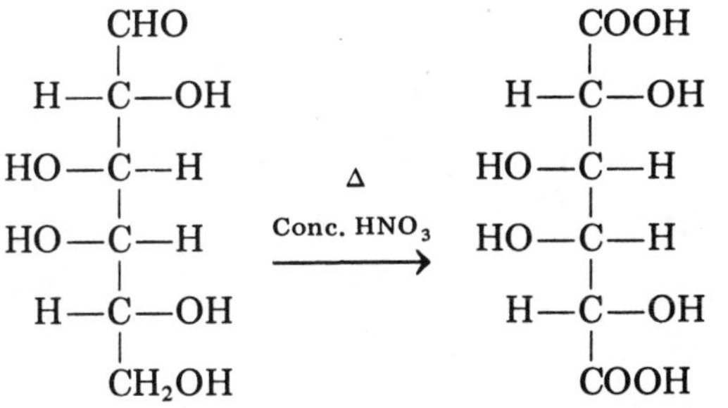

$$
\begin{array}{c}
\mathrm{CHO} \\
| \\
\mathrm{H{-}C{-}OH} \\
| \\
\mathrm{HO{-}C{-}H} \\
| \\
\mathrm{HO{-}C{-}H} \\
| \\
\mathrm{H{-}C{-}OH} \\
| \\
\mathrm{CH_2OH}
\end{array}
\quad \xrightarrow[\Delta]{\text{Conc. } HNO_3} \quad
\begin{array}{c}
\mathrm{COOH} \\
| \\
\mathrm{H{-}C{-}OH} \\
| \\
\mathrm{HO{-}C{-}H} \\
| \\
\mathrm{HO{-}C{-}H} \\
| \\
\mathrm{H{-}C{-}OH} \\
| \\
\mathrm{COOH}
\end{array}
$$

D-Galactose

4.17

D-Galactaric acid
(Mucic acid)

4.18

Mucic acid (**4.18**) occurs naturally in ripe peaches and pears. Similarly, L-tartaric acid (**4.3**) occurs as the monopotassium salt in the juice of grapes. The sodium potassium salt is known as Rochelle salt.

The ketoses, when oxidized, break at the carbonyl group to form two acids. Fructose (**4.19**) for example, yields glycolic acid and trihydroxybutyric acid.

Reduction. The reduction of the carbonyl group of a monosaccharide leads to the corresponding polyol. This may be accomplished by electrolysis of an acidified solution of the sugar or with sodium amalgam and water. The polyols are formed from the hexoses as follows:

$$\begin{array}{c} H-C=O \\ | \\ H-C-OH \\ | \\ HO-C-H \\ | \\ H-C-OH \\ | \\ H-C-OH \\ | \\ CH_2OH \end{array}$$

D-Glucose

4.6

$$\begin{array}{c} CH_2OH \\ | \\ C=O \\ | \\ HO-C-H \\ | \\ H-C-OH \\ | \\ H-C-OH \\ | \\ CH_2OH \end{array}$$

D-Fructose

4.19

$$\begin{array}{c} H-C=O \\ | \\ HO-C-H \\ | \\ HO-C-H \\ | \\ H-C-OH \\ | \\ H-C-OH \\ | \\ CH_2OH \end{array}$$

D-Mannose

4.20

$$\begin{array}{c} H-C=O \\ | \\ H-C-OH \\ | \\ HO-C-H \\ | \\ HO-C-H \\ | \\ H-C-OH \\ | \\ CH_2OH \end{array}$$

D-Galactose

4.17

$$\begin{array}{c} CH_2OH \\ | \\ H-C-OH \\ | \\ HO-C-H \\ | \\ H-C-OH \\ | \\ H-C-OH \\ | \\ CH_2OH \end{array}$$

D-Sorbitol

4.21

$$\begin{array}{c} CH_2OH \\ | \\ HO-C-H \\ | \\ HO-C-H \\ | \\ H-C-OH \\ | \\ H-C-OH \\ | \\ CH_2OH \end{array}$$

D-Mannitol

4.22

$$\begin{array}{c} CH_2OH \\ | \\ H-C-OH \\ | \\ HO-C-H \\ | \\ HO-C-H \\ | \\ H-C-OH \\ | \\ CH_2OH \end{array}$$

D-Dulcitol

4.23

Fructose forms both sorbitol and mannitol because of the formation of an additional asymmetric carbon atom during the reduction process. Thus, sorbitol and mannitol are epimers; i.e., they are optical isomers having different configurations on the second carbon atom.

The polyols are crystalline solids, soluble in water and ranging in taste from faintly sweet to very sweet. Their distribution in nature is limited to plants and they occur in both free and combined form. Glycerol, the polyol formed from glyceraldehyde (**4.1** and **4.2**) occurs in fats and other lipids. Erythritol occurs combined as esters in algae and grasses. Ribitol occurs universally in the form of its derivatives—riboflavin, flavin mononucleotide, and flavin adenine dinucleotide.

D-Sorbitol (glucitol, **4.21**) is one of the most widespread of the naturally occurring polyols. It is found in many fruits including apple, pear, cherry, plum, and peach. D-Mannitol (**4.22**) is widely distributed in plant tissues but rarely occurs in fruits. Dulcitol (**4.23**) is found in Madagascar manna. Sorbitol and mannitol are prepared in large quantities and have many commercial applications. Sorbitol is used in the manufacture of sorbose, ascorbic acid, and detergents (Tweens and Spans). It is used as a humectant in foods and in candy manufacture.

Medically, it is used as a diuretic, as a cathartic, and as a sugar substitute for diabetics. Mannitol is important as a pharmaceutical, especially the nitrate esters.

Closely related to the glycitols are the cyclic inositols. The most important compound of this group is **myo**-inositol, which is widely distributed in plants and animals. It often occurs as the hexaphosphoric acid ester (phytate) in seeds.

Condensation with Phenylhydrazine. Aldoses and ketoses react with phenylhydrazine in two steps, the first step being a condensation with the carbonyl group to form a phenylhydrazone (**4.24**):

```
   H—C=O                    H                      H
     |                      |                      |
   H—C—OH                   N—NH2   ⟶    H—C=N—N—C6H5
     |           +          |               |
  HO—C—H                   C6H5           H—C—OH
     |                                      |
   H—C—OH                                HO—C—H
     |                                      |
   H—C—OH                                 H—C—OH
     |                                      |
     CH2OH                                H—C—OH
                                            |
                                            CH2OH
```

D-Glucose   Phenylhydrazine   D-Glucose phenylhydrazone

4.24

Heating with an excess of reagent rapidly converts a hydrazone into a yellow osazone (**4.25**):

```
             H                                          H
             |                                          |
   H—C=N—N—C6H5          H                   H—C=N—N—C6H5
     |                   |                     |        H
   H—C—OH                N—NH2   ⟶             |        |
     |           + 2     |                     C=N—N—C6H5
  HO—C—H                C6H5                   |
     |                                      HO—C—H
   H—C—OH                                      |
     |                                       H—C—OH
   H—C—OH                                      |
     |                                       H—C—OH
     CH2OH                                     |
                                               CH2OH
```

D-Glucose phenylhydrazone   Phenylhydrazine   D-Glucose phenylosazone

4.25

The conversion of an aldose to a ketose is a synthetic reaction of some importance. This conversion is affected by hydrolysis of the osazone to yield an osone (**4.26**) with subsequent reduction of the aldehyde group to form a ketose. The conversion of D-glucose osazone to D-fructose is an example of this reaction:

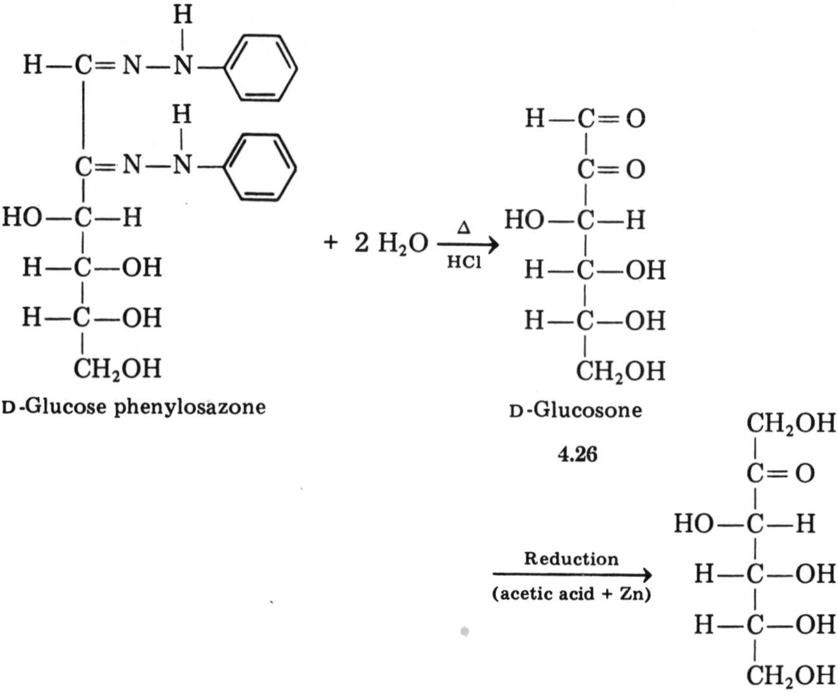

```
          H
          |
H—C=N—N—C6H5
  |
  |       H
  |       |
  C=N—N—C6H5
  |
HO—C—H
  |
H—C—OH
  |
H—C—OH
  |
  CH2OH
```

D-Glucose phenylosazone

$+\ 2\ H_2O \xrightarrow[\text{HCl}]{\Delta}$

```
H—C=O
  |
  C=O
  |
HO—C—H
  |
H—C—OH
  |
H—C—OH
  |
  CH2OH
```

D-Glucosone

4.26

$\xrightarrow[\text{(acetic acid + Zn)}]{\text{Reduction}}$

```
  CH2OH
  |
  C=O
  |
HO—C—H
  |
H—C—OH
  |
H—C—OH
  |
  CH2OH
```

D-Fructose

In osazone formation only the first two carbon atoms of the sugar are involved. The osazones of D-glucose, D-mannose, and D-fructose are identical, indicating that these sugars have the same arrangement on the last four carbon atoms; but the osazone of D-galactose differs from that of the other hexoses because its configuration on carbon 4 is different (see page 114).

If methyl phenylhydrazine is substituted for phenylhydrazine in the preparation of osazones, only ketoses will react. This reaction often is used to distinguish between an aldose and a ketose.

Isomerization, Cleavage, and Rearrangement. Three types of reactions can take place in alkaline solutions of sugars: (1) isomerization, (2) cleavage into smaller fragments, and (3) internal oxidation–reduction and rearrangement.

Reducing sugars tautomerize to form an enediol (**4.27**) salt when allowed to stand for several hours in a dilute alkaline solution (0.05 *N*).

$$
\begin{array}{c}
\mathrm{H{-}C{=}O} \\
| \\
\mathrm{H{-}C{-}OH} \\
| \\
\mathrm{HO{-}C{-}H} \\
| \\
\mathrm{H{-}C{-}OH} \\
| \\
\mathrm{H{-}C{-}OH} \\
| \\
\mathrm{CH_2OH} \\
\text{D-Glucose}
\end{array}
\rightleftharpoons
\begin{array}{c}
\mathrm{H{-}C{-}OH} \\
\| \\
\mathrm{C{-}OH} \\
| \\
\mathrm{HO{-}C{-}H} \\
| \\
\mathrm{H{-}C{-}OH} \\
| \\
\mathrm{H{-}C{-}OH} \\
| \\
\mathrm{CH_2OH} \\
\text{1,2-Enediol} \\
\mathbf{4.27}
\end{array}
+ \mathrm{NaOH}
\rightleftharpoons
\begin{array}{c}
\mathrm{H{-}C{-}ONa} \\
\| \\
\mathrm{C{-}OH} \\
| \\
\mathrm{HO{-}C{-}H} \\
| \\
\mathrm{H{-}C{-}OH} \\
| \\
\mathrm{H{-}C{-}OH} \\
| \\
\mathrm{CH_2OH} \\
\text{Enediol salt}
\end{array}
+ \mathrm{H_2O}
$$

Enediol formation destroys the asymmetry at carbon 2; hence, glucose, mannose, and fructose all form the same enediol salt because the last four carbon atoms of these sugars have the same configuration. If the enediol salt is acidified, a mixture of all three sugars will be obtained. The rearrangement of related sugars in dilute alkaline solutions is referred to as the Lobry de Bruyn–van Ekenstein reaction. The epimerization of glucose into mannose and the rearrangement of the 1,2-enediol into fructose occur as follows:

$$
\begin{array}{c}
\mathrm{H{-}C{=}O} \\
| \\
\mathrm{H{-}C{-}OH} \\
| \\
\mathrm{HO{-}C{-}H} \\
| \\
\mathrm{H{-}C{-}OH} \\
| \\
\mathrm{H{-}C{-}OH} \\
| \\
\mathrm{CH_2OH} \\
\text{D-Glucose}
\end{array}
\rightleftharpoons
\begin{array}{c}
\mathrm{H{-}C{-}OH} \\
\| \\
\mathrm{C{-}OH} \\
| \\
\mathrm{HO{-}C{-}H} \\
| \\
\mathrm{H{-}C{-}OH} \\
| \\
\mathrm{H{-}C{-}OH} \\
| \\
\mathrm{CH_2OH} \\
\text{1,2-Enediol form}
\end{array}
\rightleftharpoons
\begin{array}{c}
\mathrm{H{-}C{=}O} \\
| \\
\mathrm{HO{-}C{-}H} \\
| \\
\mathrm{HO{-}C{-}H} \\
| \\
\mathrm{H{-}C{-}OH} \\
| \\
\mathrm{H{-}C{-}OH} \\
| \\
\mathrm{CH_2OH} \\
\text{D-Mannose}
\end{array}
$$

$$
\begin{array}{c}
\Downarrow\!\Uparrow \\
\mathrm{CH_2OH} \\
| \\
\mathrm{C{=}O} \\
| \\
\mathrm{HO{-}C{-}H} \\
| \\
\mathrm{H{-}C{-}OH} \\
| \\
\mathrm{H{-}C{-}OH} \\
| \\
\mathrm{CH_2OH} \\
\text{D-Fructose}
\end{array}
$$

This type of rearrangement of a sugar into several other products is of little practical vlue, but it illustrates the detrimental effect of an alkali on a sugar. However, the isomerization of glucose ⇌ fructose and mannose ⇌ fructose are important enzymatic reactions in the intermediary metabolism of sugars.

When reducing sugars are treated with a stronger alkali (0.05 *N* and stronger), additional isomerization occurs by a continuation of the enolization process along the carbon chain. The probable process is as follows:

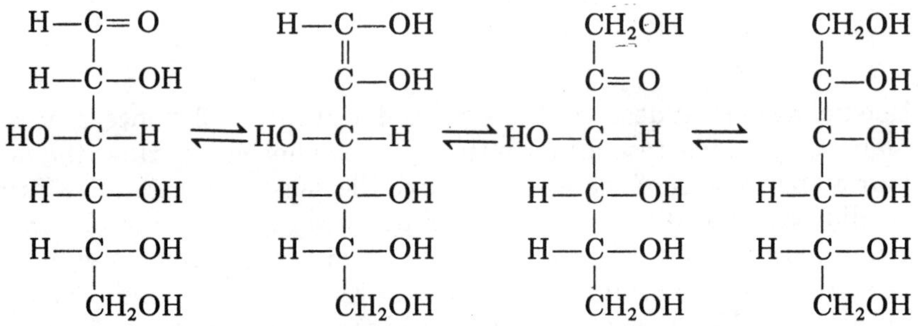

H—C=O
|
H—C—OH
|
HO—C—H
|
H—C—OH
|
H—C—OH
|
CH_2OH

D-Glucose

⇌

H—C—OH
‖
C—OH
|
HO—C—H
|
H—C—OH
|
H—C—OH
|
CH_2OH

1,2-Enediol

4.27

⇌

CH_2OH
|
C=O
|
HO—C—H
|
H—C—OH
|
H—C—OH
|
CH_2OH

D-Fructose

⇌

CH_2OH
|
C—OH
‖
C—OH
|
H—C—OH
|
H—C—OH
|
CH_2OH

2,3-Enediol

4.28

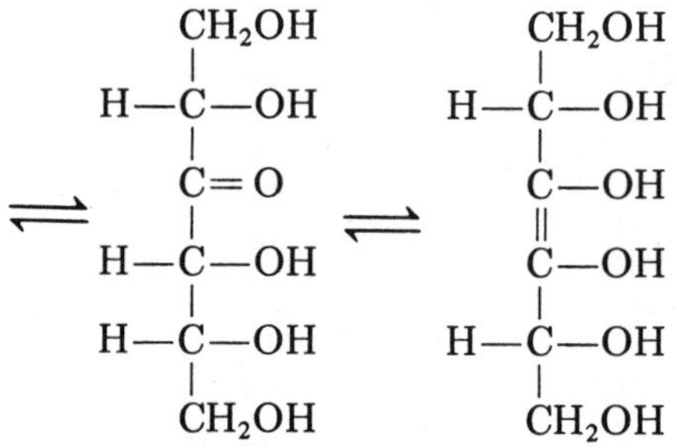

⇌

CH_2OH
|
H—C—OH
|
C=O
|
H—C—OH
|
H—C—OH
|
CH_2OH

3-Ketose

4.29

⇌

CH_2OH
|
H—C—OH
|
C—OH
‖
C—OH
|
H—C—OH
|
CH_2OH

3,4-Enediol

4.30

The enediols break at the double bonds to give a complex mixture of products. For example, cleavage between C-1 and C-2 gives rise to formaldehyde and a pentose; cleavage between C-2 and C-3 gives rise to glycolic aldehyde and a tetrose; and cleavage between C-3 and C-4 forms glyceraldehyde. The aldoses formed as a result of cleavage of the enediol may also undergo enolization and rearrangement to form a complex mixture of products. Thus, the effective reducing power of sugars and the rate at which they react with cations such as Cu^{2+} and Bi^{2+} are increased by alkali (see next section).

In addition to epimerization and cleavage into units with a smaller number of carbon atoms, sugars are converted in a strongly alkaline solution in the absence of an oxidizing agent into carboxylic acids whose overall composition does not differ from the original sugar. The acids resulting from this intramolecular oxidation–reduction and rearrangement are referred to as saccharinic acids, and several types are possible:

4.31

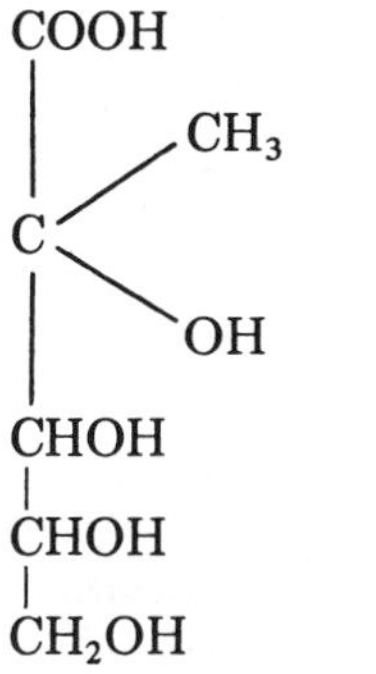

COOH
|
C (CH_3)(OH)
|
CHOH
|
CHOH
|
CH_2OH

Saccharinic acid

4.32

COOH
|
C (CH_2OH)(OH)
|
CH_2
|
CHOH
|
CH_2OH

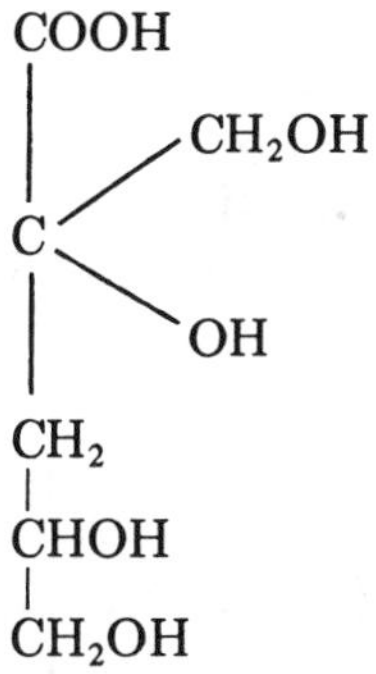

iso-Saccharinic acid

4.33

COOH
|
CHOH
|
CH_2
|
CHOH
|
CHOH
|
CH_2OH

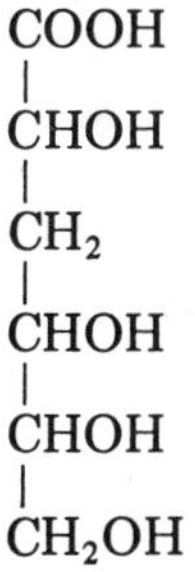

meta-Saccharinic acid

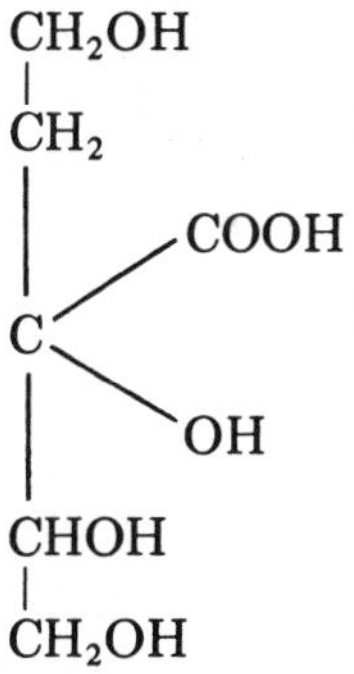

4.34

CH_2OH
|
CH_2
|
C (COOH)(OH)
|
CHOH
|
CH_2OH

para-Saccharinic acid

Carbohydrates that do not contain a free carbonyl group are not enolized by alkali and are relatively stable in alkaline solutions.

Oxidation in Alkaline Solution. All sugars that contain a free aldehyde or ketone group are classified as reducing sugars; as described in the previous section, such sugars enolize in alkaline solutions. The enediol forms of sugars are highly reactive and are easily oxidized by oxygen or other oxidizing agents. Thus, sugars in alkaline solution readily reduce oxidizing ions such as Ag^+, Hg^{2+}, Cu^{2+}, and $Fe(CN)_6^{3-}$; the sugars are oxidized to complex mixtures of acids. This reducing action of sugars is utilized for both their qualitative and quantitative determination.

Copper (Cu^{2+}) solutions, in combination with sodium citrate or sodium potassium tartrate and an alkali (sodium hydroxide, potassium hydroxide, or sodium carbonate), are the most common reagents used in sugar analysis. The citrate or tartrate prevents precipitation of cupric hydroxide by forming soluble, slightly dissociated complexes with the copper ion. However, the copper complex dissociates adequately to provide a continuous supply of cupric ions for oxidation. The alkali in the reagents enolizes the sugars and thereby causes them to break into a number of reactive fragments. For example, it is believed several reactive fragments are produced when glucose is heated with Fehling's solution, as follows:

```
                            OH                    OH                    OH
                            |                     |                     |
   H—C=O               H—C—OH              H —C—ONa              H—C—ONa
     |                    |                     |                     |
   H—C—OH   H2O        H—C—OH   NaOH        H—C—OH    -2 H2O         C—OH
     |      ——→           |     ——→             |      ——→            ||
  HO—C—H              HO—C—H               HO—C—H                   C—H
     |                    |                     |                     |
   H—C—OH              H—C—OH               H—C—OH                  C—OH
     |                    |                     |                     ||
   H—C—OH              H—C—OH               H—C—OH                H—C
     |                    |                     |                     |
    CH2OH               CH2OH                CH2OH                 CH2OH
                                                                      |
              H              H              H                         |
              |              |              |                         |
(1) CH2OH—C=    (2) =C—C=    (3) =C—C—ONa  ←—————————————————————————┘
                     |            |  |
                     OH           OH OH

              H                           H
              |                           |
(4) CH2OH—C=C—C=        (5) =C—C=C—C—ONa
            |  |              |  |  |  |
            OH H              OH H  OH OH
```

These fragments are readily oxidized and the cupric ions (Cu^{2+}) are reduced to cuprous ions (Cu^+). The course of the reaction, which is very complex, may be shown schematically as follows:

$$\text{Sugar} + \text{Alkali} \rightarrow \text{Reducing sugar fragments} \xrightarrow{+\text{Cu(OH)}_2}$$

$$\text{Cu}^+ + \text{Mixture of sugar acids} \xrightarrow{+(\text{OH})^-} \text{CuOH} \rightarrow \text{CuO}_2 + \text{H}_2\text{O}$$

As the Cu^{2+} ions (dissociated from tartrate or citrate complexes) are reduced by the sugar fragments the Cu^+ ions combine with hydroxyl ions to form yellow copper hydroxide (CuOH), which loses water in the presence of heat to give insoluble Cu_2O. The determination of the reduced copper may be accomplished by gravimetric, volumetric, colorimetric, or electrolytic methods. The Munson–Walker procedure, described later in this chapter, involves the gravimetric determination of Cu_2O.

Alkaline solutions of potassium ferricyanide are also used in the quantitative determination of reducing sugars. The reduced ferricyanide may be precipitated as the zinc salt and the excess ferricyanide determined iodometrically.

Hydroxyl Reactions

The reactions of sugars just discussed directly or indirectly involve the carbonyl groups, though a few reactions involving hydroxyl groups were considered. However, the dominant functional group of carbohydrates is the hydroxyl group, particularly if the hemiacetal group is included.

Formation of Glycosides. One of the most common reactions involving hydroxyl groups is the formation of glycosides. This is generally accomplished by treating a sugar with the appropriate alcohol and an acid catalyst.

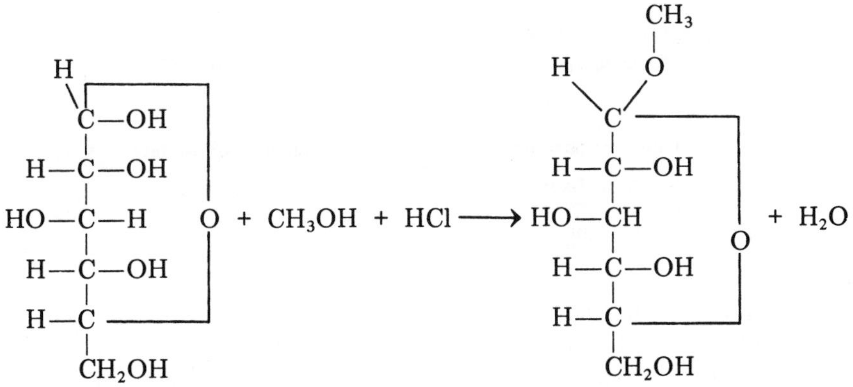

```
 H                                                   CH3
  \┌──────────┐                                        |
   C—OH      |                                   H    O
   |         |                                    \  /
 H—C—OH      |                                     C ─────────┐
   |         |                                     |          |
HO—C—H       O + CH3OH + HCl ──→                 H—C—OH       |
   |         |                                     |          |
 H—C—OH      |                                  HO—CH         O    + H2O
   |         |                                     |          |
 H—C ────────┘                                   H—C—OH       |
   |                                               |          |
   CH2OH                                         H—C ─────────┘
                                                   |
                                                   CH2OH
```

α-D-Glucose α-Methyl-D-glucoside

4.35

Both α and β anomers of methylglucoside are formed since both α and β forms exist in the original glucose. A general term for this type of compound is *glycoside*, which is defined as a derivative of a sugar in which the hydrogen atom of the potential aldehyde or ketone (hemiacetal) is replaced by an organic group to form an acetal. The glycoside, or acetal, bond is the basic linkage through which all oliogo- and polysaccharides are formed.

Aqueous solutions of glycosides are stable, exhibit mutarotation, and are nonreducing. The glycoside bond is not hydrolyzed by dilute alkali solutions but is readily cleaved in acid solutions.

Formation of Esters. The sugar esters have special significance because sugars are metabolized almost exclusively as the phosphorylated sugar. Although sugar phosphate esters can be synthesized chemically *in vitro*, the biochemical origin of sugar phosphate esters is via enzymatic synthesis. For example:

$$\alpha\text{-D-Glucose} + \text{ATP} \xrightarrow{\text{hexokinase}} \alpha\text{-D-Glucose-6-phosphate}$$

$$\text{Starch} + \text{Pi} \xrightarrow{\text{phosphorylase}} \alpha\text{-D-Glucose-1-phosphate}$$

Different phosphates of the same sugar—(e.g., glucose-1-phosphate (**4.36**) and glucose-6-phosphate (**4.37**)—behave differently in biochemical reactions.

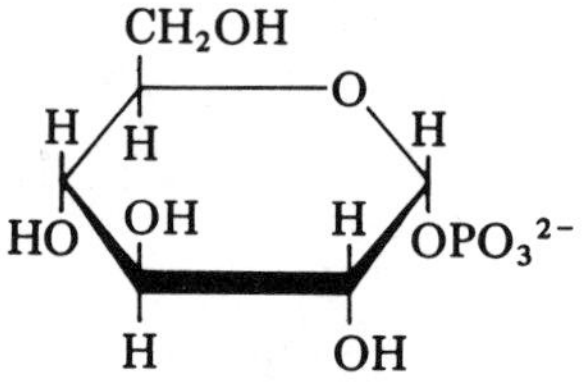

Glucose-l-phosphate

(Cori ester)

4.36

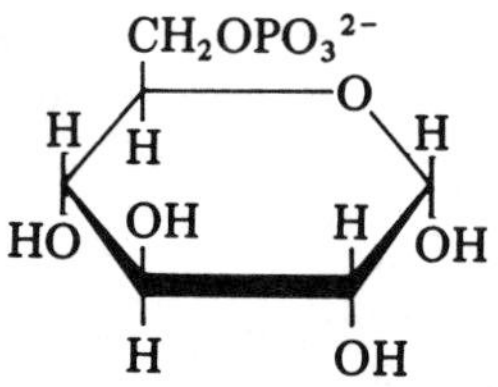

Glucose-6-phosphate

(Robison ester)

4.37

Dehydration in Acid Solution. Dilute solutions of inorganic acids have relatively little effect on the structure of monosaccharides. However, when a monosaccharide is heated in a strongly acid solution, dehydration takes place with the formation of furan derivatives. Aldopentoses are converted into furfural (**4.38**):

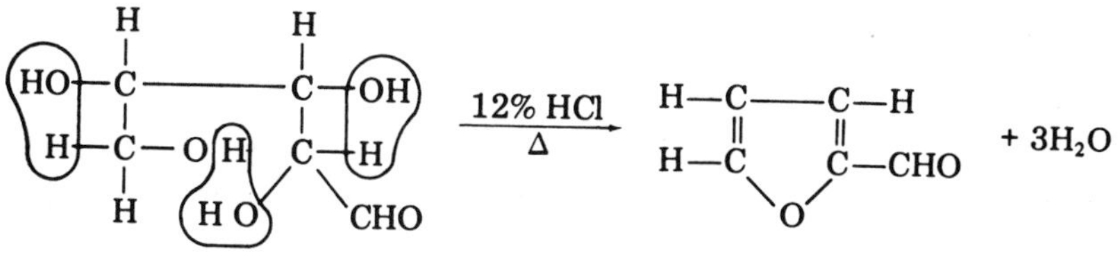

$$\text{Pentose} \xrightarrow[\Delta]{12\%\ HCl} \text{Furfural} + 3H_2O$$

Pentose

Furfural

4.38

Hexoses react in a similar way to give 5-hydroxymethylfurfural (**4.39**), which is further degraded to levulinic acid (**4.40**) and formic acid (**4.41**):

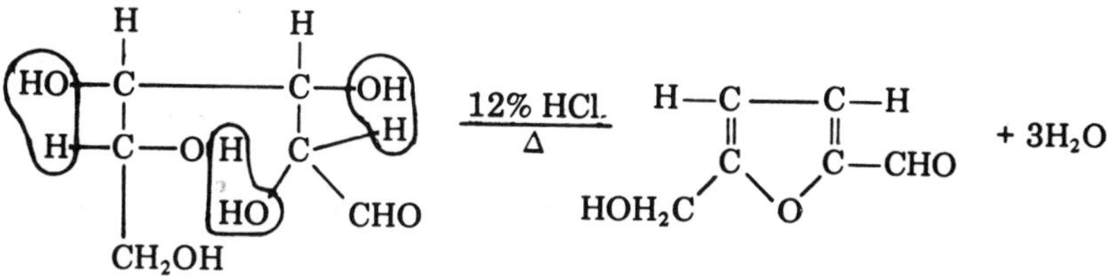

$$\text{Hexose} \xrightarrow[\Delta]{12\%\ HCl} \text{5-Hydroxymethylfurfural} + 3H_2O$$

Hexose

5-Hydroxymethylfurfural 4.39

$$CH_3{-}C({=}O){-}CH_2{-}CH_2{-}COOH + HCOOH$$

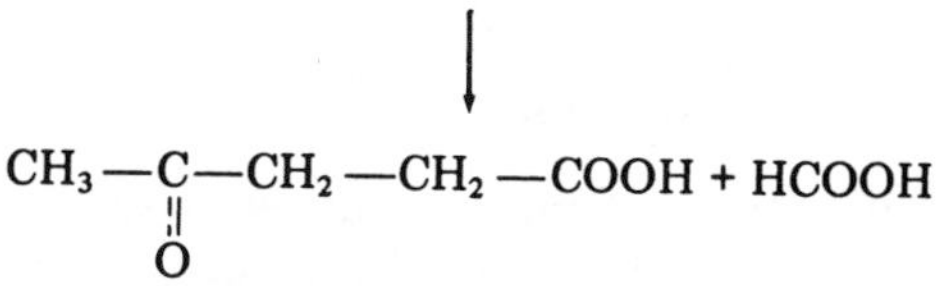

Levulinic acid

4.40

Formic acid

4.41

Ketohexoses react with acids more rapidly than aldohexoses. The furfurals readily undergo further reactions with the formation of brown colors (humins). It is probable that many of the brown colors produced in food processing result from the intermediary formation of furfurals.

Polysaccharides and compound carbohydrates are generally hydrolyzed to monosaccharides by boiling with dilute (0.05–1.0 N) mineral acids.

Oxidation. The primary hydroxyl group of aldoses may be oxidized to a carboxyl group to form uronic acids. However, it is very difficult to chemically synthesize uronic acids from aldoses because the aldehyde group readily undergoes oxidation under the same conditions. In contrast, uronic acids are synthesized in significant quantities by enzyme systems. Glucuronic acid (**4.42a**) is found as a constituent of certain polysaccharides (e.g., chondroitin sulfate). It occurs combined as monomethyl ethers in plant materials such as straws, saponins, and various woods. An important function of glucuronic acid in the body is the detoxication of substances such as phenol by the formation of glucuronides, as follows:

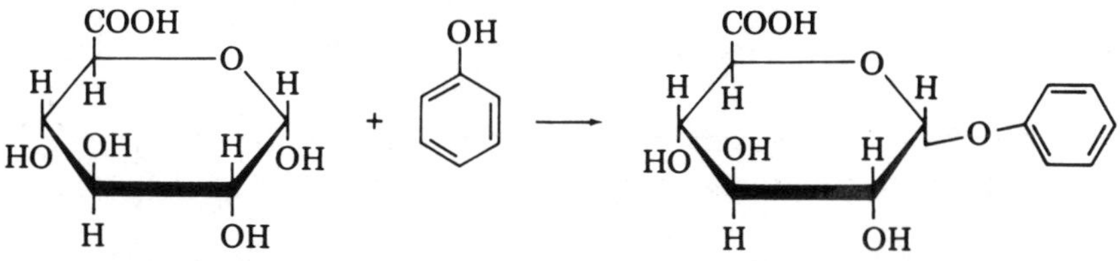

D-Glucuronic acid **4.42a** + Phenol → α-Phenylglucuronide **4.43a**

This reaction is another example of ester formation in carbohydrates. The uronic acid derived from galactose (galacturonic acid) is found as a constituent of fruit pectins, mucilages, and plant gums. D-Mannuronic acid occurs as the sole constituent of alginic acid.

Color Reactions

The action of mineral acids on sugars leads to the formation of a number of volatile products such as furfural (**4.38**) and 5-hydroxymethylfurfural (**4.39**), as already discussed. The furan derivative formed depends upon the type of sugar used, and its decomposition products condense with aromatic phenols or amines to form colored substances. This fact is the basis of several qualitative and quantitative determinations for sugars.

Color reactions of this type include the Molisch test with alcoholic α-naphthol and concentrated sulfuric acid and the anthrone reaction (anthrone + sulfuric acid). In the latter type of reaction, which can be used for the quantitative determination of sugars, ketoses and pentoses usually form colored products under conditions milder than those required for aldohexoses. The Seliwanoff reaction (resorcinol + HCl) distinguishes ketoses from aldoses. The Bial test (orcinol + HCl) and the Tollens' phloroglucinol reaction can be used to distinguish pentoses from hexoses.

All mono- and some oligosaccharides are reducing sugars. Although the tests for reducing sugars are nonspecific, they are used frequently to detect and determine sugars. The reduction of cupric ions in alkaline solution is the basis of Benedict's and Fehling's solutions.

These tests are described in more detail in the section on qualitative analysis.

☐ MONOSACCHARIDES

The monosaccharides—the simplest carbohydrates—are essentially polyhydroxy aldehydes and ketones and are classified according to the length of the carbon chain and according to the nature of the carbonyl group. Of this group only two types—the pentoses and hexoses—are of sufficient biological importance to merit detailed discussion. In addition to the pentoses and hexoses themselves, several of their derivatives—especially amino sugars and deoxy sugars—are biologically important.

Pentoses

Pentoses occur only in limited amounts as free monosaccharides, but they are widely distributed as component units of complex polysaccharides known as pentosans. On hydrolysis pentosans yield pentose sugars, such as **4.44a**, **4.45a**, and **4.46a**.

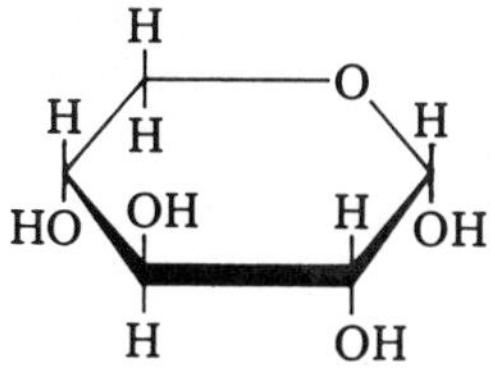

α-D-Xylopyranose
4.44a

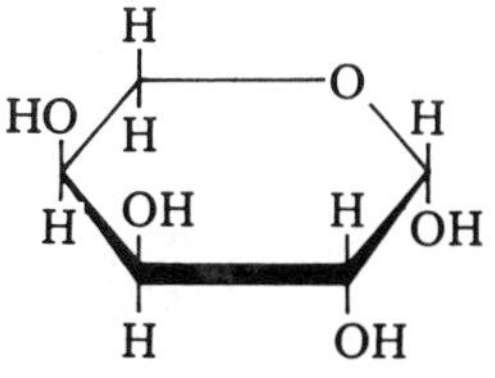

β-L-Arabinopyranose
4.45a

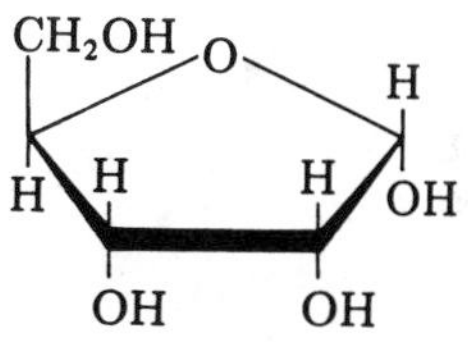

α-D-Ribofuranose
4.46a

These sugars are not fermented by yeasts and are utilized only to a limited extent by most mammals. Therefore, they are of little value as a source of energy in humans.

D-Xylose (Wood Sugar)

D-Xylose sugar is found as a pentosan (xylan) in corn cobs, straw, bran, wood gum, and the bran of seeds. The sugar can be obtained from any of the above materials by dilute acid hydrolysis. Xylose has been identified in fruits such as cherry, peach, pear, and plum. Industrially, the xylose in waste materials such as corn cobs are converted into furfural (**4.38**) through dehydration with concentrated sulfuric acid.

L-Arabinose (Pectin Sugar)

L-Arabinose occurs as a component of many gums, pectins, mucilages, and hemicelluloses. It is obtained principally by hydrolysis of gum arabic, cherry gum, mesquite gum, or beet pulp with dilute sulfuric acid. L-Arabinose has been identified in a number of fruits including apple, fig, grapefruit, lime, and some varieties of grape.

D-Ribose

D-Ribose occurs as a component of nucleic acids, nucleotide coenzymes, and riboflavin (vitamin B_2). A very important derivative of ribose is 2-deoxyribose, which is a part of deoxyribonucleic acid (DNA).

Hexoses

Six-carbon monosaccharides occur naturally in many types of plant materials, and one of them, glucose, is the circulating sugar of both plants and animals. At least four hexose sugars merit special consideration: glucose, fructose, galactose, and mannose. The cyclic formulas are shown as

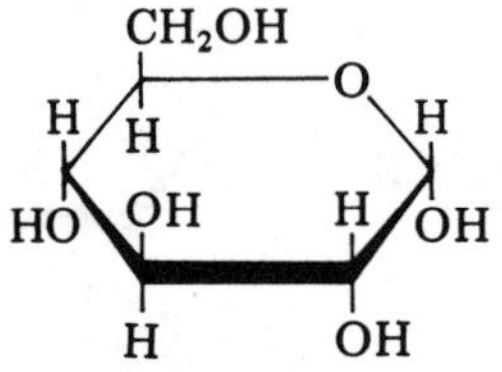

CH_2OH O H H H HO OH H OH H OH

α-D-Glucose
4.9a

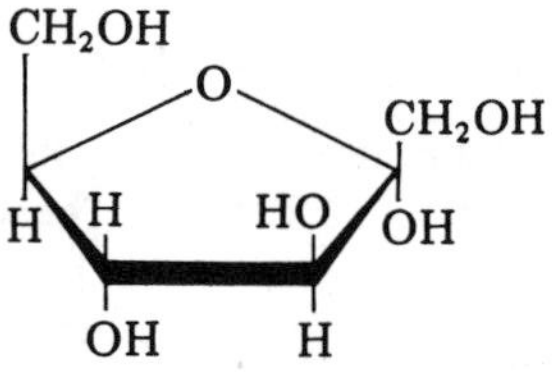

CH_2OH O CH_2OH H H HO OH OH H

α-D-Fructose

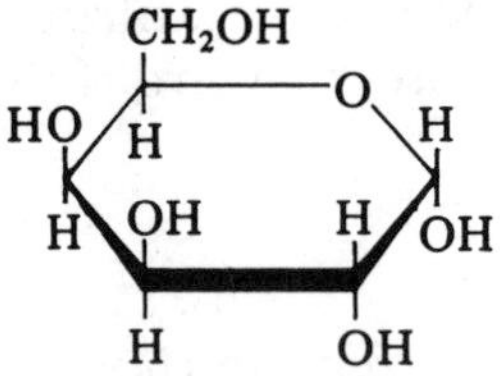

CH_2OH O HO H H H OH H OH H OH

α-D-Galactose

4.47a

CH_2OH O H H H HO OH HO OH H H

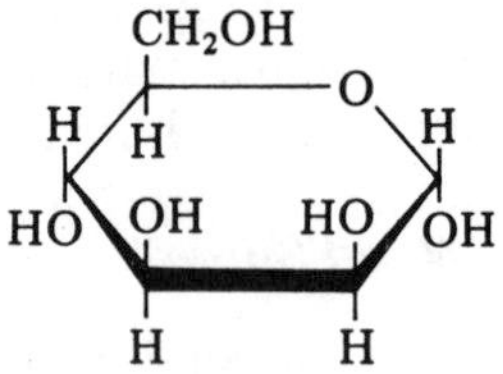

α-D-Mannose

4.48a

D-Glucose

The most widely distributed sugar in nature, because of its role in biochemical processes, is D-glucose. In the free state, it occurs in ripe fruits, flowers, leaves, roots, and the sap of plants and is the principal sugar in the blood of most animals. In the combined state, it forms a part of or the whole of a number of oligosaccharides and polysaccharides (e.g., maltose, sucrose, lactose, raffinose, dextrins, starch, cellulose, and glycogen). Glucose also occurs in the combined form as glucosides (e.g., salicin and arbutin).

D-Glucose is commonly prepared by the hydrolysis of starch. Crude dextrose, refined dextrose (Cerelose), and corn syrup are commercial glucose products. They may contain other carbohydrates as impurities, e.g., corn syrup contains, in addition to glucose, considerable amounts of oligosaccharides including maltose.

D-Fructose (Levulose)

Fructose, a ketose, is present in syrups, honey, molasses, and ripe fruits. It also occurs in small amounts in blood. Fructose is obtained from the hydrolysis of the polysaccharide inulin. It is the only ketose sugar of importance in foods.

In most fruits the glucose concentration exceeds that of fructose. However, in apples and pears the fructose concentration may exceed the glucose concentration, and in oranges, grapes, and strawberries there are equal amounts of both sugars.

Fructose exists in two structural forms, depending upon whether it is in the free state or combined. In the free state it exists principally as a pyranose structure, while in the combined form (e.g., sucrose, inulin, and several phosphate esters) it exists as a furanose.

D-Galactose

Galactose occurs most frequently as a constituent of oligosaccharides and polysaccharides rather than as a monosaccharide. Galactose has been reported to be present in grapes, olives, and possibly in peaches and pears. It may be obtained by hydrolysis of lactose, raffinose, gums, and mucilages. Legumes, impure pectin, and agar also contain galactose. It is a constituent of galactolipids found in the white matter of brain and myelin sheiths of nerve cells.

Galactose is generally made from lactose by heating with 2% sulfuric acid. α-D-Galactose is the stable isomer obtained under most conditions, whereas the β form may be obtained by crystallization from cold alcoholic solution.

D-Mannose

Mannose has been detected in oranges, olives, germinating seeds, and sugar cane molasses. It is more widely distributed in the polymeric form, as mannosans. Polysaccharides containing mannose are found in yeast and other microorganisms. In animals, mannose is a component of glycolipids, glycoproteins, and serum albumin.

Mannitol (**4.22**), the alcohol derived from mannose, is widely distributed in nature. It has been identified in the green bean, cauliflower, onion, pineapple, and in the bark and leaves of many trees.

Amino Sugars

The amino sugars are formed by the substitution of an amino group for a hydroxyl group, usually on the C-2 carbon. Two amino sugars occur frequently in organisms: 2-amino-2-deoxy-D-glucosamine and 2-amino-2-deoxy-D-galactosamine or their *N*-acetyl derivatives.

Glucosamine occurs in mucoproteins and mucopolysaccharides. It is the chief component of fungi cell walls and the shells of insects and crustaceans (lobster, crabs, etc.). Galactosamine occurs as a constituent of the mucopolysaccharide, chondroitin sulfate.

Sialic acids are naturally occurring amino-sugar derivatives consisting of a six-carbon amino sugar linked to pyruvic acid or lactic acid. Two of the more common of these are *N*-acetylneuraminic (**4.49**) acid and *N*-acetylmuramic acid (**4.50**).

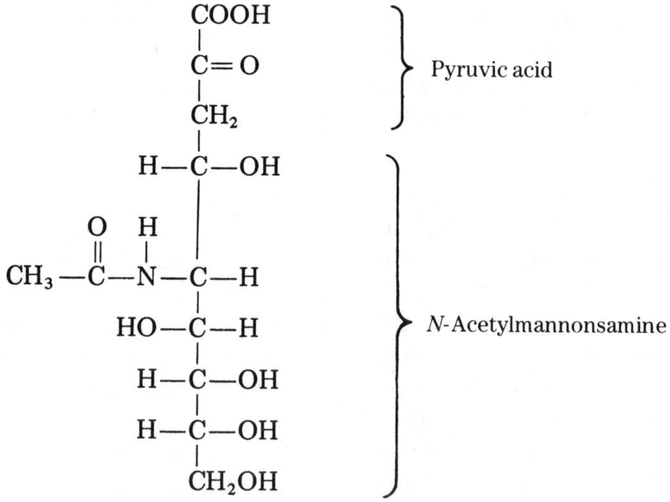

```
                    COOH        )
                     |          )
                    C=O         }  Pyruvic acid
                     |          )
                    CH2         )
                     |
                 H—C—OH         )
                     |          )
         O   H       |          )
         ‖   |       |          )
CH3—C—N—C—H                     }  N-Acetylmannonsamine
                     |          )
               HO—C—H           )
                     |          )
                H—C—OH          )
                     |          )
                H—C—OH          )
                     |          )
                  CH2OH         )
```

N-Acetylneuraminic acid

4.49

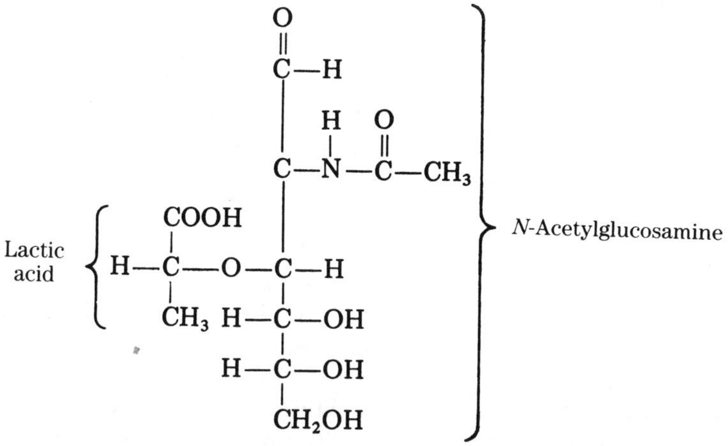

```
                         O
                         ‖
                         C—H
                         |
                         |     H   O
                         |     |   ‖
                         C—N—C—CH3
         ⎧      COOH       |                  ⎫
 Lactic  ⎨      |          |                  ⎬ N-Acetylglucosamine
 acid    ⎩ H—C—O—C—H                  ⎭
                |          |
              CH3   H—C—OH
                           |
                    H—C—OH
                           |
                         CH2OH
```

N-Acetylmuramic acid

4.50

These acids are important constituents of structural polysaccharides of cell walls and membranes.

Deoxy Sugars

Deoxy sugars are most important in biochemical processes due to their incorporation into deoxyribonucleic acids. 2-Deoxy-D-ribose (**4.51a**) is the outstanding deoxy sugar of this group.

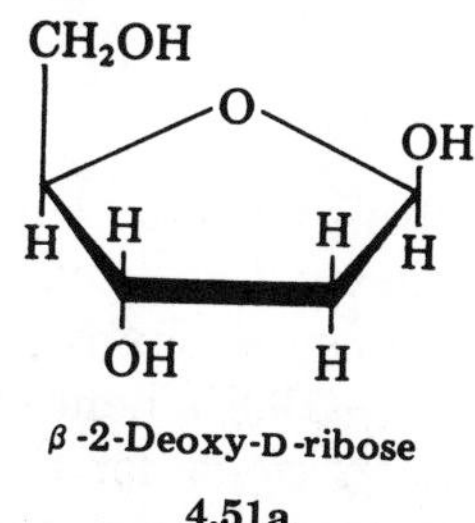

β-2-Deoxy-D-ribose

4.51a

Other deoxy sugars include L-rhamnose and L-fucose, which are important constituents of bacterial cell walls.

☐ OLIGOSACCHARIDES

As noted earlier, oligosaccharides contain two up to ten monosaccharides linked together by glycosidic bonds. However, only certain disaccharides and trisaccharides are of biological importance.

Disaccharides

Hydrolytic cleavage of the glycosidic bond in disaccharides may be brought about by acids or by enzymes which act specifically on each sugar. Of the disaccharides found in foods, sucrose, maltose, and lactose are the most important.

Sucrose

Saccharose, cane sugar, beet sugar, or simply sugar all refer to sucrose (**4.52a**), the most widely distributed of all disaccharides. Sucrose ranges in amounts from 0.1 to 25% of the fresh material; ripe fruits are especially rich in sucrose. It is obtained commercially from sugar cane and sugar beets. The sugar of commerce, granulated sugar, is available as a relatively pure food product.

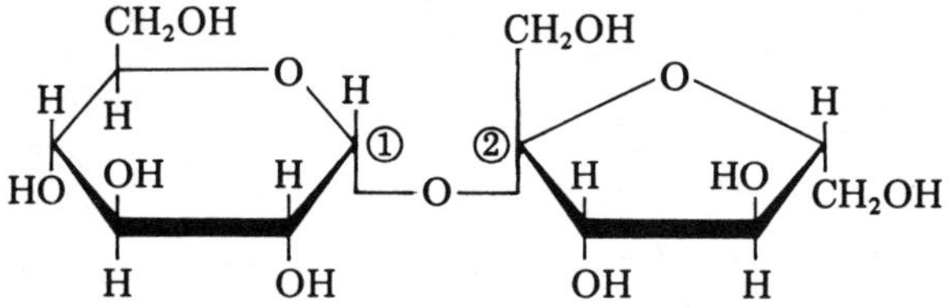

Sucrose (α-D-glucopyranosyl-(1 → 2)-β-D-fructofuranoside)

4.52a

Sucrose upon mild hydrolysis yields an equimolar mixture of glucose and fructose. The hydrolysis reaction is catalyzed by enzymes (invertases) and dilute acids. Sucrose is dextrorotatory (+66.5°), whereas invert sugar is levorotatory (−19.8°); hence the term *inversion* for the hydrolysis reaction and *invert sugar* for the product obtained. Since fructose is sweeter than sucrose, invert sugar is commonly used in candies and similar food products. Invert sugar is the principal component of honey.

Sucrose is a nonreducing sugar because the anomeric carbonyl groups of both glucose and fructose are used in the formation of a glycosidic linkage. Sucrose does not exhibit mutarotation and is unaffected by boiling sodium hydroxide. Sucrose loses water when heated to 210°C and forms a brown syrup known as caramel. It is readily fermented by yeast.

Maltose

Maltose (**4.53a**) derived its name from the fact that it occurred in the extracts from sprouted barley or other cereals (malt liquors). Maltose is composed of two glucose units and is formed when the enzyme amylase hydrolyzes starch. The enzyme maltase, which is specific for the α-linkage, splits maltose into two glucose units.

Maltose is an important constituent of corn syrups, which are prepared by the partial hydrolysis of starch with acids or enzymes. The principal saccharides in commercial starch hydrolysates are D-glucose, maltose, trisaccharides, and higher saccharides. Starch hydrolysates are used in the production of many foods, such as soft drinks, bread, confections, coffee substitutes, and infant foods.

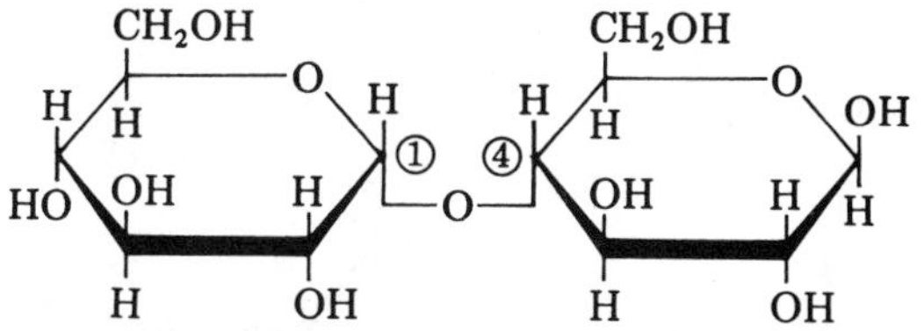

CH_2OH CH_2OH O H H O OH H H ① ④ H HO OH H O OH H H H OH H OH

β-Maltose (O-α-D-glucopyranosyl-(1 → 4)-β-D-glucopyranose)

4.53a

Technology is now available for preparing starch hydrolysates that contain 90% or more maltose (Hodge *et al.* 1972). This process involves the use of a multiple-enzyme process [i.e., isoamylases (amylo-α-1,6-glucosidases) which debranch the amylopectin fraction of gelatinized starch to linear segments] couples with β-amylase.

Maltose is a reducing sugar and exhibits mutarotation. Consequently, the hemiacetal group of one glucose unit is involved in the glycoside linkage while the hemiacetal linkage of the other glucose unit is free. Maltose is obtained as a monohydrate of the β isomer (mp 102°–103°C; $[\alpha]_D^{20} = +111.7° \rightarrow +130.4°$). The structure of α-maltose differs from **4.53a** only in the position of the groups on the unsubstituted hemiacetal group.

Lactose

Lactose consists of one glucose unit and one galactose unit. The concentration of lactose, which occurs in the milk of all mammals, varies among species. For example, cow's milk contains about 4.5% lactose while human milk contains about 6.0%.

Lactose may be hydrolyzed by dilute solutions of strong acids and the enzyme β-D-galactosidase (lactase). Lactose exists in an α form (**4.54a**) and β form; thus it exhibits mutarotation. It is a reducing sugar and is decomposed by alkali. The common crystalline form of lactose is α-lactose monohydrate, or simply alpha hydrate, which is prepared by allowing a supersaturated solution of lactose to crystallize at a temperature below 93.5°C. The alpha-hydrate loses water of hydration at about 130°C and caramelizes at 160°–180°C. Other crystalline forms may be prepared, but they change to the hydrate form in the presence of a small amount of water at temperatures below 93.5°C. The α and β isomers have the following physical properties: monohydrate of α isomer, mp 202°C and $[\alpha]_D^{20} = +85.0° \rightarrow +52.6°$; anhydrous α isomer, mp 223°C and $[\alpha]_D^{20} = +90.0° \rightarrow +55.3°$; anhydrous β isomer, mp 252°C and $[\alpha]_D^{20} = +34.9° \rightarrow +55.3°$.

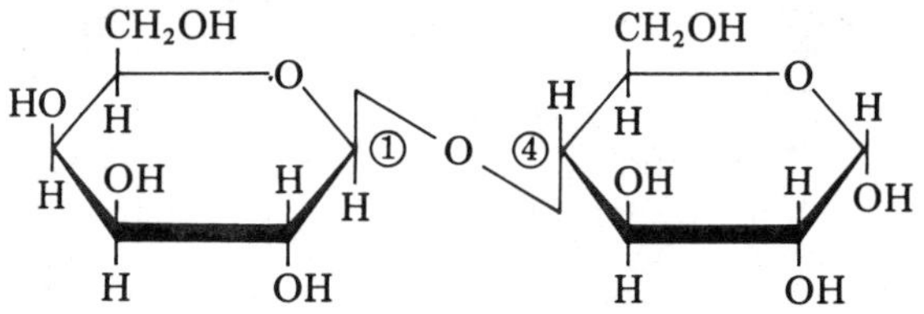

CH₂OH CH₂OH
HO H O H H O H
① O ④
H OH H H OH H OH
H OH H OH

α-Lactose (O-β-D-galactopyranosyl-(1 → 4)-α-D-glucopyranose)

4.54a

Of all the common sugars, lactose has the lowest relative sweetness and is the least soluble (17 g/100 g at 20°C). Ingestion of lactose in moderate amounts is tolerated by most people, but excessive amounts may result in gastrointestinal disturbances in certain individuals. Moreover, there are some individuals who cannot tolerate lactose and are made ill whenever they eat it.

The unique chemical and physical properties of lactose are used to advantage in the food industry. Because lactose readily absorbs flavors, aromas, and coloring materials, it is used as a carrier for such substances. Lactose is added to biscuit and other baking mixes because it readily reacts with proteins via the Maillard reaction to form the golden brown color desirable in the crusts. Lactose is not fermented by yeast so its functional properties are effective throughout the baking process, i.e., its emulsifying properties promote greater efficiency from the shortening. Lactose is used in infant foods as a coating agent for foods, and in the production of lactic acid. It is also used as a preservative to maintain the flavor, color, and consistency of meat products.

Trisaccharides

Raffinose

Several trisaccharides occur in nature but raffinose (**4.55a**) is the most important. It occurs in small quantities in sugar beets, cottonseed meal, soybeans, and coconut meats. It is a nonreducing sugar and has a specific rotation of +104°. Raffinose contains fructose, glucose, and galactose and it can be hydrolyzed by strong acid into these three monosaccharides. Hydrolysis by weak acids yields fructose and a disaccharide called melibiose (galactose + glucose). The enzyme maltase, an α-glucosidase, hydrolyzes raffinose to form galactose and sucrose. Raffinose contains water of hydration, which it loses at 100°–110°C. The disaccharide components of raffinose are indicated in **4.55a**.

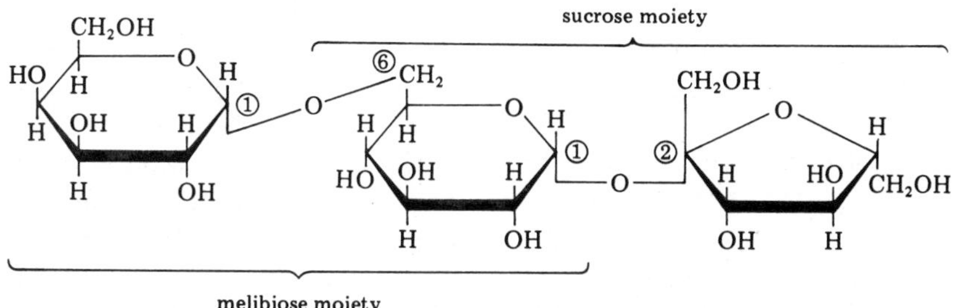

Raffinose (α-D-galactopyranosyl-(1 → 6)-α-D-galactopyranosyl-(1 → 2)-β-D-fructofuranoside)

4.55a

☐ POLYSACCHARIDES

Carbohydrates composed of ten or more monosaccharide units are referred to as polysaccharides. These substances are amorphous, colorless, and almost tasteless. Some polysaccharides possess low molecular weights, but the majority of these substances are large molecules composed of several hundred or even thousands of monosaccharide units.

Polysaccharides form colloidal solutions rather than true solutions, as the sugars do. It is for this reason that they are purified with much difficulty. For example, the empirical formula for the hexosans is written $(C_6H_{10}O_5)_x$ to indicate that the number of monosaccharide units in the condensation polymer is not known.

Polysaccharides are an important group of carbohydrates. Some are constituents in structural tissues (e.g., cellulose in plants, chitin in marine life, muramic acid in bacterial cell walls), while others constitute storage reserves (e.g., starch in plants, glycogen in animals).

The chemical structure of polysaccharides permits the following classifications: (1) homopolysaccharides, which are composed of many units of a single monosaccharide and include cellulose, starch, and glycogen; (2) heteropolysaccharides (analogous to copolymers), which are composed of two or more different components and include hemicelluloses, pectins, mucilages, and resins; and (3) conjugated compounds composed of saccharides plus lipids or proteins.

Homopolysaccharides

Cellulose

Cellulose is the major constituent of the structural tissues of plants, being present in the walls of cellular tissue combined with xylans and lignins. Cellulose differs in toughness and strength, depending on the age and type of plant. Wood fiber, flax, and cotton are excellent sources of cellulose; for example, cotton contains more than 90% cellulose.

Cellulose is insoluble in water and is quite resistant to the action of most enzymes, dilute acids, and dilute alkalies. Cellulases are absent in the digestive enzymes of animals; as a consequence, cellulose cannot be used directly by animals. However, herbivorous animals can utilize appreciable quantities of cellulose because their rumen contains microflora that can hydrolyze cellulose. The indigestible carbohydrate of foods is often referred to as "crude fiber."

Cellulose is a chain structure consisting of glucose units linked by β-(1 → 4) glycosidic linkages; in contrast starch contains α-(1 → 4) glycosidic linkages (see next section). The repeating cellobiose structural units in cellulose are shown in **4.56a**.

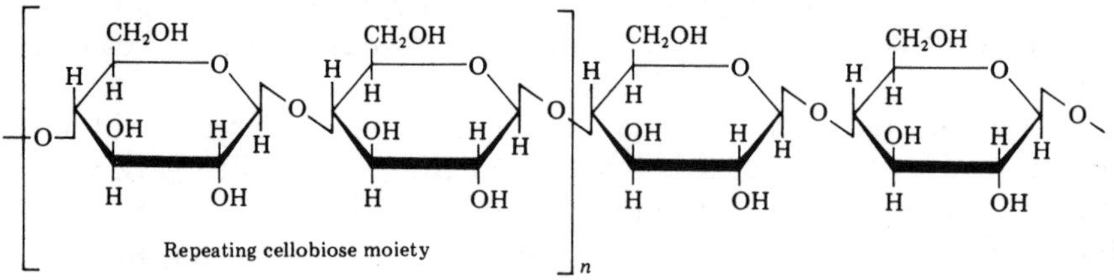

Repeating cellobiose moiety

Cellulose type structure

4.56a

Evidence indicates that cellulose is not branched but is arranged in bundles of long parallel chains. The chains are held together horizontally by hydrogen bonds between the hydroxyl groups. The molecular weights of native cellulose may vary between 100,000 and 2,000,000.

Starch

Starch occurs as the main reserve carbohydrate in most plants, being found in roots, tubers, seeds, stems, and in many fruits. Starch is found in layers within a granule that is surrounded by a thin protein layer. Although the starches are polymers of glucose, the granules of different plant species differ in size, shape, and other physical characteristics. Thus, starch grains from a given source can be identified by microscopic examination (Fig. 4.1). Starch granules are insoluble in cold water, but they will absorb water and swell slightly. The swelling is reversible, i.e., the granules will shrink on drying. In contrast, when starch granules are treated with boiling water, the granules swell until ultimately they rupture, collapse, and yield a paste. This process of swelling is referred to as gelatinization. The dispersion or paste contains two kinds of starch, amylose and amylopectin. Partial acid hydrolysis

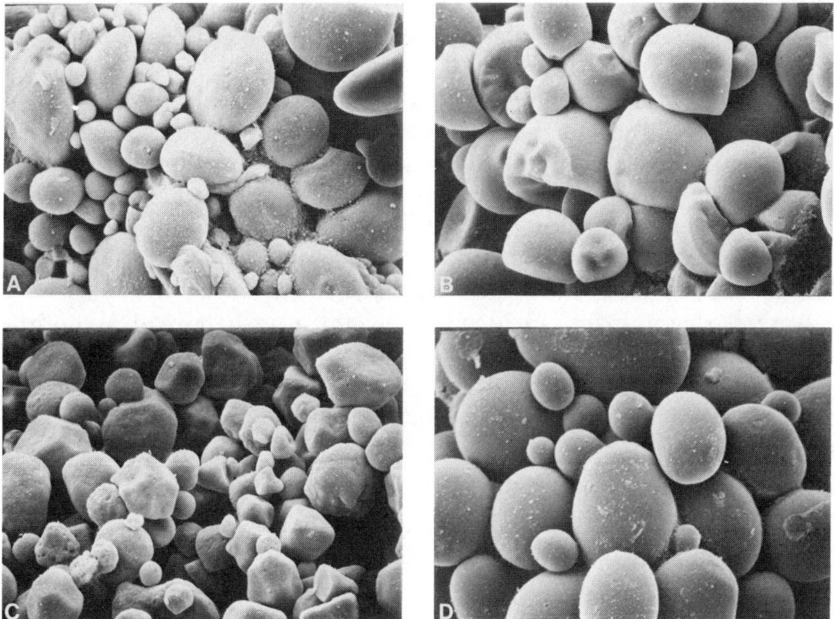

Figure 4.1. Starch granules as seen with the scanning electron microscope. A—wheat (1100×); B—tapioca (1100×); C—dent (700×); and D—potato industrial (700×).

of starch yields a complex mixture of dextrins (higher saccharides), maltose, and glucose. Complete acid hydrolysis yields D-glucose. Colloidal dispersions of starch are hydrolyzed by amylases to maltose and glucose units. Amylose and amylopectin can be distinguished by chemical methods.

Amylose. Amylose is a linear polymer consisting of 250–2000 glucose residues linked by α-D-(1 → 4) glucosidic bonds. The repeating α-maltose units are shown in **4.57a**.

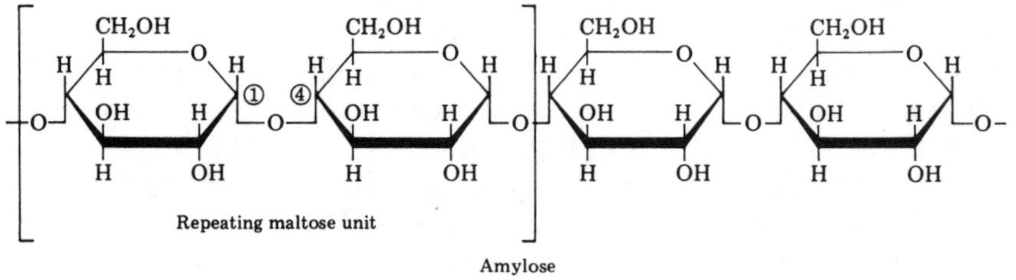

Repeating maltose unit

Amylose

4.57a

Common starches (corn, wheat, potato) contain 10–30% amylose. Starches from certain varieties of corn, sorghum, and rice contain no amylose and are known as waxy starches. On the other hand, some recently developed corn varieties (amylomaize) have starches that contain as much as 80% amylose.

In aqueous solutions, amylose readily associates to form an insoluble precipitate. This occurs because the linear molecules have a tendency to become oriented parallel to one another to such an extent that association occurs through hydrogen bonding with the net result that the affinity for water decreases, the aggregate size increases, and ultimately a precipitate is formed. This phenomenon is called retrogradation and the precipitate is called retrograded starch.

Amylose complexes with iodine by forming a helical structure around the iodine; in this form, iodine exhibits a strong absorption of light (intense blue). Amylose also complexes with fatty acids, surfactants, and polar agents such as butyl alcohol, amyl alcohol, thymol, and nitroparaffins. The latter complexing property provides a method for separating amylose from amylopectin.

Amylopectin. The other component of grain starch, amylopectin, makes up 70–100% of various starches. It is a highly branched polymer consisting of glucose units joined to each other by α-glucosidic linkages. Each branch contains about 15–25 glucose units occurring periodically in the chain. The residues are joined to each other by α-(1 → 4) linkages, with the branch point formed by α-(1 → 6) linkages, as shown in **4.58a**.

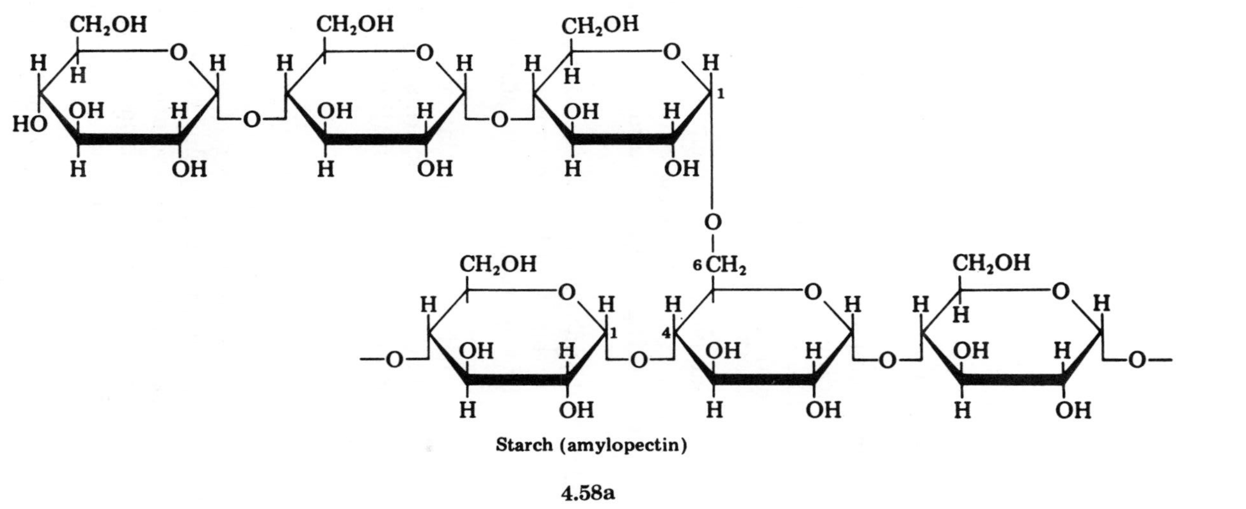

Starch (amylopectin)

4.58a

Amylopectin is usually larger than amylose, with a minimum molecular weight of approximately 1,000,000. In contrast to amylose, amylopectin gives a red-violet color with iodine and retrogradation from aqueous solution is very slow. Consequently, amylopectin starches may be characterized as being resistant to gelling and changes in water-holding properties.

Dextrins. The action of enzymes, acid, and heat upon starch produces partial degradation products called dextrins. Three kinds of dextrins are recognized as intermediate products in the hydrolysis of starch. The most complex degradation product is referred to as amylodextrin, or soluble starch, which gives a blue color with iodine. Further hydrolysis yields a less complex product called erythrodextrin, which gives a red color with iodine. Finally, one obtains achroödextrin, which gives no color with iodine. Dextrins may be made readily from starch by heating. A commercial starch gum is made by heating dry starch at 230°–260°C.

Dextrins are water soluble and are precipitated from aqueous solution by adding alcohol. All dextrins have free carbonyl groups and, as a result, reduce Fehling's solution and undergo other sugar reactions.

Dextrins occur in fruit and vegetable juices and in the leaves of many plants. These substances represent the intermediates in the synthesis of starch from glucose or the breakdown from starch to glucose. Dextrins also occur in honey.

Dextrins are used extensively in encapsulating water-soluble flavor oils for use in dry powders, mixes, etc. Dextrins are also used extensively as adhesives and are important constituents of food products such as corn syrups.

Inulin

Inulin is a linear polyfructosan composed of D-fructose units joined by β-(2 → 1) glycosidic linkages. It is only slightly soluble in hot water and does not give a blue color with iodine solution. It is a white amorphous powder, which is readily hydrolyzed by acids. In contrast to dextrins, inulin is not hydrolyzed by any of the enzymes in the gastrointestinal tract and cannot be used as a nutrient. Although inulin has no value as a nutrient, it can be used as a source of D-fructose, the sweetest simple sugar.

Inulin is found in Jerusalem artichoke and dahlia bulbs, in chicory and dandelion roots, and in the bulb of garlic and onion.

Glycogen

The structure of glycogen is similar to that of amylopectin (**4.58a**) except it is more branched and is higher in molecular weight (Fig. 4.2). Glycogen from different sources shows different degrees of branching and different molecular weights. Like amylopectin, glycogen consists of repeating glucose units joined to each other by α-(1 → 4) linkages with the branched point joined by α-(1 → 6) linkages. Glycogen has shorter linear chains (12–18 units) than amylopectin and there is more frequent branching.

Glycogen, one of the most important biochemical substances in the body, is the principal storage form of carbohydrates in animals (it also occurs in some plants, yeasts, oysters, and clams). It is found principally in the liver and muscles. Liver glycogen is the principle reserve for maintaining the normal blood level of glucose, while muscle glycogen is a source of energy for the contraction of muscle fibers. Glycogen is broken down to glucose, which can pass into the bloodstream. The glucose units on the branch exhibit a more rapid turnover than the glucose units in the interior portion of the chain.

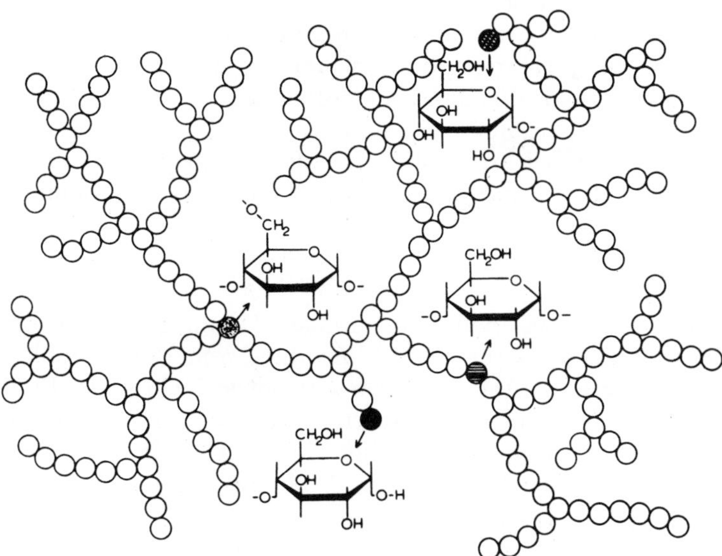

Figure 4.2. The structure of glycogen has both α-(1 → 4) and α-(1 → 6) linkages. The branch points are joined by α-(1 → 6) bonds.

Heteropolysaccharides

Mucilages

Mucilages are well known for their properties of forming gels. Agar is an example of a sulfated polysaccharide occurring in marine plants. The structure of this substance, although not completely known, is composed of D-galactose, 3,6-anhydro-L-galactose, and small amounts of an ester sulfate. Agar is insoluble in cold water and alcohol. On heating, agar solutions become viscous; on cooling, they form a gel. A firm gel is formed with as little as 1.5 g of agar per 100 ml of water. The free sugar acid is a relatively strong acid (a 1% solution having a pH 2.0), and it will not form a gel. In the presence of specific metal ions, such as Na^+, K^+, Ca^{2+}, and Mg^{2+}, agar forms thermally reversible gels. Agar is nondigestible and is at times used as a bulk laxative. It is also used as a gelling agent in meat products and in dairy products.

Carrageenan (Irish moss), which is similar to agar in composition and properties, has many applications in the food industry. It is used in chocolate milk to suspend cocoa particles; in soups and sauces to improve body; and in pie fillings as a gelling agent.

Gums

The plant gums are polysaccharides containing hexoses and pentoses in combination with a uronic acid. For example, gum arabic is a complex calcium, magnesium, and potassium salt of arabic acid, which on acid hydrolysis yields D-galactose, L-arabinose, L-rhamnose, and D-glucuronic acid. Gum arabic is soluble in water and its aqueous solutions are slightly acidic within a pH range of 4.5–5.5. It has many commercial applications such as an adhesive in icings and toppings and as a foam stabilizer in beer.

Hemicelluloses

The hemicelluloses include a number of complex heteropolysaccharides that can be extracted by dilute alkali from the insoluble plant residue that remains after extraction with hot and cold water. Evidence indicates that the hemicelluloses occur in combination with lignin and cellulose in the cell walls of fibrous tissues. The hemicelluloses are polyuronides (D-glucuronic acid) in combination with xylose, arabinose, glucose, mannose, and galactose.

Pectins

Pectins are structural polymers that occur in combination with cellulose in the cell walls of fruits and vegetables. Chemically, the pectins are linear polymers of D-galacturonic acid (**4.59a**) joined by α-(1 → 4) glycosidic linkages in which the carboxyl groups of the galacturonic acid are partly esterified with methanol (**4.60a**) and the remaining free carboxyl groups are partly neutralized (**4.60b**). Some of the hydroxyl groups on C-2 and C-3 may be acetylated. Many pectins have neutral sugars covalently linked to them as side chains. Included among the sugars are arabinose, galactose, and to a lesser extent, glucose, xylose, and rhamnose. The molecular weight of pectic substances ranges from 10,000 to 400,000.

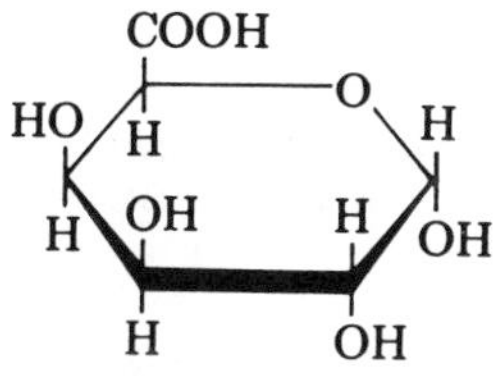

D-Galacturonic acid

4.59a

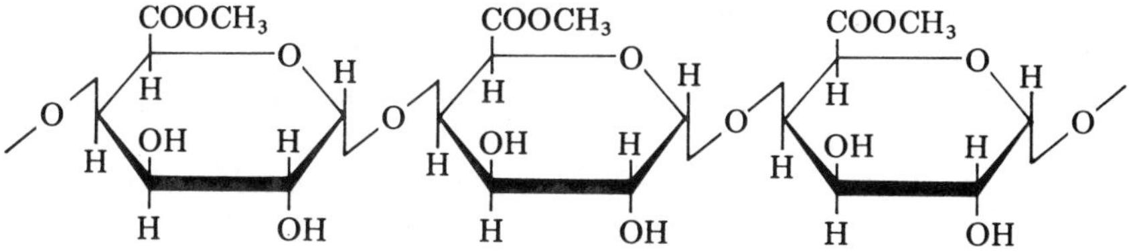

Portion of the pectin molecule

4.60a

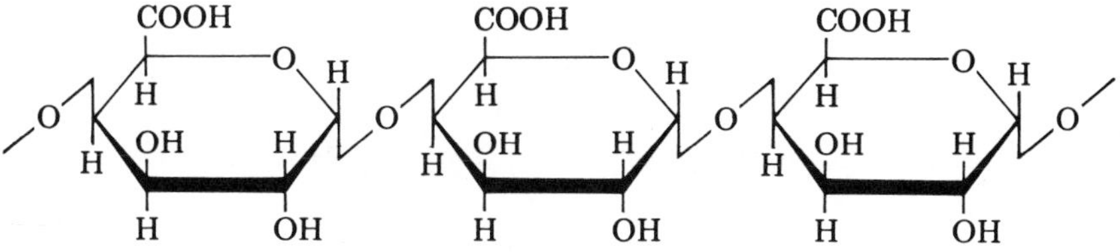

Pectic acid

4.60b

Various terms have been used to identify pectic substances and their derivatives. The nomenclature adopted by the American Chemical Society (Kertesz 1951) has been widely used to define the pectic substances:

Pectic substances. Pectic substances is a group designation for those complex, colloidal carbohydrate derivatives which occur in or are prepared from plants and contain a large proportion of anhydrogalacturonic acid units which are thought to exist in chain-like combination. The carboxyl group of these polygalacturonic acids may be partly esterified by methyl groups and partly or completely neutralized by one or more bases.

Protopectin. The term protopectin is applied to the water-insoluble parent pectic substance which occurs in plants and which, upon restricted hydrolysis, yields pectinic acids.

Pectinic acids. The term pectinic acids is used for colloidal polygalacturonic acids containing more than a negligible proportion of methyl ester groups. Pectinic acids, under suitable conditions, are capable of forming gels (jellies) with sugar and acid or, if suitably low in methoxyl content, with certain metallic ions. The salts of pectinic acids are either normal or acid pectinates.

Pectic acid. The term pectic acid is applied to pectic substances mostly composed of colloidal polygalacturonic acids and essentially free from methyl ester groups. The salts of pectic acid are either normal or acid pectates.

The following discussion relates only to pectic substances as components of foods and food products. Many excellent texts on pectins are available. The books by Kertesz (1951) and Doesburg (1965) adequately cover the various aspects of pectin chemistry.

Pectins are soluble in water. They may be precipitated with polyvalent cations, proteins, and water-miscible organic solvents. Dilute acids hydrolyze the ester and the glycosidic linkages of pectin. At low temperatures, the removal of the methyl ester groups yields pectic acid. At high temperatures, depolymerization takes place. Alkali hydrolyzes the methyl ester groups. However, with increasing temperature, the glycosidic linkage in the β-position to the ester group of pectin is cleaved following the formation of a double bond between C-4 and C-5, as shown below:

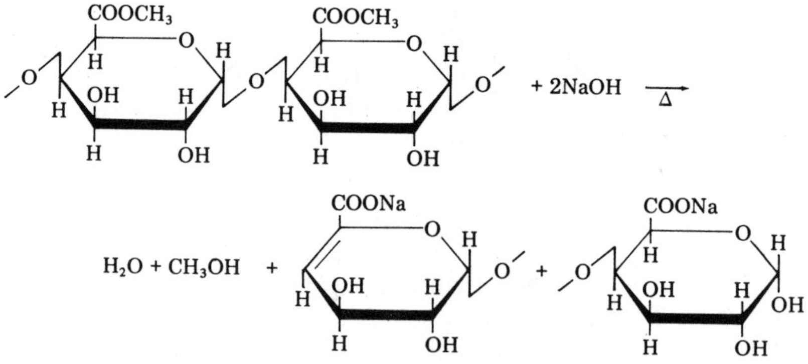

COOCH$_3$ COOCH$_3$ O H O H H H O O O + 2NaOH $\xrightarrow{\Delta}$ OH H OH H H H H OH H OH

COONa COONa O H O H H H$_2$O + CH$_3$OH + O + O OH H OH H H H OH H OH H OH H OH

This reaction is referred to as transeliminative cleavage of pectin and can also be catalyzed by enzymes (lyases).

High viscosity is a characteristic of pectin solutions. There are several factors that influence the viscosity: molecular weight of the pectin, the degree of esterification, the electrolyte concentration, and pH. Thus, the higher the molecular weight and the greater the degree of esterification, the higher the viscosity. The addition of salts of calcium and aluminum increases viscosity. The pH optimum for highest viscosity is related to the degree of esterification. The ability of pectins to form stable gels makes them an important additive to jams, jellies, and marmalades. There are two types of pectin gels: high sugar-pectin-acid gels and low-sugar gels (calcium pectinate gels). The first type of gel contains a pectin with a high methoxyl content (60–75% esterification), whereas the latter gel contains a pectin of low methoxyl content (20–45% esterification). In the high sugar-pectin-acid gel, the three components can replace each other within limits. The lower limit of sugar is approximately 55° Brix; the upper pH limit is determined by degree of esterification. Gel strength is also influenced by the molecular weight and concentration of pectin. The reactivity of low-methoxyl pectin with Ca^{2+} is used to form a low-sugar gel. This type of gel is obtained by adding calcium salts (soluble) under boiling conditions and then allowing the solution to cool. This kind of gel is used commercially to make low-calorie jams and jellies.

Commercial pectins are standardized in terms of an arbitrary unit called *jelly grade*. The unit may be defined as the proportion of sugar that one part of solic pectin or pectin extract is capable of turning, under prescribed conditions, into jelly with suitable characteristics. Thus, if 1 lb of pectin is capable of turning 100 lb of sucrose under prescribed conditions into a standard jelly, it is 100-grade pectin.

Changes occur in the pectin content and structure during growth, maturation, handling, and storage of plant material as well as in preserved products. There is limited information regarding the synthesis of pectic substances; however, information relative to the action of pectic enzymes on its substrate during processing is well-known.

Two groups of enzymes catalyze changes in pectic substances: depolymerizing enzymes and pectin esterases (Doesberg 1965). Depolymerizing enzymes are hydrolases that split the α-(1 $\rightarrow$ 4) glycosidic linkages in pectin and pectic acids. These enzymes, referred to as polygalacturonases, may be subdivided into enzymes that act primarily on pectin (polymethylgalacturonases, pectin-trans-eliminases) and enzymes that act on pectic acid (polygalacturonases and pectic acid-trans-eliminases). The pectin esterases are highly specific enzymes that hydrolyze methanol from esterified groups of pectinic acids. These enzymes do not hydrolyze the methyl ester of galacturonic acid and esters of the dimers and trimers of galacturonic acid.

Pectic enzymes are widely used in the food industry (e.g., to clarify fruit juices, especially those used for jelly manufacture and wine making). However, pectic enzymes may have an adverse effect as well as a beneficial effect. Changes in texture of a good product may be related to the activity of the pectic enzymes on their substrates.

Conjugated Polysaccharides

Glycoproteins

The glycoproteins are composed of simple proteins covalently bonded to carbohydrate moieties. These substances are widely distributed in nature and include soybean hemagglutenen, human transferrin, gamma globulin, egg ovalbumin, hormones, and ribonuclease B.

The structural characteristics of the glycoproteins include the following:

1. Seven sugars comprise the dominant portion of the carbohydrate moiety: mannose, *N*-acetylglucosamine, galactose, fucose, *N*-acetylneuraminic acid, *N*-acetylgalactosamine, and glucose.
2. The carbohydrate moieties are generally linked to the protein via the amide nitrogen of asparagine. However, there are some glycoproteins in which the carbohydrate is linked to the protein through the hydroxyl group of serine, threonine, or hydroxyproline.
3. The carbohydrate chain length is relatively short, ranging from 6 to 29 residues per mole of glycopeptide.
4. Molecular weights of the glycoproteins range from 660 to 4500.

Acid Mucopolysaccharides

Chondroitin sulfate, keratin sulfate, hyaluronic acid, and heparin are mucopolysaccharides that occur in connective tissue covalently bonded to protein. These substances are commonly referred to as glycoproteins or protein–polysaccharides and differ from typical glycoproteins in that they contain greater carbohydrate chains.

Enzymatic Degradation of Polysaccharides

In plants and animals, polysaccharides often act as storage compounds. They are better suited for this function than low-molecular-weight substances because they are less soluble and when they are in solution their osmotic pressure is so low their accumulation within the cell does not result in hypertonicity. However, before certain polysaccharides can be utilized by the body, they must first be converted into monosaccharide units.

Cleavage of glycosidic linkages is carried out in biological systems by two general mechanisms. One mechanism involves the hydrolytic cleavage of the glycosidic bond, while the second mechanism involves the *phosphorolysis* of the glycosidic bond.

Hydrolysis

The process by which polymeric starch is reduced to its component units is referred to as hydrolysis or digestion. Virtually all of these changes are brought about by the α- and β-amylases. The designation α- or β- does not refer to the configuration of the glycosidic bond that is hydrolyzed; both kinds of amylases hydrolyze α-(1 → 4) glucosidic linkages.

The α-amylases occur in many plant tissues and in the ptyalin of saliva and amylopsin of pancreatic juices of animals. The initial products resulting form α-amylase activity are oligosaccharides with six or seven glucose units. Thus, the cleavage occurs at the glycosidic linkage in the interior of the chain (Fig. 4.3). This shortening of chain length is manifested by a rapid loss in the viscosity of a colloidal starch solution; the iodine color reaction also disappears without the appearance of reducing sugars. When acting on the amylose fraction of starch (**4.57**), the initial fragments are further degraded largely to maltose. Since α-amylases cannot cleave α-(1 → 6) glycosidic linkages, the final products of starch hydrolysis are maltose, small amounts of higher

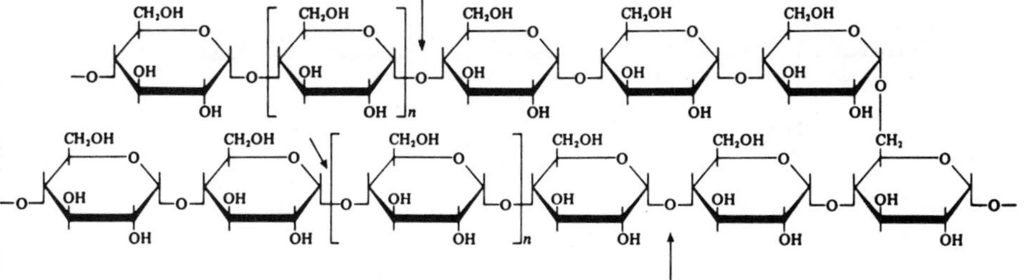

Figure 4.3. Schematic diagram of the enzymatic hydrolysis of starch by α-amylase with points of hydrolysis indicated by arrows. The α-(1 → 6) bonds are not attacked by this enzyme.

saccharides (malto-oligosaccharides), glucose, and panose (trisaccharide). The latter sugar contains the α-(1 → 6) linkage of the branching of amylopectin (**4.58**).

The β-amylases occur in higher plants, particularly in cereals and potatoes. These enzymes attach starch from the nonreducing end of the polymeric chain, cleaving alternate α-(1 → 4) glycosidic linkages to yield maltose in its β-form (Fig. 4.4). This hydrolytic cleavage is practically quantitative and negligible amounts of dextrin are formed from the straight-chain amyloses. Approximately 50–60% of branched-chain amylopectins is hydrolyzed, since enzymatic cleavage stops at the points of branching [α-(1 → 6) glycosidic linkage], leaving the *beta-limit dextrins*. Hydrolysis will cease unless debranching enzymes [α-(1 → 6) glucosidases] are present. Breakdown occurs only if an α-amylase is active at the same time.

The amylases have a wide variety of uses in the food industry (e.g., in the baking industry, the brewing industry, dextrose manufacture, sugar products, and starch removal from fruit extracts). An extremely important use for amylases is in the production of syrups that have better flavor and more sweetness than regular acid-conversion syrups. Technology has developed to the point whereby preparations of the different amylases (α-amylase, β-amylase, and glucoamylase) can be used with either acid-liquified or amylase-liquified corn starch to produce corn syrups of widely differing composition to meet the demands of different food industries (Table 4.1). For example, commercial processes are now available for producing syrups with a maltose content of about 52%, or a fructose content of 42%, or a glucose content of 39%. These corn syrups possess a number of properties, other than sweet-

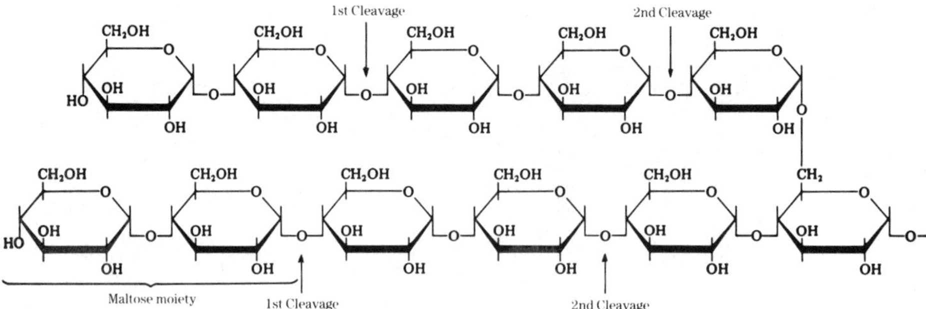

Figure 4.4. Schematic diagram of the enzymatic hydrolysis of starch by β-amylase with points of hydrolysis indicated by arrows. This enzyme cleaves alternate α-(1 → 4) bonds to yield β-maltose.

Table 4.1. Relative Sweetness and Carbohydrate Content of Several Corn Syrups

				Carbohydrate composition (dry basis)				
				Monosaccharides		Disaccharides		
Product[a]	Baumé	Relative sweetness (dry basis)	Pounds/pound dry substance	Dextrose	Fructose	(Maltose)	Tri-saccharides	Higher saccharides
28 DE Corn syrup	42°	30%	1.29	6%	—	9%	12%	73%
36 DE Corn syrup	43°	45%	1.25	15%	—	12%	11%	62%
42 DE Corn syrup	43°	50%	1.25	19%	—	14%	12%	55%
52 DE Corn syrup	43°	60%	1.23	28%	—	17%	13%	42%
62 DE Corn syrup	43°	70%	1.22	39%	—	28%	14%	19%
High-maltose 42 DE corn syrup	43°	55%	1.25	6%	—	45%	15%	36%
High-maltose 50 DE corn syrup	43°	65%	1.24	9%	—	52%	15%	24%
Glucose	—	80%	1.09	99.5%	—	—	—	—
High-fructose corn syrup	37°	100%	1.41	50%	42%	—	—	—
Liquid sucrose	36.2°	100%	1.49	—	—	100%	—	—

Source: Clinton Corn Processing Co., Clinton, IA.
[a] DE = dextrose equivalent = total reducing value expressed as glucose.

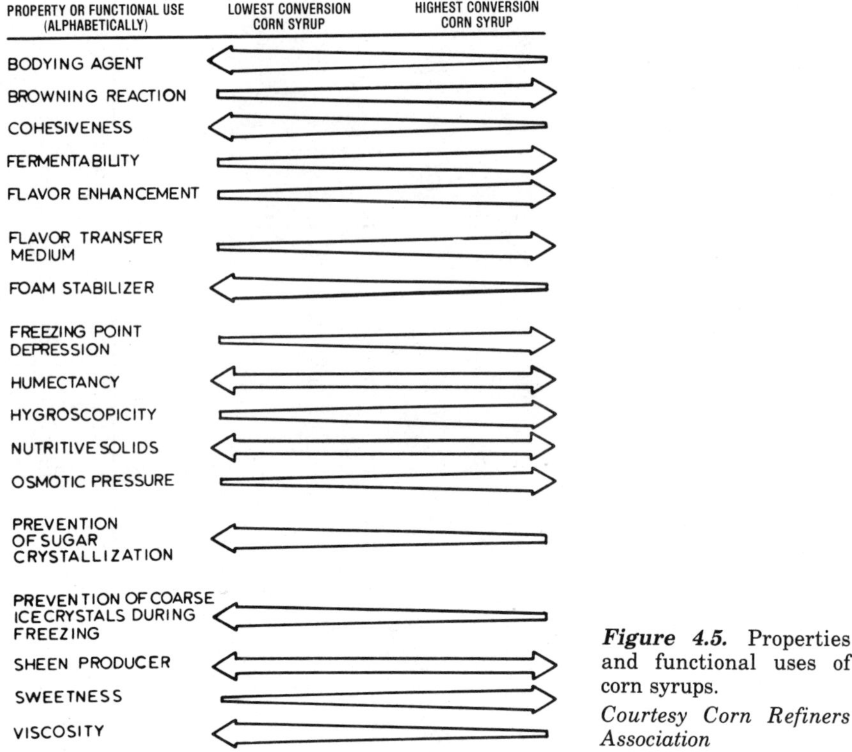

Figure 4.5. Properties and functional uses of corn syrups.
Courtesy Corn Refiners Association

ness, that represent a specific contribution to many foods. Some of the properties or functional uses of corn syrups are illustrated in Fig. 4.5.

Maltase is an intestinal α-glucosidase that splits maltose to form glucose and thereby completes the digestion of starch.

Phosphorolysis

Intracellularly, the storage polysaccharides are reduced to their component units by a process called phosphorolysis. In this reaction, catalyzed by phosphorylase, the terminal (nonreducing) glucose group of a polysaccharide chain is transferred to an inorganic phosphate group (Fig. 4.6). The chain is thus shortened and glucose-1-phosphate is formed. By this process, amylose is converted completely into glucose-1-phosphate; amylopectin is converted only to dextrin because the reaction stops near the branching point. Another enzyme, amylo-1,6-glucosidase (debranching enzyme), splits off the α-(1 → 6) linkage by simple hydrolysis, forming a molecule of glucose instead of glucose-1-

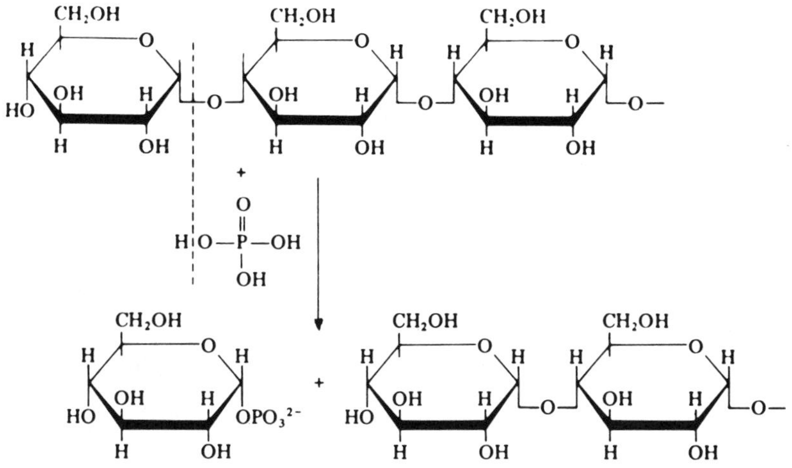

Figure 4.6. Schematic diagram of the action of phosphorylase at the nonreducing end of a polysaccharide.

phosphate. The action of the two enzymes in sequence yields a final product that is about 29% glucose-1-phosphate and 8% glucose.

Phosphorylase is found in animals, plants, and microorganisms. In animals it occurs in muscle, liver, heart, and brain. Peas, beans, and potatoes are good sources of phosphorylase in higher plants.

☐ BIOSYNTHESIS

Monosaccharides

The biosynthesis of monosaccharides in animals occurs mainly by hydrolysis reactions (e.g., starch → glucose + maltose; sucrose → fructose + glucose) and glycolysis (gluconeogenesis) in which glucose-6-phosphate and fructose-6-phosphate are key intermediates. Plants produce monosaccharides by the photosynthetic process utilizing ribulose-1,5-diphosphate, which accepts CO_2 and forms 3-phosphoglyceric acid. Glucogenesis following this carboxylation leads to the formation of glucose-6-phosphate, which can be converted to a variety of monosaccharides. Disaccharides are formed by enzyme-catalyzed linkage of the necessary monosaccharides.

Two types of enzymes can catalyze the interconversion of monosaccharides. The *isomerases* reversibly convert aldoses to ketoses. For example:

Glucose-6-phosphate ⇌ Fructose-6-phosphate

Ribose-5-phosphate ⇌ Ribulose-5-phosphate

Glyceraldehyde-3-phosphate ⇌ Dihydroxyacetone phosphate

Mannose-6-phosphate ⇌ Fructose-6-phosphate

The *epimerases* function mainly by utilizing uridine diphospho-sugars, which are formed by the general reaction UTP + sugar-1-phosphate ⇌ UDP-sugar + PPi. The UDP derivative is then epimerized at one of the asymmetric carbon atoms. For example:

UDP-glucuronate ⇌ UDP-galacturonate

UDP-D-xylose ⇌ UDP-L-arabinose

Epimerization of the sugar phosphate xylulose-5-phosphate can occur without being mediated by UDP-sugars: Xylulose-5-phosphate ⇌ Ribulose-5-phosphate.

Disaccharides

Lactose

The synthesis of lactose in the mammary gland involves several steps. The galactose moiety is formed by the following two reactions:

$$\text{Glucose-1-phosphate} + \text{UTP} \xrightleftharpoons{\text{transferase}} \text{UDP-glucose} + \text{PPi}$$

$$\text{UDP-glucose} \xrightleftharpoons{\text{epimerase}} \text{UDP-galactose}$$

Since galactose and glucose differ only at C-4, these sugars are interconvertible. Glucose-1-phosphate from metabolism serves as a ready source of galactose. UDP-galactose may also be formed as follows:

Galactose + ATP → Galactose-1-phosphate

Galactose-1-phosphate + UTP ⇌ UDP-galactose

The reaction linking galactose and glucose to give lactose is catalyzed by lactose synthetase.

UDP-galactose + Glucose ⇌ Lactose + UDP

Sucrose

The synthesis of sucrose in plants is similar to that of lactose. Fructose-6-phosphate, which is formed in glycolysis, reacts with UDP-glucose to form sucrose-6-phosphate. Sucrose phosphate is then hydrolyzed to sucrose.

$$\text{Fructose-6-phosphate} + \text{UDP-glucose} \xrightarrow{\text{synthetase}} \text{Sucrose-6-phosphate} + \text{UDP}$$

$$\text{Sucrose-6-phosphate} \xrightarrow{\text{phosphatase}} \text{Sucrose} + \text{Pi}$$

Sucrose is also synthesized by the reaction of UDP-glucose and fructose to give sucrose and UDP.

Maltose

The biosynthesis of maltose occurs mainly by the degradation of starches, which has been discussed previously (see Figs. 4.3 and 4.4).

Polysaccharides

Glycogen

Glycogen is a highly branched polymer of glucose, with each unit joined by an α-glycosidic linkage either to C-4 or C-6 of another unit (see Fig. 4.2). The synthesis of glycogen begin with the conversion of glucose-6-phosphate to glucose-1-phosphate, catalyzed by the enzyme phosphoglucomutase. The mechanism of this reaction involves a glucose diphosphate intermediate.

$$\text{Glucose-6-phosphate} + \text{Phosphoenzyme} \rightleftharpoons \text{Glucose-1,6-diphosphate} + \text{Dephosphoenzyme}$$

$$\downarrow\uparrow$$

$$\text{Glucose-1-phosphate} + \text{Phosphoenzyme}$$

The concentration of glucose-1,6-diphosphate is less than that of glucose-1-phosphate, which in turn is lower than that of glucose-6-phosphate.

It was once thought glycogen phosphorylase catalyzed both the synthesis and degradation of glycogen in reactions involving glucose-1-phosphate. This enzyme catalyzes the making and breaking of α-1,4 but not α-1,6 glucosidic bonds. It is now known that the biosynthetic pathway for glycogen involves UPD-glucose, which is formed from glucose-1-phosphate (**4.61a**) and uridine triphosphate (UTP, **4.61a**) in a reaction catalyzed by uridine diphosphoglucose pyrophosphorylase.

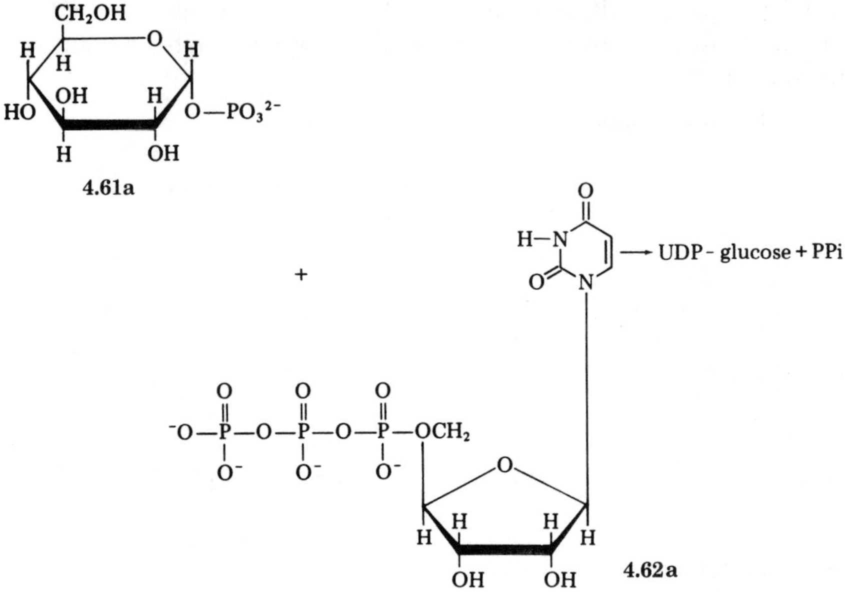

4.61a

+ → UDP - glucose + PPi

4.62a

In the second step of glycogen synthesis, the glucosyl group of UDP-glucose (**4.63a**) is transferred to C-4 of the terminal glucose unit at the nonreducing end of the straight chain in a reaction catalyzed by glycogen synthetase.

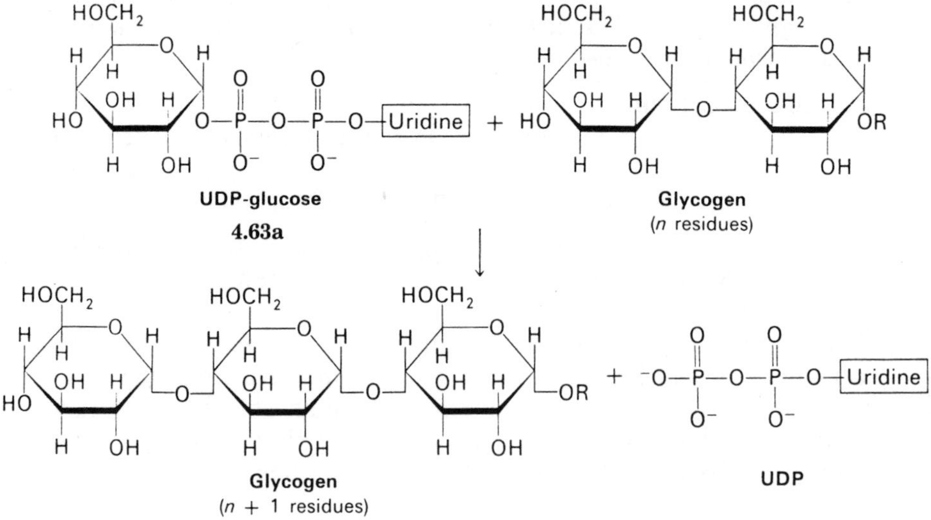

UDP-glucose

4.63a

+

Glycogen
(*n* residues)

↓

Glycogen
(n + 1 residues)

+

UDP

UDP is converted to UTP by reaction with ATP (thus the UTP/UDP ratio is maintained in the cell).

So far, only the formation of the α-1,4-glucosidic linkages in glycogen has been considered. A glycogen-branching enzyme, amylo-(1,4 → 1,6)-transglycosylase, has been isolated from liver and other tissues and has been shown to catalyze the formation of α-1,6 bonds. In this reaction there is a transfer of an oligosaccharide fragment from the end of a chain to the 6-hydroxyl group of a glucose residue of the same or another glycogen chain. The action of the branching enzyme on glycogen can be represented as follows:

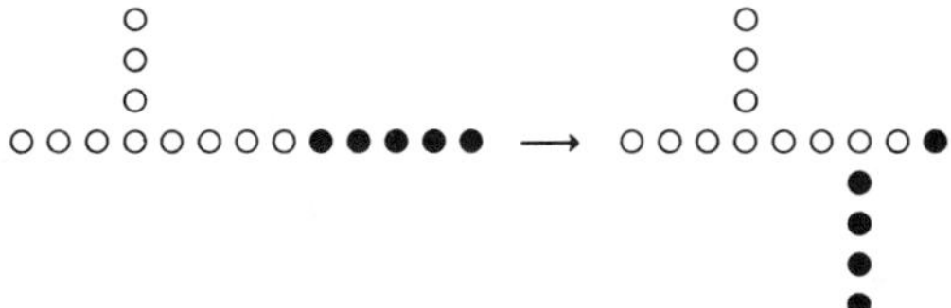

In all likelihood, intramolecular rearrangements occur with the extension of the branches by transglycosylation.

Starch

The synthesis of starch in plant tissue follows a pathway analogous to the synthesis of glycogen. In plants, the glycosyl donor is adenosine diphosphoglucose (ADP-glucose, **4.64a**), which is formed from ATP and α-glucose-1-phosphate by the action of the enzyme adenine diphosphoglucose pyrophosphorylase.

$$\text{glucose-1-phosphate} + \text{ATP} \rightleftharpoons \text{ADPG} + \text{PP}_i$$

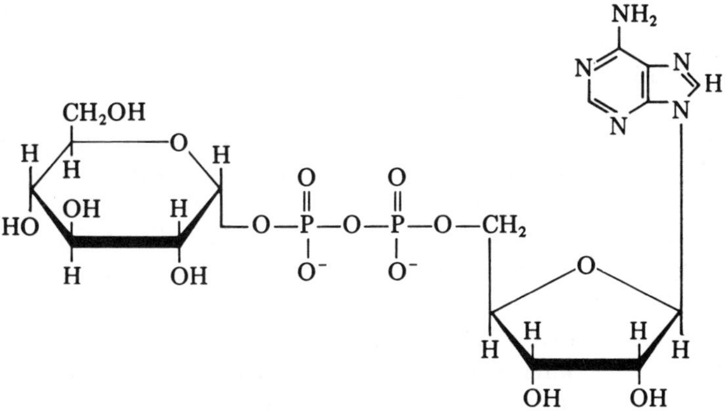

Adenosine diphosphoglucose

4.64a

This enzyme is specific for starch synthesis and is not active in glycogen synthesis. The second step of starch synthesis involves glycosyl transfer catalyzed by amylose synthetase, as follows:

$$\text{ADP-glucose} + \text{Acceptor (glucose)}_n \rightarrow \text{ADP} + \text{Starch (glucose)}_{n+1}$$

As in glycogen synthesis, the synthetase only catalyzes formation of α-1,4 glycosidic linkages. The formation of the branched structure of the starch molecule is accomplished by a branching enzyme, amylo-(1,4 → 1,6)-transglycosylase.

Cellulose

Each of the major homopolymers of glucose has as a precursor a characteristic nucleoside diphosphate glucoside. In glycogen synthesis, the glycosyl donor for the formation of the α-1,4 bonds is UDP-glucose; in starch synthesis, the donor is ADP-glucose. Similarly, the glycosyl donor in the synthesis of cellulose is guanosine diphosphoglucose (UDP-glucose).

In contrast to starch and glycogen, which both have α-1,4 linkages and are branched, cellulose (**4.40**) has α-1,4 linkages and is linear. The general reaction for the synthesis of cellulose is represented by the following formula:

$$\text{GDP-glucose} + \text{Acceptor (glucose)}_n \rightarrow \text{GDP} + \text{Cellulose (glucose)}_{n+1}$$

Other structural polysaccharides are synthesized by similar reactions. For example, xylans (α-1,4 linkages) are formed from UDP-D-xylose.

☐ RELATIONSHIP OF STRUCTURE TO SWEETNESS

Acceptance of food by an individual is influenced by such aesthetic properties as taste, color, odor, and texture. Our perception of flavor depends on both taste and odor. We are reminded of this when a head cold reduces our sense of smell so much that food seems almost tasteless. Without the sense of smell, food tastes flat because the sense of taste alone can only distinguish between sweet, salty, sour, and bitter or combinations of these four basic tastes.

The raised portions of the tongue (the papillae) are the organs of taste. The taste buds, containing the receptors, are located in the papillae on the tongue and a few may be in the mucosa of the soft palate.

Different tastes are sensed at different places on the tongue. A sweet taste is sensed at the tip of the tongue, a salty taste at the tip and edge of the tongue, a sour taste at the edge of the tongue, and a bitter taste at the back of the tongue.

The salt taste and the sour taste are produced by well-defined compounds. The salt taste is due to soluble inorganic salts, while the sour taste is related to the hydrogen-ion concentration of a solution. In contrast, sweet and bitter tastes cannot be associated with any single class of compounds. In fact, sweet and bitter responses may be found in the same class of compounds. The sweet taste is produced by a variety of organic compounds, including alcohols, glycols, sugars, and sugar derivatives.

The mechanism of sweet-taste response is not well understood. However, minor changes in the structure of a molecule can result in a new compound that has a quite different taste. For example, *p*-ethoxyphenylurea (Dulcin, **4.65**) is very sweet and *o*-ethoxyphenylurea (**4.66**) is tasteless; replacement of the oxygen of the urea group by sulfur yields a bitter compound (*p*-ethoxyphenylthiourea, **4.67**).

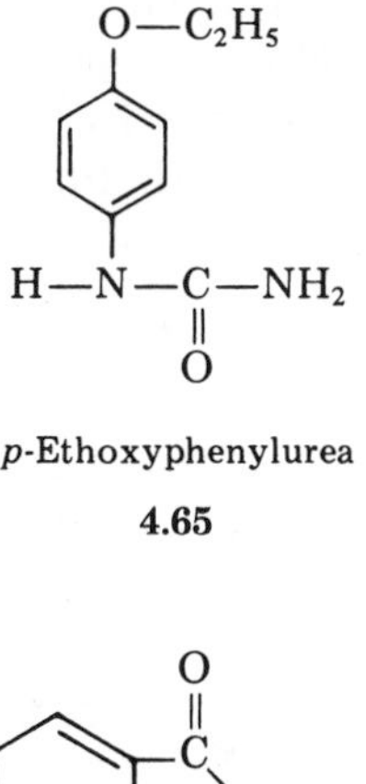

p-Ethoxyphenylurea

4.65

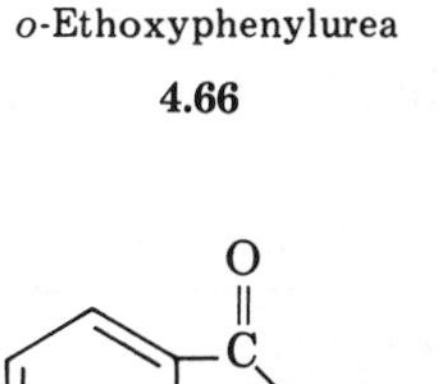

o-Ethoxyphenylurea

4.66

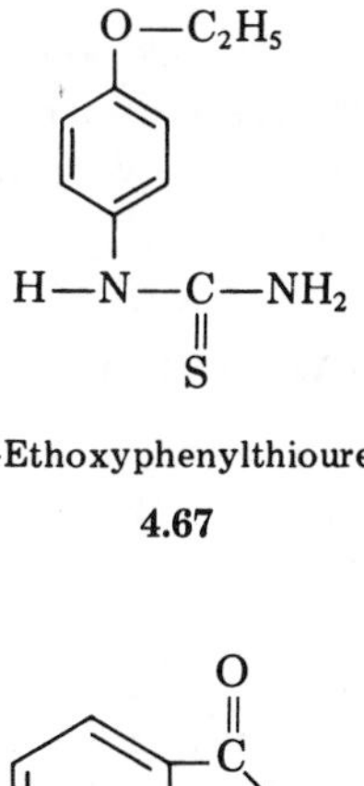

p-Ethoxyphenylthiourea

4.67

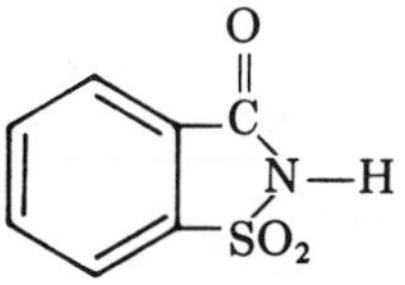

Saccharin

4.68

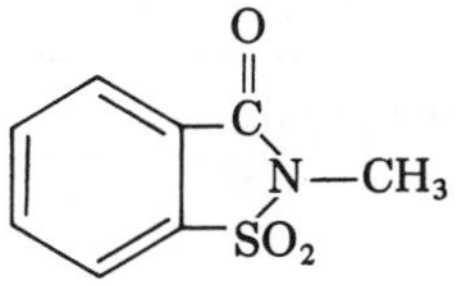

N-Methylsaccharin

4.69

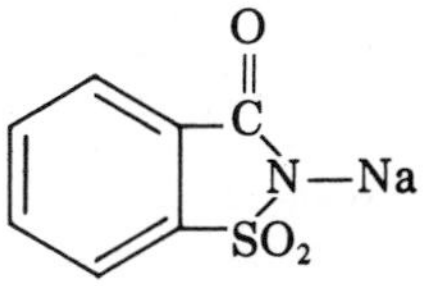

Sodium saccharin

4.70

Similarly, saccharin (**4.68**) is very sweet, but if a methyl group replaces the hydrogen of the imido nitrogen, tasteless *N*-methylsaccharin (**4.69**) results. Sodium saccharin (**4.70**) is very sweet. Stereoisomers and anomers may have different tastes. L-Glucose has a slight salty taste, while D-glucose is sweet. Similarly, the two anomers of D-mannose have different taste: α-D-mannose is sweet, whereas β-D-mannose is bitter. Many of the D-isomers of amino acids are sweet, whereas the L-isomers generally are tasteless (Table 4.2). These differences in taste are due to changes in molecular structure and indicate that discrimination between tastes reflects in part a recognition of the spatial structure of molecules.

The African fruit known as the miracle berry contains the flavor principle miraculin, which has been isolated and identified as a glycoprotein. The sweetness induced by this unusual substance abolishes the taste response to acidic materials.

Sweetness of Sugars

Since sweetness is of primary interest in sugars, a knowledge of their relative sweetening ability is important. The intensity of sweetness cannot be determined quantitatively in absolute physical or chemical terms. Sweetness can only be measured by subjective sensory methods and must always represent averages of opinion. Sucrose is usually accepted as the standard. Other sweeteners, either natural or nonnutritive, are then compared with sucrose and their sweetening power is expressed with sucrose as a base. Glucose, for example, is assigned a relative sweetness value of 69 (Table 4.3) when compared with 10% sucrose, while D-fructose has a relative sweetness value of 114.

Synthetic Sweeteners

The sweet taste is produced by a wide variety of chemical compounds. Some of the compounds that have been reported to be sweet-tasting

Table 4.2. Taste of Stereoisomers of Amino Acids

Amino acid	L Isomer	D Isomer
Glutamic acid	Unique	Almost tasteless
Phenylalanine	Faintly bitter	Sweet with bitter aftertaste
Leucine	Flat, faintly bitter	Strikingly sweet
Valine	Slightly sweet, yet bitter	Strikingly sweet
Serine	Faintly sweet, stale aftertaste	Strikingly sweet
Histidine	Tasteless to bitter	Sweet
Isoleucine	Bitter	Sweet
Methionine	Flat	Sweet
Tryptophan	Flat	Very sweet
Tyrosine	Bitter	Sweet

Source: Berg (1953).

Table 4.3. Relative Sweetness of Various Sugars

Sugar	Relative sweetness
Lactose	39
Maltose	46
D-Mannose	59
Galactose	63
Invert sugar	65
D-Xylose	67
α or β-D-Glucose	69
Sucrose	100
D-Fructose	114

Source: Nieman (1958); Amerine *et al.* (1965).

have been summarized and discussed by Moncrieff (1967). The relative sweetness of some synthetic sweeteners, relative to sucrose, is presented in Table 4.4. One of the best-known synthetic sweeteners is saccharin, which is 200–700 times as sweet as sucrose. Other sweet substances are sodium cyclamate (cyclohexylsulfamate), Dulcin (*p*-ethoxyphenylurea), and the alkoxy-2-amino-4-nitrobenzenes. Cyclamate is 15–30 times as sweet as sucrose, and the *n*-propyl derivative of 2-amino-4-nitrobenzene is over 4000 times as sweet as sucrose. A dipeptide sweetener, L-aspartyl-L-phenylalanine methyl ester, has been found to be 250 times as sweet as sucrose (Nutrasweet). The dihydrochalcones of the naturally occurring flavanones (prunin, naringin, and neohesperidin) are intensely sweet.

In addition to these sweeteners, other intensely sweet-tasting substances have been isolated from plant materials. Glycyrrhizin is a sweetener present in licorice root (*Glycyrrhiza glabra*). It is a triter-

Table 4.4. Relative Sweetness of Synthetic Substances Compared with Sucrose

Sweetener	Relative sweetness (weight basis)
Sucrose	1
Cyclohexylsulfamate (sodium cyclamate)	15–31
p-Ethoxyphenylurea (dulcin)	70–350
2,3-Dihydro-3-oxobenzisosulfonate (saccharin)	240–350
L-Aspartyl-L-phenylalanine methyl ester	250
1-*n*-Propoxy-2-amino-4-nitrobenzene	4100

Source: Inglett (1970).

penoid glycoside and its relative sweetness is about 15,000 compared that of 10% sucrose. Another naturally occurring, intensely sweet, organic compound is steviosid, which is found in the leaves of a small shrub native to Paraguay. It is a steroid glycoside that has a relative sweetness of 30,000 compared with 10% sucrose.

☐ NUTRITIONAL VALUE OF CARBOHYDRATES

Some populations in the world tend to consume large amounts of grain products and therefore derive a high percentage of calories from carbohydrates. In many parts of the world where the income is low, cereal grains supply 75–80% of the total calories. In some Asian countries carbohydrate intake is high because the diets are rice-based. This is especially true in Japan because of dietary traditions. On the other hand, the diets of Eskimos and meat-eating tribes in Africa are low in carbohydrates. The intake of starches and other complex carbohydrates from cereals, roots, tubers, and legumes decreases and the intake of sugar increases as incomes rise. In the United States the total available carbohydrate in the food supply has declined about 20% since the early twentieth century. During the same period consumption of grain products decreased about one-half, while sugar consumption increased approximately one-third. Most of the carbohydrate is furnished by flour and cereal products and by sugar and syrups. The sugar consumption in the United States has gradually increased except for a slight drop during the mid-1930s and in the early 1940s due to rationing during World War II. Because of a decrease in availability of sugar and an increase in cost, consumption per person declined after 1973. In 1976 and 1977, according to the USDA, 46% of the total calories in U.S. diets were supplied by carbohydrates compared with 56% in 1909.

Even though the importance of carbohydrate in the metabolic processes has been established, no standard for a dietary intake has been set. The daily intake of carbohydrates of most Americans amounts to 200 g or more. An inadequate intake of carbohydrates may result in malnutrition, low blood glucose levels, and a variety of metabolic disorders. But an excess of carbohydrate, especially in the form of sugar, also has adverse effects on nutritional status.

Some investigators have suggested that blood lipids are higher on diets containing sugars, especially sucrose, than on diets containing starch. These studies suggest that serum triglyceride levels are affected more by the type of carbohydrate ingested than are cholesterol levels. Other studies have shown that the type of carbohydrate has little effect on blood lipids. Although excess intake of carbohydrate is not the sole cause of obesity or diabetes mellitus, it is related to both disorders to some extent. The ingestion of more calories than is needed for energy expenditure, whether from carbohydrate, fat, or protein, results in weight gain. An excess intake of carbohydrate may cause diabetes in those who are genetically susceptible to the disorder.

A high intake of sweet foods also has an indirect effect on the nutritional adequacy of the diet. Excessive intake of cookies, cakes, candies, and soft drinks especially between meals usually results in a reduced intake of other foods that supply vitamins, minerals, and protein. As a result, malnutrition or one of the deficiency diseases may develop.

Digestion

Carbohydrates are broken down mechanically by mastication (chewing), which cuts and grinds the food and mixes it with saliva in preparation for swallowing. The muscular contractions of the stomach and walls of the intestine further mix and subdivide the food into smaller pieces. The chemical breakdown of carbohydrates to their simple sugars is brought about by the action of starch-splitting enzymes (the amylases) and also by disaccharide-splitting enzymes (sucrase, maltase, and lactase). Some enzymatic digestion of starch dextrins into maltose occurs in the mouth through the action of salivary amylase. This process continues in the stomach until amylase is inactivated by hydrochloric acid in the gastric juice. Upon acidification of the food mass, some disaccharides are broken down into monosaccharides by acid hydrolysis.

The small intestine is the principle site of carbohydrate digestion. Through the action of pancreatic and intestinal amylases and of intestinal sucrase, maltase, and lactase carbohydrates are degraded to their monosaccharide units. The disaccharide-splitting enzymes are part of the mucosal cells of the membrane lining the small intestine. Disaccharides are absorbed into these cells where they are broken down by hydrolysis into simple sugars. Monosaccharides require no digestion.

Absorption

Carbohydrates are absorbed in the small intestine. Most of the sugars in the form of monosaccharides are absorbed in the duodenum and the jejunum; small amounts are absorbed in the ileum. There are two absorptive processes involved. One is passive diffusion, which occurs when the concentration of sugars in the intestine is higher than that in the circulating blood. The other process is active transport, which occurs when the concentration of sugars in the intestine is the same or lower than that in the blood. In active transport, which requires energy supplied by ATP, sugar molecules are attached to substances called carriers, thought to be proteins or lipoproteins. The same carriers attach to sodium ions. Sugar then is transported through the intestinal wall as a sugar–sodium–carrier complex. When the complex enters the bloodstream, the sugar is transported in the portal vein to the liver. Fructose is absorbed only by passive diffusion, while glucose and galactose are absorbed by passive diffusion and active transport.

Glucose and galactose are absorbed at about the same rate, and fructose only about one-half as rapidly. Mannose and the pentoses are poorly absorbed. Factors that influence the absorption of carbohydrates include the condition of the mucous membrane of the small intestine; the secretions of the adrenal, anterior pituitary, and thyroid glands; and the intake of vitamins, particulary the vitamin B complex.

Metabolism

Carbohydrates are metabolized primarily in the liver. The simple sugars are carried in the portal vein to the liver. Galactose and fructose are converted into glucose in the presence of ATP and enzymes specific for each reaction. Glucose is phosphorylated to glucose-6-phosphate, which is converted to glucose-1-phosphate and finally to glycogen for storage in the liver. A small amount of glucose (about 100–150 g) is stored as glycogen. The rest is released into the bloodstream and carried to all tissues. Within the cells most of the glucose is converted to CO_2 and water, releasing energy for the many needs of the body. Some of the glucose is converted to fatty acids, and some combines with free amino groups to form amino acids.

For carbohydrate metabolism to proceed, several B vitamins as well as magnesium and phosphorus are required. Niacin, riboflavin, thiamin, vitamin B_6, and pantothenic acid function as coenzymes, and magnesium and phosphorus function as cofactors, in the reactions involved in the metabolism of carbohydrates.

The hormones thyroxine, glucagon, and epinephrine exert a control on both the level of sugar in the blood and the glycogen content of the liver. The hormone insulin, secreted by the islands of Langerhans in the pancreas, is essential for maintaining a normal level of blood sugar.

It is necessary for the entry of glucose into the cells, the conversion of glucose into glycogen, and glycogen into glucose, the oxidation of glucose, and the synthesis of fat and protein from glucose.

Glucose is released from the liver into the bloodstream at a rate that maintains a level of about 100 mg glucose/100 ml blood. During fasting the normal blood glucose level is 70–90 mg/100 ml. After a meal the level of glucose rises but returns to normal in a relatively short time. When a test dose of glucose is administered, the blood glucose level rises sharply for about 0.5–1 hr. If the subject has a normal glucose tolerance, the blood glucose level will return to a normal level by the end of 2 hr.

☐ COMMERCIAL SUGAR AND SUGAR PRODUCTS

Sucrose (**4.52a**) is produced from the juice of sugar cane (*Saccharum officinarum*) and from sugar beets (*Beta vulgaris*). Sugar cane is a tall grasslike plant, which stores sugar in its stalk; on the other hand, the sugar beet stores sugar in a long whitish root. The preparation of sugar-bearing juices from sugar cane and sugar beets differs considerably. However, once the juices are concentrated, the preparation of other sugar products is essentially the same.

Raw Cane Sugar Production

Mills are used to crush the stalks of sugar cane to express the sugar-bearing juices. A series of mills may consist of a shredder, or a crusher, and several three-roller mills, which extract the juice under heavy pressure. Either dilute juice or hot water is usually applied to the bagasse (the woody residue of the cane) as it leaves each of the roller mills. This process dilutes the juice but allows for extraction of the maximum amount of sugar in the stalks. The thin juice that is formed (called mixed juice) is strained and limed to alkalinity and then heated. Impurities in the juice settle to the bottom and the clear purified juice is decanted. The juice is concentrated in evaporators and then boiled under a vacuum to form sugar crystals suspended in molasses (called massecuite). The massecuite is then centrifuged to separate the sugar crystals. In most cases this raw sugar is shipped to refineries.

Raw cane sugar delivered to refiners usually contains about 97% sucrose; the remainder consists mainly of ash, invert sugar, and organic nonsugars, as well as moisture. The first step in the refining process involves mixing the raw sugar with hot saturated syrup. This serves to soften the impurities that surround each raw sugar crystal. This mixture is centrifuged with the use of spray water in the final stages to remove a portion of the softened impurities. The washed raw sugar is then dissolved in hot water, filtered, clarified, and then filtered through columns of bone char, which removes the color and most of the remaining impurities. The filtered liquor is evaporated, crystalized under vacuum, and centrifuged again. The result is a granulated sugar, which is dried and screened for crystal size and then packaged. The final product contains 99.95% sucrose if the sugar is graded medium granulated. A fine granulated sugar contains 99.9% sucrose.

Beet Sugar Production

Beets for processing into sugar are thoroughly washed to remove trash and other materials that would cause difficulty in the slicers. The beets are weighed and then sliced into cossettes whose appearance might be compared to shoestring potatoes. The slices are passed into a diffusion battery where the sugar is extracted in hot water. The underlying principle in this process is dialysis. The cell walls of the beets act as membranes through which the sugar passes. Calcium hydroxide is added to the juice and carbon dioxide is introduced until all the lime is precipitated as calcium carbonate. This treatment effects the chemical and mechanical purification of the juice. Following filtration, the juice is treated again with carbon dioxide for the complete removal of lime. The juice is then treated with sulfur dioxide, and filtered for the third time into evaporators.

The evaporators are essentially a large boiling vessel heated by exhaust steam. Connected in a series, each cell of the evaporator has a higher vacuum pressure than the preceding one, so that the juice boils progressively at lower temperatures. The evaporator process results in concentration of the liquid to about 60–70% solids. The juice is filtered again and is boiled in vacuum pans until crystals are formed. The material discharged from the vacuum pans consists of crystals of sugar suspended in a syrup. The syrup is removed by centrifugation, and the granulated sugar is washed and dried, screened for size, and packaged.

The remaining syrup is recycled for recovery of the sugar. The liquid product of the third centrifugation is usually treated to the Steffen's process, in which lime is added to precipitate sugar as calcium sucrate. This filtrate is a source of monosodium glutamate. Carbon dioxide is used to liberate the sugar, and the juice is returned to the process.

Sucrose Products

A number of different sugar products are made from cane sugar and beet sugar or the juices and syrups formed in their production.

Granulated sugar is refined crystallized sucrose obtained from sugar cane or sugar beets. It is passed over sieves so that various sizes of crystals are obtained. The crystal sizes range from ultra-fine to coarse.

Powdered sugar, also called *confectioner's sugar*, is graded from ultra-fine (10× types), to very fine (6× types), to fine (4× types).

Brown sugars are composed of fine sugar crystals covered with a film of dark-colored, cane molasses-flavored syrup. This syrup is responsible for the flavor and color of brown sugar. Light brown sugar is desirable in baked goods (e.g., as icings and glazes for doughnuts) and in butterscotch and other candies. Dark brown sugar is used in gingerbread, mincemeat, baked beans, and other dark-colored foods.

Invert sugar is obtained by inversion of a sucrose solution. This process is effected by heating the solution, treating it with an enzyme, or passing it over an acid ion-exchange bed. The two sugars that are formed are D-glucose and D-fructose, sometimes called levulose. The mixture of the two is called invert sugar. Invert sugar has a tendency to retain moisture, which makes it useful for prolonging the shelf life of many foods.

Liquid sugars are clear, practically colorless solutions of sucrose in a highly purified form. These sugars may be partially or completely inverted. Liquid brown sugars are used in baked goods, table syrups, cough drops, and other items. Liquid invert sugars are used in making beverages, baked goods, icings, and syrups. Light straw sugar and water white sugars, both of which are uninverted, are used in many products including canned foods, dairy products, frozen fruits, and ice cream.

Refiner's syrup is the residual liquid product obtained in the refining of raw cane sugars. It varies in color from dark brown to almost colorless. The better grades of refiner's syrup are used in the food processing industry.

Edible molasses is a group of products containing sugar as well as other naturally occurring substances in sugar-bearing plants. In an older method of making molasses, cane juice is boiled until a large part of the water is evaporated. This molasses is called open-kettle; since sulfur dioxide is not used in its preparation, it is sometimes called unsulfured molasses. In other methods, part or most of the crystallizable sugar is recovered from concentrated cane juice. The resulting products are used in better grades of table syrups. Molasses is used in gingerbread, spice and fruit cakes, cookies, and candies.

Maple syrup was probably first produced by American Indians, who concentrated the sap from maple trees by freezing or by dropping hot stones into bark vessels containing the sap. Later they used metallic vessels to boil down the sap. The early settlers learned the art of making maple syrup and sugar from the Indians. Nowadays, maple tree sap is concentrated from about 2% solids to 66% solids by boiling off water in open pans or kettles. The concentration process not only removes water but also produces other effects: The proteins are coagulated and skimmed off; certain of the mineral components precipitate out; and a small amount of sucrose inversion likely occurs with the production of invert sugar. The characteristic maple flavor develops as a result of heating the sap components. Most of the maple syrup that is sold wholesale is converted into blended syrups containing one part of maple to four to six parts of cane or corn sugar.

By-Products of the Sugar Industry

Commercially important by-products of the sugar industry include black strap molasses, bagasse, beet pulp, and beet tops. In the past black strap molasses was used exclusively in the distillation of alcohol and solvents. Today its primary use is in cattle feed. Bagasse was once used as fuel for raw sugar mills; however, in many parts of the world this product now is used in the production of paper and wall board. Beet pulp as well as beet tops are used as a cattle feed.

☐ CORN SWEETENERS

Corn sweeteners may be grouped into several types: corn syrups, corn syrup solids, crude corn sugar, and dextrose. The use of corn syrups and sugars has increased over the past several years due to the greater production and consumption of processed foods, and the expanded use of corn syrup and dextrose by food processors. Confectioners and chewing gum manufacturers consume more corn syrup than any other industry. Large amounts of corn syrup are used in making table syrups and syrups for soda fountain use. The largest user of refined corn sugar, or dextrose (D-glucose), is the baking industry. Corn sweeteners are also used in ice cream, canned and frozen fruits, jams and jellies, beer and ale, soft drinks, and numerous other food products.

Corn syrups are produced by hydrolysis of starch, either by acid or by acid–enzyme conversion. Hydrolysis is usually halted before it is complete, resulting in a mixture of dextrose and maltose. A higher proportion of maltose can be obtained with the acid–enzyme method than by acid hydrolysis alone. The dextrose–maltose ratio can be varied depending on the type of enzyme used and the extent of preliminary acid conversion. The acid hydrolysis process is stopped by raising the pH with sodium carbonate. Fatty substances are skimmed off or removed by centrifugation or filtration. Suspended solids are removed by filtration in vacuum filters. The filtrate is evaporated to a density of about 60% dry substance. Further clarification is accomplished using bone char or other carbon filters, resulting in a clear, practically colorless syrup. Mineral substances may be removed by an ion-exchange process. Following the final filtration, evaporation is completed in a vacuum pan at relatively low temperatures to prevent damage to the syrup. In the acid-enzyme process, the enzyme is added after neutralization, clarification, and concentration of the syrup. The enzyme is inactivated when hydrolysis has progressed to the desired degree.

Corn syrup solids are produced from completely refined corn syrup that is spray- or drum-dried to a moisture content of less than 3.5%. Since corn syrup solids are hygroscopic, the material is packed in multiwalled paper bags that contain moisture barriers.

Crude corn sugars are prepared by a more complete acid hydrolysis of corn starch than is used for syrup production. The crude sugar liquor that is formed is neutralized, filtered, clarified, and concentrated. Fine sugar from previous batches is added to the concentrated liquor, which is placed in pans and allowed to crystallize. This product is sold in slabs or billets, or is chipped and packaged for industrial and food processing uses.

Dextrose results from the complete hydrolysis of the starch. The hydrolyzate is then purified and concentrated as in the manufacture of corn syrup. The refined dextrose liquor is seeded with dextrose crystals from a previous batch. Crystallization occurs in large crystallizers equipped with agitators. Cooling rates are carefully controlled. Crystallization requires about 100 hr.

☐ QUALITATIVE ANALYSIS

It is sometimes necessary to identify the particular carbohydrates present in a food before making a quantitative analysis. The kind of food will, in general, give some indication of the various carbohydrates to anticipate. For example, one would expect fruit juices and honey to contain fructose and sucrose in uncombined forms; milk products to contain lactose and, if sweetened, also sucrose; sweet potatoes to contain starch, sucrose and glucose.

Specific carbohydrates may be detected by qualitative tests that depend mainly on differences in chemical structure. However, when employing these specific tests, the analyst frequently must separate the specific carbohydrate being tested for from other carbohydrates in order to prevent interference with the test or arriving at erroneous conclusions due to the presence of other substances of a similar nature. This is particularly true for those tests involving the reducing action of the carbonyl group of sugars. In contrast, the starch-iodine sorption test is not appreciably affected by other substances; as a consequence it is a useful test for starch in the presence of large amounts of other substances. A number of general tests for carbohydrates are described in this section, as well as paper chromatographic analysis of sugar mixtures.

Molisch's Test

Molisch's test is based on the hydrolyzing and dehydrating action of concentrated sulfuric acid on carbohydrates. In the test the acid hydrolyzes any glycosidic bonds present and dehydrates the monosaccharide to its corresponding furfural derivative (p. 122). These furfurals then condense with α-naphthol to give a colored product.

Reagent

Dissolve 10.0 g α-naphthol in 100 ml of 95% ethanol.

Procedure

Place 5 ml of each carbohydrate solution to be tested in a test tube, add 1–2 drops of Molisch's reagent, and mix. Incline the test tube and slowly run 5 ml of concentrated sulfuric acid down the side of the tube so that it forms a layer at the bottom of the tube without mixing. The formation of a reddish-violet ring at the interface of the two liquids indicates the presence of a carbohydrate.

Benedict's Test

All monosaccharides and most disaccharides act as reducing agents because they contain a free (or potentially) free aldehyde or ketone group. Thus, these sugars can reduce various cations and certain organic compounds under different conditions. Such reactions are the basis for Benedict's and Barfoed's tests.

Reagent

Dissolve 173.0 g sodium citrate and 100.0 g sodium carbonate in about 800 ml of water. Pour into a 1000-ml volumetric flask. Dissolve

17.3 g copper sulfate in approximately 100 ml of water. Add the copper sulfate solution to the flask with constant stirring. Then make to volume. Reagent does not deteriorate on long standing.

Procedure

Place 1 ml of each carbohydrate solution to be tested in a test tube and add 5 ml of Benedict's reagent. Mix. Boil in a water bath for 2 min. A yellow or red precipitate of cuprous oxide indicates the presence of a reducing sugar.

Barfoed's Test

The reagent used in this test is not reduced appreciably by disaccharides (lactose and maltose) but is reduced by monosaccharides. Thus, the test is useful in distinguishing monosaccharides in the presence of disaccharides.

Reagent

Dissolve 13.3 g of neutral, crystallized copper acetate in 200 ml of water. Filter if necessary, and add 1.8 ml of glacial acetic acid.

Procedure

Mix 5 ml of Barfoed's reagent with 1 ml of the carbohydrate solution to be tested and place in a boiling water bath for 3–4 min. Examine for a red precipitate of cuprous oxide. (Dissaccharides may cause reduction if too much sugar or acid is present, or if the heating is too prolonged.)

Seliwanoff's Test

Keto sugars, especially fructose, and aldo sugars such as glucose and lactose can be distinguished by Seliwanoff's test.

Reagent

Mix 30 ml of water, in a 100-ml volumetric flask, with 60 ml of concentrated hydrochloric acid, add 0.5 g of resorcinol, and dilute to volume.

Procedure

To 5 ml of the carbohydrate solution add 5 ml of the resorcinol reagent and mix. Boil in a water bath for 20 min, cool quickly, and examine after 2 min. A red color or red precipitate is a positive test for keto sugars.

Iodine Test

Many polysaccharides react with iodine to form a blue-black starch–iodine complex. Starch, most dextrins, amylodextrins, and glycogen give a positive iodine test.

Reagent

Prepare a 2% solution of potassium iodide and add sufficient iodine to color it a deep yellow.

Procedure

Place 2 ml of each carbohydrate solution in a test tube and add 3 drops of the iodine solution. Unbranched macromolecules (amylose moiety) give a blue color and branched macromolecules (amylopectin) give a reddish-black color.

Bial's Test

Pentoses and carbohydrates capable of yielding pentoses give a positive reaction with Bial's test. Hexoses do not react.

Reagent

Dissolve 1.5 g of orcinol in 500 ml of hydrochloric acid and add 20 drops of a 10% ferric chloride solution.

Procedure

Heat 5 ml of the reagent to boiling, remove from the flame, and add a few drops (less than 1 ml) of the carbohydrate solution to be tested. The presence of a pentose is evidenced by a vivid green color, which develops almost immediately.

Paper Chromatographic Analysis

Paper chromatography is a useful method for qualitatively identifying sugars, especially when the amounts of sugar present in the solution are relatively small. There are a variety of solvents for developing the chromatograms. Most sugars can be separated on filter paper bya phenol–water solvent. In general, the order of R_f values is pentoses > hexoses > disaccharides > trisaccharides. Paper chromatography also may be used for quantitative analysis of sugars.

Materials and Reagents

1. Phenol–water developing solution—80 parts phenol mixed with 20 parts water (80:20).

2. Whatman No. 4 filter paper.
3. Solutions (0.1 M) of glucose, ribose, raffinose, fructose, xylose, sucrose, and maltose.
4. Spray indicators—(1) Prepare Tollens' reagent for reducing sugars by mixing equal parts of 0.1% silver nitrate and 5 N ammonium hydroxide; (2) anisidine hydrochloride for nonreducing sugars prepared by dissolving 3 g of *p*-anisidine hydrochloride in 100 ml *n*-butanol.
5. Chromatogram developing jar (or 4-liter beaker).

Procedure

Draw a pencil line 2 cm from and parallel to the edge of a 20 × 20 cm piece of Whatman No. 4 filter paper. Along this line, at intervals of 2 cm, mark a small dot. Using a 20-μl micropipet, spot quantities of sugar solution on the dot, taking care that the spot remains small and discrete. Allow to dry for 10 min. Form the dried filter paper into a cylinder and staple it top and bottom, taking care to see that the edges of the sheet are not touching. A paper clip is used to join the center portion of the cylinder. The cylinder is now ready for development.

Add solvent to the developing jar to a depth of ½ cm. Place the paper cylinder in the developing chamber and close the developing chamber with a cover plate. (Saran wrap and a rubber band may be used to close the tank if a cover plate is not available). The solvent is allowed to proceed up the paper (ascending chromatography) to within 2 cm of the top. This should take approximately 6–9 hr.

Following removal of the chromatogram, the solvent boundary on the upper portion is rapidly marked and the paper is allowed to dry under a hood. If the boundary cannot be adequately defined, allow the chromatogram to dry and view under ultraviolet light. The boundary will be perceptible.

Spray the dried chromatogram lightly with Tollen's reagent, then warm it for a few minutes. Do *not* overheat or the paper will turn black. This spray is suitable for identifying reducing sugars. Alternatively, the chromatogram may be sprayed with *p*-anisidine hydrochloride and warmed 5–10 min at 105°F. This spray will detect both reducing and nonreducing sugars.

☐ QUANTITATIVE ANALYSIS

Four general types of methods are used for the quantitative analysis of sugars in foodstuffs: reduction methods, refractometric methods, polarimetric methods, and densimetric methods. The analysis of maple syrup and starch are used to illustrate these methods in the following discussion. Maple syrup, obtained by concentrating the sap from hard (rock) maple trees, is required to contain no more than 35% water and must weigh at least 11 lb/gal at 60°F. The principal sugar of maple syrup is sucrose (1–6%), but it also contains organic acids, mineral matter, proteins, and flavoring materials. The quantitative analysis of starch depends on its complete hydrolysis and determination of the resulting glucose.

Before any of these quantitative methods are used, representative samples have to be obtained and the sugar-containing solution clarified to remove substances that interfere with the analysis. The interfering substances are soluble pigments, optically active substances (amino acids, etc.), phenolic constituents, lipids, and protective colloids (protein). Such interfering substances may be separated by decolorization, ion-exchange resin treatment, or clarification with various clarifying agents (alumina cream, lead acetates, phosphotungtic acid, and animal charcoal).

In addition to the methods described in this section, sugars can be quantitatively determined by the gas chromatographic analysis of their trimethylsilyl (TMS) ether derivatives. In this procedure interfering substances are not a problem; furthermore, mixtures of sugars can be separated and analyzed by this method. A full description of this method is presented in the section on gas chromatography in Chapter 3.

Refractometric Method

The refractive index of a sugar solution is a direct measure of its concentration. Solutions of different sugars of equal concentration have approximately the same refractive index. The speed and ease with which the refractive index of a sugar solution can be determined makes this a convenient method for determining the sugar content, and indirectly the water content, of sugar solutions. Consequently, the refractometer is widely used for quality inspection in the manufacture of syrup, jams, fruit juice, and other food products.

To analyze maple syrup, determine the refractive index of the syrup at 20°C with either an Abbe or hand refractometer. Obtain the corresponding percentage of soluble solids (as sucrose) from the table in AOAC (1980) relating refractive indices to sucrose percent.

Densimetric (Hydrometer) Method

The specific gravity of a sugar solution is a function of the sugar concentration (solute) at a definite temperature. Although the presence of other soluble materials affects the specific gravity of the solution, the densimetric method gives a fair approximation of the amount of sugar present in relatively pure sugar solutions. The specific gravity is measured with a hydrometer, which is also widely used for the determination of soluble solids. A special hydrometer, reading in percentage of sugar directly at 20°C (degrees Brix), has been developed for sugar work. The solution to be analyzed (at 20°C) is placed in a tall cylinder and the Brix reading obtained after the hydrometer is spun and comes to rest.

Polarimetric Method

Polarimetric analysis depends on the fact that sugars have the ability to rotate a plane of polarized light through an axis parallel to its direction of propagation (see Optical Activity section, p. 102). The angle through which this rotation occurs is directly proportional to the concentration of the sugar, the length of the tube through which the light passes, and the specific rotating power of the sugar.

A polarimeter may be used for the quantitative determination of sugars but in actuality a saccharimeter is more commonly employed. The essential differences between these two instruments are that a polarimeter employs monochromatic light and reads in angular degrees, whereas a saccharimeter employs white light and reads sugar concentration directly, provided a single sugar is present and a normal weight of sugar is used for the reading. A normal weight of sugar is defined as the weight that, when made to volume of 100 ml and viewed in a 200-mm tube at 20°C, will give a reading of 100.

Determination of Sucrose in Maple Syrup

The principal sugar of maple syrup is sucrose. However, a small amount of invert sugar (an equimolar mixture of glucose and fructose) is normally present due to the hydrolysis of a portion of the sucrose. The Clerget–Herzfeld saccharimetric method may be used for the determination of sucrose in maple syrup. This method involves two readings: The first reading, based on a normal weight of sample, is due to the sucrose and invert sugar in the sample; the second reading of a normal weight, after inversion of the sample, is due to the invert sugar derived from sucrose plus the invert sugar originally present (see Sucrose section). The change in rotation following hydrolysis is a function of the amount of sucrose present in the sample; values of this change in rotation can be found in the literature. Use of this value in the proper equation then allows one to calculate the amount of sucrose in the sample.

Sample Preparation

Rapidly weigh to the nearest 0.005 g, 52.00 g of the maple syrup sample (twice the normal weight) in a dish and transfer with water to a 200-ml volumetric flask. Add 2–5 ml of a saturated solution of neutral lead acetate solution, dilute to volume, and mix. Filter and discard the first 25 ml of filtrate. Remove the excess lead from the remaining filtrate by adding anhydrous sodium carbonate, a little at a time, avoiding an excess; mix well and filter again, discarding the first 25 ml of filtrate.

Direct Reading

Pipet a 50-ml aliquot of the lead-free filtrate into a 100-ml volumetric flask, add 2.315 g NaCl, dilute to volume with water at 20°C, and mix well. Fill a 200-mm saccharimeter tube with this solution and obtain a reading in the saccharimeter at 20°C. Since twice the normal weight in a 2000-ml volume is equivalent to a normal weight in a 100 ml volume, a 50-ml aliquot diluted to 100 ml represents a sample of one-half the normal weight. Thus the saccharimeter reading obtained when multiplied by two will give a reading based on a normal weight. (Save the remainder of the solution if one wishes to determine the percentage of invert sugar by the Munson–Walker method.)

Invert Reading (Inversion At Room Temperature)

Pipet a 50-ml aliquot of the filtrate from the original 200 ml of clarified and deleaded solution into a 100-ml volumetric flask, add 20 ml H_2O and 10 ml HCL (sp gr 1.103 at 20°C) and let stand for 24 hr at 20°C. If the temperature is above 28°C, a 10-hr digestion is sufficient. Dilute to volume and determine the saccharimeter reading at 20°C. Again, the 50-ml aliquot represents a sample of one-half the normal weight; and a consequence, the reading obtained must be multiplied by 2. This will give the saccharimeter reading based on a normal weight.

Calculations

The sucrose present may be calculated from the formula

$$S = \frac{100(P - I)}{132.56 - 0.0794(13 - m) - 0.53(t - 20)}$$

where S = % sucrose; P = direct reading on a normal solution; I = invert reading on a normal solution; m = grams of total solids from the original sample in 100 ml of the inverted solution; and t = temperature at which readings were made.

Reduction Methods

The reducing action of sugars that contain free carbonyl groups is the basis of various quantitative methods, as well as numerous qualitative tests discussed earlier.

Munson–Walker Determination of Invert Sugar

The Munson–Walker method for determining invert sugar involves formation of a Cu_2O precipitate. From the weight of this precipitate, the amount of invert sugar in a sample can be determined. Since sucrose is nonreducing, this method can be used to determine the percentage of invert sugar in maple syrup. Thus, by combining the polarimetric determination of sucrose with the Munson–Walker determination of invert sugar, the proportions of sucrose and invert sugar in maple syrup can be determined.

Reagents

1. Fehling's A solution—Dissolve 34.64 g of $CuSO_4 \cdot 5H_2O$ in water and dilute to 500 ml.
2. Fehling's B solution—Dissolve 173 g of potassium sodium tartrate (Rochelle salts) and 50 g NaOH in water and dilute to 500 ml.

Procedure

Shake amphibole asbestos with distilled water until a fine pulp is obtained. Prepare a Gooch crucible by forming a mat of asbestos about $\frac{1}{4}$ inch thick. Wash with 10 ml alcohol and then with 10 ml ether. Dry 30 min at 100°C, cool in a desiccator, and weigh.

Prepare the sample solution as described under *Sample Preparation* and *Direct Reading* in the Determination of Sucrose in Maple Syrup. Pipet 25.0 ml of this solution into a 400-ml beaker. Add 25 ml of water and 25 ml each of Fehling's A and B solutions. Cover beaker with a watch glass, and heat it on asbestos gauze over a Bunsen burner. Regulate flame so that boiling begins in 4 min and continue boiling exactly 2 min. Filter hot solution at once through a prepared Gooch crucible, using suction. Wash precipitate of Cu_2O thoroughly with water at about 60°C, then wash with 10 ml alcohol, and finally with 10 ml ether. Dry in oven at 100°C for 30 min, cool, and weigh.

Prepare a blank in the same manner, using 50 ml H_2O in place of the sample solution.

Calculations

Subtract the weight of the blank from that of the precipitate formed with the sample solution. To correct for the presence of sucrose in the sample, refer to the Hammond table in AOAC (1984, pp. 1076–1084) or to the abridged table in Chapter 12 (Table 12.7). Calculate the percentage of invert sugar present in the original syrup on the basis of the aliquot used.

Determination of Corn Starch by Hydrolysis

The amount of starch present in a corn sample can be determined by hydrolyzing the starch with enzymes and then reacting the resulting glucose with 3,5-dinitrosalicylate reagent to give a colored solution. The absorbance of this solution at 540 nm is directly related to the concentration of glucose.

Reagents and Equipment

1. pH Meter—Standardize the pH meter each time you check the pH of different solutions. Use standard pH solutions to standardize the meter. The pH of the buffers in this procedure is very critical. Unless the exact pH specifications are adhered to, the enzymes will not breakdown starch to glucose.
2. Hot plate and mechanical stirrer.
3. Phosphate buffer (pH 6.9)—Dissolve 170 mg Na_2HPO_4 (anhydrous), 140 mg $NaH_2PO_4 \cdot H_2O$, and 30 mg NaCl in 100 ml H_2O. Determine the pH of the resulting solution with a properly standardized pH meter; adjust the pH to exactly 6.9 by adding NaOH or HCl.
4. Acetate buffer (pH 4.3)—Add 29.4 ml of glacial acetic acid (17 *N*) to 470.6 ml of H_2O. Add sodium hydroxide to this solution until an exact pH of 4.3 has been reached.
5. α-Amylase—Dissolve 100 mg of commercial enzyme in 2.0 ml of pH 6.9 phosphate buffer. The enzyme should have an activity such that 1 g will digest 50 g of starch in 30 min under standard conditions.
6. Amyloglucosidase—Dissolve 100 mg of commercial enzyme in 2.0 ml of pH 4.3 acetate buffer. One unit of activity will yield 1.0 mg of glucose from starch in 3 min at 55°C and pH 4.5. The commercial enzyme preparation contains 1200–3000 units/g of enzyme.

7. 2 *N* HCl—To prepare 200 ml, mix 33 ml concentrated HCl with 167 ml H_2O.
8. 3 *N* NaOH—To prepare 200 ml, mix 24 g NaOH with 200 ml H_2O.
9. 3,5-Dinitrosalicylate reagent—Dissolve with warming, 5 g 3,5-dinitrosalicylic acid in 100 ml 2 *N* NaOH (8 g NaOH in 100 ml H_2O). Add 150 g sodium potassium tartrate to 25 ml H_2O, and warm to dissolve. Mix the two solutions, and make up to 500 ml with H_2O.
10. Glucose standards—Prepare solutions containing 1, 2, 3, 4, and 5 mg glucose/ml water.

Preparation of the Corn Samples and Acid Pretreatment

Grind the corn in a Wiley mill or mortar. Weigh out exactly 0.80 g of the ground corn and place in a 250-ml beaker. Add 25 ml of 2 *N* HCl and approximately 25 ml of H_2O to permit more dissolution. With a mechanical stirrer agitate the solution for 20 min. Place on a preheated hot plate (around 200°–300°C) and bring to boil. After 20 min remove the corn solution and cool to room temperature. Using a pH meter slowly add 3 *N* NaOH until a pH of exactly 6.9 is reached. If the pH goes too high, slowly add 2 *N* HCl to adjust it back to pH 6.9. The volume of this solution should now be approximately 100 ml. Place solution in 100-ml volumetric flask and add H_2O to volume. From this solution (which contains some cellular matter that has little effect upon the final results), a 0.4-ml aliquot is withdrawn for each subsequent analysis.

Enzyme Hydrolysis

Add exactly 0.75 ml of the α-amylase reagent to a 0.4-ml aliquot of the acid hydrolysate. (Avoid contact of the enzyme reagent with the upper regions of the test tube.) Allow the mixture to react for at least 3 min (no more is needed but 3 min is necessary). Then add exactly 0.10 ml of the amyloglucosidase reagent. The resulting pH should then be exactly 4.5. At this pH the maximum starch hydrolysis is obtained from amyloglucosidase. The samples are then placed in a water bath at 55°C (±1°C) for 10 min.

Color Development

Add 1.0 ml of the 3,5-dinitrosalicylate reagent. (Again avoid contact with the sides of the tube.) Prepare a blank by adding 1.0 ml of the 3,5-dinitrosalicylate reagent to 1.25 ml of water. Place the test tubes into a boiling water bath for exactly 5 min. Remove the tubes and cool to room temperature. Dilute the contents with H_2O to 20 ml and mix. Read the absorbance at 540 nm against the blank.

Standard Glucose Curve

A standard curve of absorbance versus glucose concentrations is essential in determining the number of milligrams of starch present in any sample of corn. Prepare glucose standards in range from 1 mg to 5 mg in water. Treat standards as indicated in the preceding Color Development.

$$\% \text{ Starch} = \frac{\text{Wt glucose} \times 0.9 \times 100}{\text{Wt sample}}$$

SELECTED REFERENCES

ACS. 1955. Use of Sugars and Other Carbohydrates in the Food industry. Advances in Chemistry Series 12. American Chemical Society, Washington, DC.

ACS. 1966. Flavor Chemistry. Advances in Chemistry Series 56. American Chemical Society, Washington DC.

AMERINE, M. A., PANGBORN, R. M. and ROESSLER, E. B. 1965. Principles of Sensory Evaluation of Food. Academic Press, New York.

AOAC. 1984. Official Methods of Analysis. 14th ed. S. Williams (editor). Assoc. of Official Analytical Chemists, Arlington, VA.

ASPINALL, G. O. 1969. Gums and mucilages. Adv. Carbohydr. Chem. Biochem. *24*, 333–376.

AXELROD, B. 1965. Mono- and oilgosaccharides. *In* Plant Biochemistry. J. Bonner and J. E. Varner (editors). Academic Press, New York.

BEITNER, R. 1985. Regulation of Carbohydrate Metabolism. CRC Press, Boca Raton, Fla.

BERG, C. 1953. Physiology of D-amino acids. Physiol. Rev. *33*, 145–189.

CAHN, R. S., INGOLD, C. and PRELOG, V. 1966. Specification of molecular chirality. Angew. Chem. *5*, 385–415.

CAMERON, A. T. 1947. The Taste Sense and the Relative Sweetness of Sugars and Other Sweet Substances. Sci. Rept. *9*. Sugar Research Foundation, New York.

CHURMS, S. C. 1982. Carbohydrates. CRC Press, Boca Raton, Fla.

CLAMP, J. R., HOUGH, L., HIXON, J. L. and WHISTLER, R. L. 1961. Lactose. *In* Adv. Carbohydr. Chem. *16*, 159–206.

DOBY, G. 1965. Plant Biochemistry. John Wiley & Sons, London.

DOESBURG, J. J. 1965. Pectic Substances in Fresh and Preserved Fruits and Vegetables. Communication *25*. Institute for Research on Storage and Processing of Horticultural Produce, Wageningen, Netherlands.

DYKE, S. F. 1960. The Carbohydrates, John Wiley & Sons, London.

FURIA, T. E. 1986. Handbook of Food Additives. Chemical Ruber Co., Cleveland, OH.

GINSBURG, V. and NEUFELD, E. F. 1969. Complex heterosaccharides of animals. Ann. Rev. Biochem. *38*, 371–378.

GUTHRIE, R. D. and HONEYMAN, J. 1968. An Introduction to the Chemistry of Carbohydrates. 3rd ed. Clarendon Press, Oxford.

HASSID, W. Z. 1967. Biosynthesis of complex saccharides. *In* Metabolic Pathways, 3rd ed., Vol. 1. D. M. Greenberg (editor). Academic Press, New York.
HEUSER, E. 1944. The Chemistry of Cellulose. John Wiley & Sons, New York.
HODGE, J. E., RENDLEMAN, J. A. and NELSON, E. C. 1972. Useful properties of maltose. J. Am. Assoc. Cereal Chem. *17*, 180–184.
HORSTEIN, T. and TERANISHI, R. 1967. The chemistry of flavor. Chem. Eng. News *45*(17), 92–108.
INGLETT, G. E. 1970. Natural and synthetic sweetners. J. Hort. Sci. *5*, 139–141.
JONES, J. K. N. and SMITH, F. 1949. Plant gums and mucilages. Adv. Carbohydra. Chem. *4*, 243–291.
KERTESZ, Z. D. 1951. The Pectic Substances. John Wiley & Sons, New York.
LINEBACK, D. R. and INGLETT, G. E. 1982. Food Carbohydrates. AVI Publishing Co. Westport, CT.
MAHLER, H. R. and CORDES, E. H. 1971. Biological Chemistry. 2nd ed. Harper and Row, New York.
MANNERS, D. J. 1957. The molecular structure of glycogens. Adv. Carbohydr. Chem. *12*, 262–298.
MARGOLIS, R. U. and MARGOLIS, R. K. 1979. Complex Carbohydrates of Nervous Tissue. Plenum Press, New York.
MONCRIEFF, R. W. 1967. The Chemical Senses. 3rd ed. L. Hill Books, London.
NIEMAN, C. 1958. Relative sweetness of various sugars. Zucker-u. Sussenwarenwirtsch. *11*, 420–422; 465–467; 505–507; 632–633; 670–671; 974; 1051–1052; 1088–1089.
PANCOAST, H. M. and JUNK, W. R. 1980. Handbook of Sugars. AVI Publishing Co., Westport, CT.
PIGMAN, W. 1967. The Carbohydrates: Chemistry, Biochemistry, Physiology. Academic Press, New York.
SCHALLENBERGER, R. S. 1982. Advanced Sugar Chemistry. AVI Publishing Co. Westport, CT.
SCRUTTON, M. C. and UTTER, M. F. 1968. The regulation of glycolysis and gluconeogenesis in animal tissues. Ann. Rev. Biochem. *37*, 249–302.
SHARON, N. 1966. Polysaccharides. Ann. Rev. Biochem. *35*(2), 485–520.
SPIRO, R. G. 1970. Glycoproteins. Ann. Rev. Biochem. *39*, 599–638.
WEST, E. S. and TODD, W. R. 1954. Textbook of Biochemistry. Macmillan Co., New York.
WHISTLER, R. L. and SMART, C. L. 1953. Polysaccharide Chemistry. Academic Press, New York.
WOODS, A. E. and AURAND, L. W. 1977. Laboratory Manual in Food Chemistry. AVI Publishing Co., Westport, CT.
ZAPSALIS, CHARLES and BECK, ANDERLE R. 1985. Food Chemistry and Nutritional Biochemistry. John Wiley & Sons, New York.

5

Lipids

□ INTRODUCTION

Lipids have been defined as a heterogeneous group of naturally occurring substances that are insoluble in water but soluble in organic solvents such as ether, chloroform, benzene, and acetone. All lipids contain carbon, hydrogen, and oxygen, and some also contain phosphorus and nitrogen. Most lipids are soft solids or liquids at room temperature and are difficult to crystallize.

The major classes of lipids can be summarized as follows:

- Simple Lipids (esters of fatty acids and alcohols)
  - Fats and oils (esters of glycerol and fatty acids)
  - Waxes (esters of long-chain monohydroxy alcohols and fatty acids)
- Compound Lipids (simple lipids conjugated with nonlipid molecules)
  - Phospholipids (esters containing phosphoric acid in place of one mole of fatty acid)
    - Phosphoglycerides
    - Sphingolipids
    - Inositol phosphatides

Glycolipids (compounds of carbohydrate, fatty acids, and sphingosinol)
Lipoproteins (complexes of various lipids and proteins)

- Derived Lipids (products of hydrolysis of lipids)
  Fatty acids
  Alcohols (long chain or cyclic, including sterols)
  Hydrocarbons (carotenoids)
  Fat-soluble vitamins

Edible fats are complex mixtures of triglycerides and small amounts of other substances. They occur naturally or are derived through processing and storage of the fats. The natural fats are made up mostly of mixed triglycerides with only trace amounts of mono- and diglycerides and little or no free fatty acids. In contrast, processed fats may contain up to 20% of mono- and diglycerides. The other substances associated with natural fats include the following:

Phospholipids	Hydrocarbons
Sterols	Oxidation products
Fat-soluble vitamins	Trace metals
Pigments	Water

These associated substances are important for several reasons. For example, the fat-soluble vitamins, sterols, and phospholipids are of nutritional importance; the free fatty acids are an index of the degree of hydrolysis of a triglyceride; and the presence of peroxides, aldehydes, and ketones are indicative of the amount of oxidative deterioration that has taken place in a fat. Furthermore, certain of the sterols, phospholipids, carotenoid pigments, and metallic impurities may contribute to the oxidative deterioration of fat. Thus, it is necessary to understand the composition and structure of lipids in order to know their role in the biological chemistry of foods.

☐ SIMPLE LIPIDS AND THEIR CONSTITUENTS

The simple lipids can be classified into two groups: (1) fats and oils and (2) waxes. Both contain esters of fatty acids, which are monocarboxylic acids; however, the alcohol portion of the ester differs.

Fats and Oils

Triglycerides

The major components of fats and oils are triglycerides **(5.1)**, which are formed by esterification of glycerol (a trihydroxy alcohol) and fatty acids, as follows:

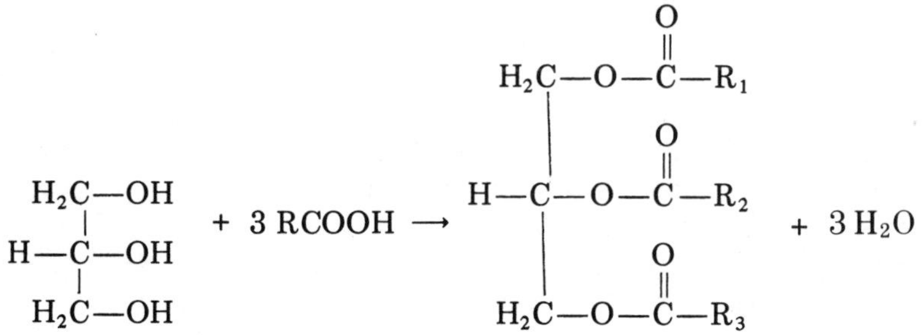

$$\begin{array}{l} H_2C{-}OH \\ H{-}C{-}OH \\ H_2C{-}OH \end{array} + 3\,RCOOH \longrightarrow \begin{array}{l} H_2C{-}O{-}\overset{O}{\overset{\|}{C}}{-}R_1 \\ H{-}C{-}O{-}\overset{O}{\overset{\|}{C}}{-}R_2 \\ H_2C{-}O{-}\overset{O}{\overset{\|}{C}}{-}R_3 \end{array} + 3\,H_2O$$

Glycerol Fatty acids Fat or oil

5.1

Naturally occuring fats are always mixtures of different triglycerides. In some triglycerides, all three acid fatty acids are the same (i.e., R_1, R_2, and R_3 in **5.1** are identical). These are known as simple triglycerides; tributyrin **(5.2)**, which contains three molecules of butyric acid, is an example. Other triglycerides contain two different fatty acids **(5.3)**, and some contain three different fatty acids **(5.4)**. These are known as mixed glycerides.

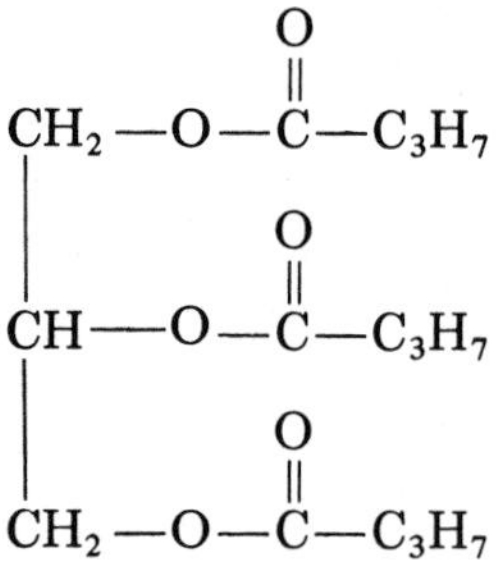

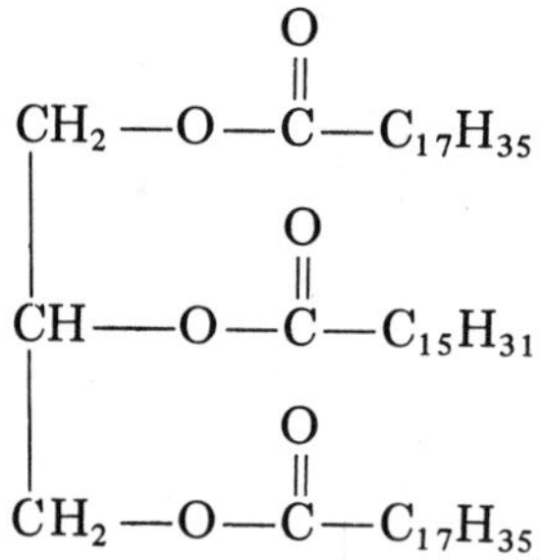

$$\begin{array}{l} CH_2{-}O{-}\overset{O}{\overset{\|}{C}}{-}C_3H_7 \\ CH{-}O{-}\overset{O}{\overset{\|}{C}}{-}C_3H_7 \\ CH_2{-}O{-}\overset{O}{\overset{\|}{C}}{-}C_3H_7 \end{array} \quad \begin{array}{l} CH_2{-}O{-}\overset{O}{\overset{\|}{C}}{-}C_{17}H_{35} \\ CH{-}O{-}\overset{O}{\overset{\|}{C}}{-}C_{15}H_{31} \\ CH_2{-}O{-}\overset{O}{\overset{\|}{C}}{-}C_{17}H_{35} \end{array} \quad \begin{array}{l} CH_2{-}O{-}\overset{O}{\overset{\|}{C}}{-}C_5H_{11} \\ CH{-}O{-}\overset{O}{\overset{\|}{C}}{-}C_{11}H_{23} \\ CH_2{-}O{-}\overset{O}{\overset{\|}{C}}{-}C_{17}H_{33} \end{array}$$

Tributyrin **5.2** β-Palmityl-α,α'-distearin **5.3** α-Caproyl-β-lauryl-α'-olein **5.4**

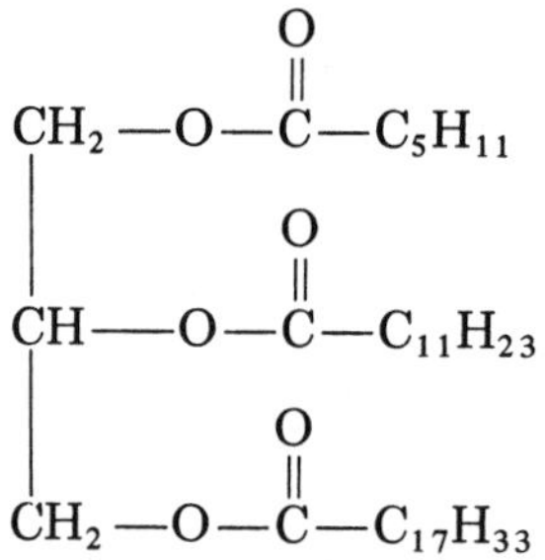

Simple triglycerides usually bear the name of the component acid with a prefix of *tri-* and a suffix of *-in* replacing the terminal *-ic* of the acid (e.g., tributyrin). With mixed triglycerides in which there are three different fatty acids (**5.4**), the names of the first two single acids end with the suffix *-yl* and the third single acid ends with the suffix *-in*, e.g., α-caproyl-β-lauryl-α′-olein. If the mixed triglyceride has two molecules of the same acid (**5.3**), the prefix *di-* and suffix *-in* are given to this acid (e.g., β-palmityl-α,α′-distearin). The glycerol carbons may be numbered as 1, 2, 3 or be designated as α, β, α′. Thus, β-palmityl-α,α′-distearin also could be designated 2-palmityl-1,3-distearin.

Mixed glycerides can exist in several isomeric forms. For example, a triglyceride containing two different fatty acids could have four structures, **5.5**, **5.6**, **5.7**, and **5.8**.

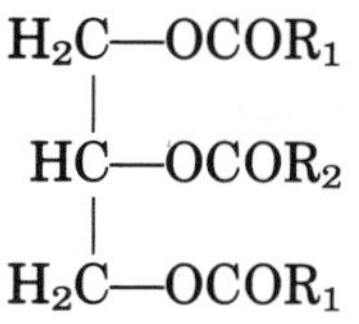

$$\begin{array}{cccc}
H_2C{-}OCOR_1 & H_2C{-}OCOR_1 & H_2C{-}OCOR_2 & H_2C{-}OCOR_2 \\
| & | & | & | \\
HC{-}OCOR_2 & HC{-}OCOR_1 & HC{-}OCOR_2 & HC{-}OCOR_1 \\
| & | & | & | \\
H_2C{-}OCOR_1 & H_2C{-}OCOR_2 & H_2C{-}OCOR_1 & H_2C{-}OCOR_2 \\
\mathbf{5.5} & \mathbf{5.6} & \mathbf{5.7} & \mathbf{5.8}
\end{array}$$

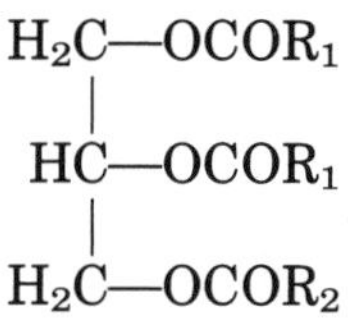

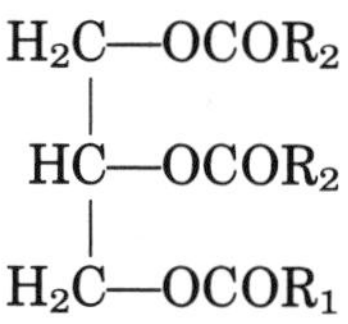

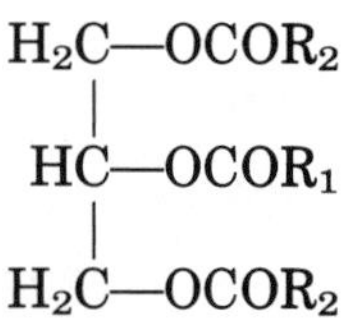

In **5.6** and **5.7**, the β carbon atom is asymmetric because two different acyl groups are attached to the α and α′ carbon positions. As a consequence, optical isomers of these two compounds exist. However, the naturally occurring fats do not exhibit an observable optical rotation.

Di- and Monoglycerides

Triglycerides, on standing, may undergo partial hydrolysis to form diglycerides and monoglycerides. The extent of hydrolysis is dependent upon the presence of water, heat, and hydrogen or hydroxide ions. Partial hydrolysis of a glyceride can proceed as follows to form various monoglycerides:

$$\begin{array}{l} CH_2-O-\overset{O}{\overset{\|}{C}}-R_1 \\ R_2-\overset{O}{\overset{\|}{C}}-O-CH \\ CH_2-O-\overset{O}{\overset{\|}{C}}-R_3 \end{array}$$

$-R_1COOH$ (left branch); $-R_2COOH$ (right branch)

$$\begin{array}{l} CH_2-OH \\ R_2-\overset{O}{\overset{\|}{C}}-O-C-H \\ CH_2-O-\overset{O}{\overset{\|}{C}}-R_3 \end{array} \qquad \begin{array}{l} CH_2-O-\overset{O}{\overset{\|}{C}}-R_1 \\ HO-C-H \\ CH_2-O-\overset{O}{\overset{\|}{C}}-R_3 \end{array}$$

$-R_2COOH$; $-R_3COOH$; $-R_1COOH$

$$\begin{array}{l} CH_2-OH \\ HO-C-H \\ CH_2-O-\overset{O}{\overset{\|}{C}}-R_3 \end{array} \qquad \begin{array}{l} CH_2OH \\ R_2-\overset{O}{\overset{\|}{C}}-O-C \\ CH_2OH \end{array} \qquad \begin{array}{l} CH_2-OH \\ HO-C-H \\ CH_2-O-\overset{O}{\overset{\|}{C}}-R_3 \end{array}$$

Complete hydrolysis yields glycerol and free fatty acids.

Triglycerides in contact with plant and animal lipases (esterases) may undergo appreciable hydrolysis. In contrast to nonenzymatic hydrolysis, lipases catalyze the hydrolysis of the ester linkage at the primary hydroxyl groups of the glycerol moiety of the molecule (α or α′ positions). Thus the initial product is an α,β-diglyceride. The latter is hydrolyzed to a monoglyceride, the β-monoglyceride. Most lipases do not hydrolyze the acyl group from the β position. The β-monoglycerides rapidly isomerize to give a mixture of approximately 12% β-monoglyceride and 88% α-monoglyceride. The reaction sequence for complete enzymatic hydrolysis of a triglyceride is as follows:

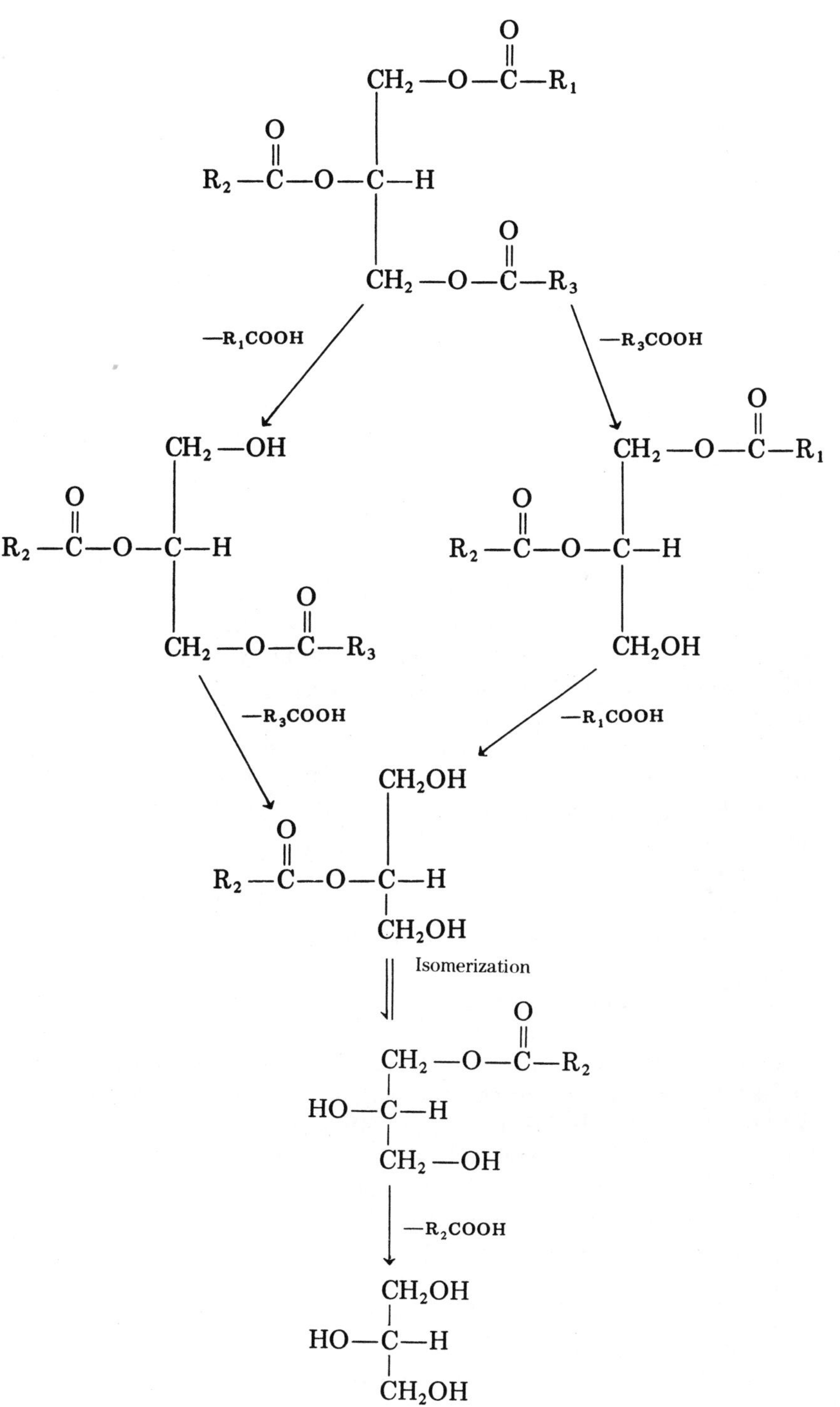

—R₁COOH
—R₃COOH
—R₃COOH
—R₁COOH
Isomerization
—R₂COOH

Mono- and diglycerides may be prepared by interesterification of fats with glycerol and an alkaline catalyst. The interesterification process is generally carried out at atmospheric pressure under a blanket of nitrogen or other inert gases to protect fats from oxidation. Heating is continued to a temperature of 200°–235°C for 1–2 hr. Various alkaline catalysts (caustic soda, caustic potash, sodium alcoholates, trisodium phosphate, and lime) in concentrations of 0.05–0.20% are employed. The reaction mixture consists of mono-, di-, and triglycerides together with unreacted glycerol. This mixture, which is commonly referred to as monoglycerides, contains 50–60% monoglycerides and 30–45% diglycerides. Molecular distillation of the technical monoglycerides yields a distillate containing 90–95% monoglycerides.

Fatty Acids

Most fatty acids are unbranched monocarboxylic acids varying in length and degree of saturation or unsaturation (presence of double bonds). There are a limited number of fatty acids containing cyclic groups, hydroxyl groups, and branched chains. As a consequence, the fatty acids may be divided into classes according to their structure, as shown in Table 5.1.

The vast majority of the naturally occurring fatty acids have an even number of carbon atoms in the molecule. The physiological processes involved in the catabolism and anabolism of fatty acids in plants and animals are so constituted that only this type of acid is produced in major quantities.

The saturated fatty acids range in chain length from C_4 to C_{30}, but fatty acids of greater than C_{20} are comparatively rare. The most common saturated fatty acids found in animal fats are palmitic ($C_{15}H_{31}COOH$) and stearic acid ($C_{17}H_{35}COOH$). Saturated fatty acids with a chain length of less than 16 carbons are frequently found in plant fats, but they occur only in small quantities in animal fats. Higher fatty acids such as arachidic acid ($C_{19}H_{39}COOH$) may also occur in animal fats, but branched-chain acids are uncommon.

The C_4 through C_8 saturated acids are liquid at ordinary temperatures (20°C); C_{10} and higher fatty acids are solids. Butyric acid (C_4) is miscible with water in all proportions. As the molecular weight of the fatty acids increases, solubility diminishes rapidly: caproic (C_6), caprylic (C_8), and capric acids (C_{10}) are slightly soluble in water, whereas lauric acid (C_{12}) and higher homologues are water insoluble.

In addition to saturated fatty acids, unsaturated fatty acids are found in the fats of animals, plants, and marine animals. Those occurring in animal fats are principally monoethenoic acids (oleic and palmitoleic) because animal cells are unable to synthesize unsaturated fatty acids containing more than one double bond. In contrast di and triethenoic acids (linoleic and linolenic) are found primarily in plant fats and oil, while polyunsaturated acids (arachidonic) are found in fish oils.

Unsaturated fatty acids are more chemically reactive than saturated fatty acids. The addition of hydrogen atoms to the double bond(s), in the presence of a suitable catalyst (palladium, platinum, nickel, or copper), gives the corresponding saturated fatty acid. Thus, oleic, linoleic and linolenic acids (all C_{18}) give stearic acid upon hydrogenation. Since the melting point of a fat is increased by this procedure, vegetable oils can be hydrogenated to yield semisolid creamy products that are used extensively as shortening agents for pies, cakes, and other bakery products. The hydrogenation process must be controlled in order to yield the type of product desired.

In addition to hydrogenation, the double bond(s) of unsaturated fatty acids are susceptible to oxidation. Exposure to oxygen of the air causes fat-containing polyunsaturated fatty acids to undergo gradual formation of peroxides together with a mixture of volatile aldehydes, ketones, and acids. The reaction is catalyzed by trace metals or the enzyme lipoxidase.

Unsaturated fatty acids show two types of isomerism: geometric (cis, trans) and positional (difference in position of the double bonds in polyunsaturated acids).

Because of the restricted rotation about the double bond in unsaturated fatty acids, geometric isomers can occur when two different atoms or groups are attached to the carbon atoms involved in the double bond. The terms *cis* and *trans* refer to the geometry of the groups (alkyl or others) attached to the carbons of a double bond. Since the carbons of the double bond are not free to rotate on their axis, a group attached to one of the carbons of the double bond can be on the same side of the double bond as a group on the other carbon; this arrangement is the cis form (e.g., **5.9**). When the groups are on opposite sides, they are trans to each other (e.g., **5.10**). Fatty acids that contain more than one double bond can exist in more than two geometrical isomeric forms. For example, linoleic acid can exist in four stereoisomeric forms: cis-cis, cis-trans, trans-cis, and trans-trans.

Most naturally occurring unsaturated fatty acids exist in the cis form; however, ruminant fats and commercial hydrogenated fats contain considerable amounts of trans acids. Upon heating with selenium or nitrous acid, the cis forms of unsaturated fatty acids are converted to a mixture of cis and trans isomers. For example, a portion of oleic

Table 5.1. Classification of Fatty Acids Present as Glycerides in Food Fats

Common name	Systematic name	Formula	Common source
		I. Saturated Fatty Acids	
A. Straight-Chain Series			
Butyric	Butanoic	$CH_3(CH_2)_2COOH$	Butterfat
Caproic	Hexanoic	$CH_3(CH_2)_4COOH$	Butterfat, coconut, and palm nut oils
Caprylic	Octanoic	$CH_3(CH_2)_6COOH$	Coconut & palm nut oils, butterfat
Capric	Decanoic	$CH_3(CH_2)_8COOH$	Coconut & palm nut oils, butterfat
Lauric	Dodecanoic	$CH_3(CH_2)_{10}COOH$	Coconut & palm nut oils, butterfat
Myristic	Tetradecanoic	$CH_3(CH_2)_{12}COOH$	Coconut and palm nut oils, most animal and plant fats
Palmitic	Hexadecanoic	$CH_3(CH_2)_{14}COOH$	Practically all animal and plant fats
Stearic	Octadecanoic	$CH_3(CH_2)_{16}COOH$	Animal fats and minor component of plant fats
Arachidic	Eicosanoic	$CH_3(CH_2)_{18}COOH$	Peanut oil
Behenic	Docosanoic	$CH_3(CH_2)_{20}COOH$	Mustard, peanut, and rapeseed oil
Lignoceric	Tetracosanoic	$CH_3(CH_2)_{22}COOH$	Small amounts in peanut oil and most natural fats
Cerotic Acid	Hexacosanoic	$CH_3(CH_2)_{24}COOH$	Wool fat
B. Branched-Chain Series			
Isovaleric	3-Methylbutanoic	$(CH_3)_2CHCH_2COOH$	Dolphin and porpoise fat
	11-Methyldodecanoic	$(CH_3)_2CH(CH_2)_9COOH$	Butterfat
	13-Methyltetradecanoic	$(CH_3)_2CH(CH_2)_{11}COOH$	Butterfat
		II. Unsaturated Fatty Acids[a]	
A. Monoethenoic Acids			
Caproleic	9-Decenoic	$C_9H_{17}COOH$	Milk fat
Lauroleic	9-Dodecenoic	$C_{11}H_{21}COOH$	Milk fat
Myristoleic	9-Tetradecanoic	$C_{13}H_{25}COOH$	Animal fat, milk fat
Physeteric	5-Tetradecenoic	$C_{13}H_{25}COOH$	Sardine and dolphin oils
Palmitoleic	9-Hexadecenoic	$C_{15}H_{29}COOH$	Marine animal oils, minor component plant and animal fats
Oleic	*cis*-9-Octadecenoic	$C_{17}H_{33}COOH$	Plant and animal fats

Vaccenic	*trans*-11-Octadecenoic	$C_{17}H_{33}COOH$	Minor component of animal fats & hydrogenated plant oils
Vaccernic	12-Octadecenoic	$C_{17}H_{33}COOH$	Minor component of hydrogenated plant oils
Gadoleic	9-Eicosenoic	$C_{19}H_{37}COOH$	Fish and marine animal oils
Cetoleic	11-Docosenoic	$C_{21}H_{41}COOH$	Marine oils
Erucic	13-Docosenoic	$C_{21}H_{41}COOH$	Rapeseed, mustard, and fanweed oils
Selacholeic	15-Tetracosenoic	$C_{23}H_{45}COOH$	Marine animal and fish liver oils
B. Diethenoid Acids			
Linoleic	9,12-Octadecadienoic	$C_{17}H_{31}COOH$	Peanut, linseed, and cottonseed oils
C. Triethenoid Acids			
Linolenic	9,12,15-Octadecatrienoic	$C_{17}H_{29}COOH$	Linseed and other seed oils
Eleostearic	9,11,13-Octadecatrienoic	$C_{17}H_{29}COOH$	Peanut seed fats
D. Tetraethenoid Acids			
Moroctic	4,8,12,15-Octadecatetraenoic	$C_{17}H_{27}COOH$	Fish oils
Arachidonic	5,8,11,15-Eicosatetraenoic	$C_{19}H_{31}COOH$	Traces in animal fats
E. Polyethenoid Acids			
Clupanodonic	4,8,12,15,19-Docosapentaenoic	$C_{21}H_{33}COOH$	Fish oil
Nisinic	4,8,12,15,21-Tetracosahexaenoic	$C_{23}H_{35}COOH$	Sardines and other fish oils
III. Unsaturated Monohydroxy Fatty Acids			
Ricinoleic	12-Hydroxy-*cis*-9-octadecenoic	$HO—C_{17}H_{32}COOH$	Peanut seed oils and castor oil
IV. Cyclic Fatty Acids			
Lactobacillic acid	ω-(2-*n*-Octylcyclopropyl)-octanoic acid		Microorganisms
Sterculic acid	ω-(2-*n*-Octylcycloprop-1-enyl)-octanoic acid		Plant seed oil
Malvalic acid	ω-(2-*n*-Octylcycloprop-1-enyl) heptanoic acid		Plant see oil

[a] In designating the position of the double bond it is customary to number the fatty acid chain in accordance with I.U.P.A.C. system beginning with the C of the carboxyl group and to give only the lower number of each pair of carbon atoms that are in the unsaturated linkage. The symbol Δ is occasionally used with a superscript denoting both carbon atoms formed by the double bond. In the nomenclature, cis-trans configuration is usually not indicated and in these cases the cis form is implied.

acid (cis form), when heated with nitrous acid, is converted to elaidic acid (trans form).

Cis

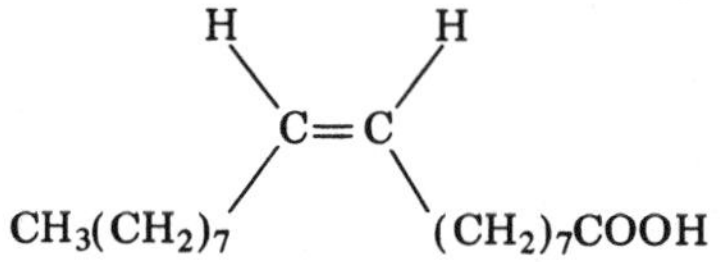

$$\begin{array}{ccc} H & & H \\ & C=C & \\ CH_3(CH_2)_7 & & (CH_2)_7COOH \end{array}$$

Oleic acid (mp 14°C)

5.9

Trans

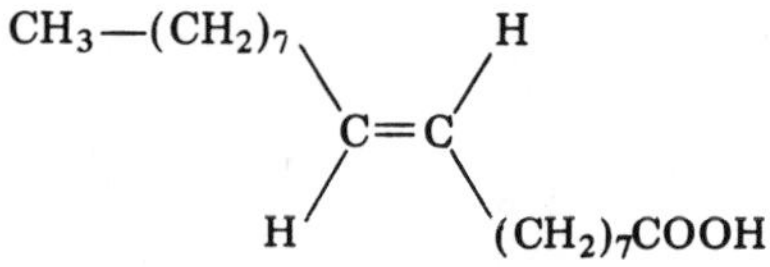

$$\begin{array}{ccc} CH_3{-}(CH_2)_7 & & H \\ & C=C & \\ H & & (CH_2)_7COOH \end{array}$$

Elaidic acid (mp 44°C)

5.10

Positional isomerism occurs in only a few of the naturally occurring unsaturated fatty acids, but it occurs in many derivatives of such acids as well as in synthetic acids. This type of isomerism depends on the relative position of the double bonds in the carbon skeleton. The two most common arrangements are (1) the *conjugated* system, with alternate single and double bonds (—CH═CH—CH═CH—) and (2) the *nonconjugated* system, with two double bonds separated by one or more methylene groups (—CH═CH—CH_2—CH═CH—).

The naturally occurring polyunsaturated fatty acids are nonconjugated; however, upon heating with alkali the nonconjugated form (1,4-system) rearranges to the conjugated form (1,3-system). A similar shift may occur when a nonconjugated system is heated to high temperatures or during autoxidation. This reaction is irreversible, and the conjugated form is the more stable form. Consequently, the reaction can be made to go to completion. The conjugated system is usually more readily oxidized than the nonconjugated system. Polyunsaturated fatty acids with a conjugated system occur in vegetable glycerides. Eleosteric acid, the principal acid in tung oil, is one of the most common fatty acids with a conjugated system.

Cyclic Fatty Acids. Several acids containing cyclic groups have been isolated from natural sources. The cyclopropenoid fatty acids, malvalic (**5.11**) and sterculic (**5.12**) acids, must be removed from cottonseed meal that is to be fed to laying birds, otherwise the acids will produce a pink-white disorder in eggs. Lactobacillic acid (**5.13**), which is extracted from different species of bacteria, contains a cyclopropane ring.

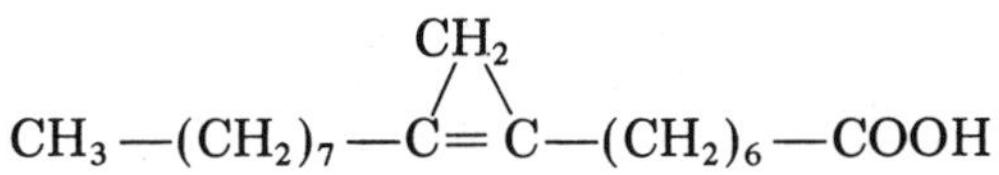

$$CH_3{-}(CH_2)_7{-}\overset{\overset{CH_2}{\wedge}}{C{=}C}{-}(CH_2)_6{-}COOH$$

Malvalic acid

5.11

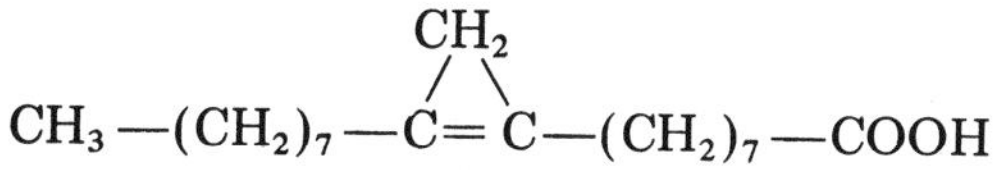

$$\begin{array}{c} CH_2 \\ / \quad \backslash \\ CH_3—(CH_2)_7—C{=}C—(CH_2)_7—COOH \end{array}$$

Sterculic acid

5.12

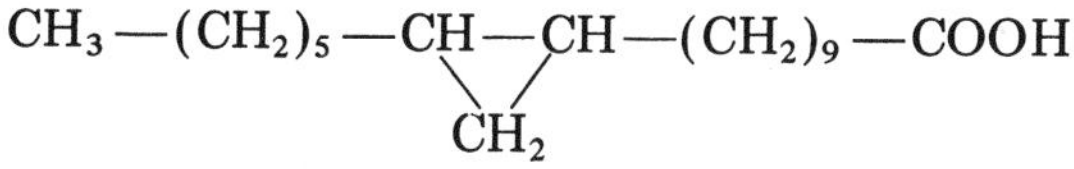

$$\begin{array}{c} CH_3—(CH_2)_5—CH—CH—(CH_2)_9—COOH \\ \backslash \quad / \\ CH_2 \end{array}$$

Lactobacillic acid

5.13

Glycerol

The common constituent of all fats and oils is glycerol (**5.14**). Since glycerol is a trihydric alcohol containing primary and secondary alcohol groups, the compound shows all the chemical reactions of alcohols. For example, on oxidation some of the compounds formed are dihydroxyacetone (**5.15**), glyceric aldehyde (**5.16**), glyceric acid (**5.17**), and tartronic acid (**5.18**).

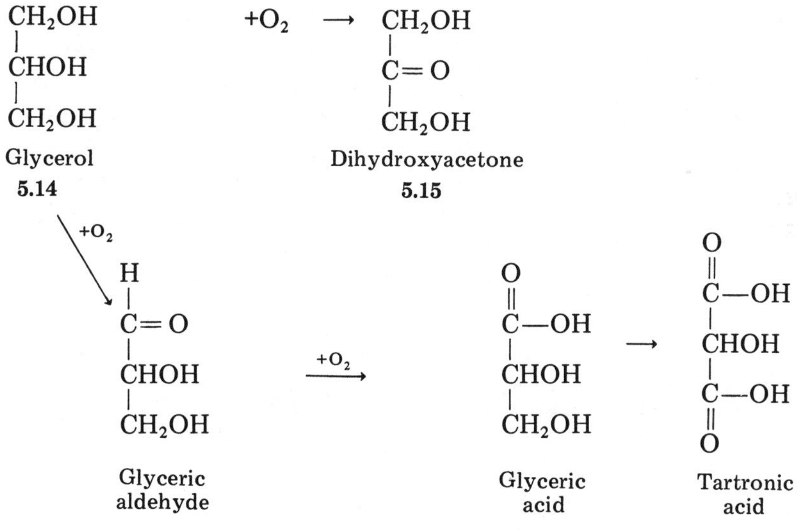

$$\begin{array}{c} CH_2OH \\ | \\ CHOH \\ | \\ CH_2OH \end{array} \quad +O_2 \longrightarrow \quad \begin{array}{c} CH_2OH \\ | \\ C{=}O \\ | \\ CH_2OH \end{array}$$

Glycerol **5.14**

Dihydroxyacetone **5.15**

$+O_2$ (from glycerol 5.14)

$$\begin{array}{c} H \\ | \\ C{=}O \\ | \\ CHOH \\ | \\ CH_2OH \end{array} \quad \xrightarrow{+O_2} \quad \begin{array}{c} O \\ \| \\ C—OH \\ | \\ CHOH \\ | \\ CH_2OH \end{array} \quad \longrightarrow \quad \begin{array}{c} O \\ \| \\ C—OH \\ | \\ CHOH \\ | \\ C—OH \\ \| \\ O \end{array}$$

Glyceric aldehyde **5.16**

Glyceric acid **5.17**

Tartronic acid **5.18**

When glycerol is heated to high temperatures, as in the case of fats spilled on a hot stove, it loses two molecules of water forming acrolein (**5.19**), a compound with a very pungent odor and irritating action.

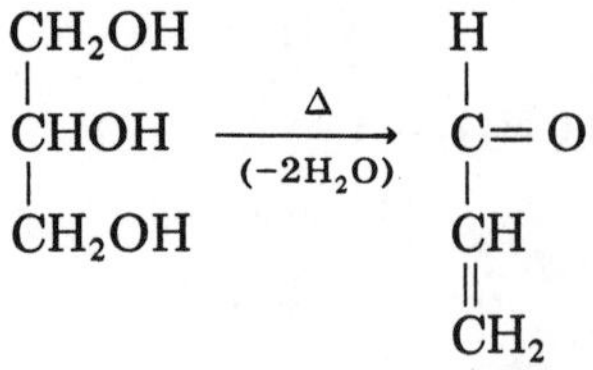

$$\begin{array}{l} CH_2OH \\ | \\ CHOH \\ | \\ CH_2OH \end{array} \xrightarrow[(-2H_2O)]{\Delta} \begin{array}{l} H \\ | \\ C=O \\ | \\ CH \\ \| \\ CH_2 \end{array}$$

Glycerol Acrolein

5.19

Glycerol, which forms esters with inorganic, as well as organic, acids is released during saponification of fats and oils. It is used commercially as a humectant in some food products. Glycerol forms a trinitrate when heated with concentrated nitric and sulfuric acid. This compound is used medically as a vasodilator to increase the flow of blood through the coronary arteries.

Waxes

In contrast to the triglycerides, waxes are esters of long-chained monohydroxy alcohols and fatty acids (e.g., **5.20** and **5.21**). Waxes are more resistant to hydrolysis than fats, requiring higher temperature and stronger alkali. Natural waxes also contain paraffins, hydroxylated and unsaturated fatty acids, secondary alcohols, and ketones. All of these compounds have high molecular weights and similar physical properties. They are widely distributed in nature, but as a general rule they never occur abundantly. In animals, they cover the surfaces of hair, wool, and feathers; in plants, they cover the surface of stems, leaves, and fruits. Fruit waxes often contain cyclic compounds of the triterpinoid type; for example, ursolic acid is found as the white coating on the surface of apples, grapes, etc.

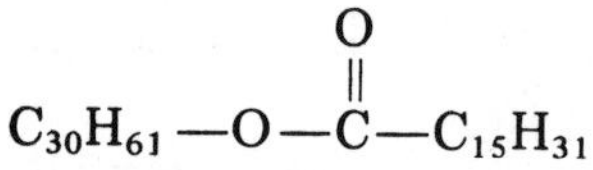

$$C_{30}H_{61}—O—\overset{\overset{\large O}{\|}}{C}—C_{15}H_{31}$$

Beeswax (myricyl palmitate)

5.20

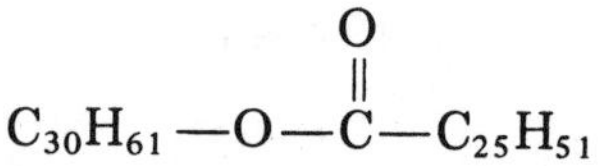

$$C_{30}H_{61}—O—\overset{\overset{\large O}{\|}}{C}—C_{25}H_{51}$$

Carnauba wax (myricyl cerotate)

5.21

☐ COMPOUND LIPIDS

Phospholipids

The phospholipids (phosphatides) are diglycerides containing phosphoric acid; many also contain a nitrogenous base, most commonly choline, ethanolamine, or serine. The phospholipids occur in varying amounts in vegetable and animal fats. Phospholipids make up about 1–2% of many crude vegetable oils and higher percentages of animal fats. Egg yolk contains approximately 20% phospholipids. Only small amounts of phospholipids are present in processed fats because they are largely removed in the refining of the crude product.

Phospholipids have several important biological functions: (1) as essential structural elements in living cells; (2) as intermediates in the transport, absorption, and metabolism of fatty acids; (3) as a storage form for fatty acids and phosphates; (4) as essential components in biological oxidations; and (5) as intermediaries in the transport and utilization of sodium and potassium ions. Phospholipids also are involved in the process of blood clotting.

Phosphoglycerides

The phosphoglycerides include lecithins, cephalins, and plasmalogens. These compounds differ in the nature of the groups attached to the glycerol moiety.

Lecithins. Lecithins from various sources have the same general structure **(5.22)**. Theoretically, two isomers of lecithin may occur depending upon whether the phosphorylcholine group is attached to the α or β carbon of glycerol. The naturally occurring lecithins are primarily of the α type. Also, since the C-2 of the glycerol molecule is asymmetric, there are two possible optical isomers; however, the lecithins found in nature all are L isomers.

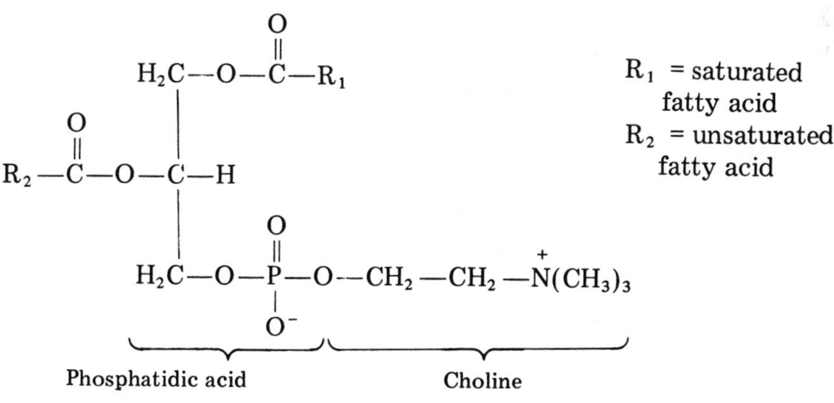

L-α-Lecithin (phosphatidyl choline)

5.22

Numerous lecithins are possible because of the wide variety of fatty acids that may be attached to the glycerol moiety. As a general rule, the fatty acid in the α position (C-1) is saturated (palmitic or stearic acid) while that in the β position (C-2) is unsaturated (oleic, linoleic, linolenic, and arachidonic acid).

Choline is an essential component of the diet. It is a quaternary ammonium base bearing a positive charge; since a negative charge may appear on the phosphate, depending on pH, lecithin may be a dipolar ion. In addition to the importance of choline in the lecithin molecule, choline is also required for synthesis of acetylcholine, a compound involved in the transmission of nerve impulses. The absence of choline in the diet leads to a fatty infiltration of the liver.

Lecithins are waxy, colorless solids, that turn yellow and then brown on exposure to light. In general, they are soluble in the usual fat solvents, are insoluble in acetone or methyl acetate, and dissolve in aqueous medium if bile salts are present. The lecithins are surface tension-active substances because of the presence of the strongly polar choline as well as the nonpolar fatty acids.

Lecithins occur in nervous tissue, egg yolk, liver, soybeans, and many crude vegetable oils. They function in tissues to keep nonpolar molecules, such as sterols, in an emulsified state. Lecithins are used commercially as emulsifiers and antioxidants in food products.

Cephalins. The cephalins differ from the lecithins only in that ethanolamine or serine replaces the choline. The distribution of fatty acids in phosphatidyl ethanol-amine **(5.23)** shows an almost equal distri-

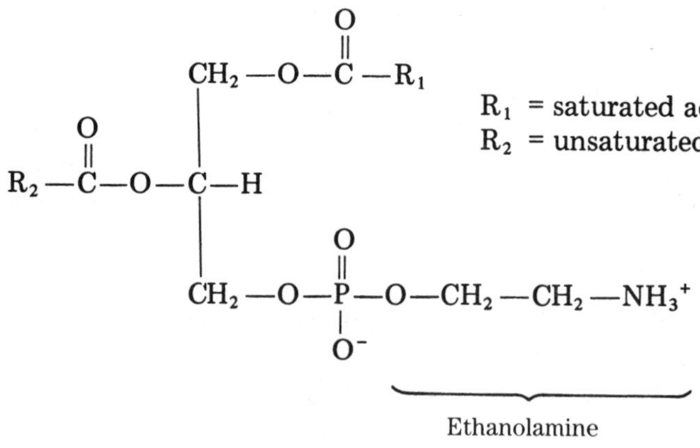

$$
\begin{array}{l}
CH_2-O-\overset{\overset{\displaystyle O}{\|}}{C}-R_1 \\
\quad | \\
R_2-\overset{\overset{\displaystyle O}{\|}}{C}-O-C-H \\
\quad | \\
CH_2-O-\underset{\underset{\displaystyle O^-}{|}}{\overset{\overset{\displaystyle O}{\|}}{P}}-\underbrace{O-CH_2-CH_2-NH_3^+}_{\text{Ethanolamine}}
\end{array}
$$

R_1 = saturated acid
R_2 = unsaturated acid

L-α-Cephalin (phosphatidyl ethanolamine)
5.23

bution between saturated and unsaturated acids; stearic acid is the only saturated fatty acid. Phosphatidyl serine (**5.24**) is also classified as a cephalin. Decarboxylation of phosphatidyl serine yields phosphytidyl ethanolamine.

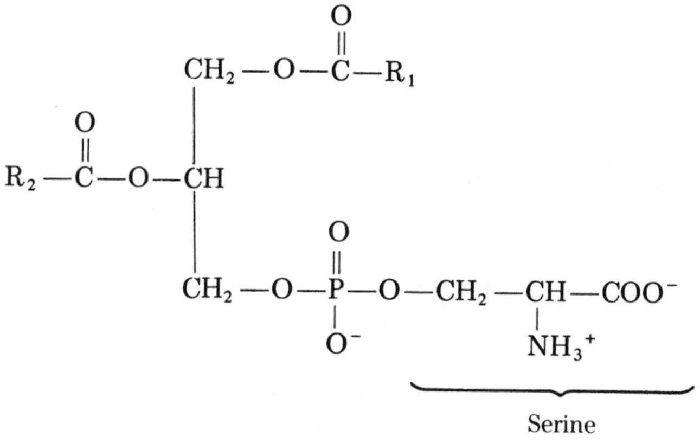

$$
\begin{array}{l}
\mathrm{CH_2{-}O{-}\overset{\displaystyle O}{\overset{\|}{C}}{-}R_1} \\
\quad | \\
\mathrm{R_2{-}\overset{\displaystyle O}{\overset{\|}{C}}{-}O{-}CH} \\
\quad | \\
\mathrm{CH_2{-}O{-}\overset{\displaystyle O}{\overset{\|}{P}}{-}O{-}CH_2{-}CH{-}COO^-} \\
\qquad\quad |\qquad\qquad\quad | \\
\qquad\quad \mathrm{O^-}\qquad\quad\ \ \mathrm{NH_3^+}
\end{array}
$$

Serine

L-α-phosphatidyl serine

5.24

Plasmalogens. Another group of phosphatides is the plasmalogens, which differ from lecithins and cephalins in that the fatty acid in the α position is replaced by an α,β-unsaturated ether. These substances are present in the membranes of muscle tissue, brain, and heart.

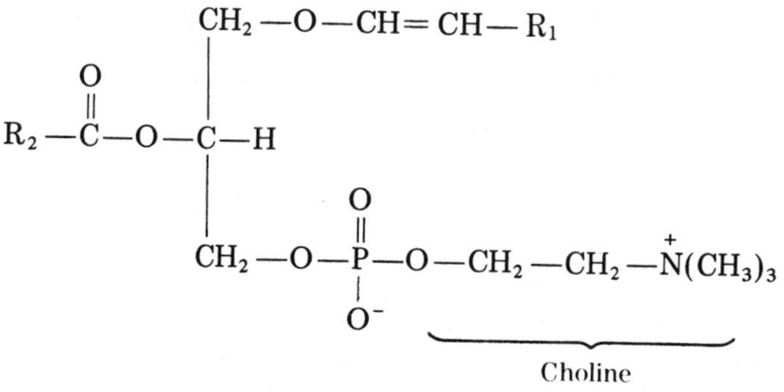

$$
\begin{array}{l}
\mathrm{CH_2{-}O{-}CH{=}CH{-}R_1} \\
\quad | \\
\mathrm{R_2{-}\overset{\displaystyle O}{\overset{\|}{C}}{-}O{-}C{-}H} \\
\quad | \\
\mathrm{CH_2{-}O{-}\overset{\displaystyle O}{\overset{\|}{P}}{-}O{-}CH_2{-}CH_2{-}\overset{+}{N}(CH_3)_3} \\
\qquad\quad | \\
\qquad\quad \mathrm{O^-}
\end{array}
$$

Choline

Plasmalogen (phosphatidyl choline)

5.25

Nonnitrogenous Phosphoglycerides. Also of biochemical importance are phosphatidyl glycerol and cardiolipin (diphosphatidyl glycerol), neither of which contains a nitrogenous base. The monophosphatidyl glycerols (**5.26**) are found in the cell membranes of plants and bacteria; diphosphatidyl glycerols (**5.27**) have been isolated from heart tissue.

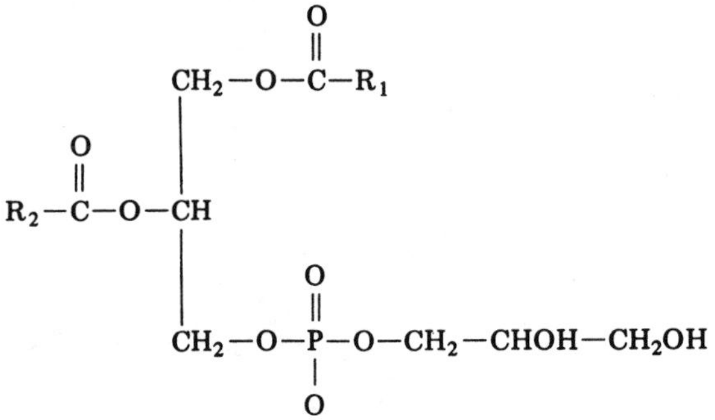

```
                 O
                 ‖
           CH2—O—C—R1
            |
     O      |
     ‖      |
R2—C—O—CH
            |
            |       O
            |       ‖
           CH2—O—P—O—CH2—CHOH—CH2OH
                    |
                    O
```

Phosphatidyl glycerol

5.26

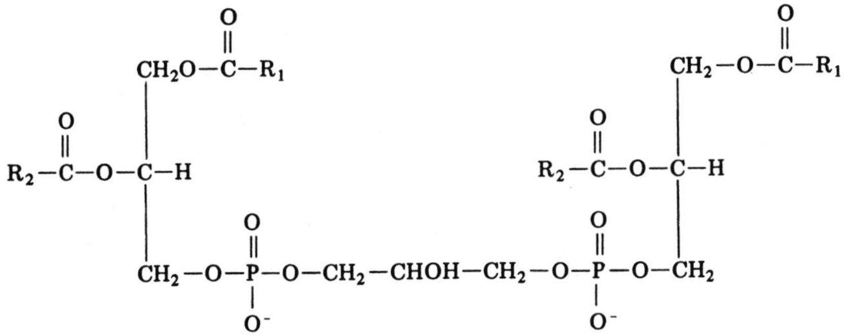

```
             O                                                        O
             ‖                                                        ‖
     CH2O—C—R1                                              CH2—O—C—R1
       |                                                      |
  O    |                                          O           |
  ‖    |                                          ‖           |
R2—C—O—C—H                                    R2—C—O—C—H
       |                                                      |
       |      O                                   O           |
       |      ‖                                   ‖           |
     CH2—O—P—O—CH2—CHOH—CH2—O—P—O—CH2
              |                                   |
              O⁻                                  O⁻
```

Diphosphatidyl glycerol

5.27

Sphingolipids

The distinguishing characteristic of the sphingolipids is that they contain sphingosine, a long-chain unsaturated amino alcohol, rather than glycerol, which is present in the other phosphatides. Sphingolipids occur in the membranes of plants and animals. Sphingomyelin

(5.28), the most common sphingolipid, is a component of nerve and brain tissue.

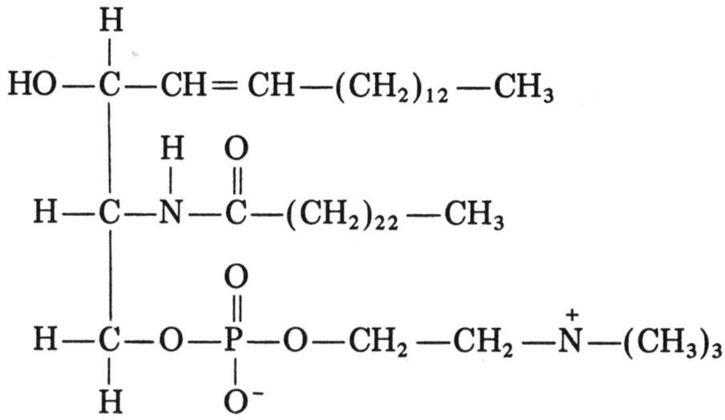

```
       H
       |
HO—C—CH=CH—(CH2)12—CH3
       |
       |   H  O
       |   |  ||
   H—C—N—C—(CH2)22—CH3
       |
       |        O
       |        ||             +
   H—C—O—P—O—CH2—CH2—N—(CH3)3
       |        |
       H        O⁻
```

Sphingomyelin

5.28

Inositol Phosphatides

Like the phosphatidyl glycerols (**5.26** and **5.27**), the inositol phosphatides (**5.29**) contain no nitrogenous base. Instead, they contain inositol (a cyclic hexahydroxy alcohol) attached to the phosphate. A relatively pure "inositide" can be prepared from soybean oil, in which as much as 16% inositol is present. These substances are found in brain tissue and there is evidence that they have an active role in transport processes in the cell.

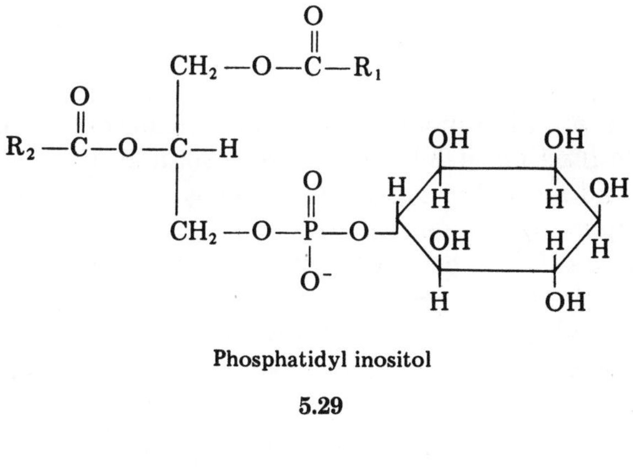

```
                      O
                      ||
              CH2—O—C—R1
   O           |
   ||          |
R2—C—O—C—H                     OH        OH
               |         O    H  H        H   OH
               |         ||
              CH2—O—P—O—         OH       H   H
                         |
                         O⁻            H        OH
```

Phosphatidyl inositol

5.29

Glycolipids (Cerebrosides)

The glycolipids are sphingolipids (**5.28**) that contain a carbohydrate, usually galactose, in place of phosphorylcholine. Thus, they contain no phosphorus. These substances are abundant in the membranes of brain and nerve cells, but they are also found in the liver, kidney, spleen, and adrenal glands. A common glycolipid is the cerebroside **5.30**.

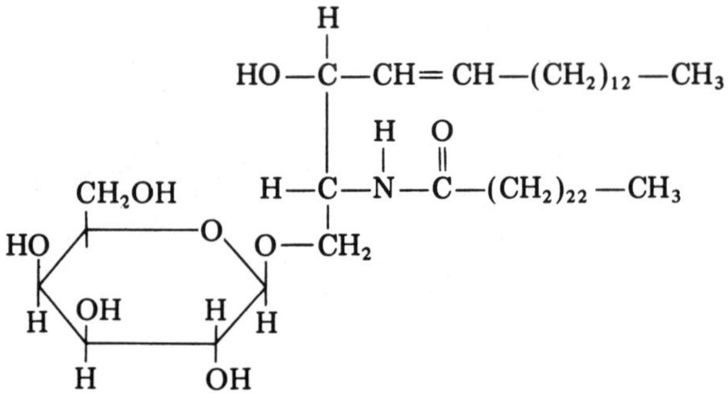

$$\begin{array}{c} \text{HO—CH(—CH=CH—(CH}_2\text{)}_{12}\text{—CH}_3\text{)—CH(—NH—CO—(CH}_2\text{)}_{22}\text{—CH}_3\text{)—CH}_2\text{—O—(}\beta\text{-galactose)} \end{array}$$

Cerebroside (β-galactolipid)

5.30

DERIVED LIPIDS

Two of the more important classes of derived lipids are the sterols and the carotene pigments. Both of these contain compounds with important biochemical functions.

Sterols

The sterols are high-molecular-weight alcohols occurring in the unsaponifiable fraction of fats. They are insoluble in water, sparingly soluble in cold alcohol or petroleum ether, and readily soluble in fats and the common fat solvents.

The fundamental carbon skeleton of the sterols is the cyclopentanoperhydrophenanthrene ring. The formula of cholesterol (**5.31**) shows the system used for designating the various rings and carbon atoms. Stereoisomerism is an important property of the sterols since they contain a number of asymmetric carbon atoms; cholesterol contains an asymmetric carbon atom at position 3, 8, 9, 10, 13, 14, 17, and 20.

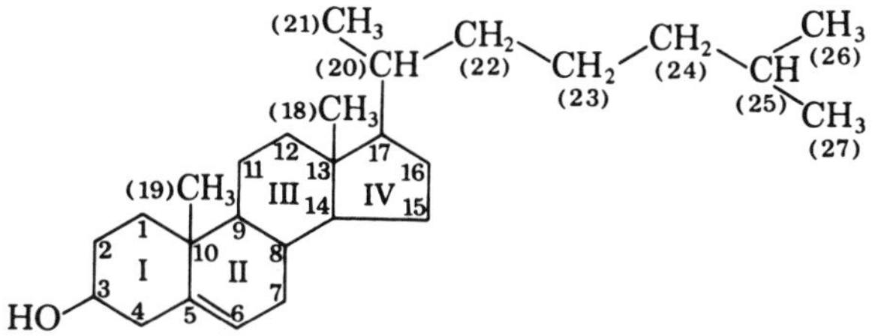

Cholesterol

5.31

Sterols occur in the fats of plants and animals. The sterols are frequently classified on the basis of their origin: (1) plant sterols are referred to as phytosterols; (2) sterols from animals, as zoosterols; and (3) sterols from lower plants (e.g., fungi), as mycosterols. A common sterol found in peanuts is ergosterol **(5.32)**.

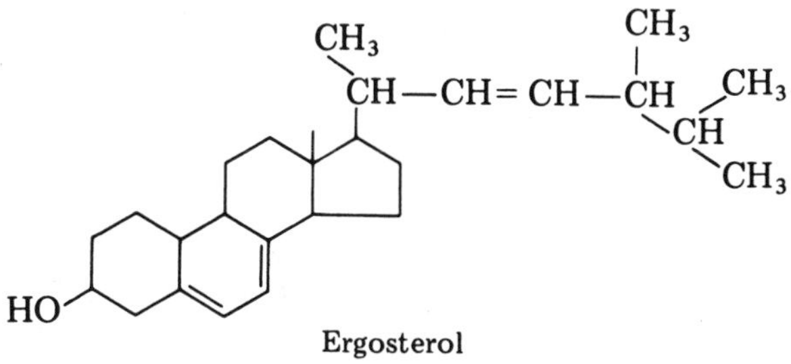

Ergosterol

5.32

The Liebermann–Burchard test is a very sensitive color test for sterols. A bluish-green to green color is obtained when a chloroform solution of the sterol is treated with acetic anhydride and concentrated sulfuric acid. The color varies in intensity with the amount of sterol present, and is the basis of a quantitative estimation.

The most common sterol in animals is cholesterol, which is present in all animal cells and has several important biological functions. It serves as the precursor of 7-dehydrocholesterol (vitamin D activity); as a part of the bile acids, it aids in the emulsification of dietary fats. Cholesterol is utilized in the biosynthesis of the adrenocortical hormones, which are important in the development of secondary male and female sex characteristics.

Pigments

Carotene and carotenoid pigments, present in the unsaponifiable fraction of most fats and oils, are responsible for the yellow-red color of many fats and oils of plant origin. The biochemical role of carotenes as provitamin A is discussed in Chapter 8.

☐ NUTRITIONAL VALUE OF FATS AND FAT PRODUCTS

Dietary fat provides a highly concentrated form of energy for the human body. On a gram for gram basis, lipids contain more than twice the energy of either carbohydrate or protein (lipids, 9.3 kcal/g; carbohydrate and protein, 4.1 kcal/g). Ethyl alcohol, with 7.0 kcal/g, almost approaches the energy value of fat. In the United States, the average dietary intake of fat has increased from 124 g/person/day in 1910 to 163 g/person/day at present. About two-thirds of the dietary lipids come from animals, and the remaining from vegetables. In addition to serving as an important energy source, dietary fat serves as a carrier for fat-soluble vitamins and provides essential fatty acids. These needs can be met by a diet containing 15–25 g of food fats.

Since the discovery of the essential fatty acids by Burr and Burr (1929), numerous studies have demonstrated that probably all animal species may develop symptoms of essential fatty acid deficiency if raised on fat-free diets (Holman 1971). Weanling rats on a fat-free diet grow poorly, showing deficiency signs such as dermatitis, poor reproduction, lowered caloric efficiency, impairment of lipid transport, and decreased resistance to stress. When linoleic acid is present in the diet, the symptoms associated with fat deficiency do not develop. Arachidonic acid and linolenic acid have also been shown to prevent these symptoms. Undoubtedly, mammals are unable to synthesize linoleic acid (18:2 *cis*-Δ^9,Δ^{12}) and linolenic acid (18:3 *cis*-$\Delta^9,\Delta^{12},\Delta^{15}$), both of which contain unsaturated double bonds. Since these two acids cannot be synthesized in mammalian tissues, but are required in the diet, they are called essential fatty acids.

Metabolism of essential fatty acids at the tissue level gives rise to the hormone-like prostaglandins. All the prostaglandins that occur naturally are a result of cyclization of C_{20} fatty acids such as arachidonic acid, which is formed from linoleic acid. The different prostaglandins are among the most potent biologically active substances. Their activities include lowering of blood pressure and constriction of smooth muscle. Other probable functions of prostaglandins include relief of nasal congestion and asthma, prevention of gastric ulcers, termination of pregnancy, and induction of labor at term. There is considerable literature on the relationship of prostaglandins to immunity (Pelus and Strausser 1977).

Within recent years it has been shown that excessive consumption of fats by humans is a factor in causing atherosclerosis. However, it has still not definitely been proven that the level of blood lipids is a primary factor in atherogenesis.

□ COMMERCIAL FATS AND FAT PRODUCTS

The edible fats and oils may be used for food, whereas the inedible fats are used in the production of soap and lubricants. Edible fats and oils, which are derived from animals and vegetables, are generally processed into lard, butter, margarine, shortening, cooking oils, and salad oils.

Natural fats and oils, which consist mostly of mixed triglycerides, also contain varying proportions of monoglycerides, diglycerides, and free fatty acids as well as small quantities of nonglyceride materials. The source of vegetable oils is the seed and seed coats of plants. Cottonseed and soybean oil accounted for 9840 and 9640 million pounds, respectively, of edible oils produced in the United States in 1976. Lard production during the same period was 1043 million pounds. The fatty acid composition of natural fats and oils is summarized in Table 5.2.

Processing

Plant Oils

The oils are usually removed from plant tissues by three methods. In the first method, cells containing oil are ruptured by heat and mechanical methods. In the past, cottonseeds free of lint were hulled, then flaked between rollers, and cooked with live steam prior to hydraulic pressing to separate the oil. A second method, called screw pressing, involves heating the flakes or cracked meats, followed by passage through close-fitting cages of screws to press out the oil. The third method involves solvent extraction using petroleum hydrocarbons at 60°–70°C. This third method, which is a continuous one, has a capacity of processing of hundreds of tons of oil per day. The by-products of these methods are proteins, which are used in animal feeds.

Table 5.2. Composition of Some Natural Fats and Oils[a]

	Bubassa	Coconut	Corn	Cottonseed	Grapeseed	Linseed	Olive	Palm	Soybean	Butter	Lard
Butyric										3	
Caproic		t[b]								1	
Caprylic	6	6								1	
Capric	4	6		t						3	
Lauric	45	44		t						4	t
Lauroleic										0.5	
Myristic	17	18		1			t	1	t	12	3
Myristoelie				t						2	
Palmatic	9	11	13	29	9	6	14	48	11	29	24
Palmatoleic				2		t	2			4	3
Stearic	3	6	4	4	4	4	2	4	4	11	18
Oelic	13	7	29	24	20	22	64	38	25	25	42
Linoleic	3	2	54	40	67	16	16	9	51	2	9
Linolenic		t				2	2		9		
Arachidic										2	1
Arachidonic										1	

[a] Values expressed as percentage of total fats and oils.
[b] t = trace.

Animal Fats

Animal fats are separated from fatty tissues by wet or dry rendering. In wet rendering, the fatty tissue is heated under steam pressure, thus rupturing the cells and liberating the fat. In dry rendering, the fatty tissue is heated in jacketed drums with agitation until the fat is released. Presently, centrifuges are used to separate the fat from water and protein.

Hydrogenation

Hydrogenation of unsaturated fats and oils increases their melting point and hardness. This process often is used in the production of shortenings, which are generally defined as plastic materials made wholly from fats and oils. Hardness can be controlled by varying the ratio of solid to liquid glyceride. The shortenings are made by blending the desired oils and/or fats, deodorizing the mixture, chilling, and finally packaging.

Interesterification

When fats and oils are heated in the presence of certain catalysts, the fatty acids attached to glycerol rearrange in a process called interesterification. While vegetable oils are randomly distributed, animal fats generally are not. For example, at a level of 0.4% tin in stannous hydroxide added to fat at 140°C, then heated to 225°C for 90 min in a vacuum, the distribution pattern of fatty acids will become random. In a similar process, the melting point of soybean oil can be increased from −7° to +5.5°C, and its softening from −13° to −0.5°C. Lard in its natural state is very grainy due to palmitic acid in the β position of the glyceride molecule. Following interesterification the melting point changes very little, but the graininess is removed. Other catalysts such as sodium methoxide and metallic sodium can be used to cause interesterification. However, with these catalysts the oil must be dried under vacuum and heated at lower temperatures. Tests are required to determine if interesterification has occurred since no color change is produced.

Interesterification is used in the industry to produce standard oils, which may be blended with others for use in the margarine and cooking fat trade. During interesterification, if the temperature is lowered, a certain amount of the higher melting triglycerides crystallize out. This has a dramatic effect on the remaining portion of the liquid oil and alters the course of esterification. This is called directed interesterification.

Uses of Fat and Fat Products in Food

Fats and fat products may consist of (1) a fat or oil, (2) a fat plus an emulsifying agent, or (3) fat emulsions such as butter and margarine. Fat products are used as shortenings, spreads, solid oils, cooking and frying fats and oils, and in the preparation of confectionery and icings.

Shortenings

The term *shortening* had its origin in the United States and referred to a preparation, originally developed from cottonseed oil, that was used to "shorten" the preparation time of shortbread and cakes. Shortenings consist entirely of fat and contain no moisture. The traditional shortening is lard.

Domestic shortenings generally fall into one of two categories—moulded products (10% air) and liquid-filled products (10–35% air). Moulded products have good cake-making properties. Liquid-filled products are more expensive but are easier to use.

Shortenings are matured (tempered) by holding at an elevated temperature (25°–30°C) for up to 48 hr. This maturation causes a change in crystal structure such that when the product is cooled it has a plastic texture. This process is accomplished with scraped-surface heat exchangers.

High-ratio shortenings allow a higher ratio of sugar to flour to be used in cakes due to the emulsifying properties of the shortenings. Emulsifiers are usually mono- or diglycerides.

Spreads

Butter and margarine are the main fat products used as spreads. Both of these products are water-in-oil emulsions.

Butter. Butter usually contains about 80% fat. In England, butter must not contain more than 16% water. Other constituents of butter include protein (1%), lactose (0.4%), mills ash (0.15%), and salt. The natural color of butter is due to carotene and other fat-soluble pigments.

Margarine. Margarine was developed after Napoleon III offered a prize for a process that would produce a butter substitute. Megè-Mouriès was awarded the prize in 1870 for his product oleo-margarine; within a few months, it was known as margarine.

Megè-Mouriès first prepared his margarine by the emulsification of beef oleo with milk. Later, other edible animal and vegetable fats were found to be satisfactory for the preparation of this product. In the United States, margarine was first made largely from oleo oil, but by 1933 more than 60% of the margarine was made from coconut oil. From 1934 on, coconut oil was replaced in increasing proportions by hydrogenated domestic oils (principally cottonseed and soya oils); at the present time practically no coconut oil is used in margarine production in the United States. Some margarine manufacturers use all-hydrogenated vegetable oils to produce their product, while others use blends of all-hydrogenated plus nonhydrogenated vegetable oils; these latter products have a greater temperature range of plasticity.

Margarine is made by thoroughly mixing, or churning, melted fat with cultured skim milk. After cooling to solidify the margarine, it is kneaded and blended into a homogeneous mass. Salt is added, and the product is then packaged. Optional ingredients include emulsifying agents, vitamins A and D, sodium benzoate, and color. Practically all margarine produced in the United States now contains vitamin A to the extent of 9,000 to 15,000 USP units.

Margarine is subject to legal regulation. It must not contain more than 16% moisture, nor more than 10% butterfat. It may not contain preservatives, but may contain coloring matter, emulsifiers, and antioxidants that are permitted by the regulations. All table margarines must contain 760–940 IU/oz of vitamin A and 80–100 IU/oz of vitamin D.

Salad Oils

Olive, corn, cottonseed, soybean, sunflower, and sesame seed oils are used as salad oils. Olive oil has been used for years as a salad oil and is particularly prized for its characteristic flavor. A salad oil must not solidify at refrigerator temperatures and must not show "clouding" (precipitation) when exposed to a temperature of 0°C for 5½ hr. Refined corn oil meets these requirements and most of the corn oil produced in this country is used as a salad oil. Cottonseed oil must be wintered in order to produce a good salad oil, and soybean oil must also be processed to make it acceptable as a salad oil. Sunflower and sesame seed oils are good natural salad oils, but they are not produced in great quantities at the present time.

Cooking Fats

Any edible fat or oil may be used for cooking. For deep-fat frying, the most desirable fat is one with a bland flavor and a high smoking temperature. Since temperatures in this type of frying may reach as high as 244°C, the fat selected should allow for heating above this point without smoking. When smoke is given off by a heated fat, it is an indication that the fat is undergoing decomposition, with the production of disagreeable odors and flavors. Fats suited for deep-fat frying include the all-hydrogenated fats such as those used for shortenings but not containing emulsifiers (mono- and diglycerides tend to decompose at high temperatures); vegetable oils (except olive oil); and high-quality lards.

For pan frying, the choice of fats is not limited to those with high smoking temperatures since the cooking is done at a lower temperature than in deep-fat frying. Sometimes fats (butter, bacon drippings, chicken fat, etc.) are used for the flavor they impart to the fried product. Fats used for pan frying include butter, margarine, olive oil, chicken fat, bacon drippings, and fats suitable for deep-fat frying. Coconut oil is used in large quantities for nut frying and popcorn popping.

☐ DETERIORATION OF FATS

Deterioration of fats, or rancidity, constitutes one of the most important technical problems in the food industries. All of the chemical reactions involved can be explained on the basis of the ester linkage and the nature of fatty acid glycerides. Deterioration occurs through hydrolysis of the ester linkage by lipases and moisture (hydrolytic rancidity); through the autoxidation of unsaturated fatty acid glycerides in atmospheric oxygen (oxidative rancidity); through the enzymatic oxidation of unsaturated fatty acid glycerides (lipoxidase rancidity); or through the enzymatic oxidation of certain saturated fatty acid glycerides (ketonic rancidity).

Hydrolytic Rancidity

In hydrolytic rancidity the off-flavor is due to the hydrolysis of a fat with the liberation of free fatty acids. Hydrolytic rancidity is extremely noticeable in fats such as butter because the volatile fatty acids released have a disagreeable odor and taste. In contrast, hydrolytic changes in fats that contain few volatile fatty acids are not accompanied by a bad odor and taste.

Oxidative Rancidity

Oxidation is the most important cause of fat spoilage because all edible fats, as such or as components of foods, contain unsaturated triglycerides. Oxidative deterioration of fat results in the destruction of vitamins (A, D, E, K, and C), destruction of essential fatty acids, and the development of a pungent and offensive off-flavor.

Experimental evidence indicates that hydroperoxides are the predominant, but not exclusive, primary products of autoxidation of unsaturated fats. Hydroperoxides are relatively unstable at or above 80°C, whereas at room temperatures they are relatively stable; therefore, different end products may be produced under different reaction temperatures. The chain or cyclic nature of autoxidation is well established (Swern 1960). The polyethenoic acids are more readily oxidized than the monoethenoic acids. The mechanism involves three types of reactions: (1) initiation of the chain reaction, (2) chain propagation, and (3) chain termination.

A mechanism for the autoxidation of a methylene-interrupted unsaturated system (e.g., linoleate) was proposed by Holman (1954). The various reactions that may occur are presented in Fig. 5.1. In the initiation step, a hydrogen atom (H) is abstracted from the methylenic carbon atom adjacent to a double bond. This reaction results in the formation of a free radical (II). The reaction is catalyzed by catalysts (trace metals, oxygen, light, etc.) or enzymes (lipoxidases). Catalysts

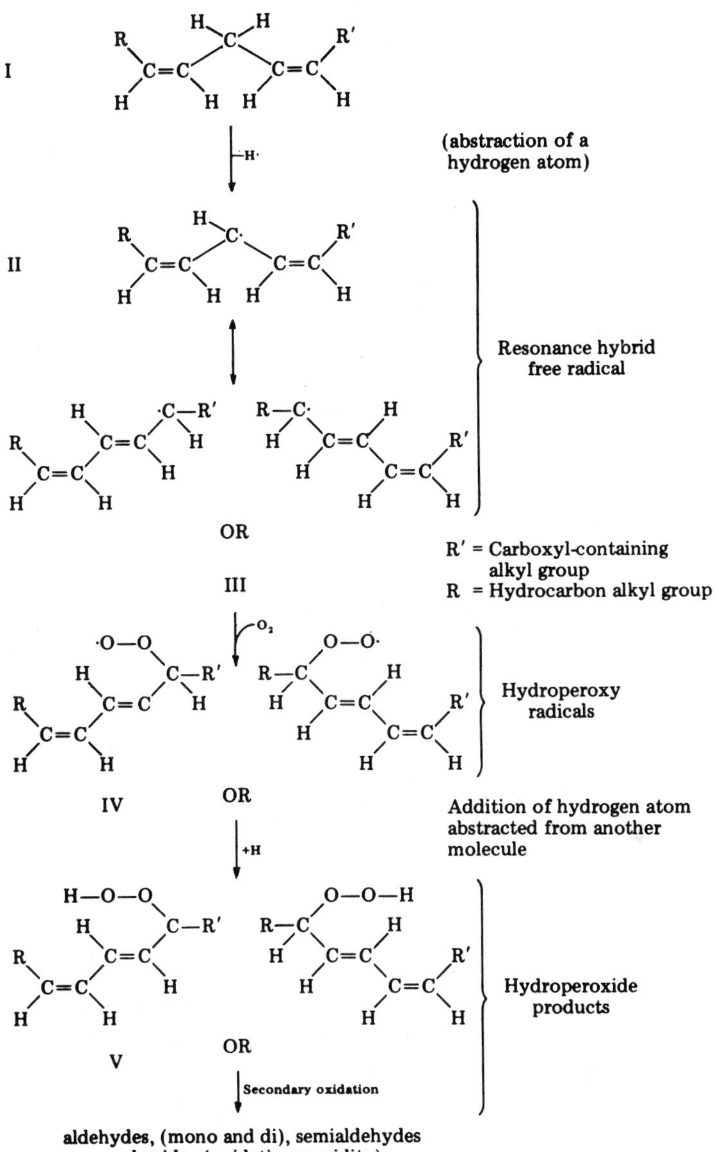

Figure 5.1. Reaction mechanisms for autoxidation of an unsaturated fatty acid.

are necessary; otherwise the reaction would be too slow to be significant. The free radical (II) is a resonance hybrid, the two extreme forms are shown as III. Molecular oxygen adds to the resonating radical, predominantly at the ends, to yield two types of hydroperoxy radicals (IV). These free radicals can accept hydrogen atoms from other molecules (linoleate) to become isomeric conjugate *cis*, *trans*-hydroperoxides (V) and in so doing propagate the chain reaction. Theoretically, only a single molecule need be involved in order to initiate the chain reaction of autoxidation. Thus, the amount of hydroperoxide needed to initiate the oxidative reaction is extremely small. The reaction can be terminated by the collision of radicals II, III, and IV with each other, as follows:

$$ROO\cdot + R\cdot \rightarrow ROOR$$

$$R\cdot + R\cdot \rightarrow RR$$

$$ROO\cdot + ROO\cdot \rightarrow ROOR + O_2$$

The hydroperoxide products (V) are unstable and undergo decomposition to form short-chain acids, alcohols, aldehydes, and ketones. These end products (secondary oxidation products) are responsible for the development of the odor and flavor of oxidized fats.

Factors Affecting Autoxidation of Fats

One of the factors that affects the autoxidation of common food fats is the total number of unsaturated linkages in the sample. However, the total amount of unsaturation may not be as important as the degree of unsaturation within a given molecule. A fat high in linolenic acid (three double bonds) would be more susceptible to oxidation than one containing a similar amount of oleic acid (one double bond).

Oxygen is necessary for autoxidation of fats. At very low oxygen pressures, the rate of oxidation is approximately proportional to the pressure. Therefore, the removal of atmospheric oxygen from a fat or food product exerts a protective effect. This is accomplished in the food industry by vacuum packing or packaging under an inert gas such as nitrogen.

All forms of light radiation from the ultraviolet to the infrared region are conducive to fat oxidation. Ultraviolet light has a more pronounced effect than visible light due to the higher energy of ultraviolet. Packaging in lightproof containers or amber-colored containers is the usual means of controlling the destructive effects of light.

Temperature also has a marked effect on the rate of autoxidation.

At ordinary temperatures, the effect of increasing temperature on the rate of autoxidation is slightly greater than for most chemical reactions because increasing temperatures accelerates both the chain propagation reactions and peroxide decompositions. Low-temperature storage helps to minimize temperature-induced oxidation.

Moisture appears to prevent or inhibit fat autoxidation, perhaps by inhibiting the absorption of oxygen. In the preparation of dehydrated foods, it has been observed that moisture levels for optimal stability vary with the type of product. These levels range from 6% moisture for starchy foods to trace moisture for high-sugar foods.

Various trace metals, especially copper and iron, act as pro-oxidants in fat. Thus, in processing packaged fats or food products containing fats, every precaution must be taken to eliminate contamination with these trace metals.

Antioxidants

Antioxidants act primarily as hydrogen donors or free radical acceptors. The reaction of an antioxidant (AH_2) may be represented as follows:

$$ROO\cdot + AH_2 \rightarrow ROOH + AH\cdot$$

$$AH\cdot + AH\cdot \rightarrow A + AH_2$$

Thus, the primary role of antioxidants is to break the chain reaction of autoxidation by reacting with hydroperoxy radicals.

The antioxidants may be classed as natural and synthetic. Many fats and oils, particularly in the unrefined form, are quite stable to oxidative rancidity because they contain natural antioxidants. The tocopherols (α, β, and γ) are the most important of the natural antioxidants and are widespread in both plant and animal tissues. Vegetable oils contain much higher concentrations of tocopherols than animal fats. The tocopherols are effective antioxidants for animal fats, but have little antioxidant effect when added to vegetable fats. The tocopherols are readily oxidized to tocoquinones, which have no antioxidant properties. In addition, the tocopherols are readily destroyed by heat, particularly at temperatures employed in the refining and processing of fats and oils.

Naturally occurring antioxidants exhibit relatively weak antioxidant properties. As a consequence, synthetic antioxidants have been developed for use in foods. These substances, to be allowed in foods, must have a low order of toxicity, should be effective in low concentrations in a wide variety of fats, should contribute no objectionable flavor, odor or color to the product, and should have approval by the Food and Drug Administration (see discussion about food additives in Chapter 1).

Since antioxidants vary in their effectiveness to stabilize fats or fat products, and in their mode of action, combinations of antioxidants are often used to stabilize fatty foods. The first satisfactory antioxidant combination was a mixture of butylated hydroxyanisole (BHA), propyl gallate (PG), and citric acid; this was used to stabilize lard for shortening purposes. At the present time BHA, butylated hydroxytoluene (BHT), and PG constitute the major proportion of the total antioxidants used. Other antioxidants are used in only limited amounts. Citric acid or other metal chelators are usually added as insurance against trace quantities of metals often present in fats and oils.

Certain acidic compounds (e.g., ascorbic, citric, and phosphoric acids) have been found to exert a synergistic effect when added along with polyphenolic antioxidants; consequently, they have been called synergists. These acidic compounds are effective metal-chelating agents, and it was at first thought that their synergistic effect was due only to their metal-chelating activity. However, when citric acid was added alone, it exhibited no antioxidant activity. Privett (1961) suggested the following series of reactions that explain the general characteristics of this synergism:

Absence of Synergist

$$2AH + 2ROOH \rightarrow 2A\cdot + RO\cdot + H_2O + R\cdot + H_2O_2$$

$$2A\cdot + 2ROOH \rightarrow O_xA + RO\cdot + H_2O + R\cdot + H_2O$$

Presence of Synergist

$$2AH + 2ROOH \rightarrow 2A\cdot + RO\cdot + H_2O + R\cdot + H_2O_2$$

$$A + S \rightarrow A\cdot S$$

$$A\cdot S + ROOH \rightarrow \text{No reaction or very slow}$$

$$A\cdot S + ROO\cdot \rightarrow O_xA + ROOH + S$$

where AH = original antioxidant; S = synergist; ROOH = hydroperoxide; A·S = antioxidant radical–synergist complex; A· = antioxidant radical; ROO· = peroxide radical formed in autoxidation; and O_xA = oxidized antioxidant.

According to these reactions, the synergist forms an antioxidant radical–synergist complex (A·S), and this chemical association suppresses the antioxidant's catalysis of peroxide composition. By suppressing the catalysis, additional chain formation is stopped and consequently the antioxidant is spared from its function in stopping such chains.

Tests for Oxidative Rancidity

Several tests have been developed to indicate oxidative rancidity in fats. Some of these tests are purely qualitative, whereas others give a quantitative indication of the degree of rancidity. The incipient stages of rancidity can be detected by these tests before the spoilage can be detected organoleptically.

Kreis Test. The Kreis test depends upon the formation of a red color when an oxidized fat is treated with concentrated hydrochloric acid and a solution of phloroglucinol in ether. The compound in rancid fats responsible for the color reaction is epihydrin aldehyde (2,3-epoxy-propionaldehyde). Due to the high sensitivity of the Kreis test, colors produced by fresh oils tend to give misleading results. All oxidized fats respond to the Kreis test, and the intensity of the color produced is roughly proportional to the degree of oxidative rancidity.

To perform the test, thoroughly mix 1 ml of melted fat and 1 ml of concentrated HCl in a test tube, add 1 ml of a 1% solution of phloroglucinol in diethyl ether (freshly prepared), and again mix thoroughly. If a pink color develops, the fat is slightly oxidized; if a red color develops, the fat is definitely oxidized.

Thiobarbituric Acid Test. The oxidation of fats produces compounds that react with 2-thiobarbituric acid to give red-colored products. Essentially the method involves dissolving the fat sample in an organic solvent such as benzene, chloroform, or carbon tetrachloride and extracting the reactive material with an acetic acid–thiobarbituric acid–water solution. The aqueous extract on heating will develop a red color if the fat is oxidized. The method, while empirical, may be made quantitative if the color intensity is measured by means of a spectrophotometer; a thiobarbituric acid value can be calculated from the absorbance.

This comparatively new test may give a more accurate picture of fat oxidation than the older color tests. For example, it was found that in samples of cottonseed and soybean oils oxidized to the same peroxide value, the soybean oil gave higher thiobarbituric acid values than the cottonseed oil. This is noteworthy because soybean oil has a greater tendency to develop oxidized flavor than does cottonseed oil.

Peroxide Value. Since peroxides are the first compounds formed when a fat oxidizes, all oxidized fats give positive peroxide tests. The peroxide value of a fat is easily determined by dissolving a weighed amount of fat in an acid–chloroform solution, adding potassium iodide, and titrating the iodine liberated with standard sodium thiosulfate solution. The value is expressed as milliequivalents of peroxide per 1000 g of sample.

☐ ANALYSIS

Sample preparation for physical and chemical analysis of fats usually involves melting solid fats followed by filtration using a hot water funnel. Oils that are not clear also should be filtered. Rancidity may be retarded if samples are stored in a cool place protected from light and air.

Physical Analyses

Moisture and Volatile Matter

A softened sample (not melted) is weighed to 5 $\pm$ 0.2 g in an aluminum moisture dish filled with a light slip-over cover. The sample is dried to constant weight in a vacuum oven at a uniform temperature 20°–25°C above the boiling point of water at working pressure ($\leq$ 100 mm Hg), then is cooled in a dessicator for 30 min and weighed. Constant weight is achieved when successive 1-hr drying periods show an additional loss in weight of $\leq$ 0.05%. The percentage loss in weight is reported as moisture and volatile matter.

Specific Gravity

A pycnometer that has been standardized is filled with the sample previously cooled to about 20°C. The pycnometer is placed in a constant-temperature bath for 30 min at 25°C, then removed from the bath, wiped dry, and weighed to constant weight as described in the previous determination. Calculate specific gravity at 25°/25°C as follows:

$$\text{Sp gr} = \frac{\text{Wt oil-filled pycnometer} - \text{Wt empty pycnometer}}{\text{Wt } H_2O \text{ at } 25°C}$$

Index of Refraction

Any standard instrument may be used to determine the index of refraction. The temperature of oils should be 20° or 25°C and that of fats 40°C. The instrument is placed so that diffused daylight or artificial light can be used for illumination. Water at a constant temperature is circulated through the prisms. Approximate corrections of butyrorefractometer readings may be made by the following formula:

$$R = R_1 + K(T_1 - T)$$

where R = reading reduced to standard temperature; R_1 = reading obtained at temperature T_1; T = standard temperature; and $K = 0.55$ for fats and 0.58 for oils.

Melting Point

The melting point of fats and fatty acids is usually determined by the capillary tube method. In this procedure, approximately 10 mm of melted and filtered fat is drawn into a thin-wall capillary tube (1-mm i.d.). The end of the tube is sealed with a small flame; be careful not to burn the fat. The tubes containing fat are stored at 4°–10°C over night (approximately 16 hr).

The tube is attached to an accurate thermometer, graduated to 0.2°C, so that the lower end is even with the bottom of the Hg bulb. The thermometer is suspended in a 600-ml beaker half filled with water to about 300 mm below the surface. Starting at 8°–10°C below the melting point of the sample, heat is applied to increase the temperature at about 0.5°C/min as the water is stirred. The melting point is when the substance becomes transparent. An average of three tests should agree within 0.5°C. The melting points of natural fats and oils are given in Table 5.3.

Table 5.3. Analytic Constants for Some Natural Fats and Oils

Fat/Oil	Iodine Value	Suponification value of oils	Melting point (°C)
Bubassa	14–18	247–251	24–26
Coconut	7.5–10.5	250–264	23–26
Corn	103–128	187–193	−12– −10
Cottonseed	99–113	189–198	−2–2
Grapeseed	135	190	−10
Linseed	155–205	188–196	−20
Olive	80–88	188–196	−3–0
Palm	44–54	195–205	27–50
Soybean	120–141	189–195	−23– −20
Butter	25–42	233–240	28–35
Lard	53–77	190–202	33–46

Saponification Value and Equivalent

The saponification value (Koettstorfer number) is defined as the milligrams of potassium hydroxide required to saponify 1 g of fat. The general reaction for the saponification of a triglyceride can be written as follows:

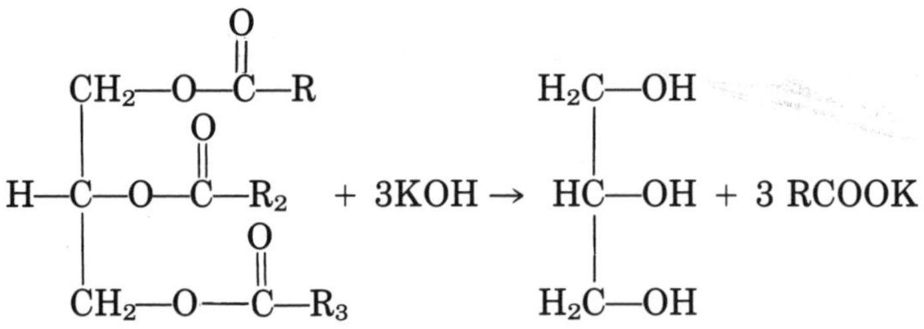

$$\begin{array}{l} CH_2{-}O{-}\overset{\displaystyle O}{\overset{\|}{C}}{-}R \\ | \\ H{-}C{-}O{-}\overset{\displaystyle O}{\overset{\|}{C}}{-}R_2 \\ | \\ CH_2{-}O{-}\overset{\displaystyle O}{\overset{\|}{C}}{-}R_3 \end{array} + 3KOH \rightarrow \begin{array}{l} H_2C{-}OH \\ | \\ HC{-}OH \\ | \\ H_2C{-}OH \end{array} + 3\ RCOOK$$

The saponification value is an index of the mean molecular weight of the glycerides comprising a fat. If the fatty acids present in the glycerides are of low molecular weight (short-chain acids), there will be more glyceride molecules per gram of fat than if the acids are of high molecular weight (long-chain acids). Thus, since each glyceride molecule requires three potassium hydroxide molecules for saponification, fats containing glycerides of low molecular weight have correspondingly higher saponification values. For example, β-oleyl-α,α′-dipalmitin (**5.33**) has a molecular weight of 833.3 and a saponification value of 202.0, whereas α,α′-dicaproyl-β-butyrin (**5.34**), with a molecular weight of 358.5, has a saponification value of 469.5. Fats such as tallow and lard, composed primarily of high-molecular-weight fatty acid glycerides, have saponification values around 195, whereas fats

$$\textbf{5.33}\quad \begin{array}{l} H_2C{-}O{-}\overset{\displaystyle O}{\overset{\|}{C}}{-}C_{15}H_{31} \\ | \\ HC{-}O{-}\overset{\displaystyle O}{\overset{\|}{C}}{-}C_{17}H_{33} \\ | \\ H_2C{-}O{-}\overset{\displaystyle O}{\overset{\|}{C}}{-}C_{17}H_{33} \end{array}$$

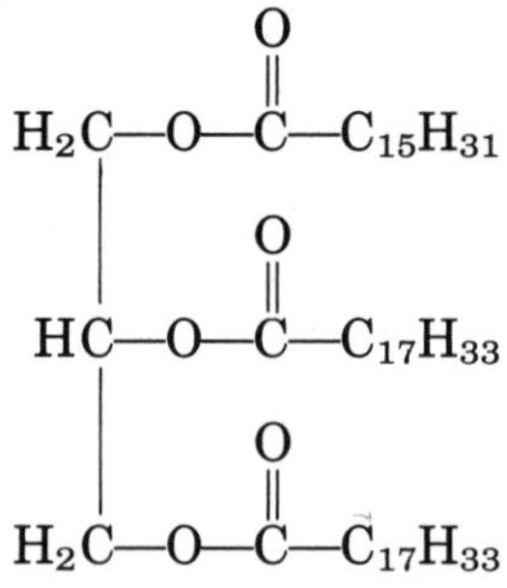

β-Oleyl-α,α′-dipalmitin
Molecular weight = 833.3
Saponification value = 202.0

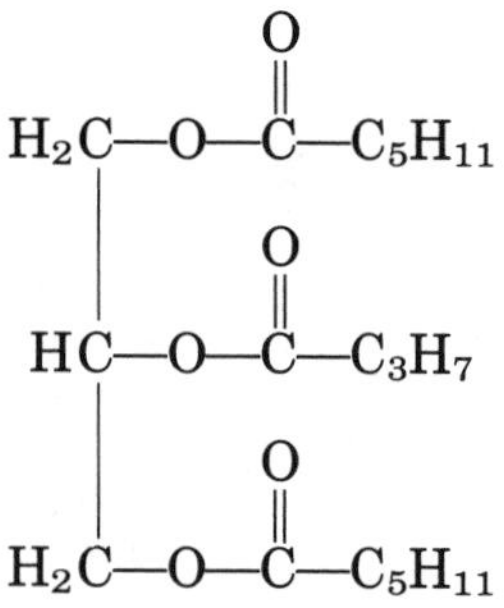

$$\begin{array}{l} H_2C{-}O{-}\overset{\displaystyle O}{\overset{\|}{C}}{-}C_5H_{11} \\ | \\ HC{-}O{-}\overset{\displaystyle O}{\overset{\|}{C}}{-}C_3H_7 \\ | \\ H_2C{-}O{-}\overset{\displaystyle O}{\overset{\|}{C}}{-}C_5H_{11} \end{array}\quad \textbf{5.34}$$

α,α′-Dicaproyl-β-butyrin
Molecuar weight = 358.5
Saponification value = 469.5

such as butterfat and coconut oil, containing appreciable quantities of low-molecular-weight fatty acid glycerides, have saponification values around 230–250. The saponification values for many natural fats and oils are listed in Table 5.3.

Fats containing appreciable quantities of unsaponifiable matter, such as sterols, have low saponification values. On the other hand, fats containing unsaturated fatty acid glycerides have saponification values that are higher than those of the corresponding saturated fats. The reason for this is that the oxidation of unsaturated fatty acid glycerides causes the unsaturated acid to split at the point of oxidation with the formation of carboxyl groups. Each carboxyl group formed in this way reacts with potassium hydroxide in the saponification reaction and hence causes an increased saponification value over that normal for the fat.

To determine the saponification value, a weighed sample of the fat is treated with a measured volume of alcoholic KOH solution and saponified by heating. The excess potassium hydroxide is backtitrated with standard acid solution. A blank determination is run simultaneously with the sample to determine the concentration of the alcoholic potassium hydroxide solution. From the weight of sample and the titration values of the blank and samples, the saponification value may be calculated.

Reagents

1. Aldehyde-free alcoholic KOH—To prepare aldehyde-free alcohol, reflux a mixture of 1200 ml of 95% ethanol, 10 g of KOH, and 6 g of aluminum foil for 30 min. Distill and collect 1 liter of the ethanol after discarding the first 50 ml coming over. Dissolve 40 g of high-grade KOH (low in carbonate) in a liter of aldehyde-free ethanol, maintaining the temperature below 15°C during the solution of the alkali. The solution should remain clear. It is approximately 0.7 *N*.
2. Standard 0.5 *N* hydrochloric acid.
3. Indicator solutions—(a) To prepare phenolphthalein indicator, dissolve 1 g of phenolphthalein in 50 ml of 95% ethanol and dilute to 100 ml with water. (b) To prepare alkali-blue 6B indicator, boil 2 g of the sodium salt of triphenyl-pararosanilinetrisulfonic acid in 100 ml of ethanol under reflux for 2 hr and then filter.

Procedure

Place the liquid fat to be analyzed in a small weighing bottle equipped with a medicine dropper. Weigh the bottle and fat, then transfer by means of the medicine dropper approximately 5 g of the fat to a 300-ml Erlenmeyer flask; reweigh the weighing bottle and remaining fat. The weight of the fat sample is obtained by difference. Approximately 40–50 drops of fat will be required to give a weight of 1 g. Determinations should be repeated on each sample.

Accurately measure 50 ml of the alcoholic KOH solution by means of a transfer pipette or burette and add to the fat sample in the Erlenmeyer flask. Attach an air condenser to the top of the flask and reflux on a steam bath for 30 min. Complete saponification of fats usually occurs within 15 min; hence, a 30-min heating period is normally sufficient. However, if it is known that the material is difficult to saponify, a longer heating period may be necessary. Complete saponification has occurred when the solution is entirely homogeneous. Cool the flask.

Add 1 ml of phenolphthalein indicator solution, and titrate the excess KOH with the standard 0.5 *N* HCl. If the solution becomes reddish-brown during the saponification, it is helpful to add 1 ml of the alkali-blue 6B solution in addition to the phenolphthalein before titrating. Phenolphthalein and alkali-blue 6B have the same color (red) in alkali solutions; in acid solutions, however, phenolphthalein is colorless, whereas alkali-blue 6B is blue.

Two blank determinations are run simultaneously with the samples. The blank determinations should contain the same volumes of alcoholic KOH solution measured in the same way and refluxed for the same length of time as for the samples. The titration values obtained on the blanks determine the concentration of the KOH solution and thus are as essential as the samples.

Calculation

From the weight of sample and the titration values for the blank and the sample, the saponification value may be calculated. To be acceptable, the values obtained should check within three saponification values. With care and experience, greater accuracy than this can be expected.

When exactly the same volume of alcoholic KOH solution is used in the blank as in the sample, the following formula may be used to calculate the saponification value (S.V.):

$$\text{S.V.} = \frac{\begin{array}{c}(\text{Blank titration} - \text{sample titration})\\ \times \text{Normality of acid} \times 56.1\end{array}}{\text{Sample weight in grams}}$$

Remarks

Aldehyde-free alcohol is used for the preparation of the alcoholic KOH solution so that a minimum amount of resinification with attendant color formation occurs during the saponification of the fat. High-grade KOH, low in carbonate, is used in the preparation of the alkali solution, since the presence of carbonates in the solution interferes with the titration when phenolphthalein is used as indicator. Furthermore, the saponification and titration should be done with as little access of air as possible to prevent absorption of carbon dioxide by the alkali solution.

An air condenser is used to prevent the loss of alcohol during saponification. At the end of the titration, the alcohol solution should still be approximately 50% to prevent hydrolysis of the potassium soaps present. If alcohol is lost during saponification, sufficient neutral aldehyde-free alcohol should be added to give the required 50% alcohol concentration at the end of the titration.

Saponification Equivalent (Mean Molecular Weight)

The saponification equivalent offers a means for expressing directly the mean molecular weight of a fat or fatty acid ester, whereas the saponification value is useful merely as an index of the molecular weight.

In calculating the saponification equivalent of a fat, the assumption is made that it consists entirely of triglycerides free from unsaponifiable matter and other impurities. The following formulas may be used to calculate the saponification equivalent (S.E.) of a fat or of a simple ester from data obtained in the determination of the saponification value:

$$\text{S.E. of a fat} = \frac{(\text{Sample weight} \times 3000)}{(\text{Blank titration} - \text{sample titration}) \times \text{Normality of acid}}$$

$$\text{S.E. of a simple ester} = \frac{(\text{Sample weight} \times 1000)}{(\text{Blank titration} - \text{sample titration}) \times \text{Normality of acid}}$$

If the saponification value has been calculated, the saponification equivalent may be obtained upon dividing the weight of the moles of potassium hydroxide required for saponification by the saponification value expressed in grams (S.V. $\times$ 0.001). For example, α-,α'-dicaproyl-β-butyrin (**5.34**) has a saponification value of 469.5; since it is a fatty acid triglyceride, three moles of potassium hydroxide (168.3 g) are required to saponify one mole of **5.34**. Thus, in this case 168.3 ÷ 0.4695 = 358.5, the saponification equivalent or molecular weight of triglyceride **5.34**. A simple ester, such as methyl palmitate, would have required only one mole of potassium hydroxide (56.1 g) for saponification. These relationships can be simplified to the following formulas:

$$\text{S.E. of a fat} = \frac{168{,}300}{\text{S.V.}}$$

$$\text{S.E. of a simple ester} = \frac{56{,}100}{\text{S.V.}}$$

Iodine Value

The iodine value is defined as the number of centigrams of iodine absorbed by 1 g of fat. Fats containing unsaturated fatty acid glycerides will, under proper conditions, quantitatively add halogen, calculated as iodine, to the unsaturated linkages. Saturated fatty acid glycerides will not add halogen. Thus, the iodine value, the quantity of halogen added, is a measure of the amount of unsaturation (number of double bonds) in a fat.

Since the unsaturated fatty acids commonly occurring as components of edible fats are liquid at room temperature, the iodine value can be related to the melting point or hardness of a fat. For example, corn oil, which contains about 83% unsaturated fatty acids, has a melting point of −10° to −12°C and an iodine value of 103–128. In contrast, lard with about 54% unsaturated fatty acids has a melting point of 33°–46°C and an iodine value of 53–77. There is also some evidence that extremely hard fats (predominantly long-chain saturated glycerides) are not digested to the same degree in the body as the softer fats containing an appreciable amount of unsaturated glycerides. To some degree there also is a relationship between fat spoilage due to oxidative deterioration and the iodine number of a fat. The more highly unsaturated glycerides containing fatty acids with two and three double bonds are highly susceptible to oxidative deterioration; glycerides with one double bond are much less susceptible; and the saturated acid glycerides are not at all susceptible to this type of spoilage. The iodine values for many fats and oils are presented in Table 5.3.

To determine the iodine value, a weighed sample of fat is dissolved in chloroform and treated with a measured volume of halogen solution in a special glass-stoppered iodine flask. The flask and contents are stored for sufficient time to permit the addition of halogen to the unsaturated glycerides, and then the excess halogen is back-titrated with standard sodium thiosulfate solution. Blank determinations are run simultaneously with the samples to determine the concentration of the halogenating agent. The amount of halogen reacting with the unsaturated glycerides is calculated as iodine.

Because iodine (I_2) itself is not reactive enough to add quantitatively to the unsaturated bonds in fats, iodine monochloride (ICl) and iodine monobromide (IBr) are used as halogenating agents in two common methods of determining the iodine value. Even though it is likely that ICl or IBr add to the double bonds in the reaction, the total halogen added may be expressed as iodine. It is immaterial, as far as the calculations are concerned, whether the halogenating solution contains ICl, IBr, or I_2. Treatment with KI just before the back-titration of the excess halogen releases an equivalent amount of iodine from ICl or IBr:

$$ICl + KI \rightarrow I_2 + KCl$$

$$IBr + KI \rightarrow I_2 + KBr$$

The Wijs method employs ICl and the Hanus method employs IBr as the halogenating agent.

Wijs Method

Because the Wijs ICl solution is somewhat more reactive than the Hanus IBr solution, results obtained by the Wijs method are in better agreement, particularly for oils with high iodine values. Hence, the American Oil Chemists Society has adopted the Wijs method as the standard procedure for determining iodine values.

Reagents

1. Glacial acetic acid (C.P.)—To test the purity of the acetic acid, dilute 2 ml of the acid with 10 ml of distilled water and add 0.1 ml of 0.1 *N* potassium permanganate solution. The pink color must not be discharged completely within 2 hr.
2. Potassium iodide solution—Dissolve 15 g of KI in 85 ml of water.
3. 0.1 *N* Sodium thiosulfate solution standardized against potassium dichromate as a primary standard.
4. Starch indicator solution—Make a starch suspension by stirring 1 g of soluble starch in 10 ml of water. Pour this well-mixed suspension into 100 ml of boiling distilled water, stir thoroughly,

and continue the heating for 2 min. Cool, and if a precipitate settles out, decant the supernatant liquid for use as the indicator solution.

5. Wijs iodine solution—Dissolve 13 g of resublimed iodine in 1 liter of glacial acetic acid using heat, if desired, to effect a more rapid solution of the iodine. Cool the solution and titrate a 20-ml portion with standard 0.1 *N* sodium thiosulfate solution. Pour a small portion of the iodine solution (100–200 ml) into a beaker and pass washed and dried chlorine gas into the main portion of the solution until the original titration figure is doubled or slightly exceeded. Pour the small portion of the original iodine solution that was set aside into the chlorinated solution; this should reduce the chlorine content of the entire solution to slightly less than that of the iodine content. It is desired in the chlorination that the halogen content be almost, but not quite, doubled. An excess of iodine is necessary in the solution, but an excess of chlorine must be avoided. Store the Wijs solution in a glass-stoppered amber bottle, being sure that it is tightly stoppered. Record the date of preparation on the bottle and do not use the solution after it is more than 1 month old.

Procedure

Weigh accurately about 0.5 g of fat or 0.25 g of an edible oil, or 0.1–0.2 g of a drying oil into a 300-ml glass-stoppered iodine flask. Dissolve the fat sample in 15 ml of chloroform. Add 25 ml of the Wijs iodine solution from a glass-stoppered burette. Stopper the flask, rotate gently to mix the solution, and place in the dark for 30 min. At the end of this time, loosen the stopper and wash it off into the flask with 20 ml of the KI solution followed by 100 ml of water.

Titrate the iodine solution with standard 0.1 *N* sodium thiosulfate solution until the yellow iodide color has almost disappeared. Add a few drops of starch indicator solution and continue the titration to the disappearance of the blue color. Near the end of the titration, stopper the flask and shake vigorously after each small addition of the standard thiosulfate solution to get all the iodine held in the chloroform layer into the water phase, where it can react with the thiosulfate.

Two blank determinations should be run simultaneously with the samples. The same quantities of reagents are used in the blanks as in the samples and the volumes are measured with the same volumetric apparatus. The average value obtained for the blank determinations is used to ascertain the halogen concentration of the iodine solution.

Calculation

When exactly the same volume of iodine solution is used in the blank as in the sample, the following formula may be used to calculate the iodine value (I.V.):

$$\text{I.V.} = \frac{\begin{array}{c}\text{(Blank titration} - \text{sample titration)} \\ \times \text{ Normality of } Na_2S_2O_3 \times 12.69\end{array}}{\text{Sample weight in grams}}$$

Remarks

In order that halogen will add quantitatively to each unsaturated linkage present in the fat, it is necessary that at the completion of the addition reaction, a 100% excess of halogen remain in the solution. This may be accomplished by keeping the fat sample constant and increasing the volume of halogen reagent when the fat is more highly unsaturated, or by keeping the volume of halogen reagent constant and decreasing the weight of the sample as the degree of unsaturation increases. The latter method, given in the directions above, is the one commonly used.

The fat and halogenating agent are stored in the dark for exactly 30 min to ensure complete addition of halogen to the unsaturated linkages without any substitution taking place. Light catalyzes the substitution reaction and hence precautions are taken to eliminate it.

Since the iodine solution consists of halogens dissolved in glacial acetic acid, which has a high thermal coefficient of expansion, blank determinations are run at regular intervals along with the samples so that changes in laboratory temperature, which would affect the volume and hence the concentration of the iodine solution, will be taken into consideration.

The titrated solution, when allowed to stand, frequently reverts to a blue color, likely due to the splitting off of halogen from the addition compound. The end point taken should be the first one observed.

Hanus Method

The determination is conducted in exactly the same manner as in the Wijs method except that the Hanus iodine solution (IBr) is used in place of the Wijs iodine solution (ICl). The Hanus solution is prepared as described in the Wijs method except that bromine gas is substituted for chlorine gas. The calculation of the iodine value is the same as for the Wijs method. Rasults by the Hanus method are about 2–4% lower than those obtained by the Wijs method.

Effect of Fatty Acid Structure on Iodine Value

The iodine value normally gives a true indication of the relative amount of unsaturation in a fat. However, if the unsaturated linkages are found in abnormal places in the fat acid, or if they are in a conjugated system, true iodine values may not be obtained.

The effect of the place of the double bond in the fatty acid molecule on the iodine value obtained by the usual procedure is well illustrated by comparison of the theoretical and actual iodine values of four isomeric oleic acids:

Acid	*Formula*	*Theoretical Iodine Value*	*Actual Iodine Value*
2-Oleic acid	$CH_3(CH_2)_{14}CH{:}CHCOOH$	90.07	9.04
3-Oleic acid	$CH_3(CH_2)_{13}CH{:}CHCH_2COOH$	90.07	16.27
4-Oleic acid	$CH_3(CH_2)_{12}CH{:}CH(CH_2)_2COOH$	90.07	26.90
9-Oleic acid	$CH_3(CH_2)_7CH{:}CH(CH_2)_7COOH$	90.07	90.07

The closer the double bond gets to the carboxyl group, the lower the actual iodine value. Fortunately, the 9-oleic acid is normally present in the glycerides of the naturally occurring fats, so that the iodine value gives a true picture of the amount of unsaturation.

The effect of conjugated double bonds on the iodine value is well illustrated in the case of elaeostearic acid. On the basis of iodine value, this acid was originally considered to have two double bonds. Later it was discovered that this acid had three double bonds and that iodine did not add to one of the double bonds in the conjugated system present in the acid. Some idea of the amount of conjugation present in a fat or fatty acid may be obtained from a determination of the diene number, maleic acid values, or by the ultraviolet spectrophotometric method. Conjugated unsaturation is not likely to occur in the fatty acid glycerides of edible oils.

Thiocyanogen Value

The thiocyanogen value may be defined as the centigrams of thiocyanogen, expressed in iodine equivalents, absorbed by 1 g of fat. Iodine values indicate the total amount of unsaturation in fats. However, there is still the question as to whether the unsaturation is due to oleic, linoleic, or linolenic acids in the common edible glycerides, which may contain any or all three of these unsaturated fatty acids in their structure. The thiocyanogen value in conjunction with the iodine value permits calculation of the amounts of each of these unsaturated acids present.

Thiocyanogen $(CNS)_2$ exhibits halogenating activity between that of bromine and iodine. This reagent does not react with the saturated fatty acids but adds one molecule of thiocyanogen to every molecule of oleic acid (one double bond), as is also true of iodine; one molecule of thiocyanogen to every molecule of linoleic acid (two double bonds), whereas iodine adds two molecules; and two molecules of thiocyanogen to every molecule of linolenic acid (three double bonds), to which iodine adds three molecules. Thus, the amounts of oleic and linoleic acids may be determined in a mixture of the two, if the iodine and thiocyanogen values are known. Furthermore, the amounts of oleic, linoleic, and linolenic acids can be determined if, in addition to the iodine and thiocyanogen values, the amount of saturated acids present is obtained by a separate method. The amounts of these various unsaturated acids may be calculated when present as a mixture of free acids or combined as glycerides.

Acetyl and Hydroxyl Values

The acetyl value is defined as the number of milligrams of KOH required to neutralize the acetic acid obtained by saponifying 1 g of an acetylated fat. It is best determined by the Andre–Cook method, which involves a calculation based upon the increase, after acetylation, in the saponification value of a fat containing hydroxy groups.

The hydroxyl value is defined as the number of milligrams of KOH equivalent to the hydroxyl content of 1 g of fat. In the method developed by Roberts and Schuette, the sample is acetylated in a sealed tube in the presence of an excess of acetic anhydride. The excess acetic anhydride is then hydrolyzed and titrated with a standard KOH solution, and the hydroxyl value is calculated.

The acetyl and hydroxyl values are a measure of the hydroxy groups present in a fat. The hydroxy groups are present either in hydroxy fatty acid glycerides, such as ricinoleic acid glycerides in castor oil, or in sterols, which are high-molecular-weight alcohols constituting the bulk of the unsaponifiable matter of fats. Hydroxy fatty acids are not common in the naturally occurring fats; hence, the acetyl values of most of the common fats are due to their sterol content. Since partial oxidation of unsaturated fatty acid glycerides may form hydroxy acid glycerides, the acetyl value of an oxidized fat may be slightly greater than that of the normal fat. The presence of mono- and diglycerides in a fat also increases the acetyl value.

The acetyl and hydroxyl values are determined primarily for castor oil and certain lubricating oils; consequently they are of little interest to the food analyst. They have been used on occasion as a measure of the hydroxy content (mainly sterols) in edible fats.

Esterification and Gas–Liquid Chromatography of Neutral Lipids

Lipids are a class of heterogeneous compounds. As a consequence no single solvent is suitable for extracting lipids from naturally occurring materials. Solvents of low polarity, such a carbontetrachloride, diethyl ether and liquid hydrocarbons, will extract glycerides, sterols and small amounts of complex lipids, whereas use of mixtures of the above solvents and polar solvents, such as alcohol, will extract the majority of complex lipids. Examples of the latter mixtures are ethyl ether/ ethanol and chloroform/methanol. Extraction is often performed in a nitrogen atmosphere to prevent oxidation of unsaturated fatty acids of the glyceride.

In this experiment the glycerides and sterols are extracted with ether. The glycerides are separated from the sterols by saponification, the glycerides are esterified, and analyzed by gas–liquid chromatography.

Reagents

1. Diethyl ether.
2. Chloroform.
3. 95% Ethyl alcohol.
4. Sodium chloride.
5. 5% HCl in methanol.
6. Petroleum ether, 30°–60°C.
7. 100/200 mesh Unisil (Carlson Chemical Co., Williamsport, Pa.).
8. Soxhlet extraction apparatus.
9. Peanut butter.
10. Peanuts or nuts of any kind.
11. Gas Chromatograph.

Procedure

Extraction—Weigh a 1.5–2.0 g sample of peanut butter on a piece of Whatman No. 1 filter, then fold the filter paper and place in an extraction thimble which, in turn, is placed into a Soxhlet extraction tube. Attach the tube to a weighed flask. Half-fill the flask with ether and extract the mixture for about 16 hr, warming the flask gently on a heated mantle. Rearrange the apparatus for distillation and evaporate ether to dryness. Redissolve the residue in 5 ml of chloroform.

Separation of neutral lipids—A 100/200 mesh Urisil silicic acid column is used to separate neutral lipids from other lipid material (e.g., phospholipids, etc.). Mix 7 g Urisil with a small volume of chloroform and add to a 1 × 15 cm column. Transfer the above lipid material onto the top of the column. The neutral lipids are recovered by evaporating the solvent on a rotary evaporator to 10 ml (at room temperature). Transfer to a small flask and evaporate to dryness under a stream of nitrogen. Redissolve the residue in 5 ml chloroform and transfer to a 6-in. test tube. Add 3 ml of 5% HCl in methanol to the flask and cap with a small funnel and a marble. (The funnel and marble act as a crude reflux condenser reducing the amount of alcohol lost.) Place the test tube in a 75°C water bath and allow to reflux for 1 hr. If too much methanol evaporates during reflux, add a little more to the flask. Remove, cool and add an equal volume of water to the test tube. Saturate with sodium chloride and add 2 ml of petroleum ether. Cap, invert, and shake. The petroleum ether layer contains the methyl esters of fatty acids.

Gas Chromatography

Analyze the esters by gas–liquid chromatography to ascertain the composition of the sample.Compute peak areas by triangle technique. Determine the identity of your sample by comparison with standards.

GC Apparatus and Conditions

1. Column: 5–10 ft × $\frac{1}{8}$ in. or $\frac{1}{4}$ in.; glass or stainless steel or aluminum or copper; packed with polyester liquid phase on acid-washed Chromosorb W.
2. Column temperature: 170°–210° ± 1.0°C; inlet port approximately 50°C higher than column temperature.
3. Carrier gases: Helium (minimum purity 99.95 mole %) for thermal conductivity detector; helium, nitrogen, or argon (minimum purity 99.95 mole %) for flame ionization detector (FID).
4. FID gas: Hydrogen (minimum purity 99.95 mole %) and breathing-quality dry air (< 2 ppm hydrocarbons equivalent to CH_4) for the flame of the FID.

Halphen Test

A positive Halphen test is usually considered specific for the presence of cottonseed oil. However, kapok and baobab oils give a reaction similar to cottonseed in the Halphen test. Also, lard or butterfat obtained from animals fed on cottonseed meal also gives a positive Halphen test.

Furthermore, the absence of a positive test does not necessarily preclude the presence of cottonseed oil, since this oil, upon oxidation or hydrogenation, loses its ability to react in the Halphen test. Heating the oil for 10 min at 250°C also destroys its ability to give the color reaction.

Procedure

Mix equal volumes of amyl alcohol and a 1% solution of sulfur in carbon disulfide. Add 3–5 ml of this reagent to an equal volume of the oil sample in a test tube and heat in a bath of boiling saturated NaCl solution for 1–2 hr. Place a control tube containing cottonseed oil plus the reagent in the bath for comparison. A reddish color should develop if cottonseed oil is present, and the intensity of the color is somewhat proportional to the amount of the oil present.

Furfural Test for Sesame Oil

Sesame oil gives a somewhat specific color reaction with furfural that can be used to detect sesame oil in the presence of other oils. The intensity of color formation is roughly proportional to the amount of sesame oil present. Hydrogenated sesame oil responds to this test, but the test is more sensitive with unhydrogenated oil. As little as 0.25–0.5% sesame oil can be detected by the furfural test.

Procedure

To 10 ml of a liquid sample add an equal volume of concentrated HCl solution and mix. Add 0.1 ml of a furfural solution (prepared by adding 2 ml of furfural to 100 ml of ethanol), shake well for 15 sec, and let stand until the emulsion breaks. Look for color in the lower layer when the emulsion breaks. Absence of a pink color indicates a negative test. If a pink color is present, add 10 ml of water, shake again, and observe the color as soon as the layers separate. If the pink color persists, sesame oil is not present. It is necessary to observe the color as soon as possible so that, should a pink color be present, it is not masked by the development of other colors.

Fitelson Test for Teaseed Oil

The glycerides of teaseed oil and olive oil are quite similar in their chemical composition; consequently the two oils have similar physical and chemical characteristics. However, they do differ somewhat in the nature of their unsaponifiable matter, and the Fitelson test, which can differentiate between the two oils, is based upon this difference. The test is applicable only to mixtures of olive and teaseed oils.

Procedure

To a clear test tube, add exactly 0.8 ml of acetic anhydride, 1.5 ml of chloroform, and 0.2 ml of sulfuric acid. Mix the contents and cool in an ice–water bath to 5°C. Add 7 drops of the oil sample to be tested. If the tube contents become cloudy, add acetic anhydride, drop by drop, shaking after each addition until no turbidity is evident. Hold at 5°C for 5 min.

To the contents of the tube, add 10 ml of cold (5°C) ether, stopper with a cork, and mix by inverting the test tube. Return the tube to the ice–water bath and note if a color develops. A sample of pure teaseed oil will give an intense red color within 1 min, develop to a maximum intensity, and then fade away. In some cases olive oil will develop a faint pink color, which fades.

Remarks

The intensity of color produced is closely proportional to the amount of teaseed oil present. Thus it is possible to estimate the amount of teaseed oil present in a sample by comparing its color with that of mixtures of known composition tested simultaneously with the sample. The test is reliable for mixtures containing more than 10% of teaseed oil in olive oil.

Modified Bellier Test for Peanut Oil

The modified Bellier test can detect peanut oil in the presence of olive, cottonseed, corn, and soybean oils. In this test, the fatty acids are liberated by saponification and dissolved in a warm alcohol solution. The solution is then cooled and the temperature at which turbidity or precipitation of the fatty acids occurs is noted. If peanut oil is present, turbidity will occur at a higher temperature than would be anticipated for the other oils because of the precipitation of the relatively insoluble arachidic acid, a component of peanut oil but not of the other oils.

Reagents

1. 70% Ethanol solution—Dilute 700 ml of ethanol to a volume of 950 ml with water. Check the specific gravity or refractive index of the solution and adjust if required.
2. 1.5 *N* KOH solution—Dissolve 10 g of KOH in aldehyde-free

ethanol to a volume of 100 ml. To prepare aldehyde-free ethanol, see *Reagents* under Saponification Value and Equivalent.

3. HCl solution (sp gr 1.16)—Dilute 83 ml of concentrated HCl to 100 ml with water. Check the specific gravity with a hydrometer and adjust if required.

Procedure

Transfer a 0.92-g sample of the oil into a 125-ml Erlenmeyer flask equipped with a standard taper joint carrying an air condenser. Add 5 ml of the alcoholic KOH solution and heat on a steam bath for 5 min using the air condenser; swirl the contents of the flask once or twice during saponification. Remove the flask from the bath; add 50 ml of the 70% ethanol solution and 0.8 ml of the HCl solution. Dissolve any precipitate that may form by warming. Carefully cool the flask and contents in water, stirring continuously with a thermometer, so that the temperature falls at the rate of 1°C/min. Note the temperature at which a definite precipitate first appears as evidenced by turbidity or clouding. Turbidity can best be observed by looking through the solution toward a good light or toward a black background with good light coming from one side.

Remarks

It is essential to prevent premature formation of turbidity by stirring continuously when cooling the sample and also to ensure that the cooling water does not rise above the level of liquid in the flask, allowing local cooling to occur.

If turbidity appears above 9°C in the case of olive oil or 13°C in the case of cottonseed, corn, or soybean oils, peanut oil is present.

Bromine Test for Fish Oils and Marine Animal Oils

The fish oils and marine animal oils contain large amounts of polyunsaturated fatty acid glycerides. The formation of insoluble bromides of these polyunsaturated acids can be used to identify them in the presence of vegetable oils. Metallic salts must be absent in the sample.

Procedure

Dissolve approximately 6 g of the oil in 12 ml of a mixture of equal parts of acetic acid and chloroform contained in a test tube. Add bromine, drop by drop, until a slight excess has been added as evidenced by the color, keeping the temperature of the solution at 20°C. Permit the mixture to stand 15 min or more and then place the test tube in a boiling water bath. The solution will be clear if only vegetable oils are present but will remain cloudy if fish oils are present.

Test for Mineral Oil in Fats

This test depends on the fact that mineral oil is nonsaponifiable and forms a turbid mixture with water.

Procedure

Place in an Erlenmeyer flask, 1 ml of the oil or melted fat, 1 ml of a saturated KOH solution, and 25 ml of ethanol. Attach an air condenser and boil until saponification is complete (usually 5 min), shaking occasionally. Add 25 ml of water, mix, and note if turbidity appears. When more than 0.5% mineral oil is present, a distinct turbidity appears.

Preparation of Component Mixed Fatty Acids

At times, the iodine value, melting point, and refractive index of the mixed fatty acids derived from a fat are more significant for identification and the detection of adulteration than the characteristics of the fat itself. Thus, if sufficient sample is available, they should be considered when they add information to the particular problem.

Procedure

To 25 g of the fat in an Erlenmeyer flask, add 20 ml of an aqueous solution containing 10 g of KOH and 20 ml of ethanol. Equip the Erlenmeyer flask with a reflux condenser and heat on the steam bath until saponification is complete. Then remove the condenser and evaporate off the ethanol. Dissolve the residue in 200–300 ml of hot water. Add an excess of sulfuric acid and boil gently until the fatty acids form a clear layer on top of the liquid. Either siphon off the aqueous layer or draw it off in a separatory funnel and discard. Wash the fatty acids with hot distilled water until free from mineral acid. Separate the melted fatty acids as thoroughly as possible from the water and filter, using a hot water-jacketed funnel to obtain the acids in a dry condition. Use these acids for the desired determinations.

Uses of Analytical Results

The various fat values that have been discussed in this section are those normally used for the identification of an unknown fat, for the detection of admixture of a foreign fat, or for obtaining information relative to the industrial possibilities of a fat.

The general approach for identifying an unknown fat is to determine certain of its significant physical and chemical values, particularly its melting point, specific gravity, refractive index, saponification value, and iodine value. By comparing these values with those of known fats it is usually possible to identify an unknown fat by the similarity of its characteristics with those of a known fat. Because few fats are similar in more than one or two of their characteristic values, if several values are compared, identification should be possible.

Foreign fat added to a sample is often difficult to identify. A determination of the refractive index, iodine value, and saponification value may indicate the presence of a foreign fat. However, should these characteristics for the foreign fat not differ widely from those of the pure fat sample, considerable quantities could be added without possible detection in this manner. Furthermore, by combining several fats, a mixture could be prepared with characteristics similar to a pure fat; detection of such a mixture, when used as an adulterant of the pure fat, would be difficult. Thus, in addition to the fat's characteristics, confirmatory tests and even tests on the fatty acids derived from the fats are frequently necessary to determine the presence of a foreign fat.

The general approach to detecting the presence of a foreign fat can be illustrated by olive oil. Because it is relatively expensive, olive oil is subject to adulteration with less-expensive oils, such as cottonseed, peanut, corn, lard, teaseed, sesame seed, rape seed, poppy seed, and even coconut. The fat characteristics and the confirmatory tests helpful in detecting the added oil in each instance are as follows:

- *Cottonseed Oil.* The most useful values for detecting cottonseed oil in olive oil are the iodine number and melting point of the fatty acids. A positive Halphen test also strongly suggests the presence of cottonseed oil.
- *Peanut Oil.* Because the fat characteristics of peanut oil and olive oil are quite similar, they are not useful in differentiating the two oils. The separation and identification of arachidic acid by the modified Bellier test positively indicates peanut oil.
- *Corn Oil.* The refractive index and iodine value are significantly higher for corn oil than for olive oil. A mixture of lard oil and corn oil that has an iodine value similar to that of olive oil would still have an abnormal specific temperature value.

- *Lard Oil.* The iodine value for lard oil is lower than the average for olive oil. However, this permits the possibility of adding considerable quantities of lard oil to olive oils which have high iodine values or by blending lard oil with other oils (cottonseed, peanut, etc.) to reduce the iodine value comparable to olive oil. More significant for detecting lard oil are its odor (especially when warmed), the high melting point of its fatty acids, and especially the presence of cholesterol.
- *Teaseed Oil.* The fat characteristics of teaseed oil are identical with those of olive oil and so are useless in distinguishing the two oils. Fitelson's test can be used to detect teaseed oil in the presence of olive oil.
- *Sesame Seed Oil.* The refractive index and iodine values of sesame seed oil are appreciably higher than they are for olive oil; therefore, they would indicate the presence of sesame seed oil if added in considerable quantities. The furfural test is specific for sesame seed oil and should be made.
- *Rape Seed Oil.* The lower saponification value of rape seed oil is its most significant difference from olive oil. Adulteration with mineral oil would also lower the saponification value of olive oil but would give abnormally high unsaponifiable matter.
- *Poppy Seed Oil.* The refractive index and iodine value of poppy seed oil are considerably higher than they are for olive oil and therefore would indicate its presence.
- *Coconut Oil.* The low iodine value and high saponification value of olive oil containing coconut oil would be significant.

Information relative to the industrial uses of a fat may be obtained from certain of its characteristics. The drying property of oils is associated with the amount and, particularly, the type of unsaturation present in their glycerides. The iodine value and refractive index indicate the relative amount of unsaturation. The iodine value in conjunction with the thiocyanogen value indicates the amounts of the different unsaturated fatty acids present in the glycerides. In the most rapid drying oils, unsaturation occurs in a conjugated system. Fats used in foods should not exhibit appreciable oxidizing ability; hence the fat characteristics giving information relative to the amount and types of unsaturation may also be used to indicate the relative keeping ability of fats. When it is important to know the length of the carbon chain of the fatty acids present in the glycerides, the saponification value or saponification equivalent will give the desired information.

The melting point is related to the relative hardness of a fat in the solid state, which is related to its digestibility and also to its ability to be used as a shortening agent in baked goods. These few illustrations should indicate how the fat characteristics may be used to obtain information on the industrial uses of a fat.

SELECTED REFERENCES

AOAC. 1980. Official Methods of Analysis. 13th ed. W. Horwitz (editor). Assoc. of Official Analytical Chemists, Arlington, VA.

AOAC. 1984. Official Methods of Analysis. 14th ed. S. Williams (editor). Assoc. of Official Analytical Chemists. Arlington, VA.

BERGSTROM, S., CARLSON, L. A. and WEEKS, J. R. 1968. The prostaglandins: A family of biologically active lipids. Pharmacol. Rev. *20*, 201–210.

BLOCH, K. 1963. Lipid Metabolism. John Wiley & Sons, New York.

BOLLAND, J. L. and TEN HAVE, P. 1947. Kinetic studies on the chemistry of rubber and related materials. IV. The inhibitory effect of hydrogen on the thermal oxidation of ethyl linoleate. Trans. Faraday Soc. *43*, 201–210.

BRAVERMAN, J. B. S. 1963. Introduction to the Biochemistry of Foods. Elsevier Publishing Co., New York.

BROCKERHOFF, H. and YURKOWSKE, M. 1966. Stereospecific analysis of several vegetable fats. J. Lipid Res. *7*, 62–65.

BROWN, P. and MORTON, I. D. 1970. Fats and Fatty Acids. *In* Food Industries Manual, 20 ed. A Woollen (editor). Chemical Publishing Co., New York.

FIESER, L. F., and FIESER, A. M. 1959. Steroids. Van Nostrand Reinhold Co., New York.

FURIA, T. E. 1968. Handbook of Food Additives. Chemical Rubber Co., Cleveland, OH.

GIBSON, D. M. 1965. The biosynthesis of fatty acids. J. Chem. Educ. *42*, 236–243.

GUARNIERI, M. and JOHNSON, R. M. 1970. Nomenclature. Adv. Lipid Res. *8*, 116–118.

GURR, M. I. and JAMES, A. T. 1980. Lipid Biochemistry: An Introduction. 3rd ed. Chapman and Hall, New York.

HOLMAN, R. T. 1971. Essential fatty acid deficiency. Prog in Chem. Fats Other Lipids *9*, 275–348.

JOSLYN, M. A. 1970. Methods in Food Analysis, 2nd ed. Academic Press, New York.

KRITCHEVSKY, D. 1958. Cholesterol. John Wiley & Sons, New York.

LENNARZ, W. J. 1970. Lipid metabolism. Ann. Rev. Biochem. *39*, 359–388.

LUNDBERG, W. O. (editor). 1961. Autoxidation and Antioxidatants. Vol. I and II. John Wiley & Sons, New York.

MAJERUS, P. W. and VAGELOS, P. R. 1967. Fatty acid biosynthesis and the role of the acyl carrier protein. Adv. Lipid Res *5*, 2–30.

MARKLEY, K. S. 1961. Fatty Acids: Their Chemistry, Properties, Production and Uses. 2nd ed., Parts 1 & 2. John Wiley & Sons, New York.

PRIVETT, O. S. 1961. Some observations on the course and mechanism of autoxidation and antioxidant action. *In* Proc. Flavor Chemistry Symposium. Campbell Soup Co., Camden, NJ.

SAMUELSSON, B. 1972. Biosynthesis of prostaglandins. Federation Proc. *31*, 1442–1450.

SCOTT, G. 1965. Atmospheric Oxidation and Antioxidants. Elsevier Publishing Co., New York.

SHAPITO, B. 1967. Lipid metabolism. Ann. Rev. Biochem. *36*, 247–270.

SOBER, H. A. (editor). 1970. Handbook of Biochemistry. 2nd ed. Chemical Rubber Co., Cleveland, OH.

SONNTAG, N. O. V. 1979. Structure and composition of fats and oils. *In* Bailey's Industrial Oil and Fat Products, Vol. I, pp. 1–98. John Wiley & Sons, New York.

SWERN, D. 1960. Oxidation by atmospheric oxygen. *In* Fatty Acids, 2nd ed., Part 2. K. S. Markley (editor). John Wiley & Sons, New York.

WAKIL, S. J. (editor). 1970. Lipid Metabolism. Academic Press, New York.

WILLIAMS, K. A. 1966. Oils, Fats, and Fatty Acids. 4th ed. J. and A. Churchill-Little, London.

ZAPSALIS, C. and BECK, A. R. 1985. Food Chemistry and Nutritional Biochemistry. John Wiley & Sons, New York.

6

Proteins

☐ INTRODUCTION

Proteins are essential components of every living cell. Some are utilized in the formation and regeneration of tissue; certain specific proteins serve as enzymes and others as antibodies; still others fulfill indispensable functions in metabolic regulation and contractile processes. All things considered, proteins are concerned with virtually all physiological events.

Plants are able to synthesize proteins from inorganic sources of nitrogen, water, and carbon dioxide assimilated by their roots and leaves. In contrast, animals and man are dependent on plant and animal proteins in the diet to provide the necessary constituents for protein synthesis.

Since proteins are of such great importance to animals and man, many plants are grown because of the nutritional value of their proteins. However, not all proteins have the same biological value: some proteins are rich in certain essential amino acids, and thus have a high biological value, whereas other proteins are devoid of some of these acids or contain them in very small amounts. In those areas of the world where the main protein source is vegetable protein that lacks certain amino acids, protein deficiency diseases often occur, particularly in children.

All proteins contain nitrogen, carbon, hydrogen, and oxygen. Many also contain sulfur; some contain phosphorus, and a few contain other elements such as zinc, iron, and copper. Although proteins vary somewhat in composition, the typical elemental analysis is 16% nitrogen, 50% carbon, 7% hydrogen, 22% oxygen, and 0.5–3% sulfur.

Proteins exhibit a number of common properties that must be accounted for in any definition of these compounds:

- They are polymeric materials of high molecular weight.
- They are amphoteric, i.e., they may act chemically both as acids and as bases.
- Following complete hydrolysis of a protein, the hydrolysate consists entirely of amino acids (except that additional groups, such as heme, iron, or copper, may also be found in the case of a conjugated protein).
- In their polymeric structures, the amino acid units of proteins are joined together in definite sequences and exist in definite three-dimensional conformations.

☐ AMINO ACIDS AND THE PEPTIDE BOND

All proteins yield amino acids when hydrolyzed, and all but two of these are α-amino carboxylic acids. Proline and hydroxyproline are α-imino acids. In proteins, amino acids are united through amide linkages between the α-carboxyl and α-amino functional groups of adjacent amino acid residues. Such bonds are called *peptide bonds;* this linkage is represented as follows:

$$H_3^+N-\underset{R_1}{\overset{H}{C}}-\overset{O}{\overset{\|}{C}}-\overset{H}{N}-\underset{R_2}{\overset{H}{C}}-\overset{O}{\overset{\|}{C}}-\overset{H}{N}-\underset{R_3}{\overset{H}{C}}-\overset{O}{\overset{\|}{C}}\text{-----(etc.)---}\underset{R_n}{\overset{H}{C}}-C\begin{matrix}\nearrow O \\ \searrow O^-\end{matrix}$$

The substances resulting from the formation of peptide bonds are termed *peptides,* and the individual amino acid units of peptides are termed *residues.* Thus, a peptide containing two amino acid residues is referred to as a dipeptide; one containing many residues is referred to as a polypeptide. By convention, peptides are written with the terminal α-amino group (N terminus) to the left and the terminal α-carboxyl group (C terminus) to the right.

The peptide linkages of proteins are somewhat resistant to hydrolysis and, in the absence of enzymes, require prolonged heating with strong acids or alkalies for completion of the process. (Enzymatic hydrolysis of proteins is discussed in the Proteolysis section.) Acid hydrolysis is generally preferred because alkaline hydrolysis partly converts the optically active amino acids (L isomers) into racemic mixtures (D and L forms). During complete acid or alkaline hydrolysis some of the amino acids are destroyed: all of the glutamine and asparagine, most of the tryptophan, and some of the serine and threonine residues are destroyed by acid, whereas arginine, cystine, and cysteine, as well as asparagine and glutamine, are destroyed by alkali.

The α-amino acids, except glycine, have at least one asymmetric carbon atom and, as a consequence, may exist in optically active D and L forms or in optically inactive racemic mixtures. The configuration around the asymmetric carbon of most naturally occurring amino acids is the same as the configuration of the asymmetric carbon atom of L-glyceraldehyde; accordingly, they are classified as L-amino acids. (See Optical Activity section in Chapter 4 for more detailed discussion of this topic.)

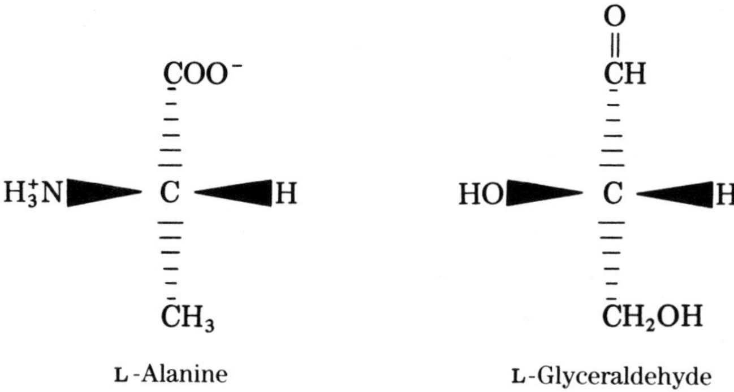

L-Alanine L-Glyceraldehyde

The amino acids have been classified in various ways. One system classifies them according to the structure of their R group and another according to the number of amino and carboxyl groups they possess. The structures of the 23 amino acids found in proteins are given in Table 6.1 in which the amino acids are classified according to both of these systems.

Ionic Properties

Typical free amino acids have at least two groups—the amino group and the carboxyl group—that can act as proton acceptors or donors.

Table 6.1. Classification of α-Amino Acids

I. Aliphatic Amino Acids

A. Monoamino-monocarboxylic acids

	Structure	pK_a at 25°
1. Glycine (Gly)	$H_3^+N—CH_2—COO^-$	2.35; 9.78
2. Alanine (Ala)	CH_3 | $H_3^+N—CH—COO^-$	2.35; 9.83
3. Valine (Val)	CH_3 CH_3 \\ / CH | $H_3^+N—CH—COO^-$	2.29; 9.72
4. Leucine (Leu)	CH_3 CH_3 \\ / CH | CH_2 | $H_3^+N—CH—COO^-$	2.33; 9.74
5. Isoleucine (Ile)	CH_3 | CH_2 | $CH—CH_3$ | $H_3^+N—CH—COO^-$	2.34; 9.76
6. Serine (Ser)	CH_2OH | $H_3^+N—CH—COO^-$	2.19; 9.21
7. Threonine (Thr)	CH_3 | H—C—OH | $H_3^+N—CH—COO^-$	2.09; 9.10
8. Cysteine (Cys)	$CH_2—SH$ | $H_3^+N—CH—COO^-$	1.71 8.39 (sulfhydryl) 10.76 (α-amino)
9. Cystine (Cys Cys)	NH_3^+ | $CH_2—CH—COO^-$ | S | S NH_3^+ | | $CH_2—CH—COO^-$	1.65; 2.26 (carboxyls) 7.85; 9.85 (α-amino)

Table 6.1. *(Continued)*

	Structure	pK_a at 25°
10. Methionine (Met)	$CH_2—S—CH_3$ $\vert$ CH_2 $\vert$ $H_3^+N—CH—COO^-$	2.13; 9.28
B. Monamino-dicarboxylic acids		
1. Aspartic (Asp)	COO^- $\vert$ CH_2 $\vert$ $H_3^+N—CH—COO^-$	2.05 (α-carboxyl) 3.87 (β-carboxyl) 10.00
2. Asparagine (Asn)	$O{=}C—NH_2$ $\vert$ CH_2 $\vert$ $H_3^+N—CH—COO^-$	2.02 8.80
3. Glutamic (Glu)	COO^- $\vert$ CH_2 $\vert$ CH_2 $\vert$ $H_3^+N—CH—COO^-$	2.16 (α-carboxyl) 4.27 (γ-carboxyl) 9.36
4. Glutamine (Gln)	$O{=}C—NH_2$ $\vert$ CH_2 $\vert$ CH_2 $\vert$ $H_3^+N—CH—COO^-$	2.17; 9.13
C. Diamino-monocarboxylic acids		
1. Arginine (Arg)	$(CH_2)_3—N(H)—C(=NH_2^+)—NH_2$ $\vert$ $H_3^+N—CH—COO^-$	1.82 8.99 (α-amino) 12.48 (guanido)
2 Lysine (Lys)	$(CH_2)_4—NH_3^+$ $\vert$ $H_3^+N—CH—COO^-$	2.18 8.95 (α-amino) 10.53 (ϵ-amino)

Table 6.1. *(Continued)*

	Structure	pK_a at 25°
3. Hydroxylysine (Hyl)	$CH_2—NH_3^+$ | CHOH | CH_2 | CH_2 | $H_3^+N—CH—COO^-$	2.13 8.62 (α-amino) 9.67 (ε-amino)
II. Aromatic Amino Acids		
A. Monoamino-monocarboxylic acids		
1. Phenylalanine (Phe)	CH_2 | $H_3^+N—CH—COO^-$	1.83; 9.31
2. Tyrosine (Tyr)	OH CH_2 | $H_3^+N—CH—COO^-$	2.20 9.11 (α-amino) 10.03 (phenolic hydroxyl)
3. Tryptophan (Trp)	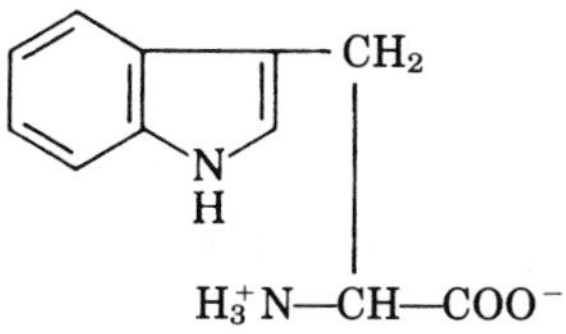 CH_2 N H $H_3^+N—CH—COO^-$	2.46; 9.41

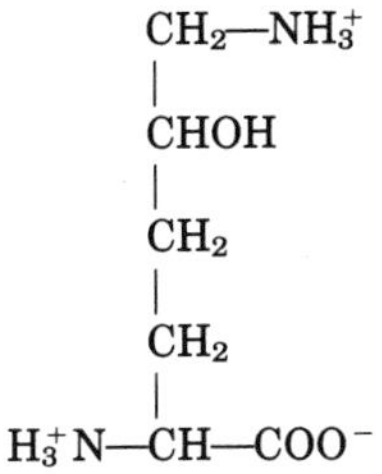

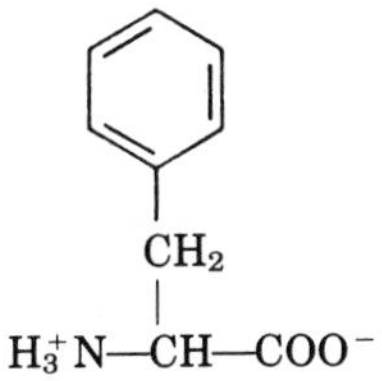

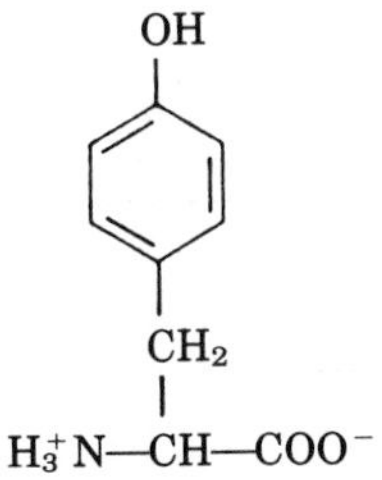

(continued)

Table 6.1. (*Continued*)

	Structure	pK_a at 25°
III. Heterocyclic Aliphatic Amino Acids		
A. Monoamino-monocarboxylic acid		
1. Histidine (His)	Imidazolium ring (HC—N(H)⁺=CH—N(H)—C=) —CH_2— H_3^+N—CH—COO^-	1.82 6.04 (imidazole) 9.17
B. α-Imino acids		
1. Proline (Pro)	Pyrrolidine ring with N^+H_2, α-H, —COO^-	1.95; 10.64
2. Hydroxyproline (Hyp)	HO-substituted pyrrolidine ring with N^+H_2, α-H, —COO^-	1.82; 9.66

The high melting points and water solubilities of amino acids suggest that they occur in the ionic form. Titration of a simple amino acid with both acid and base indicates that the α-amino group exists as an α-ammonium group (pK_a approximately 10) and the carboxyl group exists as a carboxylate group (pK_a approximately 2). The structure that is consistent with this phenomenon is commonly called the zwitterionic form. At low pH the cationic form of the amino acid is present, and at high pH, the anionic form predominates.

The relationship between pH and the ionic form of an amino acid is illustrated by the titration curve for glycine, a monoamino-monocarboxylic acid (Fig. 6.1). At a pH midway between the pK_a of the cationic

$$
\begin{array}{c}
\text{R} \\
| \\
\text{H}_3\overset{+}{\text{N}}\text{—C—COO}^- \\
| \\
\text{H}
\end{array}
$$

Zwitterionic form

$$
\begin{array}{c}
\text{R} \\
| \\
\text{H}_3^+\text{N—C—COOH} \\
| \\
\text{H}
\end{array}
$$

Cationic form

$$
\begin{array}{c}
\text{R} \\
| \\
\text{H}_2\text{N—C—COO}^- \\
| \\
\text{H}
\end{array}
$$

Anionic form

and pK_a of the anionic form is the isoelectric point (pI) of the amino acid, at which it carries no net charge, being zwitterionic.

The other functional groups of amino acids that contribute to their

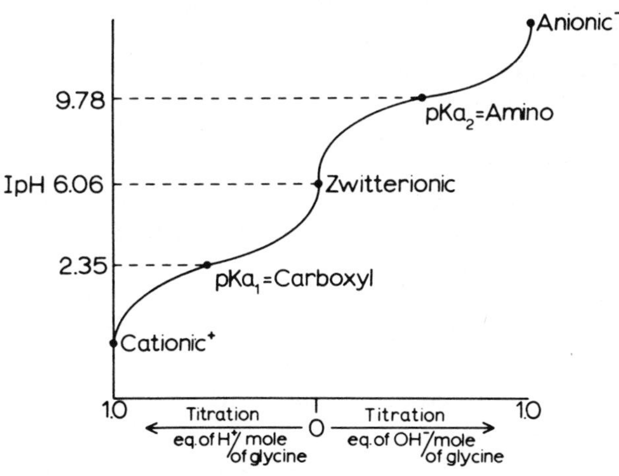

Figure 6.1. Titration curve for glycine.

acid–base properties are the following:

- Sulfhydryl of cysteine (pK_a = 8.39)
- β-Carboxyl of aspartic acid (pK_a = 3.87)
- γ-Carboxyl of glutamic acid (pK_a = 4.27)
- Guanidino of arginine (pK_a = 12.48)
- ε-Amino of lysine (pK_a = 10.53)
- Phenolic hydroxyl of tyrosine (pK_a = 10.03)
- Imidazole of histidine (pK_a = 6.04)

Titration curves of these amino acids have four inflection points, rather than the three associated with the glycine curve.

Color Reactions

Generally speaking, the color reactions of proteins or amino acids can be divided into two classes: those that involve specific reactive side chains and general reactions that occur at the amino or carboxyl groups of amino acids or at peptide bonds. Several of the color reactions are used for the qualitative and quantitative estimation of proteins and of their constituent amino acids.

The phenolic side chain of tyrosine is responsible for the red color produced by the Millon reagent and the blue color given in an alkaline solution of phosphomolybdotungstate (Folin–Ciocalteau reagent). The xanthoproteic reagent gives a yellow color with proteins containing amino acids having a benzene ring (tryptophan, phenylalanine, and tyrosine). The indole side chain of tryptophan reacts with the Hopkins–Cole reagent to give a purple color and with 2-hydroxy-5-nitrobenzylating reagents to incorporate a yellow color. The guanidino side chain of arginine reacts with α-naphthol and sodium hypochlorite (Sakaguchi test) to give a red color.

Probably the most widely used color reaction is that of ninhydrin with the free amino group of amino acids to produce various colors. The ninhydrin test may be used for the qualitative detection of proteins and their hydrolytic products and for the quantitative determination of separated amino acids. Another very common test for protein is the biuret reaction. This is a test for peptide linkages and is positive when two or more peptide linkages are present; thus, a dipeptide does not react with the biuret reagent.

The Millon, biuret, and ninhydrin reactions are described in detail in the Analysis section.

Metabolic Roles

As indicated already, the primary metabolic function of amino acids is as constitutents of proteins. In addition, however, amino acids occur in a number of important nonprotein compounds. For example, glycine is incorporated into the porphyrin portion of hemoglobin and the cytochromes; the hormones adrenaline, and thyroxine are formed from tyrosine; glutathione, an important cellular constituent, is a tripeptide composed of glutamate, cysteine, and glycine; and the posterior pituitary hormones vasopressin and oxytocin are oligopeptides, which each contain nine amino acid residues. The nucleotides are synthesized from amino acids such as glycine, aspartate, and glutamine.

□ CLASSIFICATION OF PROTEINS

There are many ways of classifying proteins based on either their chemical or physical properties. A uniform system of classification has evolved in which proteins are divided into three main groups: simple, conjugated, and derived proteins. The first two groups represent proteins as they exist in nature; the third group contains proteins that have been changed or modified by enzymes, chemical reagents, or physical agents.

Simple Proteins

Substances that on hydrolysis yield only amino acids or their derivatives are classified as simple proteins. Some of the major types of simple proteins are described briefly in this section.

Albumins are soluble in water and salt solutions, and are coagulated by heat. Common examples are egg albumin, lactalbumin, serum albumin, legumetin of peas, and leucosin of wheat.

Globulins are soluble in neutral solutions of salts of strong bases and strong acids but are only sparingly soluble in water. Serum globulin (blood), myosin (muscle), edestin (hemp seed), and legumin (peas) are typical globulins.

Glutelins are soluble in very dilute acid or base but insoluble in neutral solvents. These proteins, which occur only in plant materials, include glutelin (wheat) and oryzenin (rice).

Prolamins are soluble in 50–90% alcohol but insoluble in water, absolute alcohol, and other neutral solvents. The prolamins, like the glutelins, occur only in plants. Common prolamines are gliadin (wheat, rye), zein (corn), and hordein (barley).

Scleroproteins are characterized by their insolubility in neutral solvents. They are found only in animals. Collagen and keratin in connective tissue, bones, hair, horns, hoofs, etc., and fibroin in silk are examples of scleroproteins.

Histones, which contain a high content of basic amino acids, are soluble in water, dilute acids, and alkali but are insoluble in dilute ammonia. They are found only in animals (e.g., histones from calf thymus and pancreas.

Protamines, like histones, are strongly basic but have lower molecular weights than histones. They are soluble in water and ammonia, form stable salts with strong acids, are highly basic, and are not coagulated by heat. The protamines are rich in the basic amino acid arginine, which may account for 70–80% of the total amino acid content. The protamines occur in combination with nucleic acid in spermatazoa of salmon, herring, and sardines.

Conjugated Proteins

The structure of conjugated proteins is more complex than that of simple proteins. Molecules of complex proteins, in addition to having the typical amino acid structure of a simple protein, are combined with one or more nonprotein substances such as lipids, carbohydrates, organic prosthetic groups, and nucleic acids.

Nucleoproteins contain nucleic acids, which are strongly acidic polymers, combined with a protein moiety. Common protein components are basic proteins of the histone and protamine classes. Nucleoproteins make up the chromatin material of cell nuclei.

Lipoproteins are conjugates of protein with lipids (e.g., lecithin or cholesterol). They are found in cell nuclei, blood, egg yolk, milk, brain and nerve tissue, and cellular and intracellular membranes.

Glycoproteins contain carbohydrates as the nonprotein moiety. They are involved in intercellular contact and recognition responses.

Chromoproteins contain an organic colored prosthetic group. An example is flavoprotein in which the prosthetic group is a derivative of riboflavin. Other protein-bound pigments include chlorophyll, heme, and ferritin.

Derived Proteins

The derived proteins include all the products resulting from the partial hydrolysis of naturally occurring proteins, as well as those resulting from the action of physical agents (in particular, heat). The derived proteins can be classified according to the degree of change they have undergone: those that have been changed least are called primary derivatives; those that have undergone considerable change are known as secondary derivatives.

Primary derivatives include proteans, metaproteins, and coagulated proteins. Proteins that have been only slightly modified by water, dilute acids for alkalies, or enzymes, are classified as proteans. They are insoluble in water. Common examples are fibrin (coagulated blood), casein (curdled milk), and edestin. Metaproteins, which result from further hydrolysis, are soluble in acids and alkalies but not in neutral solvents.

Secondary derivatives include proteoses, peptones, and peptides. Proteoses are soluble in water and not coagulated by heat but are precipitated by saturated solutions of ammonium sulfate. Simpler products of protein hydrolysis are peptones. They are soluble in water, not coagulated by heat, and not precipitated by saturated solutions of ammonium sulfate. The peptides, the simplest derived proteins, are relatively low-molecular-weight combinations of two or more amino acids.

☐ STRUCTURE OF PROTEINS

Because of the complexity and high molecular weights of proteins, their structure is conveniently considered at several levels. The *primary structure* of proteins is determined by the ordered sequence of amino acids linked by covalent peptide bonds. The *secondary structure* of a protein is the three-dimensional arrangement in space that it assumes because of hydrogen bonding between peptide bonds. Many proteins have additional, *tertiary,* structure due to interactions of the amino acid side chains. Finally, *quarternary structure* refers to the association, in some proteins, of more than one polypeptide chain to form oligomeric molecules; the individual chains may be the same or different.

Primary Structure

Evidence for the peptide bond as the primary structural unit of all proteins is now unequivocal. For example, although proteins contain few titratable α-amino and α-carboxyl groups, the number of these groups increases markedly on hydrolysis. Moreover, amino acids can be isolated from protein hydrolysates, and polypeptides can be synthesized chemically from amino acids (or their derivatives). The simplest concept of a protein is that of a straight-chain (i.e., nonbranched) polymer consisting of several hundred amino acids possessing a variety of side chains. At one end of the polymeric chain is a free α-amino group, termed the N-terminal amino acid, and at the other end is a free α-carboxyl group, called the C-terminal amino acid. The physical and chemical properties of proteins depend not only on the peptide bonds but also on the side chains, which have a pronounced effect on the overall reactions and interactions of proteins.

It is obvious that if a peptide is described only in terms of its constituent amino acids, one can not be certain of the sequential order of the residues in the peptide structure. A useful method for detecting and identifying N-terminal amino acids is the Sanger reaction, in which 2,4-dinitrofluorobenzene reacts with the free amino groups of peptide molecules. Following acid hydrolysis of the treated peptide, the α-*N*-dinitrophenyl derivative of the N-terminal amino acid can be extracted with ether and identified by chromatographic procedures.

An end-group analysis, known as the Edman degradation, permits the sequential removal and identification of amino acids from the N-terminus of a protein molecule. In this method, the peptide is allowed to react with phenyl isothiocyanate in alkaline media to form an *N*-phenylthiocarbamyl peptide (PTC-peptide). Upon treatment with strong anhydrous acid, the N-terminal peptide bond is broken and the derivative cyclizes to form a thiazolone, which undergoes rearrangement to a phenylthiohydantoin of the terminal amino acid (PTH-amino acid); the cyclization is carried out in an anhydrous medium (e.g., hydrogen chloride in nitromethane) so as to prevent nonspecific acid hydrolysis of other peptide bonds. The phenylthiohydantoin may be identified directly by gas chromatography or thin-layer chromatography. The sequence of reactions on the opposite page are involved in the Edman degradation:

By careful control of the reaction conditions, the amino acids of a protein can be sequentially removed by the Edman degradation method. In this way, the primary structure—i.e., the sequence of amino acid residues linked by peptide bonds—can be determined. By convention, this sequence is written starting from the N-terminal amino acid and proceeding to the C-terminal acid. For example, in a tripeptide whose primary structure is represented by Ala→Leu→Ser, alanine (Ala) is the N-terminal acid and serine (Ser) is the C-terminal acid.

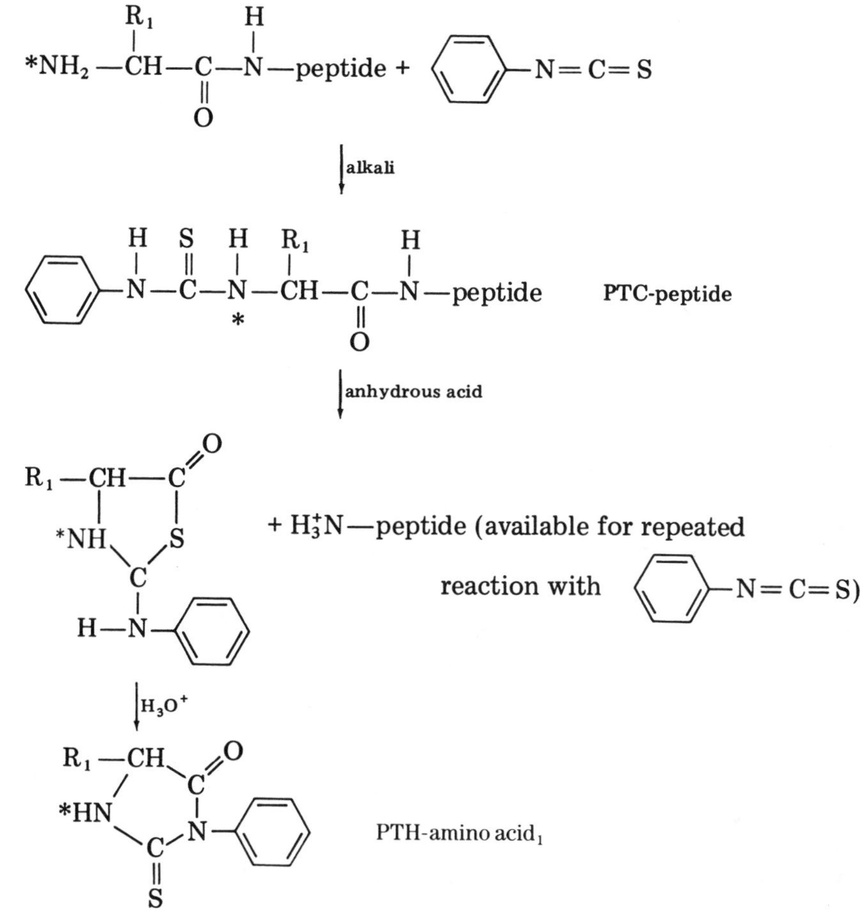

$$^{*}NH_2-\underset{R_1}{CH}-\underset{\|O}{C}-\underset{H}{N}-\text{peptide} + C_6H_5-N=C=S$$

↓ alkali

$$C_6H_5-\underset{H}{N}-\underset{\|S}{C}-\underset{H}{\overset{*}{N}}-\underset{R_1}{CH}-\underset{\|O}{C}-\underset{H}{N}-\text{peptide}$$ **PTC-peptide**

↓ anhydrous acid

$R_1-CH-C(=O)$, *NH, S, C, H—N—C_6H_5 (thiazolinone ring) + H_3^+N—peptide (available for repeated reaction with $C_6H_5-N=C=S$)

↓ H_3O^+

R_1-CH, C=O, *HN, C=S, N—C_6H_5 (phenylthiohydantoin ring) PTH-amino acid$_1$

Secondary Structure

The chemical and physical properties of a protein depend on its three-dimensional molecular shape (conformation) as well as its amino acid sequence. Specific terms are used to discuss the several levels of structural organization of proteins and large peptides. The secondary structure of a protein refers to the spatial arrangement of its polypeptide chain(s) (extended, partially extended, or helically coiled) that is conferred by hydrogen bonding between peptide bonds.

The α helix (Fig. 6.2) is the most stable of the various secondary structures that certain polypeptide chains (depending on the nature of the individual residues) might assume. There are approximately 5.4Å

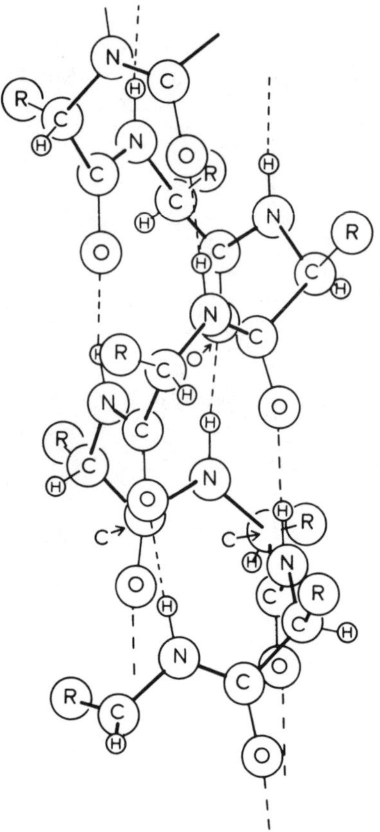

Figure 6.2. αHelix form of a protein molecule (right-handed).

per turn of the helix, with each amino acid residue contributing approximately 1.5 Å translation along the helical axis. This gives 3.6 amino acid residues per turn and, thus, 100° rotation per residue.

The secondary conformations of peptide chains are stabilized by hydrogen bonds between the peptide linkages—i.e., the attractive force between a hydrogen atom attached to the N of one peptide bond and the carbonyl O of another peptide bond. An isolated hydrogen bond is relatively weak compared with a covalent bond, but there may be hundreds of intrachain hydrogen bonds in a protein molecule, thereby resulting in overall strength. Agents such as heat, acids, bases, urea, sodium dodecylsulfate, and salts can interfere with hydrogen bonding, thus affecting the stability of the secondary conformation of proteins.

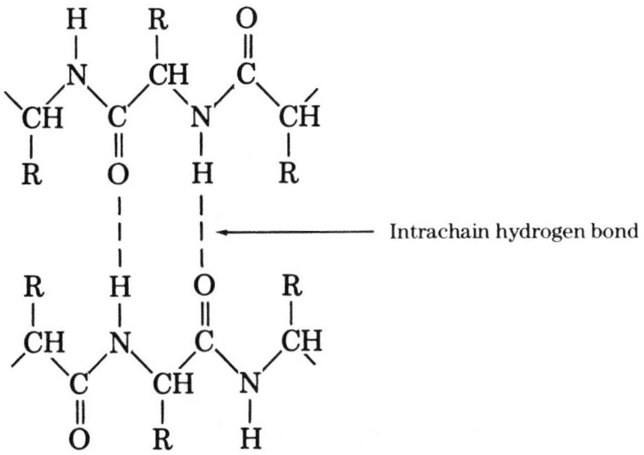

Intrachain hydrogen bond

Hydrogen bonding is particularly prevalent in fibrous proteins. Two such proteins are keratins (hair and wool) and myosin (a protein of muscle tissues). These proteins can exist in either the alpha form (helical, contracted form) or the beta form (stretched, pleated sheet form). Silk fibroin, on the other hand, naturally occurs in the pleated sheet secondary structural conformation (beta form).

Tertiary Structure

Protein molecules containing only hydrogen bonds may be extremely flexible. A stable, globular three-dimensional structure requires other forces in order to compress the polypeptide chain into a more or less spherical form. The forces that contribute to the folding and stabilization of the coiled chain (tertiary structure) are attributable to the properties of the side chains of the amino acid residues. Thus, tertiary structure may be imposed by a combination of various interactions: (1) hydrophobic interactions, which constitute the major force stabilizing tertiary structures of all proteins; (2) disulfide bonds, present in some but not all proteins; (3) hydrogen bonds between polar side chains (relatively infrequent); and (4) ionic bonds (salt linkages), which are also infrequent.

The basic amino acids (lysine and arginine) and the dicarboxylic amino acids (aspartic acid and glutamic acid) are in certain instances involved in salt linkages, but they are usually hydrated by solvent. Cysteine is the sole source of sulfhydryl groups to form disulfide bonds, which are covalent linkages. A number of amino acid residues have side chains made up of aliphatic or aromatic groups; these residues (e.g., leucine, valine, phenylalanine, and methionine) tend to be hydrophobic. When placed in an aqueous environment, the hydrophobic side chains are repelled by the polar water molecules and tend to associate together. Side-chain hydrogen bonding may occur between aspartic acid, glutamic acid, asparagine, glutamine and the phenolic hydroxyl group of tyrosine.

Quaternary Structure

Quaternary structure is the association of individual polypeptide chains (monomeric units) via noncovalent bonding to form an oligomeric protein molecule. Hemoglobin, for example, consists of four subunits—2 α and 2 β chains—each chain having a molecular weight of about 16,000, and thus possesses quaternary structure. Myoglobin, on the other hand, consists of a single polypeptide chain (molecular weight about 16,000) and thus lacks quaternary structure. Quaternary structural organization also occurs in certain enzymes that are made up of subunits. For example, the enzyme glycogen phosphorylase contains four subunits, which alone are catalytically inactive but when linked as a tetramer form the active enzyme.

Denaturation

Denaturation involves an unfolding or other significant alteration of the folded structure of a protein without breaking its covalent peptide bonds. Denaturation may be caused by heat, acid, alkali, or a variety of other chemical and physical agents.

Protein denaturation can affect both the flavor and texture of foods containing proteins and thus is an important consideration in the processing of certain foods. For example, in the dairy industry, the deciding factor in the choice of a particular temperature treatment is made on the basis of chemical changes produced by the heat treatment. Thus, temperatures in excess of 72°C for 15 sec in the pasteurization of milk are avoided; otherwise the milk serum (or whey proteins) becomes denatured and the milk acquires a cooked flavor and has a resistance to clotting by rennet. Such changes are of practical significance in the production of cheese.

The denaturation of proteins does not necessarily imply a decrease in their digestibility by proteolytic enzymes. Denaturation increases the digestibility of egg albumin by trypsin. Native hemoglobin is not digested by trypsin, whereas denatured hemoglobin is readily digested.

☐ MOLECULAR WEIGHT AND ISOELECTRIC POINT OF PROTEINS

Molecular Weight

Because proteins are large complex substances, it is extremely difficult to determine their molecular weights, and several methods for doing so are used.

When solutions of proteins are centrifuged, the rate of sedimentation of the protein molecules is determined by the molecular weight, shape, and density of the molecule, by the density and viscosity of the dispersion medium, and by the centrifugal force developed in the ultracentrifuge. Based on experience and theoretical considerations, a technique was developed to calculate the molecular weight of a protein from its sedimentation rate in an ultracentrifuge. This ultracentrifugal method has proven to be the most generally satisfactory method for the determination of the molecular weights of proteins. In addition to determining molecular weight, ultracentrifugation can be used to determine the purity of a protein solution because molecules of different molecular size sediment at different rates giving multiple boundaries.

Molecular weights can also be calculated from light scattering. This method is based on the fact that protein solutions opalesce. A measurement of the intensity of light scattered (sidewise) by a known weight of protein in solution allows computation of its molecular weight. Gel filtration chromatography also has been used extensively for determining the molecular weights of proteins. This method is simple and inexpensive, but its accuracy depends upon the particular protein being investigated.

Table 6.2 lists the molecular weights of some typical proteins. Most of these values were determined by the ultracentrifugal method.

A number of naturally occurring proteins exist in different micellar or aggregate forms. Casein has been shown to be a complex mixture of molecules varying in molecular weight from 75,000 to 375,000. Insulin is an example of a protein that naturally exists as a molecular aggregate. Its molecular weight was first found to be approximately 36,000. However, upon dilution insulin dissociates to units of 12,000, which in turn are made up of two subunits, each with a molecular weight of 6,000.

Isoelectric Point

Since the charges of the α-carboxyl groups and the α-amino groups of internal amino acid residues are removed in the formation of peptide

Table 6.2. Approximate Molecular Weights of Some Typical Proteins

Protein	Source	Approximate molecular wt
Myoglobin	Muscle	17,200
Lactalbumin	Cow's milk	17,400
Gliadin	Wheat	27,500
β-Lactoglobulin	Cow's milk	35,400
Zein	Corn	40,000
Egg albumin	Hen	44,000
Gelatins	Skins	10,000–100,000
Caseins	Cow's milk	75,000–375,000
Myosin	Muscle	1,000,000
Nucleohistone (thymus)	Calf	2,300,000

bonds, the chemical nature of the side chains determines the overall charge of a protein molecule. Carboxyl groups in the side chains of aspartic and glutamic acid residues bear negative charges, while the amino groups of lysine and arginine residues bear positive charges at physiological pH.

The behavior of a protein toward acid or base is similar to the behavior of an amino acid. At a low pH, the H^+ ions protonate the side-chain functional groups of basic amino acids and the carboxyl groups of acidic amino acids, so that the protein is positively charged. At a high pH, the protons are removed from the functional groups of basic amino acids and acidic amino acids, so that the protein becomes negatively charged. At some intermediate pH, the protein is electrically neutral; that is, it possesses exactly the same number of positive and negative charges. This particular pH is called the isoelectric point (*pI*), and it has a characteristic value for each given protein. As a general rule, most of the physical properties of proteins are minimal at or near the isoelectric point. This is true for solubility, conductivity, degree of hydration, viscosity, osmotic pressure, stability, and electrokinetic phenomena.

□ METABOLISM OF PROTEINS

In animals, dietary proteins are digested, or broken down, by a number of proteolytic enzymes into their constituent amino acids, which are absorbed in the small intestine. The absorbed amino acids undergo one of two fates: (1) they are used to build endogenous proteins and other nitrogenous metabolites essential to the organism (nucleotides, nucleic acids, porphyrins, amino sugars, etc.) or (2) they are degraded further into carbon dioxide and water or other catabolites with the elimination of nitrogen either in the form of urea or uric acid.

Proteolysis

The enzymes that catalyze protein hydrolysis to give peptides and amino acids are called proteolytic enzymes and the process of protein fragmentation is called *proteolysis*. Proteolytic enzymes may be divided into two groups, endopeptidases and exopeptidases. *Endopeptidases* attack interior, as well as terminal peptide bonds, of polypeptide chains, and are also referred to as proteases. Examples of this group of enzymes are pepsin, elastase, trypsin, and chymotrypsin (from mammals) and papain and ficin (from plants). *Exopeptidases* are enzymes that attack only the terminal peptide bonds of peptide chains so as to sequentially remove amino acids. Carboxypepidases act on the carboxyl-terminal peptide bonds of oligopeptides (tripeptides and longer) and polypeptides, and aminopeptidases act on the amino-terminal peptide bonds; there are also dipeptidases whose action is limited to dipeptides (i.e., C-terminal and N-terminal).

Many of the proteolytic enzymes have a relatively high degree of specificity although several can hydrolyze esters of L amino acids. Denatured proteins are attacked more readily than are native proteins. The proteases cleave the peptide chain only at specific points; i.e., some enzymes act selectively on peptide bonds of basic amino acyl residues, others act on aromatic or acidic residues. For example, trypsin selectively hydrolyzes peptide bonds whose carbonyl group is contributed by lysine or arginine. The specificity of α-chymotrypsin is somewhat less. Its activity is directed towards the cleavage of peptide bonds in which the carbonyl group is derived from phenylalanine, tyrosine, or tryptophan, or residues with long aliphatic side chains. Pepsin's specificity is even broader (less selective) than that of chymotrypsin.

Pepsin, the proteolytic enzyme in gastric juice, initiates the breakdown of protein entering the digestive tract of monogastric animals. This enzyme is synthesized in the form of an inactive enzyme precursor (zymogen) by the chief cells of the stomach mucosa. The zymogen, known as pepsinogen, is converted to pepsin by the action of hydrochloric acid secreted by parietal cells of the gastric mucosa, and once a small amount of pepsin is formed, it acts on additional pepsinogen molecules to generate additional pepsin. Such a process is known as *autocatalysis*. The optimum pH for peptic activity is at the normal acidity of gastric juice (pH 1.5–2.5), and the products of hydrolysis are mixtures of polypeptides.

As the contents of the stomach pass into the small intestine, they are neutralized by the pancreatic juice, which contains the zymogens (trypsinogen, three chymotrypsinogens, proelastase, and two procarboxypeptidases). Trypsinogen is activated by enterokinase, an enzyme present in the intestine, to give trypsin. Once trypsin is formed, it catalyzes the further transformation of trypsinogen into active trypsin by autocatalysis; trypsin also catalyzes activation of the other pancreatic zymogens, namely, the chymotrypsinogens and procarboxypeptidases.

The intestinal mucosa furnishes two exopeptidases: aminopeptidase, which hydrolyzes N-terminal peptide bonds, sequentially; and dipeptidase, which hydrolyzes dipeptides to form free amino acids.

Thus, through the combined actions of proteolytic enzymes, food proteins are ultimately broken down into their constituent amino acids, which are absorbed and utilized. Absorption is not simply a diffusion process, since L amino acids are absorbed more readily than are the D isomers.

Large molecules such as polypeptides and even proteins can be absorbed only to a limited extent under certain conditions. If undigested or partly digested proteins are repeatedly injected into an animal's blood stream, the animal may undergo a severe shock or even die after reinjection of the same protein. Similarly, if the absorption process of a person is such that it allows polypeptides or proteins to be absorbed, shock or death may occur upon eating of certain food proteins. Asthma, eczema, and hayfever are hyperallergenic symptoms known to be intensified by certain protein foods.

Proteolysis is not limited to the gastrointestinal tract. There are intracellular proteases that digest necrotic tissue. These enzymes, called cathepsins, are usually localized in intracellular sacs called lysosomes; as long as the cell is undamaged, the lysosomes remain intact and the cathepsins are isolated. Cell damage or cell death results in the destruction of the lysosomal membrane and release of the intracellular proteases with subsequent autolysis of the cell tissue. The increase in tenderness of beef following storage at temperatures just above freezing for 1–4 weeks is attributed to the catheptic enzymes present in the muscle. Today, proteolytic enzyme preparations (papain, bromelain, microbial proteases, etc.), added in the form of powders, are routinely used to increase the tenderness of meat, especially beef.

Microorganisms must frequently fragment proteins in their media; they accomplish this by secreting extracellular proteolytic enzymes. In general, the proteases and peptidases from microorganisms hydrolyze a wider variety of bonds than do the individual proteolytic enzymes of the gastrointestinal tract. The proteolytic enzymes of microorganisms are used extensively in the ripening of special cheeses (i.e., Liederkranz, Limburger, Camembert, and Cheddar).

Rennin is the proteolytic enzyme found in the gastric juice of calves. Its substrate is the κ-casein of milk, which is converted by the enzyme to insoluble *para*-κ-casein. The casein micelle can also be hydrolyzed by pepsin, trypsin, and other proteases. The formation of paracasein is an important step in the manufacture of cheeses.

Protein Synthesis

Living organisms contain many different kinds of proteins, which serve as catalysts in cellular reactions, as structural materials, and as antibodies for the defense of the body against intrusion of foreign organisms. The mechanism of protein synthesis must provide not only for the synthesis of the peptide bonds from amino acids, but also for the synthesis of specific protein structures (i.e., the ordered sequence of specific amino acids to give the primary structure of a protein). Furthermore, the primary structure of a given protein is species specific. In other words, even proteins that have the same function (e.g., α-amylases, which hydrolyze starch) have slightly different primary structures in different organisms. For example, the primary structures of α-amylases from humans, monkeys, and horses are not identical but are characteristic of the species from which the protein is derived. Thus, the primary structures of the proteins synthesized by an organism are genetically determined, and the process of protein synthesis is involved with the cell's genetic material—deoxyribonucleic acid.

Three of the most important macromolecules in all microorganisms, plants, and animals are deoxyribonucleic acids (DNA), ribonucleic acids (RNA), and proteins. These substances, either directly or indirectly, control essentially all life processes. The involvement of DNA in genetic duplication is one of the outstanding discoveries of all times. Subsequent studies of the involvement of DNA and RNA in protein synthesis have opened new avenues of insight into chemotherapy, nutrition, and a host of other biochemical phenomena. Before discussing the process of protein synthesis in detail, we examine the structure and function of the nucleic acids.

Structure and Function of DNA and RNA

Both DNA and RNA are made up of subunits called nucleotides, which contain a purine or pyrimidine base, a sugar (either D-ribofuranose or 2-deoxy-D-ribofuranose), and phosphate (esterified on the 5′-hydroxyl of the sugar molecule). The sugar is attached to either the N-1 position of pyrimidine moieties or to the N-9 position of purine moieties via β-*N*-glycosidic linkages, as shown in **6.1** and **6.2**.

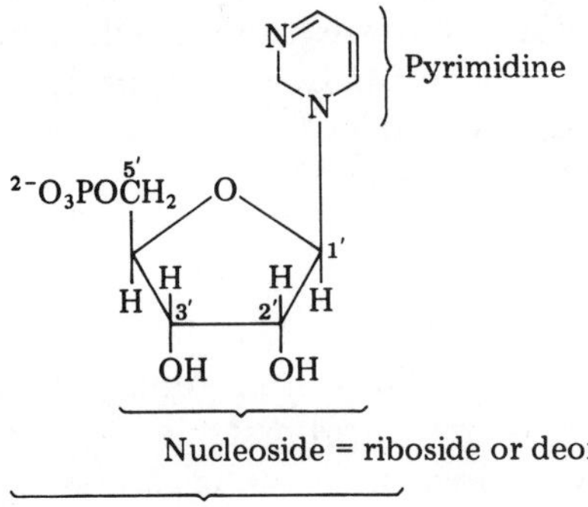

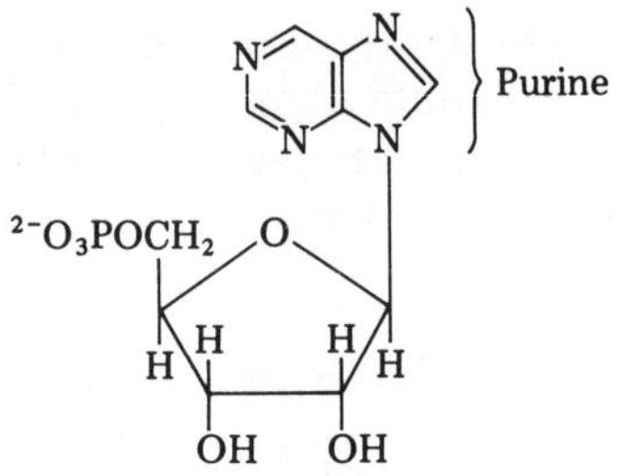

Pyrimidine

Purine

Nucleoside = riboside or deoxyriboside of purine or pyrimidine

Nucleotide = phosphate ester of nucleoside

6.1 **6.2**

Free purine or pyrimidine, or their nucleosides or nucleotides, are not found in nature; rather, purines and pyrimidines involved in living cells all contain one or more nitrogen and/or oxygen substituents.

DNA Deoxyribonucleic acids are found primarily in the nucleus of eucaryotic cells and are responsible for the "passing on" of genetic characteristics of a parent cell to the daughter cell. The structure of DNA is the well-known double helix, which consists of two strands of polydeoxyribonucleotide. The monomeric deoxyribonucleotides are predominantly deoxyadenylate (**6.3**), deoxyguanylate (**6.4**), deoxycytidylate (**6.5**), and deoxythymidylate (**6.6**).

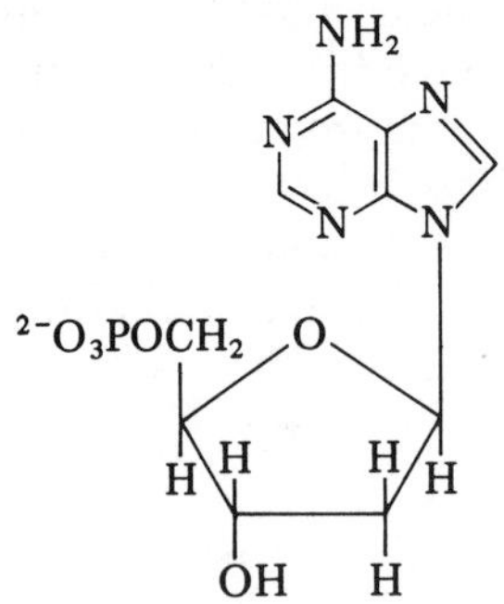

2′-Deoxyadenosine-5′-monophosphate (deoxyadenylate; dAMP)

6.3

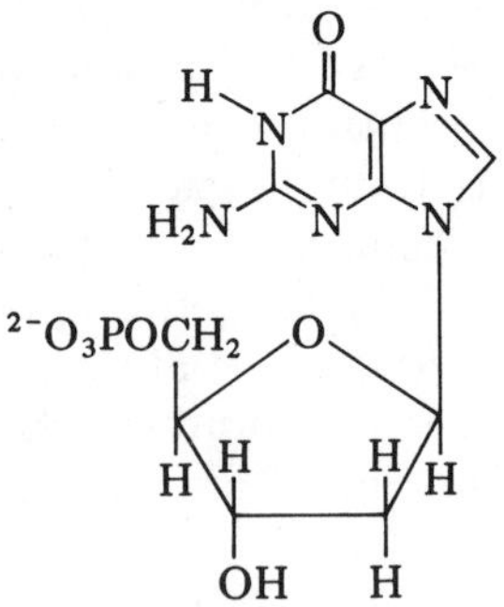

2′-Deoxyguanosine-5′-monophosphate (deoxyguanylate; dGMP)

6.4

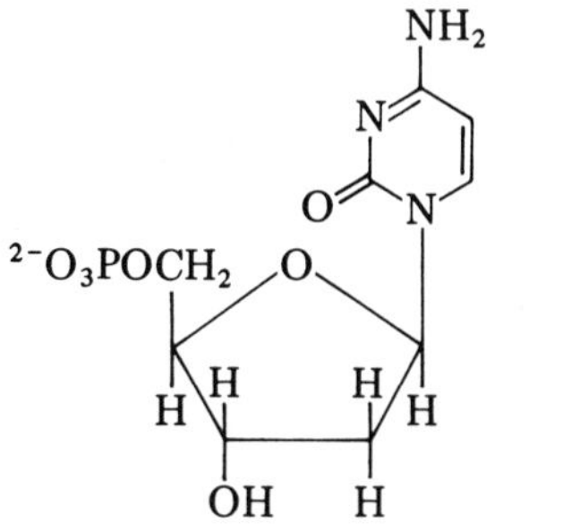

2′-Deoxycytidine-5′-monophosphate (deoxycytidylate; dCMP)

6.5

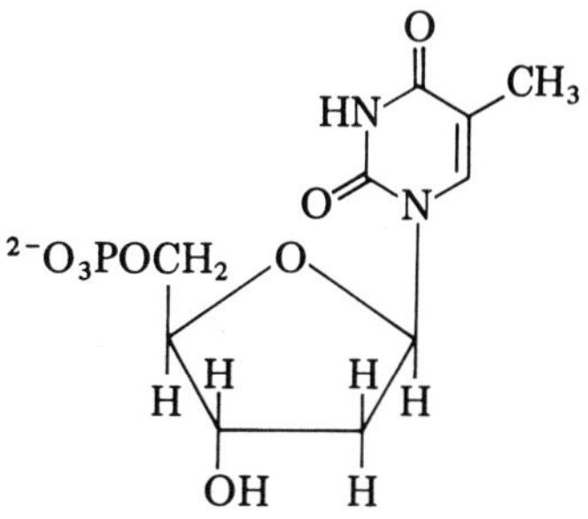

2′-Deoxythymidine-5′-monophosphate (deoxythymidylate; dTMP)

6.6

DNAs are polymers of these deoxyribonucleotides linked by phosphodiester bonds between the 3′-oxygen of one residue and the 5′-oxygen of the adjacent residue. The basic DNA structure is represented as follows:

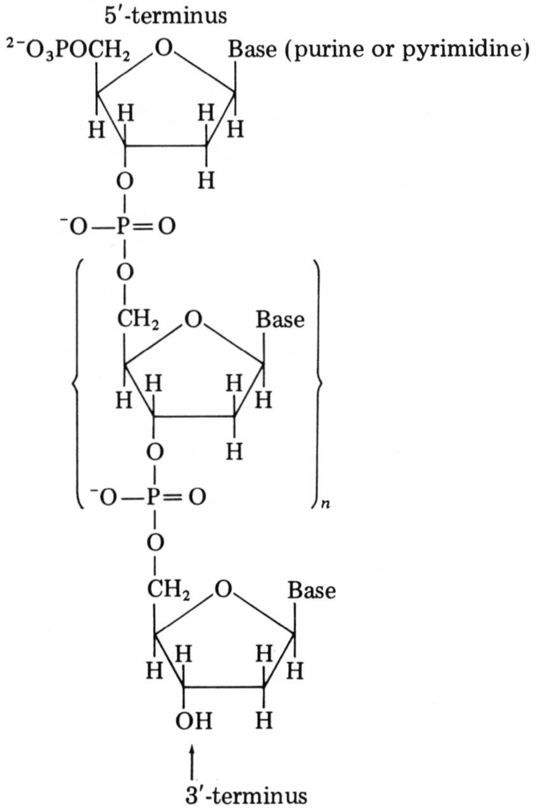

5′-terminus

$^{2-}O_3POCH_2$ Base (purine or pyrimidine)

3′-terminus

The double-stranded helical form of DNA arises from hydrogen bonding between complementary base pairs. Each adenine of one polydeoxyribonucleotide chain hydrogen-bonds to thymine of the other chain (**6.7**), and each guanine of one chain pairs with cytosine of the other chain (**6.8**). The ratios of adenine:thymine and of the guanine:cytosine in double helical DNA are thus one. This 1:1 ratio is a consequence of the complementary base pairing.

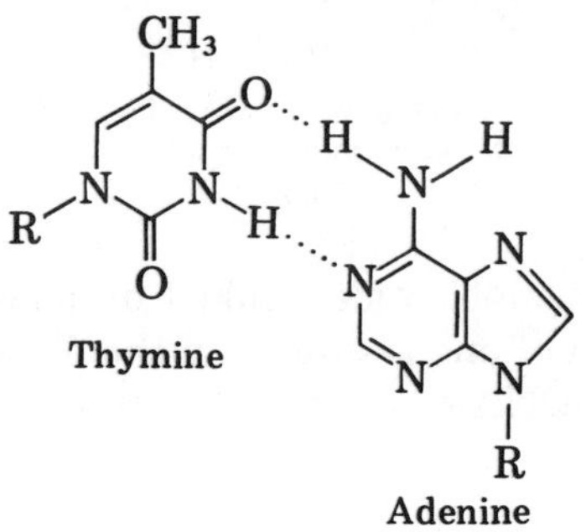

Thymine Adenine

6.7

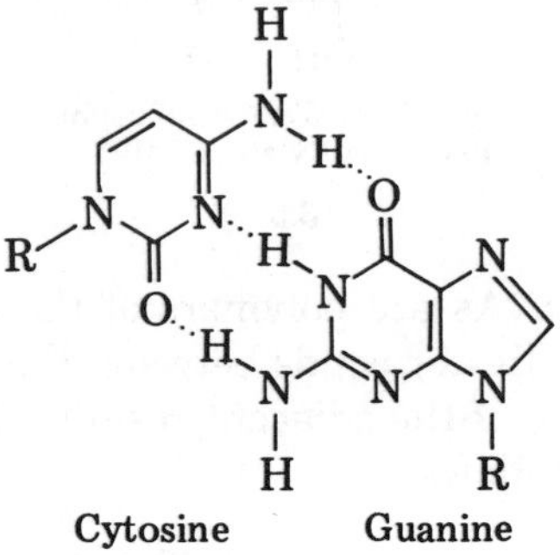

Cytosine Guanine

6.8

The DNA double helix thus consists of two polydeoxyribonucleotide strands coiled around a central axis. The strands are of opposite polarity, i.e., one strand has its 5′ terminus hydrogen-bonded to the other strand's 3′ terminus, and vice versa. There are 10 nucleotide pairs per turn of the right-handed helix with a translation of approximately 34 Å, giving 3.4 Å per base pair. The bases are arranged in a stacked fashion such that their planes are perpendicular to the axis of the DNA helix. The helix has two grooves, one shallow (minor) and one deep (major) as shown in Fig. 6.3.

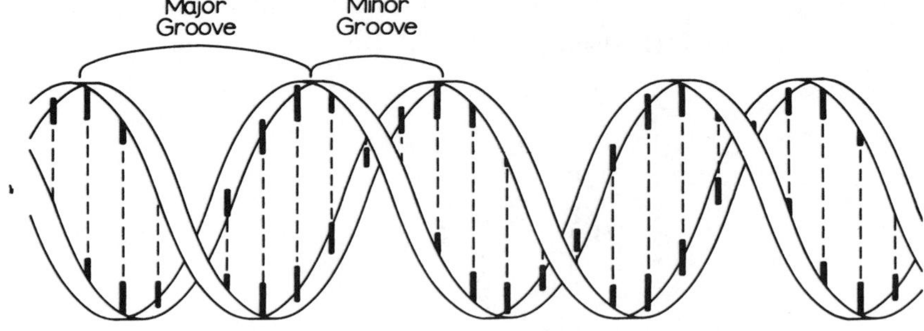

Figure 6.3. Double-helix structure of DNA.

The most notable feature of the DNA molecule is the ability of each strand of the double helix to serve as a template for replication of the complementary strand. This ability, correctly referred to as *semiconservative replication,* can be represented as shown in Fig. 6.4. This molecular property provides the chemical basis for the genetic continuity of biological systems.

The two strands of a "parent" DNA molecule separate and are duplicated, producing two double-stranded molecules, each identical to the original molecule. Thus, each daughter cell contains the same quantity and quality of genetic material as did the parent cell. These molecules can in turn replicate by the same mechanism for subsequent cell division.

Polymerization of nucleotides to form DNA is catalyzed by the enzyme DNA polymerase. The substrates for this enzyme are deoxynucleoside 5′-triphosphates (dATP, dGTP, dCTP, and dTTP). The structure of dATP is shown below. As in the case of virtually all nucleoside

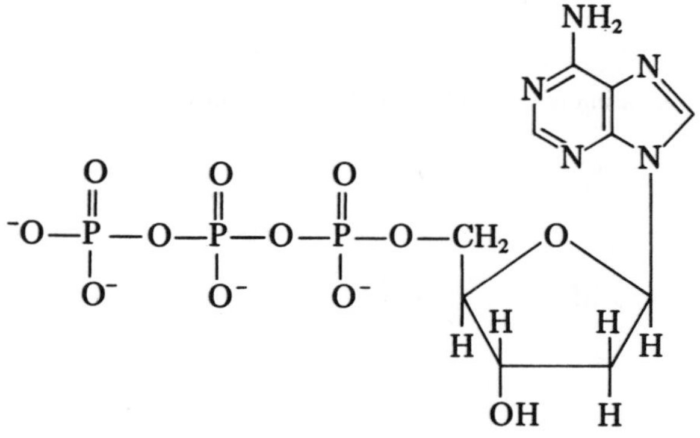

2′-Deoxyadenosine-5′-triphosphate (dATP)

triphosphate metabolic reactions, Mg^{2+} is required. The enzyme also requires an existing polynucleotide, which serves as a template. The reaction can be represented as follows:

$$n(\text{deoxynucleoside triphosphates}) + \text{DNA template strand from parent molecule} \xrightleftharpoons{Mg^{2+}} (\text{deoxynucleoside monophosphate})_n\text{-DNA template strand} + nPP_i$$

It should be realized that Fig. 6.4 is an oversimplification. Both strands of the double helix appear to be duplicated simultaneously,

Figure 6.4. Duplication of double helix of DNA. Parent DNA is shown as dark lines.

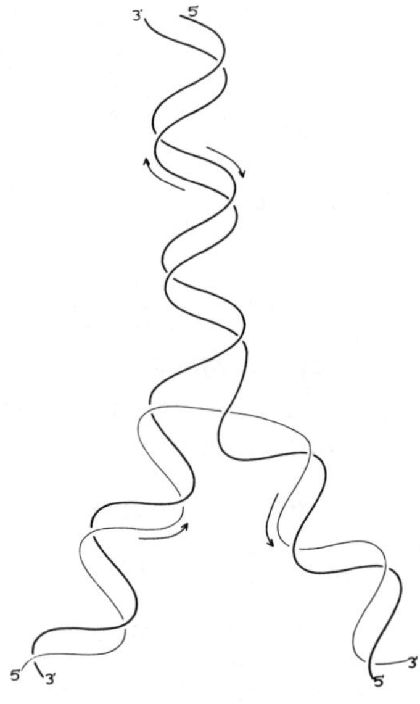

Figure 6.5. Duplicating sequence of DNA molecule. Arrows indicate polarities.

rather than splitting and then duplicating. DNA polymerase, which builds a chain in the 5′→3′ direction, is augmented by another enzyme, polynucleotide ligase, which completes the chain by catalyzing covalent bond formation between segments of DNA polymerase-synthesized chains. The partially completed duplication is shown in Fig. 6.5. The double strands of the parent molecule (dark lines) are of opposite polarity, as indicated by the arrows. Duplication begins at the 3′ end of the parent chain and proceeds along this chain adding deoxynucleoside monophosphate units having 5′-3′ phosphodiester linkages (light line). Ultimately, the duplicating chain "crosses over" to the other parent strand and moves along this chain from the 3′ to the 5′ end as in the case of the previous parent chain. It should be noted that the direction is the same in duplication of both parent strands, thus the final daughter molecules have the same polarity as the original parent molecules.

RNA. Although ribonucleic acids are found mainly in the cytoplasm, most RNA appears to be formed in nuclei of eucaryotic cells, e.g., liver. The three major classes of RNA—messenger RNA (mRNA), transfer RNA (tRNA), and ribosomal RNA (rRNA)—have different functions in the cell.

The most important features that distinguish RNA from DNA are the following:

- RNA has D-ribofuranose as the component sugar, whereas DNA has 2-deoxy-D-ribofuranose.
- RNA generally contains uracil in place of thymine, which is present in DNA. Both RNA and DNA also contain adenine, guanine, and cytosine. The structure of uridylate, the ribonucleotide containing uracil, is represented by **6.10**.

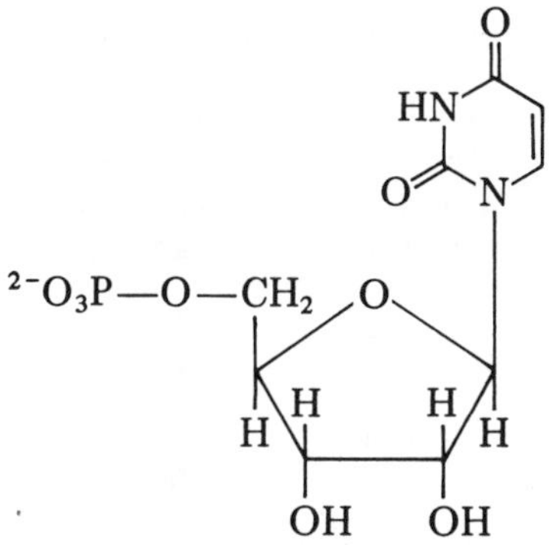

$^{2-}O_3P—O—CH_2$ O HN O N O H H H H OH OH

Uridine-5′-monophosphate
(uridylate; UMP)

6.10

- The RNA strands do not appear to associate to form a double helix as does DNA, except in certain viruses that contain double-stranded RNA. However, intrachain base pairing does occur in RNA.

Messenger RNA is synthesized by enzymatic reactions in a process referred to as *transcription*. The transcription process is dictated by a strand of DNA which, through hydrogen bonding of complementary bases, determines the sequence of ribonucleotides in the mRNA strand that is being formed. The sequence of the mRNA chain determines, ultimately, the order in which amino acids are linked when a polypeptide chain is synthesized. Each sequence of three adjacent nucleotide residues in the mRNA chain beginning at or near its 5′ terminus is called a *codon*. These three-residue units (i.e., triplet codons) determine which of the tRNA species (carrying activated amino acids) attach to the mRNA. Three complementary bases in each tRNA species are called *anticodons*. The interrelationship of codons, anticodons, and their role in protein synthesis is discussed in the next section. A list of the codons for each of the 20 activated amino acids used for protein biosynthesis is presented in Table 6.3.

Table 6.3. Genetic Code: Codons of mRNA (5′ to 3′)[a]

	U	C	A	G
	UUU Phe	UCU Ser	UAU Tyr	UGU Cys
	UUU Phe	UCC Ser	UAC Tyr	UGC Cys
U				
	UUA Leu	UCA Ser	UAA Ochre	UGA Umber
	UUG Leu	UCG Ser	UAG Amber	UGG Trp
	CUU Leu	CCU Pro	CAU His	CGU Arg
	CUC Leu	CCC Pro	CAC His	CGC Arg
C				
	CUA Leu	CCA Pro	CAA Gln	CGA Arg
	CUG Leu	CCG Pro	CAG Gln	CGG Arg
	AUU Ile	ACU Thr	AAU Asn	AGU Ser
	AUC Ile	ACC Thr	AAC Asn	AGC Ser
A				
	AUA Ile	ACA Thr	AAA Lys	AGA Arg
	AUG Met	ACG Thr	AAG Lys	AGG Arg
	GUU Val	GCU Ala	GAU Asp	GGU Gly
	GUC Val	GCC Ala	GAC Asp	GGC Gly
G				
	GUA Val	GCA Ala	GAA Glu	GGA Gly
	GUG Val	GCG Ala	GAG Glu	GGG Gly

[a] Codons in blocks are termination signals.

Transfer RNAs play the important role of mobilizing activated amino acids and mediating their transfer to the proper positions along a mRNA molecule. Each of the 20 amino acids used in the synthesis of proteins has a corresponding tRNA; some of the amino acids have more than one species of tRNA. These tRNAs are named for the amino acid with which they react. For example, the first tRNA whose complete covalent structure was determined was alanyl tRNA isolated from yeast. The covalent structure of alanyl tRNA ($tRNA_{Ala}$) is shown in Fig. 6.6. The amino acid, in this case alanine, is esterified through its carboxyl group to the 2′- and 3′-terminal hydroxyl groups of its specific tRNA chain to form an aminoacyl-tRNA (e.g., alanyl-$tRNA_{Ala}$; i.e., $tRNA_{Ala}$ "charged" with alanine). This ester linkage represents an activated form of the carboxyl group, which can thus be transferred to the free α-amino group of an adjacent activated amino acid at the site of the mRNA.

In summary, then, the genetic information contained in the nucleo-

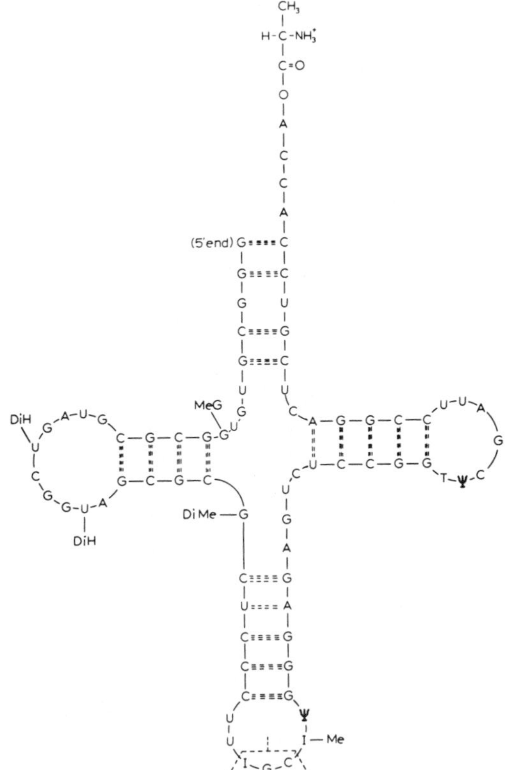

Figure 6.6. Alanine transfer RNA. Anticodon is shown inside dotted lined area. MeG = 1-methyl-guanosine; DiHU = 4,5-dihydrouridine; I = inosine; DiMeG = dimethylguanosine; MeI = 1-methylinosine; ψ = pseudoridine, and T = thymidine.

tide sequences of DNA is *transcribed* into mRNAs and ultimately *translated* into the primary structure of proteins via tRNAs.

Process of Protein Synthesis

The sites of protein biosynthesis are ribonucleoprotein particles called ribosomes. In addition to these intracellular structures, other factors essential for protein synthesis include transfer RNA, messenger RNA, aminoacyl-tRNA synthetases, ATP, Mg^{2+}, GTP, initiation factions, K^+, elongation factors, peptide synthetase, transfer factors, and release factors.

The various steps in protein synthesis can be summarized as follows:

1. Each amino acid is "activated" by ATP to form an aminoacyl-tRNA. These two-stage reactions are catalyzed by specific aminoacyl-tRNA synthetases (alanyl-tRNA synthetase, phenylalanyl-tRNA synthetase, etc.). For example:

$$CH_3—\underset{|}{\overset{NH_3^+}{CH}}—COO^- + ATP \xrightarrow[\text{Enzyme (Ala specific)}]{Mg^{2+}}$$

$$CH_3—\overset{NH_3^+}{\underset{}{CH}}—\overset{O}{\overset{\|}{C}}—AMP - \text{Enzyme} + PP_i$$

$$\downarrow tRNA_{Ala}$$

$$CH_3—\overset{NH_3^+}{CH}—\overset{O}{\overset{\|}{C}}—tRNA_{Ala} + \text{Enzyme} + AMP$$

2. As the ribosome attaches to the mRNA chain at or near its 5′ terminus, the aminoacyl-tRNA corresponding to the first codon of the mRNA chain is attached via the ribosome to the chain (Fig. 6.7). The initiation of this first step in most microorganisms is controlled by the first aminoacyl-tRNA attached to the mRNA. In bacteria *N*-formylmethionyl-tRNA is the initiating factor, and thus becomes the N terminus of the growing polypeptide chain.
3. A second aminoacyl-tRNA (AA_2tRNA_2) enters the ribosome at the mRNA codon adjacent to that which binds the first aminoacyl-tRNA (AA_1tRNA_1). This brings the two amino acids into close proximity and the elongation factor, GTP, and peptide synthetase promote the formation of a peptide bond with the release

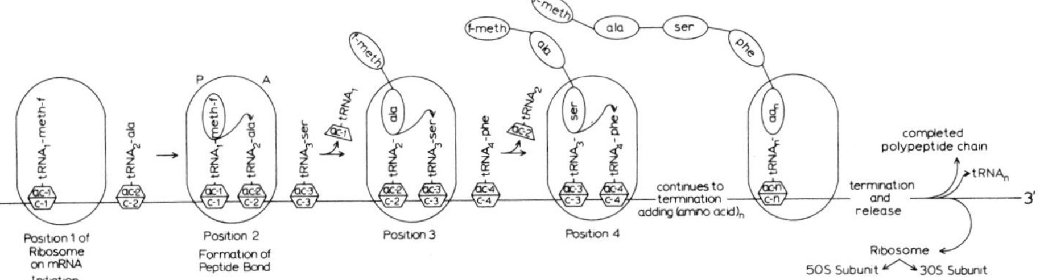

Figure 6.7. Sequence of protein synthesis. In diagram, ac = anticodon and c = codon.

of $tRNA_1$, leaving a dipeptidyl-$tRNA_2$ attached to the second codon of the mRNA chain.

4. Dipeptidyl-$tRNA_2$ remains attached to its codon and a new AA_3-$tRNA_3$ then attaches to the adjacent codon (in the direction proceeding from the 5′ end toward the 3′ end of the mRNA). Again GTP, the elongation factor, and protein synthetase produce a new peptide bond to yield a tripeptidyl-$tRNA_3$ with release of $tRNA_2$.
5. This process is repeated until the entire "message" of codons in the mRNA has been "translated"; at that point a terminating codon is reached on the mRNA chain. Thereupon, the release factor catalyzes the dissociation of the completed polypeptide chain from the terminal tRNA. The ribosome is then released from the mRNA and can be reused in the synthesis of additional polypeptide chains.

Metabolic Fate of Nitrogen

In a healthy adult animal, the breaking down of tissues (catabolism) and the building up of tissues (anabolism) are in dynamic equilibrium. Expressed in another way, the dietary nitrogen intake is equal to the amount of nitrogen excreted (in the urine and feces). During growth and convalescence from disease, increased rates of protein synthesis will depress the excretion of nitrogenous catabolites, and the animal is said to be in a state of *positive nitrogen balance.* Conversely, during starvation or wasting diseases, the reverse is true; i.e., the rate of catabolism exceeds that of anabolism. Under such circumstances, the animal is in a state of *negative nitrogen balance.*

Amino acids present at levels in excess of metabolic requirements for the synthesis of body proteins and essential nonprotein nitrogenous compounds can be catabolized for the production of energy. The amino group is generally removed from the carbon chain early in the catabolic process. Many of the resulting α-keto acids are shunted into the general metabolic pathways to be eventually oxidized to carbon dioxide and water, with the release of energy. In ureotelic animals, including man, the nitrogen appears in the urine in the form of urea, which is synthesized by an energy-requiring process from ammonia and carbon dioxide. There are three types of initial reactions involved in the catabolism of amino acids: (1) oxidative deamination to form α-keto acids; (2) transamination to form α-keto acids; and (3) decarboxylation to form primary amines.

Oxidative Deamination

In the oxidative deamination of an amino acid, two hydrogen atoms probably are removed to form an imino acid, which in turn is hydrolyzed to ammonia and a keto acid. The reaction is catalyzed by specific enzymes (L-amino acid oxidases), some of which have flavin adenine dinucleotide (FAD) as a prosthetic group to serve as the hydrogen acceptor.

$$\text{R—CH(NH}_2\text{)—COOH} \xrightarrow{-2\text{H}} \text{R—C(=NH)—COOH} \xrightarrow{+\text{H}_2\text{O}} \text{R—C(=O)—COOH} + \text{NH}_3$$

Transamination

The transamination of α-keto acids, which is possibly the most important general pathway for elimination of nitrogen, involves the transfer of an amino group from an amino acid to α-ketoglutaric acid, yielding the corresponding α-keto acid and L-glutamic acid. The transfer of the amino group is catalyzed by specific enzymes (transaminases) via the prosthetic group pyridoxal-5-phosphate. The α-ketoglutaric acid required for this reaction is readily available as one of the intermediaries of the tricarboxylic acid cycle.

$$\underset{\alpha\text{-Ketoglutaric acid}}{\text{HOOC—CH}_2\text{—CH}_2\text{—C(=O)—COOH}} + \underset{\text{Amino acid}_\text{R}}{\text{R—CH(NH}_2\text{)—COOH}} \rightarrow$$

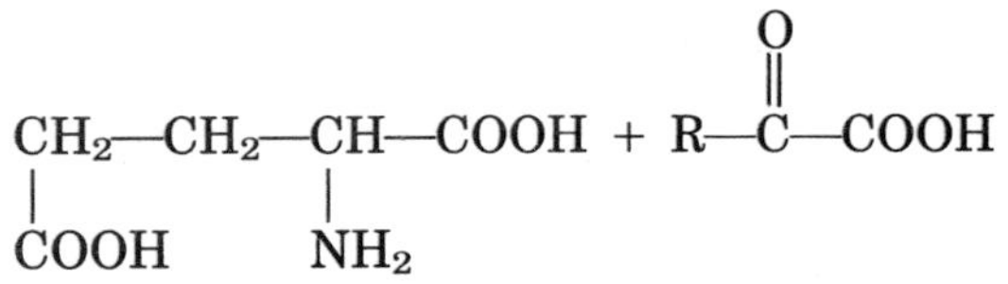

$$\underset{\displaystyle \text{COOH}}{\overset{|}{\text{CH}_2}}\text{—CH}_2\text{—}\underset{\displaystyle \text{NH}_2}{\overset{|}{\text{CH}}}\text{—COOH} + \text{R—}\overset{\displaystyle \text{O}}{\overset{\|}{\text{C}}}\text{—COOH}$$

L-Glutamic acid α-Keto acid$_R$

Most transamination reactions are metabolically reversible. Therefore, certain amino acids can be formed from the corresponding α-keto acids and L-glutamate. For example, L-alanine can be synthesized from pyruvate. However, each transaminase is highly specific with regard to amino acid substrates, and several amino acids (lysine, threonine, and others) are not transaminated. In addition, there are a number of amino acids for which the α-keto precursors or other suitable intermediates are not synthesized by an organism. As a consequence, these amino acids must be supplied to the organism; these so-called essential amino acids are discussed later in this chapter.

The amino group of the glutamic acid formed via transamination is subsequently removed through the action of glutamic acid dehydrogenase, which uses nicotinamide adenine dinucleotide (NAD^+) as the hydrogen acceptor, in the following reaction:

$$\text{Glutamate} + NAD^+ + H_2O \rightarrow NH_4^+ + \alpha\text{-Ketoglutarate} + NADH$$

The reduced pyridine coenzyme, NADH, then transfers its hydrogen and pair of electrons to other systems. The ammonia that is formed can be detoxified by biosynthesis of urea. The net reaction can be written as follows:

$$CO_2 + 2NH_4^+ \rightarrow H_2N\text{—}\overset{\displaystyle \text{O}}{\overset{\|}{\text{C}}}\text{—}NH_2 + H_2O$$

Urea

It should be recognized that the above reaction is a gross oversimplication of the actual process of urea formation. Four amino acids—ornithine, citrulline, aspartic acid, and arginine—take part in a cyclic process, in which ATP is hydrolyzed to provide the required energy, in the formation of urea. This cyclic process can be depicted as follows:

$H_2N—\overset{\overset{O}{\|}}{C}—NH_2$ L-ornithine $NH_4^+ + CO_2 + ATP$

Urea arginase Citrulline + ADP + P_i

H_2O NH_4^+ (L-aspartic acid) + ATP

L-arginine + AMP + PP_i fumarate

Decarboxylation

Decarboxylation of amino acids, which is quite common in bacteria, is catalyzed by specific enzymes (amino acid decarboxylases), which require pyridoxal phosphate; the net result is the production of carbon dioxide and a primary amine. Some of the primary amines produced by mammalian decarboxylases have important pharmacological effects (e.g., histamine, tyramine, γ-aminobutyric acid); others serve as precursors of hormones, as components of coenzymes, and in other metabolic roles. For example, the decarboxylation of serine gives rise to ethanolamine, which is found in phosphatides (it is also the parent substance of choline), and the decarboxylation of cysteine gives β-mercaptoethylamine, which is a constituent of coenzyme A.

☐ PROTEINS AND NUTRITION

The most important function of dietary protein is to supply amino acids either directly or indirectly (i.e., the protein is a source of nitrogen used by an organism to synthesize amino acids). These amino acids, derived either directly or indirectly from dietary protein, are then used by an organism to synthesize its own proteins, as already described.

Essential and Nonessential Amino Acids

Most plants and microorganisms have the ability to synthesize all the amino acids required for protein generation. In animals, some of the amino acids required for protein synthesis are formed from α-keto acids (e.g., pyruvate, oxaloacetate, and α-ketoglutarate) and some from other precursors at rates sufficient to meet metabolic needs. These amino acids have been named endogenous, or nonessential, amino acids. In contrast, there are certain amino acids that animals cannot synthesize at all or cannot synthesize at rates sufficient to meet metabolic needs. These amino acids are termed essential amino acids and must be present in the diet either as free amino acids or as constituents of the dietary proteins. Amino acid requirements vary from species to species. The essential and nonessential amino acids for man are listed in Table 6.4.

It should be stressed that the terms essential and nonessential amino acids relate only to dietary intake. However, the role of the nonessential amino acids should not be underemphasized since they participate in many diverse metabolic reactions and thus are key intermediates in both anabolism and catabolism; i.e., while they are not essential nutrients, they are essential metabolites.

Nutritive Value of Proteins

It is well-known that proteins differ in their nutritive values. Such differences in the nutritive value of proteins are due to the variability of amino acid compositions. Accordingly, it is possible to rate proteins on the basis of their constituent amino acids and to assign numbers to these ratings. One useful rating system is based on the biological values (BV) of proteins. Actually, a number of nutritional characteristics are included in this expression: the digestibility, the availability of the digested product to the organism, and the presence and amounts of the various essential amino acids. Biological values are useful for comparing proteins from a variety of sources and for judging the nutritional adequacy of a combination of proteins such as might be found in a normal diet.

Protein score, based on the amino acid composition of a protein mix-

Table 6.4. Amino Acids Required by Man for the Biosynthesis of Protein

Essential		Nonessential	
Lysine	Threonine	Alanine	Histidine[a]
Isoleucine	Tryptophan	Arginine[a]	Proline
Leucine	Valine	Aspartic acid	Serine
Methionine	Phenylalanine	Cysteine	Tyrosine
		Glutamic acid	Asparagine
		Glycine	Glutamine

[a] In the young growing child, L-histidine and L-arginine are essential amino acids, although in adults they are apparently synthesized at rates suficient to meet metabolic needs.

ture or individual protein, permits one to predict the "performance" of a dietary mixture. This method is based on the concept that the limiting factor in the quality of a protein is the concentration of an essential amino acid that falls below the requirement. The standard of reference, the amino acid pattern of the nutrient that is used in calculating the score, is still a matter of debate. Some of the proposed reference standards are whole egg, cow's milk, human milk, and a pattern based on human amino acid requirements.

Several other methods have been developed for evaluating protein quality. Two of the more widely used methods are the protein efficiency ratio (PER), which is equal to the gain in body weight per gram of protein ingested, and net protein utilization (NPU), which characterizes the relationship between the nitrogen (protein) ingested and the increase in carcass nitrogen.

In vivo assays are generally the methods of last recourse for evaluating protein quality because animal tests are both time- and material-consuming. Furthermore, the results are open to a wide range of interpretations, yielding valid information only in the hands of an expert. For these reasons *in vitro* methods—be they chemical, microbiological, or enzymic—are preferred; when used with discrimination *in vitro* methods serve as an indispensible complement to animal assays.

The problem of improving protein quality is one of major importance in the world today. A knowledge of the quality of individual proteins permits one to improve the nutritional adequacy of a given protein of low biological value by supplementing it with one or more amino acids (such as lysine) or with another protein that provides the limiting or missing essential amino acids (mutual supplementation).

Effect of Heat

Proteins are generally affected by heat; the changes they undergo may be beneficial in some respects and harmful in other respects.

The beneficial effects relate to some change in the structure of the protein molecule. These transformations, commonly involving partial denaturation, render proteins more susceptible to enzymatic digestion, reduce their solubility, and destroy (or lower) the biological activity of toxic proteins. The effect of heat in inactivating a trypsin inhibitor present in soybeans is well-known. Raw fish contains an enzyme (thiaminase) that catalyzes the destruction of vitamin B_1; heat inactivates the enzyme, thereby preventing the destruction of thiamin in diets rich in fish. The effect of heat on the curd of calcium ceseinate of milk has been used to advantage in child feeding. Raw eggs contain a protein, avidin, which combines with and inactivates another B-vitamin, biotin. Heat denatures avidin and, as a consequence, the protein cannot function as a vitamin inhibitor.

On the other hand, excessive heating of proteins may result in the destruction of certain amino acids or render them unavailable for digestion. One of the most important changes that results from heating is the interaction of certain amino acid residues in proteins with reducing sugars such as glucose. The amino acid lysine is most frequently involved in this type of reaction, but tryptophan, arginine, and histidine may also be involved. In the early stages (melanoidin condensation), a bond is formed between the sugar and the amino group which interferes with the action of proteolytic enzymes but which may be broken by boiling with acid. Thus, the reactions are in a sense reversible. If the reaction is continued, a discoloration (brownish color) begins to appear. In the last stages, i.e., when browning has become extensive, amino acids are destroyed.

Moisture content plays a very important part in the so-called *browning reaction*. Experiments with simple systems of amino acids and sugars have shown that a moisture content of 30% is most favorable for this reaction. Moisture contents above or below this value are less favorable for the reaction. The browning reaction is most likely to occur in dried and dehydrated food products because in their preparation the moisture content is reduced from a high value to a low value, usually passing through the range conducive to protein damage. Similarly, baking should promote protein damage as a result of the browning reaction. In contrast, cooking and toasting should have no adverse effect on the nutritive value of cereal protein.

Proteins

Meat and milk are the most important sources of dietary protein for most people; the proteins of these foodstuffs are discussed in detail in Chapters 12 and 13.

☐ ANALYSIS

The complex chemical composition of proteins makes them quite difficult to characterize by simple chemical or physical procedures. However, their component amino acids may be conveniently detected by various specific chemical tests. Also they may be qualitatively and quantitatively determined by first hydrolyzing the protein, then separating the amino acids by paper, ion-exchange, or thin-layer chromatography. Amino acids may also be conveniently determined by converting them to *N*-acetyl-*n*-propylesters, which can be separated, identified, and quantitized by gas chromatography.

Qualitative Methods

As noted in an earlier section, amino acids and proteins undergo a number of color-forming reactions that can be used to determine the presence of peptide bonds or specific amino acids. Three of the most common qualitative tests for proteins are described in this section.

Proteins to be tested should be essentially free of salts such as chlorides and ammonium salts. Desalting can be accomplished by passing a protein solution through a Sephadex G-25 column or by dialysis. The desalted solution may then be lyophilized and stored at low temperatures (less than 0°C). Egg albumin, lactalbumin, and lactoglobulin are all suitable proteins for trying out these qualitative tests.

Biuret Reaction

When a protein is mixed with a solution of sodium hydroxide and a weak solution of copper sulfate, a violet color is produced. This is a test for the peptide linkage and will be positive when two or more peptide linkages are present. The color is due to the presence of a coordination complex with Cu^{2+} in which the four water molecules, normally coordinated with the cupric ion, are displaced by α-amino groups. The alkali serves to convert the complex into a soluble sodium salt. The peptide linkage in proteins and peptides can lead to a stable complex with Cu^{2+} in which two five-membered rings are formed. Other compounds (e.g., urea) also give a positive biuret test, for example, urea forms biuret, as follows:

$$NH_2{-}\overset{\overset{\displaystyle O}{\|}}{C}{-}NH_2 \Rightarrow O{=}C{=}NH + NH_3$$

$$O{=}C{=}NH + NH_2{-}\underset{\underset{\displaystyle O}{\|}}{C}{-}NH_2 \Rightarrow H_2N{-}\overset{\overset{\displaystyle O}{\|}}{C}{-}\overset{\overset{\displaystyle H}{|}}{N}{-}\overset{\overset{\displaystyle O}{\|}}{C}{-}NH_2 \text{ (biuret)}$$

Procedure

Mix 1 ml of a 2% protein solution with 1 ml of 10% NaOH. Dropwise, add 0.1% $CuSO_4$ solution with mixing. A violet-pink color should develop. Repeat this reaction with a 2% glycine solution.

Millon's Reaction

When a protein solution is heated with Millon's reagent, a red color is produced. A positive test is due to the presence of phenolic groups in the protein molecule. Tyrosine is the only common amino acid that contains a phenolic group; thus, the Millon reaction is a specific test for the presence of tyrosine.

Procedure

Prepare Millon's reagent by dissolving 1 part of mercury in cold fuming nitric acid; dilute with twice the volume of water and decant the clear solution after several hours. Place a small amount of powdered protein on a spot plate. Add 5 drops of Millon's reagent. Observe the red color on standing. The test may be repeated using solid tyrosine. To 1 ml of a 2% protein solution add 5 drops of Millon's reagent and heat gently until a red color appears.

Ninhydrin Reaction

Ninhydrin (triketohydrindene hydrate) reacts with the free amino group of amino acids to produce various shades of blue or purple with the exception of proline and hydroxyproline, which give yellow products. The reaction of ninhydrin with α-amino acids is represented as follows:

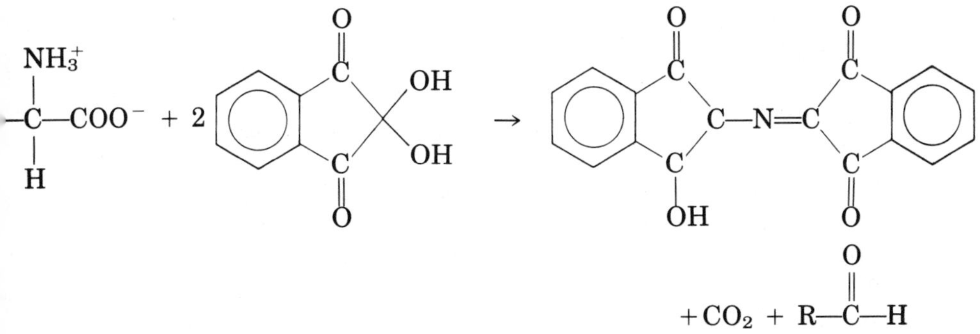

$$\begin{array}{c} NH_3^+ \\ | \\ -C-COO^- \\ | \\ H \end{array} + 2\; C_6H_4\begin{array}{c} C{=}O \\ C{=}O \end{array}\!\!>\!C\!<\!\begin{array}{c} OH \\ OH \end{array} \rightarrow C_6H_4\begin{array}{c} C{=}O \\ C{-}OH \end{array}\!\!>\!C{-}N{=}C\!<\!\begin{array}{c} C{=}O \\ C{=}O \end{array}\!\!C_6H_4$$

$$+\,CO_2 + R-\overset{\overset{\displaystyle O}{\|}}{C}-H$$

The blue-purple color may be used to quantitatively determine amino acids and certain peptides by measuring its absorbance at 570 nm. Commercial automated systems based on the ninhydrin reaction are available for amino acid determinations.

Procedure

Adjust the pH of a 0.5% solution of protein to pH 7. To 1 ml of this solution add 10 drops of 0.2% ninhydrin solution. heat in a 100°C bath for 10 min. Note the color. This test may be repeated using concentrated $(NH_4)_2SO_4$ or partially hydrolyzed protein solutions.

Paper Chromatographic Analysis

The amino acids released by hydrolysis of a protein may be separated by paper chromatography and the separated compounds visualized by spraying the chromatogram with ninhydrin. By comparing the migration pattern of a protein hydrolysate with the migrations of known amino acids, the component amino acids in a protein or peptide can be identified.

Materials and Reagents

1. Whatman No. 1 chromatography paper (46 cm × 57 cm).
2. Chromatography tank (30 cm in diameter and 60 cm high).
3. Developing solvents—(a) Butanol–acetic acid–water solvent (6:1:2 by volume) and (b) phenol (chromatography grade) saturated with water.
4. Protein hydrolysate prepared as in following quantative Ninhydrin Method.
5. Amino acid solutions—Prepare by dissolving 100 mg of known amino acids in 100 ml of water.
6. Ninhydrin spray—Prepare by dissolving 200 mg of ninhydrin in 100 ml of ethanol. Commercial aerosol spray cans of ninhydrin solution are available.

Procedure

Spot the protein hydrolysate in a corner approximately 6 cm from each edge. Roll the paper and sew the edges together using cotton (colorless) thread. (Paper staples corrode and discolor the paper.) Place approximately 2 cm of the butanol—acetic acid—water developing solvent in the chromatography tank. Place the cylinders in the tank and allow the solvent front to rise to within approximately 5 cm of the top of the paper. When the solvent front migrates the desired distance, remove the paper, mark the solvent front, and air-dry. (Be careful not to touch the paper with your fingers as they contain amino acids. Use rubber gloves or forceps.) Spray the chromatograms with ninhydrin spray and allow to air-dry. The ninhydrin reacts with the amino acids and blue-violet spots appear. Use the same procedure to chromatograph known amino acids. It is best to spot and develop the known amino acids and hydrolysate on the same paper.

Two-dimensional chromatography may be used if more positive identification is required. In two-dimensional chromatography the chromatogram is spotted as usual and developed first in the phenol–water solvent. The chromatogram is then dried and washed with ether to remove the phenol. The paper is then turned 90°, placed back in the developing tank, and developed in the other direction using the butanol–acetic acid–water solvent. Spots are visualized as described above. Known amino acids are chromatographed and their positions compared to those for the components of the hydrolysate. In this manner, it is usually possible to positively identify the amino acids of protein hydrolysates.

It should be noted that acid hydrolysis may destroy tryptophan and cause some losses of serine and threonine.

Quantitative Methods

Both the biuret and ninhydrin reactions may be used for the quantitative determination of protein. The Kjeldahl method is used to determine the nitrogen content of a sample; from this value the protein content can be calculated.

Quantitative Biuret Method

The amount of protein in a sample can be determined quantitatively by comparing the absorbance at 550 nm of the sample with that of standard protein solutions after reaction with biuret reagent.

Procedure

Prepare a standard protein solution using crystalline bovine albumin that contains 10 mg/ml. Add 0.25 ml, 0.5 ml, and 0.75 ml of this standard to three reaction tubes. Bring the final volume to 1 ml with distilled water. Add 4 ml of biuret reagent to the tubes with mixing. Read the absorbance (550 nm) at the end of 30 min against a blank containing 1 ml of water and 4 ml of biuret reagent.

Determine the protein content of an unknown protein solution provided by the instructor. This solution will contain 5–10 mg/ml; therefore, several dilutions should be prepared so that at least two would be in the range of concentration of the albumin standard curve.

Quantitative Ninhydrin Method

The amino acid content of a sample can be determined by first hydrolyzing the sample, then reacting the hydrolysate with ninhydrin reagent and comparing the absorbance at 570 nm with that of known concentrations of glycine, also treated with ninhydrin reagent.

Procedure

Using a desalted and lyophilized protein sample, prepare a hydrolysis mixture containing 100 mg/ml dissolved in 8 *N* H_2SO_4. Autoclave the sample in an appropriate container for 5 hr at 15 lb.

Neutralize the hydrolyzed sample with solid $Ba(OH)_2$. Barium sulfate precipitates as the $Ba(OH)_2$ is added. Do not add an excessive amount. Neutralize to pH 5. Centrifuge the solution, pour off the supernatant (save it), and wash the $BaSO_4$ precipitate with 10 ml of hot water. Save the washes and discard the precipitate. The final volume should be adjusted to 30 ml.

Dilute a portion of hydrolysate solution 1:100 with water. The final solution, after dilution, will contain the equivalent of 0.033 mg/ml of the original protein. Prepare a standard curve using glycine. Generally a 1-*mM* (75 μg/ml) solution of glycine may be used. Prepare assay mixtures as shown in Table 6.5. Mix the contents of the assay tubes, heat in a 100°C bath for 20 min, and then add 8 ml of 50% 1-propanol with thorough mixing. Determine the absorbance at 570 nm 10 min after addition of the alcohol.

Table 6.5. Components of Assay Tubes in Quantitative Ninhydrin Method

	Volume of components (ml)							
Components	1	2	3	4	5	6	7	8
Standard glycine	0.5	0.25	0.1	—	—	—	—	—
H_2O	—	0.25	0.4	—	0.25	0.4	0.45	2.0
Hydrolysate				0.5	0.25	0.1	0.05	—
Ninhydrin reagent	1.5	1.5	1.5	1.5	1.5	1.5	1.5	1.5
	Caution: Mix and Heat before adding 1-propanol							
50% 1-Propanol	8.0	8.0	8.0	8.0	8.0	8.0	8.0	8.0

Kjeldahl Nitrogen Determination

The standard method for determining nitrogen in organic compounds is the Kjeldahl method. The procedure described here (AOAC method 2.059) differs from that described in Chapter 2 in that the ammonia liberated from a protein sample is distilled into a standard acid solution and the excess acid is back-titrated with standard alkali solution.

In this procedure, metallic mercury or HgO is used as the catalyst in the digestion step. Sulfide or thiosulfate is added to precipitate the mercury before distillation of ammonia. Mercury forms an ammonia-mercury complex that remains even after addition of sodium hydroxide; thus if the mercury is not precipitated, some of the ammonia would not distill and low values of % N would be obtained. Since the presence of nitrate also yields ammonia under the conditions of the Kjeldahl procedure, this method is applicable only in nitrate-free samples.

Precautions

Always add H_2SO_4 to H_2O. Wear face shield and heavy rubber gloves to protect against splashes.

Mercury (Hg) is hazardous in contact with NH_3, halogens, and alkali. Its vapors are extremely toxic and cumulative. Regard Hg spills on hot surfaces as extremely hazardous and clean up promptly. Powdered sulfur sprinkled over Hg can assist in cleaning up spills. When Hg evaporation is necessary, use effective fume removal device.

Procedure

Place weighed sample (0.7–2.2 g) in digestion flask. Add 0.7 g HgO *or* 0.65 g metallic Hg, 15 g powdered K_2SO_4 *or* anhydrous Na_2SO_4, and 25 ml concentrated H_2SO_4. If a sample greater than 2.2 g is used, increase H_2SO_4 by 10 ml for each additional gram of sample. Place flask in inclined position and heat gently until frothing ceases (if necessary, add a small amount of paraffin to reduce frothing); boil briskly until solution clears and then about 30 min longer (2 hr for samples containing organic material).

Cool, add approximately 200 ml of H_2O, cool below 25°C, add 25 ml of potassium sulfide (40 grams diluted to 1 liter) or sodium thiosulfate (80 grams diluted to 1 liter), and mix to precipitate Hg. Add a few Zn granules to prevent bumping, tilt flask, and add a layer of NaOH without agitation. (For each 10 ml H_2SO_4 used, or its equivalent in diluted H_2SO_4, add 15 g solid NaOH or enough solution to make contents strongly alkaline. Thiosulfate or sulfide solution may be mixed with the NaOH solution before addition to flask.) Immediately connect the flask to the distilling bulb on a condenser; with the tip of the condenser immersed in a standard acid solution containing 5–7 drops of indicator in the receiver, rotate flask to mix contents thoroughly. Then heat until all NH_3 has distilled (greater than or equal to 150 ml of distillate). Remove receiver, wash tip of condenser, and titrate excess standard acid in distillate with standard NaOH solution. Run a blank determination on the reagents.

Calculations

The milliequivalents (meq) of NH_3 distilled over equals the difference between the meq of acid originally present in the distillation flask and the meq of acid left after distillation, which is determined by titration with standard NaOH. Thus, meq NH_3 = (ml std acid × normality acid) − (ml std NaOH × normality NaOH). Since the meq of N in the protein sample equals the meq of NH_3, the grams of N is calculated by multiplying by 0.014007 (the grams of N in 1 meq), as follows: g N = meq N × 0.014007. Thus, the percentage nitrogen can be calculated from the following expression:

$$\%\ \mathrm{N} = \frac{[(\text{ml std acid} \times \text{normality acid}) - (\text{ml std NaOH} \times \text{normality acid})] \times 1.4}{\text{g sample}}$$

As discussed in Chapter 2, the weight of nitrogen in a sample can be converted to protein by using the appropriate factor based on the percentage of nitrogen in protein. The most common nitrogen-to-protein conversion factor is 6.25, although this value varies somewhat depending on the source of the protein (see Table 2.1).

As an example of how to calculate the % protein in a 1.0-g sample of a food product, assume that 50 ml of 0.1121 *N* standard acid is originally in the NH_3 collection flask and that 20 ml of 0.1231 *N* standard NaOH is required to titrate the remaining acid after partial neutralization by NH_3. Then,

- $\text{meq NH}_3 = (50\ \text{ml} \times 0.1121\ \text{N}) - (20\ \text{ml} \times 0.1231\ \text{N})$
  $= 5.605\ \text{meq acid} - 2.462\ \text{meq base}$
  $= 3.143\ \text{meq}$
- $\text{g of N} = 3.143\ \text{meq} \times 0.014007\ \text{g/meq}$
  $= 0.0440\ \text{g}$
- $\text{g of protein} = 0.0440\ \text{g N} \times 6.25\ \text{g protein/g N}$
  $= 0.275\ \text{g}$
- $\%\ \text{protein} = \frac{0.275\ \text{g} \times}{1.0\ \text{g}} 100 = 27.5\%$

Isolation of Lactalbumin and Globulin

To prepare milk serum, measure 400 ml of fresh skim milk into a large beaker and heat to 40°C. Add 10% lactic acid solution slowly from a burette, stirring vigorously with a rubber-tipped stirring rod, until a flocculent precipitate forms. Allow the precipitate (casein) to settle and decant the supernatant fluid through a filter. The casein may be discarded. Heat the filtrate again to 40°C, add a few drops of phenolphthalein solution, and carefully neutralize to a faint pink with a thin suspension of $Ca(OH)_2$ in water. Filter off the precipitate through a large fluted filter. This precipitate is called the neutralization precipitate of milk serum and consists largely of phosphates. Add dilute (1%) HCl to the filtrate until neutral to litmus. The following procedures require 300 ml of filtrate (milk serum).

Salting Out Globulin

Dilute 50 ml of serum with an equal volume of water. Warm to 30°C; maintain at 30°C and saturate the solution with $MgSO_4$ crystals (about 40 g will be required). The precipitate that forms is globulin. All animal globulins are salted out of their neutral solutions on saturation with $MgSO_4$.

Salting Out Lactalbumin and Globulin

Dilute 50 ml of serum with equal volume of distilled water, warm to 30°C, and saturate with $(NH_4)_2SO_4$ crystals (about 80 g will be required) at this temperature. The precipitate that forms is a mixture of globulin and lactalbumin. Determine by inspection the relative amounts of protein salted out in this procedure and the previous globulin procedure.

Separating Globulin from Albumin

Dilute 50 ml of serum with an equal volume of a saturated $(NH_4)_2SO_4$ solution that has pH 7.0. Let stand for 1 hr. Filter off the precipitate, which is globulin. To the filtrate add 5% H_2SO_4 solution drop by drop with stirring until definite coagulation occurs. The coagulum is albumin.

Heat Coagulation of Serum Proteins

Dilute 50 ml of serum with an equal volume of water, add 0.3 ml of 10% acetic acid, and heat slowly to 75°C in a water bath. Record the temperature at which the first cloudiness appears and that at which coagulation occurs. The stages in the heat coagulation of proteins are demonstrated, i.e., (a) denaturation as shown by the cloudiness and

(b) agglutination of the denatured colloidal particles. Now raise the temperature of the solution to boiling and boil for 5 min. Filter off the coagulated material while the solution is still hot, in order to facilitate filtration. Test the filtrate for protein as follows: To 10 ml of water-clear filtrate add 1 ml of 10% acetic acid; then add, drop by drop, with shaking after each addition, a 5% solution of potassium ferrocyanide. A positive test for protein is shown by a distinct cloudiness.

Determination of Protein and Amino Acid Content of Sweet Potato

As discussed earlier, in the section on the nutritive value of proteins, the content of essential amino acids has been used as one measure of the quality of proteins. Such protein scores are expressed as a percentage value, i.e. the ratio of the quantity of each essential amino acid in a test protein to the quantity of each essential amino acid in a reference protein. Obviously, the choice of the reference protein affects the score of a test protein. Evaluation of protein quality by this approach requires determination of the protein and amino acid content of a material. The general procedures are described in this section, using sweet potato as an example.

Protein and Dry Matter

Samples for analyses are obtained by cutting 3-mm-diameter plugs from the mid-section (equator) of three to five sweet potatoes with a cork borer. Kjeldahl N is determined on the plugs (weighed to the nearest 0.1 mg) and reported as protein after multiplying by 6.25. Dry matter is determined by drying 6- to 8-g samples in a vacuum oven at 60°C for 16–18 hr.

Extraction of Protein

Peel and dice six to eight sweet potatoes and blend 300 g in a Waring blender with 600 ml cold water for 4 min at high speed. Filter the resulting slurry through pelon to remove cell walls and fibers. Centrifuge at 300 *g* for 25 min at 4°C; this will sediment most of the starch granules. The starch pellets are resuspended and centrifuged at 16,000 *g* for 10 min. The supernatants from each centrifugation are combined, and the proteins then coagulated by adding trichloracteic acid to 12% concentration and heating to 50°C. The coagulum is precipitated by centrifugation at 10,000 *g* for 25 min, then extracted with a mixture of acetone–ether (1:1) until the extracts are colorless. Dry the protein powders overnight at 18°–20°C in a hood and store in an evacuated desiccator. Perform Kjeldahl N analyses on the residue.

Hydrolysis of Protein

Weigh about 60 mg of the extracted protein powder into 10-ml ampoules, then add 8 ml 6 *N* HCl. Degas ampoules with nitrogen and seal under 0.5 atm of nitrogen. Place sealed ampoules in a reaction vessel and heat with refluxing toulene for 16 hr. Transfer the contents of the ampoule to centrifuge tubes and centrifuge at 15,000 *g* for 10 min. Pour the supernatant liquid into a 20-ml beaker, wash the ampoule with 2 ml distilled water, and add the washing to the supernatant liquid. Dry sample over NaOH pellets in an evacuated desiccator. The dried samples are then dissolved in 5 ml of sample dilution buffer (pH 2.2)[1] and filtered through a 0.2-micron Millipore filter.

Since tryptophan is known to be destroyed by acid hydrolysis, another set of samples must be weighed and hydrolyzed as before except 0.17 ml thioglycholic acid is added.

Ion-Exchange Cleanup

The hydrolysate is highly colored due to the presence of sugars. These interfering substances can be removed by passage of the hydrolysate through an ion-exchange column. Dowex 50 W (200–400 mesh) is prepared by three alternate washings with 1 *N* HCl and water, then dried at room temperature overnight. Columns are prepared by suspending 0.35 g of washed resin in 2 ml 1 *N* HCl and transferring the suspensions into a Pasteur pipet. Allow the acid to drain until it reaches the top of the resin, then add 1 ml water and allow it to drain to the top of the column; *do not at any time allow the top of the column to become dry*. Transfer the hydrolysate to the column; when level of hydrolysate nears the top of the resin column, wash three times with 1-ml portions of 0.5 *N* acetic acid followed by two 1-ml portions of water. The amino acids are eluted with 4 ml of an eluant containing 20% triethylamine, 10% acetone, and 70% water. The eluate is dried over sulfuric acid in a partially evacuated desiccator. (Note: Care should be taken to prevent bumping when applying vacuum to the desiccator.) After drying, dissolve the eluate in 5 ml of sample dilution buffer for the amino acid analysis.

Amino Acid Analysis

The mixture of amino acids can be resolved using an amino acid analyzer. Norleucine is added to the sample dilution buffer as an internal standard for the long column and β-guanidine-propionic acid

[1] Sample dilution buffer (pH 2.2): sodium citrate-$2H_2O$—19.6 g, conc. HCl—16.5 ml, thiodiglycol (25% solution)—10 ml, and caprylic acid—0.1 ml. Make up to final volume of 1 liter.

for the short column. The amino acids are measured as micromoles per sample and reported as moles/100 kg based upon the Kjeldahl protein content of the powder.

A mixture of amino acids is applied to an ion-exchange resin column. Buffers of carefully controlled pH and ionic strength are pumped through the column. The individual amino acids are selectively bound to the resin and then eluted on the basis of the chemical nature of the resin, the temperature, and the pH and ionic strength of the buffer used. The resolved components of the hydrolysate, after being pumped out the bottom of the column, are monitored by mixing the sequential eluate samples with ninhydrin, reacting the mixture in a heated coil, then passing the stream of colored substances through a colorimeter. A photocell at the cuvettes transmits the light response (570 nm and 440 nm) to a recorder, which prints out the chromatogram. The chromatogram is then interpreted on the basis of previously accumulated data, i.e., calibration runs with known amino acids under operating conditions exactly the same as those used with test samples.

SELECTED REFERENCES

ACKERS, G. K. 1970. Analytical gell chromatography of proteins. Adv. Protein Chem. *24,* 343–442.

ALBANESE, A. A. (editor). 1959. Protein and Amino Acid Nutrition. Academic Press, New York.

ALTSCHUL, A. M. 1958. Processed Plant Protein Foodstuffs. Academic Press, New York.

ALTSCHUL, A. M. 1966. World Protein Resources. Advances in Chemistry Series *57,* American Chemical Society, Washington, DC.

AMERICAN MEAT INSTITUTE FOUNDATION. 1960. The Science of Meat and Meat Products. W. H. Freeman and Co., San Francisco.

ANFINSEN, C. B., EDSALL, J. T. and RICHARDS, F. M. 1985. Advances in Protein Chemistry, Academic Press, New York.

ANON, 1958. Exchange of Genetic Material: Mechanisms and Consequences. Cold Spring Harbor Symposium on Quantitative Biology, Vol. 23. Cold Spring Harbor Laboratory, Long Island, NY.

ANON. 1966. The Genetic Code. Cold Spring Harbor Symposium on Quantitative Biology, Vol. 31. Cold Spring Harbor Laboratory, Long Island, NY.

ANON. 1968. Replication of DNA in Microorganisms. Cold Spring Harbor Symposium on Quantitative Biology, Vol. 33. Cold Spring Harbor Laboratory, Long Island, NY.

ANON. 1969. The Mechanism of Protein Synthesis. Cold Spring Harbor Symposium on Quantitative Biology, Vol. 34. Cold Spring Harbor Laboratory, Long Island, NY.

AOAC. 1980. Official Methods of Analysis. 13th ed. W. Horwitz (editor). Assoc. of Official Analytical Chemists, Arlington, Va.

AOAC. 1984. Official Methods of Analysis. 14th ed. S. Williams (editor). Assoc. of Official Analytical Chemists, Arlington, Va.

AURAND, L. W. and TRIEBOLD, H. O. 1963. Food Composition and Analysis D. Van Nostrand Co., New York.

AURAND, L. W. and WOODS, A. E. 1973. Food Chemistry. AVI Publishing Co., Westport, CT.
BAILEY, J. L. 1967. Techniques in Protein Chemistry. 2nd ed. Elsevier Publishing Co., Amsterdam.
BISSWANGER, H. and SCHMINCKE, E. 1980. Multifunctional Proteins. John Wiley & Sons, New York.
BLACKBURN, S. 1968. Amino Acid Determination. Marcel Dekker, New York.
CROFT, L. R. 1980. Introduction to Protein Sequence Analysis. John Wiley & Sons, New York.
GEIDUSCHEK, E. P. and HASELKORN, R. 1969. Messenger RNA. Ann. Rev. Biochem. *38,* 647–676.
HAUROWITZ, F. 1963. The Chemistry and Function of Proteins, 2nd ed. Academic Press, New York.
KLOTZ, I. M. 1967. Protein subunits: a table. Science *155,* 697–698.
KLOTZ, I. M. and DARNELL, D. W. 1969. Protein subunits: A table. Science *166,* 126–127.
KOPPLE, K. D. 1966. Peptides and Amino Acids. W. A. Benjamin, New York.
LEHNINGER, A. L. 1982. Principles of Biochemistry. Worth Publishing Co., New York.
LUNDBLAD, R. L. 1984. Chemical Reagents for Protein Modification. CRC Press, Boca Raton, FL.
LIPMAN, F. 1969. Polypeptide chain elongation in protein biosynthesis. Science *164,* 1024–1031.
MAHLER, H. R. and CORDES, E. H. 1971. Biological Chemistry. 2nd ed. Harper and Row, New York.
MEISTER, A. 1965. Biochemistry of Amino Acids. 2nd ed. Academic Press, New York.
MOLDAVE, K. 1965. Nucleic acids and protein biosynthesis. Ann. Rev. Biochem. *34,* 419–448.
NATIONAL RESEARCH COUNCIL. 1950. The Problem of Heat Injury to Dietary Protein. Reprint and Circular Series No. *131.* National Academy of Sciences, Washington, DC.
NATIONAL RESEARCH COUNCIL. 1963. Evaluation of Protein Quality. Publication *1100.* National Academy of Sciences, Washington, DC.
NESELSON, M. and STAHL, G. W. 1958. The replication of DNA. Cold Spring Harbor Symp. Quant. Biol. *23,* 9–12.
NEURATH, H. and HILL, R. 1975. The Proteins, 3rd ed. Academic Press, New York.
NOVELLI,L G. D. 1967. Amino acid activation for protein synthesis. Ann. Rev. Biochem. *36,* 449–484.
PLATT, B. S. and MILLER, D. S. 1959. The net dietary protein value: Its definition and application. Proc. Nutr. Soc. *18,* 7–8.
POMERANZ, Y. and MELOAN, C. E. 1978. Food Analysis Theory and Practice. AVI Publishing Co., Westport, CT.
SCHULTZ, H. W. (editor). 1964. Symposium on Foods, Proteins and Their Reactions. AVI Publishing Co., Westport, CT.
SCHWEET, R. and HEINTZ, R. 1965. Protein synthesis. Ann. Rev. Biochem. *34,* 723–758.
SCOPES, R. K. 1982. Protein Purification: Principles and Practice. Springer-Verlag, New York.
SOBER, H. A. (Editor). 1970. Handbook of Biochemistry. 2nd ed. Chemical Rubber Co., Cleveland, OH.
STEIN, S. and MOORE, W. 1961. The structure of proteins. Sci. Am. (Feb.), 81–82.

TANFORD, C. 1970. Protein denaturation. Adv. Protein Chem. *24,* 2–92.

VOGEL, H. J. and VOGEL, R. H. 1967. Regulation of protein synthesis. Ann. Rev. Biochem. *36,* 519–538.

WATSON, J. D. 1970. Molecular Biology of the Gene. 2nd ed. W. A. Benjamin, New York.

ZAPSALIS, C. and BECK, A. R. 1985. Food chemistry and Nutritional Biochemistry. John Wiley & Sons, New York.

7

Enzymes

□ INTRODUCTION

A catalyst is a substance that accelerates a chemical reaction without changing the nature or quantity of products formed and without itself being consumed in the process. Enzymes are biological catalysts, i.e., they catalyze reactions in animal and plant tissue that, if performed *in vitro,* would proceed at a very slow rate or would require more drastic conditions to reach a velocity equivalent to those observed in cells. Enzyme-catalyzed reactions take place readily under the mild conditions of the cellular environment; without enzymes, the biochemical reactions characteristic of living organisms could not proceed at rates sufficient to maintain life.

All enzymes are protein, but some also contain a nonprotein organic moiety—either a prosthetic group or a coenzyme. A prosthetic group is tightly bound to the protein and remains attached throughout the course of a catalytic reaction; flavin adenine dinucleotide (FAD) enzymes such as the dehydrogenases (succinic and fatty acyl) are examples. Coenzymes such as nicotinamide adenine dinucleotide (NAD^+) also bind to enzymes but do so less strongly than prosthetic groups; thus coenzymes dissociate from the enzyme during or at the termination of a catalytic reaction. Additionally, enzymes often have an obligatory requirement for metal ions. Some metal ions and other cations or anions act only as activators; that is, their presence increases the catalytic activity of the enzyme, but the enzyme can function without the activating ion.

The protein nature of enzymes confers the property of specificity on them. Thus, different enzymes are required to catalyze different reactions. Many chemical reactions are reversible and proceed to equilibrium concentrations of reactants and products, which is reflected in the equilibrium constant. The same enzyme catalyzes a reversible reaction in both directions, and the presence of the enzyme has no effect on the equilibrium constant of a reversible reaction. However, there are some reactions in which one enzyme is required for the reaction to occur in one direction and another enzyme is required for the reaction to take place in a "reverse" manner. An example is the reaction of glucose + ATP to form glucose-6-phosphate + ADP, which is catalyzed by hexokinase. Reactions of this type have an equilibrium constant that favors the formation of the product to such a degree that the reaction can be considered essentially irreversible. Indeed, the "reverse" reaction is not actually the reverse but involves the hydrolysis of glucose-6-phosphate to form glucose and phosphate, catalyzed by phosphatase. These two reactions are represented as follows:

$$\text{Glucose} + \text{ATP} \xrightarrow{\text{Hexokinase}} \text{Glucose-6-phosphate} + \text{ADP}$$

$$\text{Glucose-6-phosphate} \xrightarrow{\text{Phosphatase}} \text{Glucose} + P_i$$

Since enzymes are proteins, any agent such as heat, strong acids or bases, organic solvents, or other materials that denature proteins will also destroy the activity of an enzyme. Thus, one of the important characteristics of enzymes is their inactivation by heat. A temperature of 80°C for a few minutes is sufficient to inactivate most enzymes.

Many foods and food products contain considerable enzyme activity even after processing. As discussed later, this activity can be either beneficial or harmful to food quality. Enzyme reactions can be controlled to some degree by refrigeration but even then some reactions occur at an appreciable rate.

□ KINETICS OF ENZYME–CATALYZED REACTIONS

As noted already, the fundamental property of enzymes is to increase the rate, or velocity, at which a reaction occurs. Measurement of the effect of enzyme concentration, substrate concentration, and other variables on the rate of enzyme-catalyzed reactions (i.e., the amount of product formed or substrate consumed per unit time) provides insight into how enzymes function. Thus, much of the scientific study of enzymes is concerned with the kinetics of enzyme-catalzyed reactions.

Steady-State Kinetics

Enzyme Concentration

The rate of an enzyme-catalyzed reaction is proportional to the concentration of the enzyme. This relationship is the basis for methods used to determine enzyme concentrations in samples of unknowns. Curve a in Fig. 7.1 shows a linear relationship between the rate of the reaction and the enzyme concentration in the presence of an excess of substrate. Deviations from this linear relationship indicate some limiting factor. For example, if the concentration of an enzyme is increased sufficiently, the rate of the reaction will decrease due to substrate exhaustion as indicated by Curve b in Fig. 7.1. Deviations from linearity also occur when an enzyme preparation contains either an activator or an inhibitor. In the presence of an activator, the response curve shows an upward curvature (Curve d, Fig. 7.1), whereas in the presence of an inhibitor, the response curve shows a downward curvature (Curve c, Fig. 7.1). Curve e shows the effect of an enzyme inactivator such as Pb^{2+} or Hg^{2+}.

Substrate Concentration

The rate of an enzymic reaction also depends on the concentration of the substrate. At low substrate concentration (in the presence of excess enzyme) the relationship of reaction rate to substrate concentration is almost linear and complies with *first-order kinetics,* which

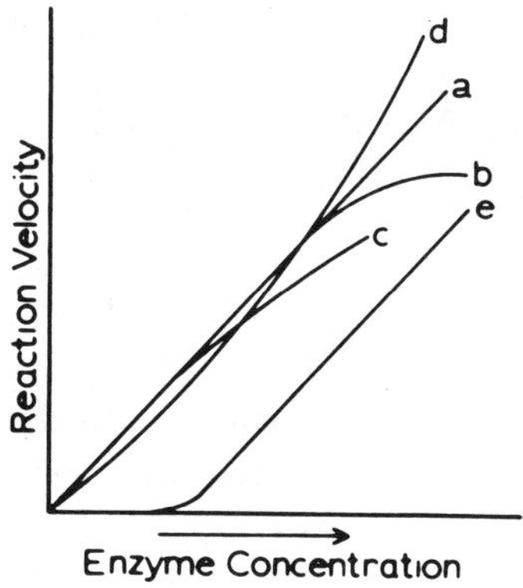

Figure 7.1. Effects of various factors on the relationship between reaction velocity and concentration of enzyme. a—Normal. b—Substrate exhaustion. c—Effect of inhibitor. d—Effect of activator. e—Enzyme inactivator.

may be expressed as

$$\frac{d[S]}{dt} = k'([S] - [P])$$

where k' = the first-order reaction constant; $[S]$ = initial substrate concentration; and $[P]$ = product concentration. Thus, the amount of remaining substrate ($[S] - [P]$) at any given time is proportional to the rate of reaction. For purposes of graphical presentation, the equation may be integrated to give

$$k' \times t = 2.3 \log \frac{[S]}{[S] - [P]}$$

where $[S]$ = initial concentration of substrate and $[P]$ = concentration of substrate converted to product after the lapse of time t. If the time (t) is plotted against the log of $[S]/([S] - [P])$, a straight line results; the slope of the line is constant and equals $2.3/k'$.

At high substrate concentration, however, the velocity of the reaction is maximum and is independent of substrate; hence, the course of reaction follows *zero-order kinetics,* which is expressed as

$$\frac{d[S]}{dt} = k^0$$

where k^0 is the zero-order reaction constant. For such a reaction the amount of end product formed is proportional to time.

Michaelis–Menten Constants

The velocity (μmoles of substrate changed per minute) of an enzymic reaction is much greater than that of organic reactions catalyzed by other types of catalysts. The effectiveness of a catalyst is described by its molecular activity (turnover number), which is defined as the number of molecules of substrate transformed per minute per molecule of catalyst. Turnover numbers for enzymes generally range from several hundred to several thousand molecules of substrate per molecule of enzyme per minute. For some enzymes, e.g., catalase, the molecular activity is as high as one million or more.

With increasing substrate concentration (at constant enzyme concentration), the rate of reaction begins to decrease until a certain substrate concentration is reached beyond which no further change in rate is observed (Fig. 7.2). This effect is easily explained if we assume the formation of an enzyme–substrate complex. If enzyme molecules combine with substrate molecules, the rate of reaction will increase with

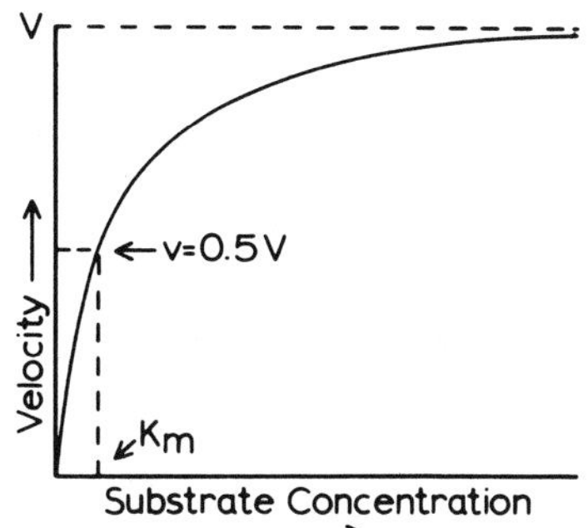

Figure 7.2. Michaelis-Menten plot of reaction velocity versus substrate concentration at constant enzyme concentration.

the addition of more substrate until sufficient substrate is present so that all enzyme molecules have combined with substrate; at this point, an increase in substrate will no longer increase the rate of reaction. This line of reasoning suggests that the enzyme binds the substrate to form an unstable enzyme–substrate complex.

The formation of the enzyme–substrate complex from its components and its breakdown to the free enzyme and product is indicated by the following equation:

$$E + S \underset{k_{-1}}{\overset{k_1}{\rightleftharpoons}} ES \underset{k_{-2}}{\overset{k_2}{\rightleftharpoons}} P + E$$

where E = enzyme; S = substrate; ES = enzyme–substrate complex; P = end products; and k_1, k_{-1}, k_2, and k_{-2} are the velocity constants of the reaction. The formation of ES from $P + E$ has a negligible effect on the initial velocity of the reaction, which is defined as the velocity before an appreciable quantity of product is formed. If $[E]$ = total molecular concentration of the enzyme, $[S]$ = molecular concentration of the substrate, $[ES]$ = concentration of the enzyme–substrate complex, and $([E] - [ES])$ = concentration of free enzyme, one can derive the following expression:

$$\frac{[S]([E] - [ES])}{[ES]} = \frac{k_{-1} + k_2}{k_1} = K_m$$

where K_m is the Michaelis–Menten constant. The K_m can be determined solely from the reaction velocities, according to the following expression:

$$v = \frac{V[S]}{K_m + [S]}$$

where v = the observed reaction velocity, V = the maximum velocity of the reaction at saturating substrate concentration, and $[S]$ = initial substrate concentration. When the actual velocity of the reaction (v) is equal to half the maximum velocity (V), K_m is equal to the substrate concentration. This is illustrated graphically in Fig. 7.2. The K_m can, therefore, be determined experimentally by plotting the reaction rates at varying substrate concentrations. The reciprocal of the Michaelis–Menten equation, known as the Lineweaver–Burk equation, can be written as

$$\frac{1}{v} = \frac{K_m}{V}\frac{1}{[S]} + \frac{1}{V}.$$

According to this useful expression, a plot of the reciprocal of v versus the reciprocal of $[S]$ will give a straight line whose intercept on the y axis is equal to $1/V$ and whose slope is equal to K_m/V, as shown in Fig. 7.3.

The Michaelis–Menten constant is an important characteristic of an enzyme. A small K_m value indicates a high affinity of the enzyme for its substrate, while a large K_m value indicates a low affinity (i.e., a

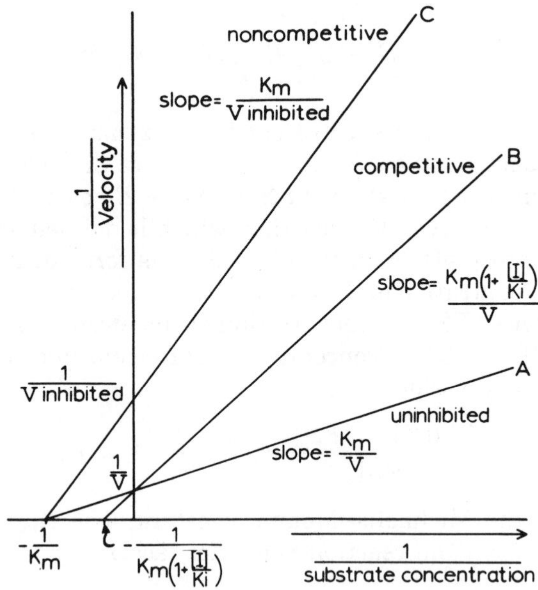

Figure 7.3. Lineweaver-Burk plot of the reciprocal of reaction velocity versus the reciprocal of substrate concentration. Uninhibited, competitively inhibited, and noncompetitively inhibited reactions are shown.

high substrate concentration is necessary to reach half-saturation). Michaelis–Menten constants are usually expressed in moles per liter. These constants usually range from 10^{-2} to 10^{-5} moles per liter.

Other Types of Kinetics

The responses of reaction velocity to changes in enzyme and substrate concentration described in the previous section are observed with many enzymes and can be described by Michaelis–Menten steady-state kinetics. However, not all enzymes exhibit such kinetics, and in recent years considerable attention has been focused on so-called sigmoidal kinetics and ping pong kinetics.

Sigmoidal Kinetics

Three types of sigmoidal response of initial velocity (v) to changes in substrate concentration ($[S]$) have been observed in enzyme-catalyzed reactions:

1. Sigmoidal response to substrate shown in plots of v vs $[S]$.
2. Sigmoidal response to substrate shown in plots of v vs $[S]$, but with additional modulation of velocity by effector.
3. Sigmoidal response to inhibitor (allosteric inhibitor) shown in plots of v/v_0 vs [I], where v = observed velocity in the presence of I and v_0 = original velocity in the absence of I.

The apparent major function of enzymes that respond sigmoidally to changing concentrations of substrate is the metabolic regulation of a physiologically important reaction or series of reactions. According to Monod *et al.* (1963):

> It would appear that certain proteins, acting at critical metabolic steps, are electively endowed with specific functions of regulation and coordination; through the agency of these proteins, a given biochemical reaction is eventually controlled by a metabolite acting apparently as a physiological "signal" rather than as a chemically necessary component of the reaction itself . . . . These proteins are assumed to possess two, or at least two, stereospecifically different, non-overlapping receptor sites. One of these, the *active site,* binds the substrate and is responsible for the biological activity of the protein. The other, or *allosteric site,* is complementary to the structure of another metabolite, the *allosteric effector,* which it binds specifically and reversibly. The formation of the enzyme–allosteric effector complex does not activate a reaction involving the effector itself: it is assumed only to bring about a discrete reversible alteration of the molecular structure of the protein or *allosteric transition,* which modifies the properties of the active site, changing one or several of the kinetic parameters which characterize the biological activity of the protein.

One of the early demonstrations of sigmoidal kinetics involved the enzyme aspartate transcarbamylase, which catalyzes the reaction of L-aspartate and carbamylphosphate to produce carbamyl aspartate; a series of further reactions ultimately results in formation of cytidine-5′-triphosphate (CTP). Thus, the aspartate transcarbamylase reaction is the first in the biosynthetic pathway for CTP.

The sigmoidal curve obtained when velocity is plotted versus aspartic concentration is theorized to be due to the effect of the first molecule of aspartate on the binding of subsequent molecules of aspartate (Fig. 7.4, curve B). If CTP is added, it causes an inhibition of the enzyme (curve C), the K_m increases and V_{max} remains unchanged but the sigmoidicity is still observed. CTP is obviously competitive with aspartate but both bind at different sites. This is evidenced by the fact that the inhibitory effect of CTP is not observed when the enzyme is "desensitized" by heating or treatment with heavy metals or urea (curve A). The desensitized enzyme follows ordinary hyperbolic kinetics (described by the Michaelis–Menten equation) indicating that the CTP site is no longer functional. Thus, CTP acts as an allosteric effector that modulates the first reaction in its biosynthetic pathway although CTP is not structurally or chemically similar to either of the substrates.

Physical chemical measurements indicate that aspartate transcarbamylase has a molecular weight of 310,000. The enzyme molecule has been shown to be composed of two types of subunits: regulatory ones (MW = 34,000) composed of two peptides of 17,000 MW and catalytic ones (MW = 99,000) made up of three subunits of 33,000 MW. Each molecule of the native enzyme has six regulatory peptide units and six

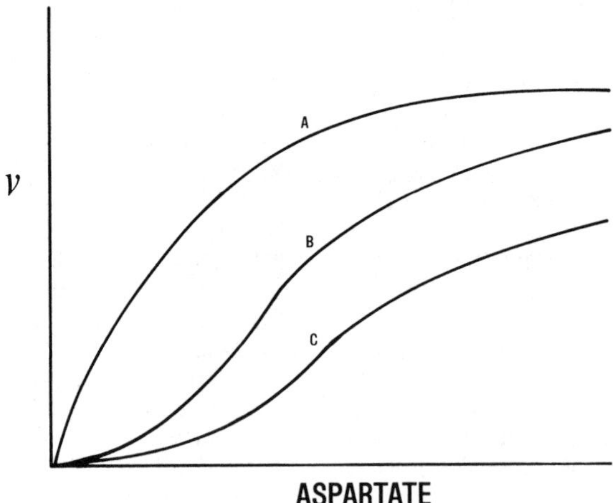

Figure 7.4. Sigmoidal kinetics exhibited by aspartate transcarbamylase. Curve A—desensitized enzyme in presence of CTP exhibits normal Michaelis Menten kinetics. Curve B—normal enzyme in absence of CTP. Curve C—normal enzyme in presence of CTP exhibits competitive inhibition.

catalytic subunits for a total molecules weight of approximately 310,000. The regulatory subunit has no catalytic activity but binds CTP, whereas the catalytic subunit is not inhibited by CTP and is catalytically active toward aspartate.

Two models have been widely accepted to explain the sigmoidal response of reaction velocity to changes in substrate concentration as illustrated in curves B and C in Fig. 7.4: These are the sequential interaction model and the concerted transition model.

Sequential Interaction Model. In this model, binding of substrate causes a conformational change in the enzyme so that the binding of a second molecule of substrate is affected. Each substrate molecule that binds makes it easier for the next substrate molecule to bind. After an initial lag, the velocity increases markedly, followed by the normal sloping off as the enzyme approaches saturation. Modification of the Michaelis–Menten equation yields the following expression, which describes a sigmoidal curve:

$$v = \frac{V_{max}\,[S]^n}{K' + [S]^n}$$

where n = number of substrate binding sites for each molecule of enzyme. In this equation, K' is numerically equal to $[S]^n$ at 50% maximum velocity and is not the same as K_m, which equals $[S]$ at 50% V_{max}.

The linear form of this equation, called the Hill equation, is patterned after the linear expression developed by Hill for the binding of oxygen to hemoglobin (which is a sigmoidal curve). The Hill equation for enzyme-catalyzed reactions is

$$\log \frac{v}{V_{max} - v} = n \log [S] - \log K'$$

where n represents only the sigmoidicity of the curve and not the absolute number of binding sites. A value of 2.5 for n may represent three fairly strong binding sites or four, five, or six weaker ones.

Concerted Transition Model. This model assumes that the identical units composing the enzyme have only one binding site for each ligand (substrate, inhibitor, or activator). The complete enzyme can exist in different conformations, which exist in equilibrium. As a particular ligand binds preferentially to a particular conformation, there is a shift in equilibrium to the form that must be replenished.

R and T have been used to designate two forms of a theoretical tetramer. T is the taut form with a lower affinity for substrate, and R is the relaxed form with a higher affinity for substrate. Thus, the equilibrium constant (L) for $R \rightleftharpoons T$ is $L = [T]/[R]$.

The dissociation constant for S binding to T is often designated K_{ST}. The dissociation constant for S binding to R is designated K_{SR}. The ratio K_{SR}/K_{ST} is designated c. The cooperativity of substrate binding depends on L and c. The velocity curves become more sigmoidal as L increases and c decreases. When L increases the $R_0 \rightleftharpoons T_0$ equilibrium shifts in favor of T_0. When c decreases, the affinity of the T state decreases relative to the affinity of the R state for S. When allosteric inhibitors bind preferentially to the T state, the $T_0 \rightleftharpoons R_0$ equilibrium shifts in favor of T_0. As the constant L increases, the velocity curves become more sigmoidal and n (Hill coefficient) approaches the actual number of binding sites.

Ping Pong Kinetics

When an enzyme-catalyzed reaction involves two or more substrates or two or more products, the reaction mechanism, or sequence, becomes more complicated than that described previously in the section on Michaelis–Menten Constants. Many multisubstrate/multiproduct reactions exhibit ping pong kinetics. For such reactions, Lineweaver–Burk plots of $1/v$ versus $1/[S]_1$ at several fixed concentrations of S_2 gives parallel straight lines (S_1 and S_2 are two different substrates).

Cleland (1967, 1970) devised a shorthand method for designating ping pong reaction sequences. A summary of Cleland's terminology, based on Plowman (1972), follows:

1. The number of kinetically important substrates or products is designated by the syllables *Uni, Bi, Ter, Quad,* etc., as they appear in the mechanism.
2. The number of reactants involved in the reaction in one direction is the reactancy for that direction. For example, a reaction with two substrates and two products is called *Bi Bi* and is said to be bireactant in both directions. Likewise, a reaction with one substrate and two products is a *Uni Bi* reaction, unireactant in the forward direction and bireactant in the reverse direction. Reactions with three substrates and four products are termed *Ter Quad* reactions.
3. Since most reactions are studied at constant pH in an aqueous medium, H^+ and H_2O are not ordinarily considered substrates.
4. Consideration of cofactors, coenzymes, or prosthetic groups as substrates depends on the circumstances.

5. A *sequential* mechanism is one in which all the substrates must be present on the enzyme before any products can leave.
6. Sequential mechanisms are designated *ordered* or *random* depending on whether the substrates add and the products release in an obligatory sequence or in a nonobligatory sequence.
7. A *ping pong* mechanism is one in which one or more products are released during the substrate addition sequence, thereby breaking the substrate addition sequence into two or more smaller segments.
8. Sequential and ping pong mechanisms are further described by a set of four or more syllables (*Uni, Bi,* etc.) to indicate the number of substrate additions and product releases in each such group.
9. The letters A,B,C, and D designate substrates in the order in which they add to the enzyme. The letters P,Q,R,S . . . designate products in their order of release from the enzyme.
10. Stable enzyme forms are designated by the letters E,F,G,H . . . , with E being the least complex, or "free", enzyme if such a distinction can be made.
11. Transitory enzyme forms that undergo or can isomerize to undergo unimolecular degradation with release of substrates or products rather than participating in bimolecular steps with substrate or product are called *central complexes*. These central complexes normally are enclosed in parentheses to distinguish them from other transitory forms.
12. For the purpose of simplifying the equations to be written, only one central complex is included for a given reaction sequence. The rationale for this convention is that steady-state kinetic studies, even in combination with many other types of data, are incapable of measuring isomerization of central complexes (as opposed to other transitory forms).
13. In mechanisms where isomerization of a stable enzyme form takes place as a part of the reaction sequence, new terms are added to the rate equation and are designated by the prefix Iso (e.g., Iso Ordered, Iso Random, and Iso Ping Pong).
14. In ping pong mechanisms, two or more stable enzyme forms may occur, leading to *Di*-Iso Ping Pong or to *Tri*-Iso Ping Pong. For greater precision the Iso prefix may be inserted at the appropriate point in the full name to indicate which stable enzyme form is isomerizing.
15. A reaction sequence is written from left to right, with a horizontal line or group of lines representing the enzyme in its var-

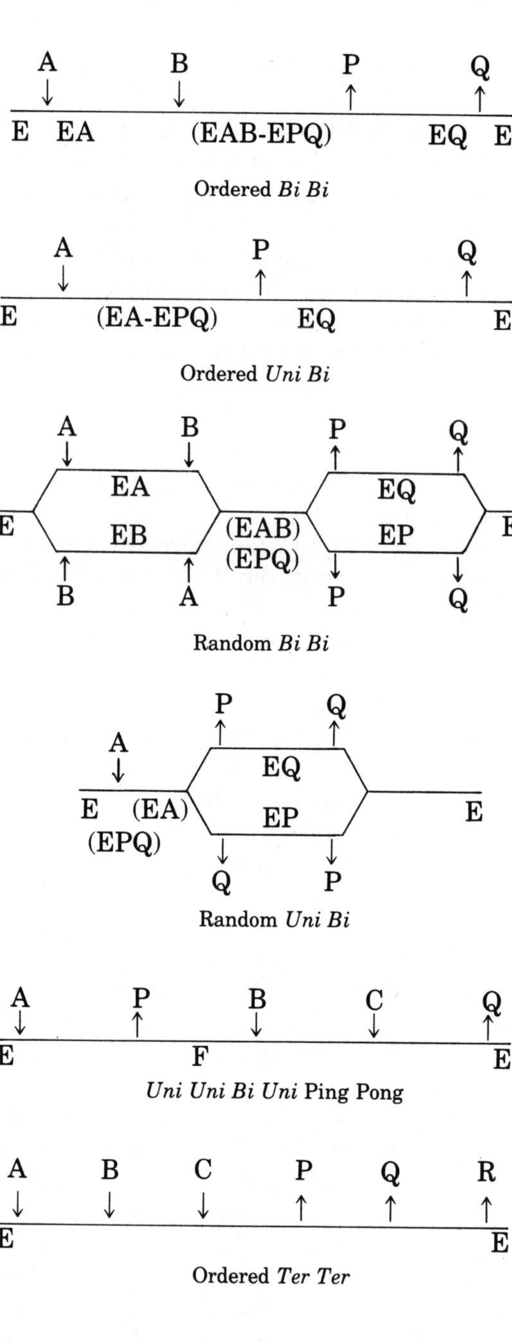

Ordered *Bi Bi*

Ordered *Uni Bi*

Random *Bi Bi*

Random *Uni Bi*

Uni Uni Bi Uni Ping Pong

Ordered *Ter Ter*

Bi Uni Uni Uni Ping Pong

ious forms. Substrate additions and product dissociations are indicated by appropriately aimed vertical arrows above the enzyme line.

Several reaction sequences written according to Cleland's terminology are illustrated in Fig. 7.5. Based on detailed kinetic studies of an enzyme-catalyzed reaction, it is possible to determine which type of reaction sequence is involved.

Effect of Temperature

An increase in temperature increases the velocity of an enzyme-catalyzed reaction only within definite limits. Initially, the reaction velocity increases with increasing temperature to an optimum, but at higher temperatures it decreases eventually to zero. The optimum temperature is dependent upon time; i.e., lengthening the time of exposure at elevated temperatures causes a fall in the apparent optimum (Fig. 7.6). This is due to two opposing effects: (1) the temperature-dependent increase in velocity of the catalyzed reaction due to a greater molecular activity and (2) an increase in the rate of destruction of the enzyme by heat denaturation. It is generally found that increases in temperature, up to about 45°C, produce corresponding increases in reaction rate (effect a). Above about 45°C, thermal denaturation of protein (effect b) becomes increasingly important, and the activity of the enzyme gradually decreases, eventually reaching zero when catalysis ceases.

Most enzymes are almost instantly inactivated by exposure to temperatures near 100°C; heating at 80°C requires a longer exposure time before complete inactivation occurs. This fact is utilized in the preparation of vegetables for canning and freezing. The vegetables are blanched by immersion in hot water or by treating with steam in order to inactivate enzymes that would otherwise adversely affect nutritive values, colors, and flavors. In some instances heat inactivation is reversible. For example, the peroxidase of peas is completely inactivated by blanching for 60 sec, but regenerates on standing and regains its activity.

The optimum temperature for an enzyme is influenced not only by the time of exposure to heat but also by pH, the effect of activators or inhibitors, the presence of protective proteins, etc.

The effect of temperature changes on the kinetics of an enzyme re-

Figure 7.5. Representative reaction sequences written according to terminology of Cleland (1967, 1970). See text for explanation of terms.

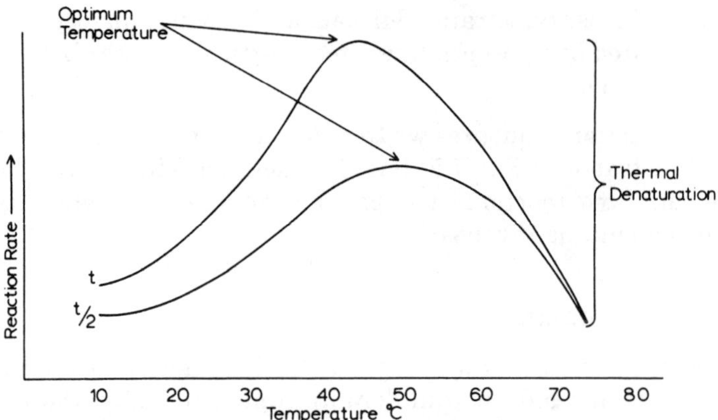

Figure 7.6. Effect of temperature on enzyme reaction rate at constant enzyme and substrate concentrations.

action is best described by the Arrhenius equation:

$$k = Ae^{-E_a/RT} \qquad \text{or} \qquad \log k = -\frac{E_a}{2.3R}\frac{1}{T} + \log A$$

where k = the rate constant for a specific rection; A = a constant for that reaction; E_a = activation energy; R = the Boltzmann constant; and T = temperature. Maximum velocity (V_{max}) can be used instead of k; thus, the Arrhenius plot for an enzyme reaction may be constructed simply by determining V_{max} at various temperatures and then plotting log V_{max} versus $1/T$ (°K) as shown in Fig. 7.7.

Curve A is the ideal-type plot and indicates that V_{max} increases with increasing temperature. However plots such as curve B occur when a different step in the reaction becomes rate limiting at a certain temperature (indicated by inflection point). Curve C is observed when inactivation of the enzyme (e.g., denaturation) occurs at a certain temperature. For enzymic reactions, the temperature coefficient (Q_{10}), which is the increase in reaction rate for each 10°C increase in temperature, generally is 1.1–5.3.

The activation energy (E_a) of a reaction can be determined from the slope of its Arrhenius plot. The activation energy (cal/mole) represents the amount of energy required to activate the molecules of reactants sufficiently for them to collide and react. Catalysts, including enzymes, reduce the activation energies of reactions and thus increase their velocities. In Table 7.1, the values of E_a for a number of reactions are listed. It is apparent that reactions catalyzed by enzymes have lower

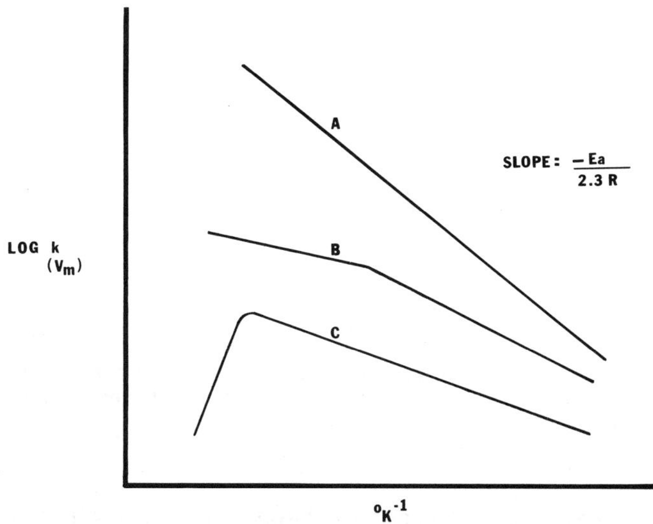

***Figure** 7.7.* Arrhenius plot of V_{max} of enzymic reaction versus of the maximum velocity (V_{max}) of an enzymic reaction versus the reciprocal of the temperature (°K). At sufficiently high temperatures, enzymes are denatured and the reaction rate falls to zero.

energies of activation than the same reactions catalyzed by inorganic catalysts. In other words, the presence of an enzyme mediates a chemical reaction at a lower energy level and at a lower temperature.

Effect of pH

The activity of an enzyme is greatly affected by the hydrogen ion concentration of its medium. It is likely that the change in enzyme

***Table** 7.1.* Examples of Activation Energies with Various Catalysts

Reaction	Catalyst	E_a (cal/mole)
Sucrose inversion	H^+ (nonenzymatic)	26,000
	Yeast invertase	11,500
Casein hydrolysis	H^+ (nonenzymatic)	20,600
	Trypsin	12,000
Ethyl butyrate hydrolysis	H^+ (nonenzymatic)	13,200
	Pancreatic lipase	4,200

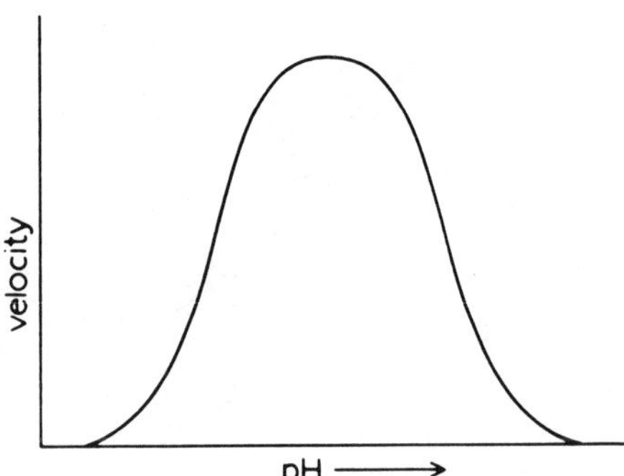

Figure 7.8. Effect of pH on enzyme reaction rate at constant enzyme and substrate concentrations. The optimal pH may vary with the source of the enzyme, temperature, and type of substrate.

activity with varying pH is due to changes in ionization of either the enzyme, the substrate, or the enzyme–substrate complex. In addition to such ionic effects, low or high pH can cause inactivation due to the denaturation of the enzyme protein.

If the rate of an enzymatic reaction, at a given enzyme and substrate concentration, is plotted against different pH values, the resulting curve usually is bell-shaped (Fig. 7.8). Such curves have a relatively small pleateau of optimum activity with sharply decreasing rates on either side of the optimal pH. The optimal pH is not constant under all conditions; it may vary with the source of the enzyme, the kind of substrate, and the temperature (Table 7.2).

Table 7.2. Effect of Enzyme Source, Substrate, and Reaction Temperature on pH Optimum

Factors	Enzyme	Optimum pH
Source		
Green bean	Pectin methylesterase	8.2
Fungae	Pectin methylesterase	5.5
Substrate		
Ovalbumin	Pepsin	1.5
Casein	Pepsin	1.8
Gelatin	Pepsin	2.0
Temperature		
25°C	Malt diastase	4.3
45°C	Malt diastase	5.0
60°C	Malt diastase	5.7

Activation

Some enzymes show an absolute requirement for a particular inorganic ion for their catalytic activity, while other enzymes show increased activity when certain cations are included in the reaction medium. Examples of enzymes with an absolute requirement are cytochrome oxidase (Fe^{2+}) and enolase (Mg^{2+} and Mn^{2+}). On the other hand, pyruvate kinase is activated by K^+, $NH_4{}^+$, and other ions but can function without them. Among the more important cation activators are K^+, Mg^{2+}, Zn^{2+}, Fe^{2+}, Ca^{2+}, and Co^{2+}. Some of the divalent cations can replace each other, but occasionally they compete with one another. For example, Mn^{2+} can replace Mg^{2+} in activating some of the phosphate-transferring enzymes, but pyruvate kinase, which is normally activated by Mg^{2+}, is inhibited by Ca^{2+}.

Relatively few enzymes are affected by the presence or absence of anions. Salivary and pancreatic amylases, which are activated by chlorides, are notable exceptions.

Cells sometimes produce proteins that are inactive forms of enzymes. This type of precursor is called a *zymogen* or *proenzyme*. An example of a zymogen is trypsinogen, which is produced by the pancreas. When the pancreatic juice containing trypsinogen reaches the small intestine, it is converted to trypsin by the enzyme enterokinase. In the conversion of the inactive form to the active form, an inhibitory peptide is removed and there is an alteration in the conformation of the protein molecule.

Another type of activation involves the use of reducing agents to maintain the integrity of sulfhydryl groups in an enzyme. These groups are essential parts of the active center of some enzymes and when oxidized to the disulfide form these enzymes are inactivated. The plant proteolytic enzymes papain, ficin, and bromelin are inactivated by exposure to oxygen; when a suitable reductant is added (cysteine, glutathione, or bisulfite), the enzyme remains in an active state.

Inhibition

Inhibitors, which lower the catalytic activity of enzymes, may be divided into two categories, competitive and noncompetitive inhibitors.

Competitive Inhibition

Competitive inhibition occurs when a compound competes with a substrate (or coenzyme) for the binding site of an enzyme. It implies a similarity in the structure of the inhibitor and the substrate. The amount of inhibition is related to (1) the concentration of substrate, (2) the concentration of inhibitor, and (3) the relative affinities (K_m)

for the enzyme–substrate and enzyme–inhibitor complex. The inhibitory effect is reversible; that is, activity can be restored by increasing the substrate concentration.

The kinetic behavior of a competitively inhibited system is shown in Fig. 7.3 (curve B). In a competitively inhibited system, the maximal velocity remains unchanged since the effect of the inhibitor is overcome at higher substrate concentrations. The slope of a Lineweaver–Burk plot is increased because the effective K_m of the enzyme is increased by the factor $(1 + [\mathrm{I}]/[K]_i)$ where $[K]_i$ is the dissociation constant of the enzyme–inhibitor complex.

Noncompetitive Inhibition

Two types of noncompetitive inhibition occur. In classical irreversible inhibition, the inhibitor binds irreversibly to the enzyme. This results in an apparent decrease in $V_{\max}$ but K_m remains constant. This type of inhibition involves the irreversible reaction $\mathrm{E} + \mathrm{I} \rightarrow \mathrm{EI}$ and cannot be reversed by increasing the substrate concentration. The amount of inhibition is dependent upon inhibitor concentration and the affinity of the inhibitor for the enzyme.

The kinetics of noncompetitive inhibition is illustrated by curve C in Fig. 7.3. The decrease in maximal velocity is due to the decrease in concentration of active enzyme. The K_m is unchanged since noncompetitive inhibitors appear kinetically only to remove active enzyme molecules from the reaction not to decrease the enzyme's affinity for substrate. The slope is increased due to the decreased velocity throughout the range of enzyme concentrations. Some irreversible inhibitors are specific in their action: iron-containing enzymes are inhibited by cyanide; enzymes containing sulfhydryl groups are inhibited by Cu^{2+}, Hg^{2+}, and *p*-hydroxymercuribenzoate; and some esterases are inhibited by certain organic phosphates (e.g., parathion and tetraethyl pyrophosphate).

In reversible noncompetitive inhibition, the inhibitor has no effect on substrate binding, nor does substrate binding have any effect on inhibitor binding. Substrate and inhibitor bind reversibly, randomly, and independently at different sites. The reaction may be represented as follows:

$$\begin{array}{ccccc} \mathrm{E} + \mathrm{S} & \rightleftharpoons & \mathrm{ES} & \rightarrow & \mathrm{E} + \mathrm{P} \\ + & & + & & \\ \mathrm{I} & & \mathrm{I} & & \\ \upharpoonleft\downharpoonright & & \upharpoonleft\downharpoonright & & \\ \mathrm{EI} + \mathrm{S} & \rightleftharpoons & \mathrm{ESI} \text{ or } \mathrm{EIS} & & \end{array}$$

However ESI is not a catalytically active and productive complex. Reversible noncompetitive inhibitors have the same effect on V_{max} and K_m as irreversible noncompetitive inhibitors.

The use of chemical enzyme inhibitors in foods is for the most part rather limited because of the toxic nature of those substances. However, living organisms produce substances that inhibit enzyme action. For example, a trypsin inhibitor, which occurs naturally in soybeans, inhibits the digestion of the protein in raw soybean meal. Antipepsin and antitrypsin protect the stomach and intestines from the action of proteolytic enzymes that are active in these organs.

☐ STRUCTURE AND ACTIVE SITE

Since all enzymes are proteins, their basic structural characteristics are the same as those described in Chapter 6 for proteins in general. As noted earlier in this chapter, many proteins also contain a non-protein component, either a prosthetic group or coenzyme.

The activity of an enzyme depends not only on its primary structure (i.e., sequence of amino acids) but also on its conformation (secondary and tertiary structure). Disruption of an enzyme's conformation, by heat or other agents, generally results in a loss of activity. In addition, as mentioned in the section on sigmoidal kinetics, some enzymes are known to contain more than one polypeptide chain and thus exhibit quarternary structure. Disruption of the quarternary structure of such enzymes (e.g., by dissociating them into their subunits) also is likely to cause a loss of activity.

Although enzymes, like all proteins, are large molecules, only a portion of an enzyme molecule—called the active site(s)—is critical for catalytic activity. Indeed, the active site(s) of an enzyme is the position or area at which the catalysis of the substrate occurs. Since a large number of enzymes contain only protein, it follows that in such cases the active site must in fact be one or more amino acid residues. Additionally, coenzymes and metals are frequently involved in the catalytic process as an integral part of the enzyme–substrate complex. Thus, there must also be binding sites for these obligatory components of the reaction. Kinetically, the coenzymes and metal ions frequently follow the same type of Michaelis kinetics as is shown by the substrate. Many enzymes are known to also undergo conformational changes when placed in contact with their substrate, cofactor, or metal ion; that is, their tertiary structure is modified during reaction.

Large portions of the protein moeity of some enzymes can be removed with little change in activity. This indicates that the portion removed plays no special role in the catalytic process and that the active site corresponds to a relatively small portion of the enzyme. It has been found that some amino acids are more likely than others to be components of the active site. Those that are frequently found at the active site are serine, histidine, glutamic acid, aspartic acid, arginine, and tyrosine.

Serine, which is especially common as an active-site amino acid, is known to be present in the active site of chymotrypsin, trypsin, phosphoglucomutase, phosphorylase, and alkaline phosphatase. A classic example of the role of serine in enzymatic catalysis occurs in the isomerization reaction of glucose-1-phosphate to glucose-6-phosphate catalyzed by phosphoglucomutase [see Walsh (1979)]. The intermediate (glucose 1,6-diphosphate) and the phosphorylated enzyme are sufficiently stable to allow isolation. The reaction sequence can be represented as follows: Serine phosphate may be isolated from the phos-

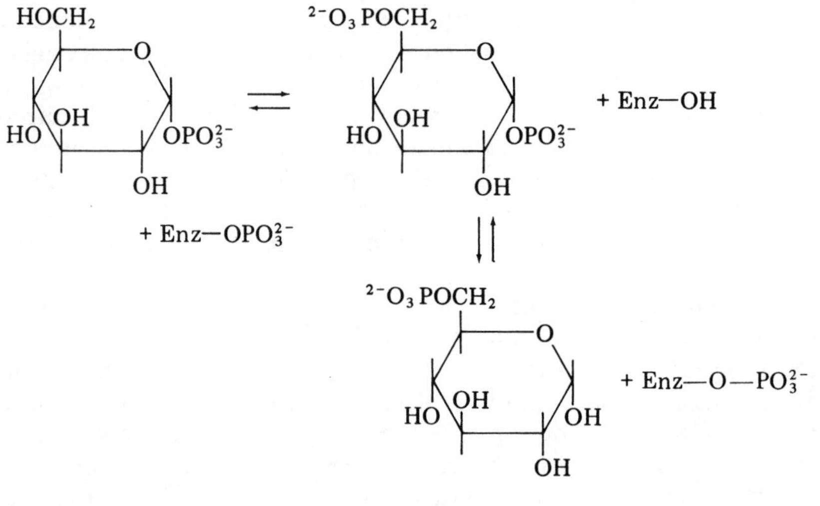

$HOCH_2$ O, HO, OH, OPO_3^{2-}, OH + Enz—OPO_3^{2-} ⇌ $^{2-}O_3POCH_2$ O, HO, OH, OPO_3^{2-}, OH + Enz—OH

⇅

$^{2-}O_3POCH_2$ O, HO, OH, OH, OH + Enz—O—PO_3^{2-}

phorylated enzyme; the amino acid residue on the N-terminal side of the serine is alanine or glycine and on the C-terminal side is histidine.

Active serine residues may be located in other enzymes by reaction with diisopropylfluorophosphate, as shown in the following. Enzymes that contain more than one serine residue appear to react with diisopropylfluorophosphate more rapidly at the active-site hydroxyl.

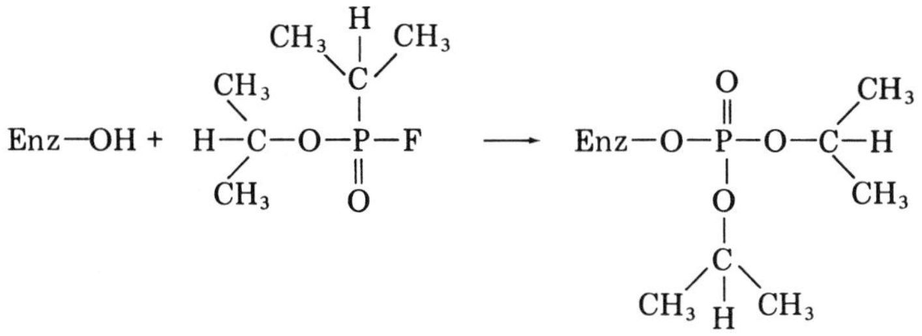

$$\text{Enz—OH} + \text{H—}\underset{\text{CH}_3}{\overset{\text{CH}_3}{\text{C}}}\text{—O—}\underset{\text{O}}{\overset{\text{CH}(\text{CH}_3)_2}{\text{P}}}\text{—F} \longrightarrow \text{Enz—O—}\underset{\text{O—CH}(\text{CH}_3)_2}{\overset{\text{O}}{\text{P}}}\text{—O—}\underset{\text{CH}_3}{\overset{\text{CH}_3}{\text{C}}}\text{—H}$$

NOMENCLATURE AND CLASSIFICATION

Enzymes are frequently named after the substrate upon which they act plus the suffix *-ase*. Thus, the term *amylase* indicates an enzyme that acts upon starch, while *lipase* is an enzyme acting upon lipids. Further definition has proven necessary in order to indicate narrower substrate specificities, e.g., acid phosphatase and alkaline phosphatase. It is clear that a nomenclature system that is based partly on trivial names, partly on names of substrates, etc., is not very satisfactory.

The International Union of Biochemistry (1978) has adopted the recommendations of its Commission on Enzymes for naming and classifying enzymes. The classification of enzymes is based on the type of reaction catalyzed by the enzyme. Six main divisions are employed in the classification:

1. Oxidoreductases
2. Transferases
3. Hydrolases
4. Lyases
5. Isomerases
6. Ligases

Oxidoreductases are enzymes that catalyze oxidation-reduction reactions; dehydrogenases, oxygenases, and peroxidases are typical enzymes of this group. Transferases are enzymes catalyzing group transfers; transaminases and kinases are typical enzymes of this group. Hydrolases have as their primary function the hydrolysis of a variety of compounds by water; lipases and peptidases are well-known enzymes in this group. Lyases are enzymes which cleave C–C, C–O, C–N, and other bonds by elimination, leaving double bonds or rings, or conversely adding groups to double bonds. Aldolases, decarboxylases, and hydratases are included in this class of enzymes. Various subclasses of lyases include pyridoxal-phosphate enzymes that catalyze the elimination of a β- or γ-substituent from an α-amino acid followed by a replacement of the substituent by some other group. Isomerases catalyze the geometric or structural changes with one molecule. This class includes racemases, epimerases, cis–trans isomerases, isomerases, tautomerases, mutases, and cyclo-isomerases.

Finally ligases catalyze the linking together of two substrate molecules with the breaking of a pyrophosphate bond; aminoacyl-tRNA synthetases are important enzymes in this group.

In the systematic nomenclature, each enzyme is assigned a unique number consisting of four parts, which are separated by decimals; the entire number is preceded by EC, which denotes the Enzyme Commission. The designation EC 1.1.1.1 would be deciphered as follows:

1 Oxidoreductase
1.1 Action on CH_2-OH
1.1.1 NAD^+ or $NADP^+$ as hydrogen acceptor
1.1.1.1 Specific enzyme—Alcohol:NAD^+ oxidoreductase

In general, the systematic name of an enzyme consists of two parts: the first part denotes the substrate and the second part indicates the nature of the reaction. To illustrate, the well-known β-amylase (trivial name) has the systematic name α-1,4-glucan maltohydrolase (EC 3.2.1.2). The name indicates that the substrate is a glucan (a glucose polymer) in which the molecules are joined by α-1,4-glucosidic linkages. The reaction is hydrolytic and it results in the splitting off of maltose. Although the use of systematic names for enzymes is recommended, in food technology trivial names are used most frequently.

☐ FOOD ENZYMES

The enzymes present in most fresh food materials have the capability of causing both undesirable and desirable changes in foods. Therefore, the management of these naturally occurring enzymes is an important consideration in food technology. The characteristics of the more important enzymes occurring in foods are summarized in Table 7.3.

Amylases are among the most widely distributed of all enzymes, being found in plants, animals, and some microorganisms. These enzymes, which hydrolyze polysaccharides to disaccharides and some monosaccharides, are used widely in the brewing and baking industries.

Most living cells contain proteolytic enzymes that can act on proteins under appropriate conditions. In living tissue these enzymes are in-

Table 7.3. Important Enzymes Occurring in Foods

Enzyme	Sources	Substrates	pH optimum	Temp optimum	Metal ions involved	Mode of action and product(s)
CARBOHYDRASES						
α-Amylase (EC 3.2.1.1)	Plants, mammalian tissue, microorganisms	Amylose, amylopectin glycogen, other α-1,4 polysaccharides	4.5–7.0 depending on source. Generally near 4.5	25°–70°C; bacterial sources may have high temp. optimum	Ca (not obligate) and Zn	Source-dependent random endodegradation produces reducing sugars. Amylose degraded in two steps: (1) amylose $\xrightarrow{\text{fast}}$ maltose and maltotriose and (2) oligosaccharides $\xrightarrow{\text{slow}}$ glycose and maltose. Amylopectin → glucose, maltose, α-limit dextrins, and oligosaccharides. (≥4) with α-1,6 glycosidic bonds. α-Amylases from different sources produce different limit dextrins. α-C-1 configuration generally retained.
α-Amylases (EC 3.2.1.2)	Higher plants (e.g., wheat, sweet potatoes, soybeans, barley malt)	Amylose, amylopectin glycogen, other α-1,4 polysaccharides	5.0–6.0	Less stable than α-amylase sweet potato β-amylase stable to 65°C	None known	Attacks α-1,4 glycosidic bonds by exodegradation starting at nonreducing end of the chains to yield maltose straight-chain molecules with an even number of glucose residues. Glucose and maltotriose are also end products when polysaccharide has an uneven number of glucose units. C-1 position undergoes inversion from α → β.
Glucoamylase (EC 3.2.1.3)	*Aspergillus* and *Rhizopus*	Saccharides with α-1,6 and α-1,4 linkages. Lack of specificity, e.g., amylose, maltose, maltotriose, nigerose, isomaltose and many others are attacked.	4–5	50°–60°C	None known	Attacks nonreducing end of polymer (starch) at α-1,3, α-1,6, or α-1,4 linkages. Yields glucose with inversion at C-1 from α to β configuration.

(continued)

Table 7.3. *(Continued)*

Enzyme	Sources	Substrates	pH optimum	Temp optimum	Metal ions involved	Mode of action and product(s)
Amylo-1,6-glucosidase (EC 3.2.1.33) and 4-α-glucanotransferase (EC 2.4.1.25)	Plants and animals	Limit dextrins modified from glycogen and/or amylopectin	4–5	50°C	—[a]	EC 3.2.1.33 causes endohydrolysis of α-1,6-glucosidic linkages at points of branching in chains of α-1,4-linked glucose residues. EC 2.4.1.25 transfers a segment of a α-1,4-glucan to a new 4-position in an acceptor, which may be glucose or a α-1,4-glucan. These two enzymes make up a system for so-called "indirect debranching of limit dextrins," where transfer precedes hydrolysis by the glucosidase. Converts branched-chain polyglucoside to straight-chain ones.
Pullunlanase (EC 3.2.1.41)	Plants (e.g., potatoes, beans, sweet corn), *Aerobactei aerogenes*, *E. intermedia*, *Streptococcus mitis*, etc.	Pullulan, α- and β-limit dextrins of amylopectin and glycogen, and amylopectin, *per se*, with some enzymes	5.0	25°C	—	Endohydrolysis of α-1,6-glucosidic linkages in pullulan and α- and β-amylase-produced limit dextrins of amylopectin and glycogen, also amylopectin, *per se*. Specifically, plant pullulanase (R-enzymes) hydrolyze α-1,6-glycosidic bonds in amylopectin and its β-limit dextrin, also cleave 1,6-bonded α-maltose and maltotriose residues in α-limit dextrin. R-enzymes do not attack glycogen. The microbial pullulanases are similar in activity and specificity to the R-enzyme.
Isoamylase (EC 3.2.1.68)	*Pseudomonas*, *Cytophaga*, and yeast	Glycogen, amylopectin, and their β-limit dextrins	5.0	25°C	—	Does not attack pullulan and weakly attacks α-limit dextrins. Generally hydrolyzes α-1,6-glucosidic branched linkages in glycogen, amylopectin and their β-limit dextrins. Bacterial enzyme action on glycogen is complete.

Invertases 1. α-glucosidase (EC 3.2.1.20)	*S. cerevisiae* *S. carlsbergensis* *K. fragilis* *S. mellis*	Sucrose, maltose, α-1,4-oligosaccharides, e.g., gluc-fruc-gluc (melezitose)	4.5	55°C	—	Hydrolyzes the terminal, nonreducing α-1,4 linked glucose residues to release α-glucose. Will hydrolyze melezitose (gluc-fruc-gluc) to glucose and turanose (fruc-gluc) but will not hydrolyze raffanose (galac-gluc-fruc).
2. β-Fructo-furanosidase (EC 3.2.1.26)		Sucrose and other terminal, unsubstituted β-D-fructofuranosyl residues, e.g., galac-glu-fruc	4–5.5 (range 3.5–5.5)	ca. 55°C. media dependent (55°–7.0°C range)	—	Will hydrolyze sucrose and oligosaccharides which contain head-terminal fructose (β-fructofuranosyl) e.g., raffinose (galac-gluc-fruc). Also shows transferase activity. Catalyzes D-fructofuranosyl from sucrose to the carbonyl group of D-glucose and D-fructose to form *o*-β-D-fructofuranosyl-(2→6)-D-glucopyranose, *o*-β-D-fructofuranose, and *o*-β-D-fructfuranosyl-(2→6)-D-fructofuranose. Methanol and ethanol may also serve as acceptors.
Lactase or β-Galactosidase (EC 3.2.1.23)	Plants (e.g., almonds, peaches, apples, apricots), microorganisms (e.g., *A. oryzae*, *A. foetidus*, *K. fragilis*, *E. coli*, *A. niger*), some yeasts mammalian intestine, primarily in animals that suckle their young.	Lactose. Some enzymes will hydrolyze α-L-arabinosides	Bacteria—7.0 Fungal—5.0 Yeast—6.0	Fungal—50°C Yeast—37°C Bacterial—37°C. Other bacterial reported to be 50°C or more.	None required. Hg^{2+}-is inhibitory as is copper and iron.	Hydrolyzes lactose → glucose and galactose. Generally hydrolyzes nonterminal β-D-galactose residues in β-galactosides. β-D-galactosyl transfer occurs preferentially at the primary alcohol of D-glucose to lactobiose, 3-*o*-β-D-galactopyranosyl-D-galactopyranose and at the nonreducing galactosyl residues of lactose or galactobiose to form lactotriose or galactotriose.

(*continued*)

Table 7.3. *(Continued)*

Enzyme	Sources	Substrates	pH optimum	Temp optimum	Metal ions involved	Mode of action and product(s)
Cellulase (EC 3.2.1.4)	Soil organisms (e.g., wood-rotting fungi) and digestive juices of cattle and invertebrates, but probably is of micoorganism origin. Specific sources are *Trichoderma viride*, *T. koningii*, *Fusarium solani*, *Chaetomium globosum*, and others.	Cellulose, lichenin and cereal β-glucans	4.5–6.5	Very stable to higher temperatures; e.g., some are stable for 10 min at 100°C. Optimum is much lower, e.g., 35°–50°C.	Some are inhibited by heavy metals.	Endohydrolysis of β-1,4-glucosidic linkages in cellulose lichenin and cereal β-glucans. Carboxymethyl cellulose generally used as the substrate for measuring activity. Enzyme preparations from *Trichoderma viride* have been found to contain three components: (1) C1 component-no action on CMC or cellobiose. (2) A component with high activity toward CMC but low activity toward cellobiose. (3) A component with a high activity toward cellobiose but a low activity toward CMC. The structure and function of C1 factor is not conclusively known.
PECTIC ENZYMES						
Pectinesterase (EC 3.1.1.11)	Plants (e.g., tomato, alfalfa, orange), fungi (*Coniophera cerebella*, and bacteria (*X. campestris* and *vasculorum*)	Methyl ester of polygalacturonic acid (pectin)	Fungal—4.5 (substrate dependent); 6.5 for nongalacturonide esters. Plant—5.5–8. Bacterial—7.5–8.0.	30°–40°C depending upon source	Plant pectinesterases activated by 0.15 *M* NaCl or 0.03 *M* $CaCl_2$	Pectinic acid (methyl ester of polygalacturonic acid) + $nH_2O \rightarrow$ polygalacturonic acid + methanol (Carbohydrate units include a chain of axial-axial α-1,4-linked D-galacturonic acid units with regions of L-rhamnose and side chains of arabinose, galactose and xylose. Approximately 75% of the carboxyl groups are esterified with methanol.)
Endopectin lyase (EC 4.2.2.10)	Fungi (*Aspergillus*, *Fusarium*, *Penicillium*, *Sclerotinia*)	Pectinic acids (does not act on de-esterified pectin)	5.1–5.9	30°–40°C	None observed	Elimination of 6-methyl-Δ-4,5-D-galacturonate residues from pectinic acids, thus bringing about depolymerization.

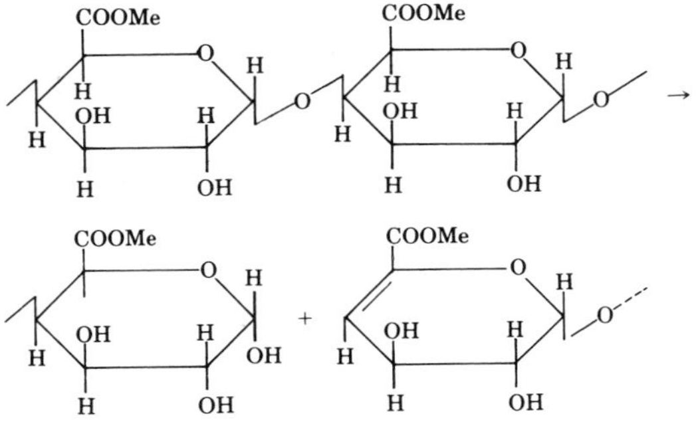

Enzyme	Source	Substrate	pH optimum	Temperature optimum	Reaction	Comments
Endopolygalacturonases (EC 3.2.1.15)	Higher plants (avocado and tomato), fungi (widely distributed between various genus) yeasts, bacteria	Pectic acid (pectate) or other galacturonans (e.g., tetra-, tri- and digalacturonic acids)	Tomato—2.5, 3.5, and 4.5 depending on mol wt of substrates Yeast (*K. fragilis*)—3.3–4.4 depending on mol wt of substrate *A. niger* PG—4.0–4.2	30°–40°C	Hydrolyzes 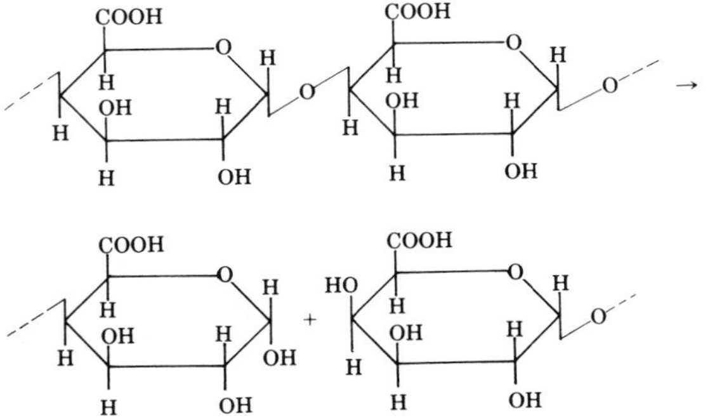 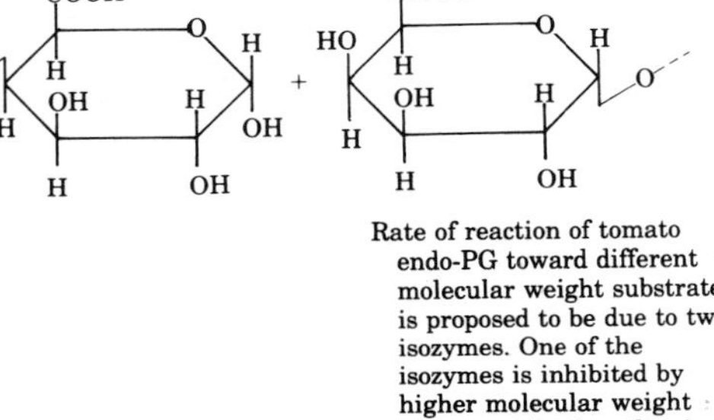	Rate of reaction of tomato endo-PG toward different molecular weight substrates is proposed to be due to two isozymes. One of the isozymes is inhibited by higher molecular weight pectate while the other is not. This would also explain the variation in pH optimum.

(*continued*)

Table 7.3. *(Continued)*

Enzyme	Sources	Substrates	pH optimum	Temp optimum	Metal ions involved	Mode of action and product(s)
Exopolygalacturonase (EC 3.2.1.67)	Carrots, *A. niger*	Same as for Endo-PG	Two forms from *A. niger*: PGI, 4.4–4.6; PGII, 5.0–5.1	30°–40°C	PGI activated by Hg^{2+}	PGI hydrolyzes pectate to monogalacturonate. PGII degrades pectic acid (pectate) 28% to galacturonic acid. Di- and trigalacturonic acids are hydrolyzed by both PGI and II
Endopectate lyases (EC 4.2.2.2)	Fungi (*Fusarium*) bacteria (*Bacillus*, *Clostridium*, and *Erwinia*), and protozoa	Pectic acid (pectate) and other polygalacturonides	Fungal—9.0–9.3; Protozoa—8.0; Bacteria—8.9–9.4	30°–45°C	Ca^{2+} required by fungal and bacterial. *Erwinia* enzyme not activated by Ca^{2+}.	Elimination of Δ-4,5-galacturonide residues from pectate, thus bringing about depolymerization.
Exopectate lyase (EC 4.2.2.9)	Bacteria (*C. multifermentans* and *E. aroideae*)	Pectic acid	Bacteria—8.5–9.4	35°C	Divalent metal ions required; Zn not an activator.	Elimination of Δ-4,5-D-galacturose-D-galacturonate residues from the reducing end of the unesterified pectic acid. (See Arch. Biochem. Biophys. *125*, 734, 1968.)
Oligogalacturonide lyase (EC 4.2.2.6)		*o*-(4-deoxy-β-L-5-threo-hexopyranos-4-enyl-uronate-(1,4)-D-galacturonate and related oligosaccharides (depolymerization)				
PROTEASES						
Papain (EC 3.4.22.2)	Papaya	Peptides, proteins (e.g., egg albumin, casein, gelatin)	5–7 depending upon substrate	30°–40°C	Inhibited by heavy metals. Free SH is required for activity.	Preferentially cleaves peptide bonds to the right of Arg-, Lys-, Phe-X residues; limited hydrolysis of native immunoglobulins.
Ficin (EC 3.4.22.3)	Figs (*Ficus carica*)	Peptides, proteins (e.g., casein, gelatin)	Stable at 3.5–9; optimum is substrate dependent	30°–40°C Time dependent; optimum as high as 63°C	Same as for papain	Preferential cleavage: Phe-, Tyr- (See papain.)
Bromelain (EC 3.4.22.5; EC 3.4.22.4)	Pineapple juice (EC 3.4.22.5) Pineapple stems (EC 3.4.22.4)	Juice enzyme rapidly cleaves benzoyl-arginine amide and also synthesizes acylamino acid anilides; acts on casein and other	6–8	30°–40°C	Inhibited by Hg^{2+}. Free SH required for activity.	The enzyme from stems and fruit hydrolyzes arginine esters and amides but not L-Leu-NH_2, Cbz-L-Glu-L-Phe, Cbz-Gly-L-Phe, or Gly-L-Phe-NH_2.

		peptides and proteins. Stem enzyme also acts on benzoyl-arginine ethyl ester; preferentially cleaves Phe^5-Ser^6 in brandykinin; acts on casein and other proteins				
Trypsin (EC 3.4.21.4)	Pancreas of mammals; also isolated from fishes, crustaceans, insects and *Streptomyces*	Preferential cleavage of peptide bonds in which Arg and Lys are the carboxyl-contributing amino acids. Endopeptidase	7–9	Assay usually performed at 30°C	Ca^{2+} is activator for autocatalysis of trypsinogen.	Pancreas cells of vertebrates produce inactive form of trypsin, (trypsinogen), which is converted, autocatalytically, to β-trypsin by cleavage of Lys^6-Ile^7 (N-terminal = 1). β-Trypsin is converted to α-trypsin by cleaving at Lys^{131}-Ser^{132}. α-Trypsin is further converted to ψ-trypsin by cleavage at Lys^{176}-Asp^{177}. All forms are active. The specificity of trypsin toward arginine and lysine is due to the ionic bonding of the basic groups of these amino acids with the carboxyl group of Asp^{177} at the active site. Classed as serine proteases.
Chymotrypsin, A & B (EC 3.4.21.1) C (EC 3.4.21.2)	Pancreas of mammals; also isolated from fishes and insects.	Preferential cleavage of peptide bonds in which Tyr, Trp, Phe, and Leu are the carboxyl-contributing amino acids.	7–9	Assay usually performed at 30°C	—	Bovine pancreas extract contains two inactive forms of chymotrypsin (chymotrypsinogen A and B). Trypsin hydrolysis is involved in the conversion of the precursors to the active enzyme. Specifically attacks peptides containing amino acids with aromatic rings (e.g., Tyr-, Phe-, and Trp-). Contains serine as active site, thus, is classed as serine protease.

(*continued*)

Table 7.3. *(Continued)*

Enzyme	Sources	Substrates	pH optimum	Temp optimum	Metal ions involved	Mode of action and product(s)
Pepsin A (EC 3.4.23.1) B (EC 3.4.23.2) C (EC 3.4.23.3)	Gastric acid of various mammals	A form (predominant enzyme) preferentially cleaves peptide bonds in which Phe and Leu are the carboxyl-contributing amino acids. B form degrades gelatin and shows little activity toward hemoglobin. C form has activity with hemoglobin as a substrate	Stable at 5–5.5, and active at 1–4 with maximum at approximately 1.8 for several proteins and at 2–4 for synthetics substrates	37°C	LiCl causes loss of activity	The three forms of the pepsin—A, B and C—differ in specificity; generally all have a broad range of specificity. Pepsin also exists as an inactive form (pepsinogen), which is converted at a pH of below 5 to pepsin. Pepsin also has esterase activity.
Rennin (EC 3.4.23.4)	Gastric juice of calf (fourth stomach). Also isolated from *Mucor pusillus*, *Endothia parasitica*, *Bacillus cereus*, *Aspergillus candidus* and *Byssochlamys fulva* (these sources have an enzyme described as "rennin like").	Casein and some modified caseins (e.g., reduced casein, rephosphorylated dephosphocasein and *p*-mercuribenzoate *S*-casein). Causes clotting.	Stable at 5.3–6.2. Prorennin is stable at 5.3–9. Optimum pH with hemoglobin is 3.7 (mammal only).	Assays are commonly performed at 37°C but hydrolytic activity can be observed at 2°C	—	Casein coagulation involves two steps: first, casein is enzymatically converted to *p*-casein, then *p*-casein is heat-coagulated with Ca^{2+}. The enzyme-catalyzed step is a hydrolysis reaction that can be separated from the coagulation step by performing the hydrolysis at 2°C. Data suggest that the amino acid residues near the rennin-sensitive linkage of *k*-casein participate in the cleavage reaction by rennin which is not surprising in light of previous reaction mechanisms proposed for proteolytic enzymes. Prorennin is converted to rennin by autocatalytical hydrolysis at pH 5; autocatalysis is minor in prorennin conversion at pH 2.

Cathepsins B (EC 3.4.22.1) C (EC 3.4.14.1) D (EC 3.4.23.5) E (EC 3.4.23.5) G (EC 3.4.21.20) L (EC 3.4.22.15)	All are found in vertebrate tissues such as liver, spleen, kidney, lung, and muscles and in leucocytes.	Proteins of muscle tissue. Specificity of B is similar to papain (possibly also similar to trypsin). Meat aging is accomplished by cathepsin activity. C and G have specificity similar to chymotrypsin while D resembles pepsin in specificity.	4.0–8.0 depending on substrate, condition, and form of enzyme.	35°–40°C	—	Cathepsin B resembles papain or trypsin in mechanism and specificity. Cathepsin C resembles chymotrypsin while D resembles pepsin. D appears to differ considerably in specificity from substrates of A, B, and C. D is classed by I.U.B. as an acid proteinase. E is classed as homologous with cathepsin D. Cathepsin A is not listed in the Enzyme Commission nomenclature. Cathepsin C, classed as a dipeptidyl peptidase, removes two residues from the N-terminal end and also actively polymerizes dipeptide amides and transfers dipeptide residues; -SH group forms part of the active center. G and L are new additions to the I.U.B. list.
Carboxypeptidase (EC 3.4.17.1)	Bovine and porcine pancreas	Peptides and proteins with C-terminal amino acids other than Arg, Lys, and Pro.	7–8	Assays are usually performed at 25°C, but appears to be stable at lower and higher temperatures.	Zn^{2+} is part of active enzyme complex.	Bovine procarboxypeptidase A isolated from the pancreas is a zymogen. The active form can be generated from the inactive form by proteolysis by trypsin. Arg^{145}, Zn, Glu^{270} and Tyr^{248} are involved most with the active site-binding of the substrate. (Hartsuck and Lipscomb, 1971).
Aminopeptidase (EC 3.4.11.1) (cylosol)	Porcine kidney bovine lens	Hydrolyzes most L-peptides, splitting off the N-terminal amino acid with free amino group. Does not act on Lys and Arg peptides.	8–9	Assays are usually performed at 25°C.	Zn^{2+} is part of active complex. Some are activated by divalent metal ions.	This exopeptidase with broad specificity releases asparagine and glutamine from peptides. Lysyl and arginyl residues are resistant to hydrolysis. The assay is leucine amide → leucine + NH_3. Other aminopeptidases in pig kidney are assayed by measuring the rate of hydrolysis of *p*-nitroanilides of alanine or leucine.

(*continued*)

Table 7.3. *(Continued)*

Enzyme	Sources	Substrates	pH optimum	Temp optimum	Metal ions involved	Mode of action and product(s)
EC 3.4.11.2	Porcine kidney and bovine lens	Splits α-amino acids (preferentially alanine not proline) from peptides	8–8.5	Assays are usually performed at 25°C.	Zn^{2+}-enzyme not activated by heavy metals	Only derivative of L-α-amino acids are hydrolyzed by particle-bound enzymes. The relative rates of hydrolysis of different L-amino acid-*p*-nitroanilides are alanine (100), phenylalanine (83), leucine (71), glycine (22). This enzyme hydrolyzes alanine or glycine peptide bonds faster than leucine aminopeptidase.
EC 3.4.11.14	Human liver, pancreas, kidney, and duodenum	Preferentially cleaves N-terminal alanine	Approximately 8–8.5	Assays are usually performed at 25°C.	Co^{2+} can replace Zn^{2+}.	This metalloenzyme with high temperature stability and broad specificity, releases all N-terminal amino acids, including arginine and lysine.
OXIDOREDUCTASES						
Xanthine oxidase (EC 1.2.3.2)	Bovine milk and liver	Xanthine, hypoxanthine, other purines and pterins, and aldehydes	8.3	Assayed at 30°–37°C	Fe and Mo	Xanthine oxidase from milk is a molybdoflavoprotein with the flavin moiety being tightly bound to the protein. The sequence of reaction where uric acid and H_2O_2 are formed from Xanthine and O_2 is postulated to be
				Xanthine → Uric acid	2Mo (VI) → 2Mo (V); 2FAD → 2FADH·	2Fe (III) → 2Fe (II); O_2 → O_2^{2-} → H_2O_2
Peroxidase (EC 1.1.1.1.7)	Most plant and animal tissue, milk, and microorganisms. Typical sources are horseradish, fish, crayfish, and peas.	ROOH where donor R may be H, alkyl, or acyl. Donor may be *p*-cresol, guaiacol, or resorcinol. Some peroxidases are also capable of	6–8; optimum is 7.	Assays usually performed at 20°–25°C	Fe in several peroxidases	There are classes of peroxidases: ferriprotoporphyrin, verdoperoxidases, and flavoperoxidases. The former two have iron porphyrin nuclei; the latter has FAD as

		halogenating phenolic compounds (e.g., tyrosine).				the prosthetic group. Four types of reactions are proposed for peroxidase: peroxidatic, oxidatic, catalactic, and hydroxylation [see Whitaker (1972)]. Several peroxidases, e.g., lactoperoxidase, are also capable of halogenation (Morrison and Schonbaum 1976). The scheme proposed is

E (peroxidase) + $H_2O_2 \rightarrow$ EO (peroxide derivative or compound I) + H_2

$$EO + I^- \text{ (halide)} \rightarrow EO/I^- \rightleftarrows EOI \rightleftarrows EOH \rightarrow E + HOI$$

(EO └─XH; EOH └─XI)

EOH
└─XI can then iodinate a phenolic compound.

Peroxidase is commonly assayed by the spectrophotometric determination at 470 nm of guaiacol oxidation to tetraguaiacol. Halogenation can be determined by the change in absorbance at 290 nm when tyrosine is converted to iodotyrosine.

Catalase (EC 1.11.1.6)	Various animals, plants, and microorganisms	H_2O_2	Wide range of pH; essentially independent of pH at 3–9.	25°C	Fe bound to porphyrin	Overall reaction (catalatic) is $2H_2O_2 \rightarrow 2H_2O + O_2$. Intermediate reactions have been proposed as shown:

$$\text{Catalase-}Fe^{3+}\cdot H_2O \xrightarrow{H_2O_2 \;\; H_2O} \text{Catalase—}FE^{2+}\cdot HOOH$$

$$\downarrow$$

HAAH AH· Compound I

AH₂ / AH· → AH· HAAH

$$\text{Catalase—}Fe^{3+}\cdot H_2O \xleftarrow{AH\cdot \;\; AH_2} \text{Compound II}$$

(*continued*)

Table 7.3. *(Continued)*

Enzyme	Sources	Substrates	pH optimum	Temp optimum	Metal ions involved	Mode of action and product(s)
Polyphenol oxidases (EC 1.10.3.1, 1.10.3.2, and 1.14.18.1)	Various animals and plants including fruits and mushrooms.	Catechol + O_2 (1.10.3.1), benzenediol + O_2 (1.10.3.2), and L-tyrosine + dihydroxy-L-phenylalanine + O_2 (1.14.18.1).	5–8 depending upon source; assays are performed near 6.5	25°C	Cu	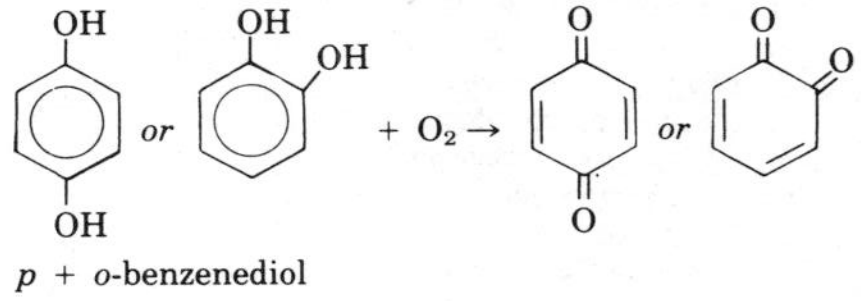 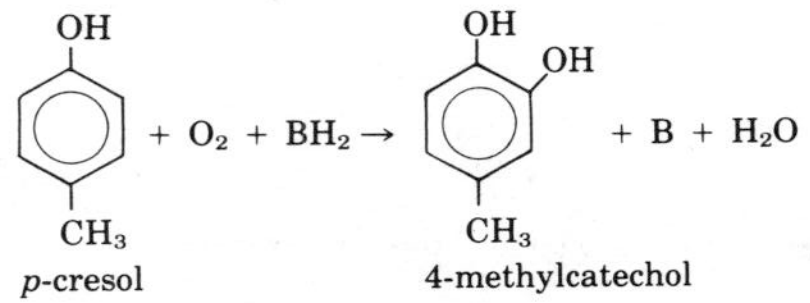 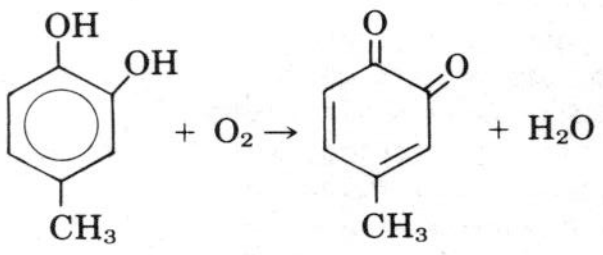 The reactions for these enzymes include 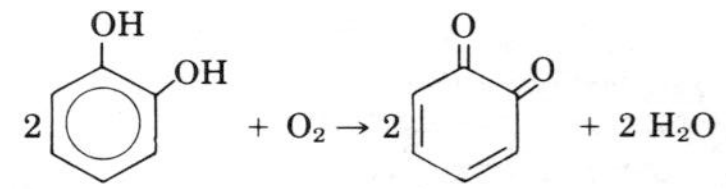 2 [catechol: OH, OH] + O_2 → 2 [o-quinone: O, O] + 2 H_2O (Oxidation) [OH, OH] *or* [OH, OH] + O_2 → [O, O] *or* [O, O] *p* + *o*-benzenediol (Oxidation) [OH, CH_3] + O_2 + BH_2 → [OH, OH, CH_3] + B + H_2O *p*-cresol → 4-methylcatechol (Hydroxylation) [OH, OH, CH_3] + O_2 → [O, O, CH_3] + H_2O (Oxidation)

HYDROLASES Glucose oxidase (EC 1.1.3.4)	*Aspergillus niger*, *Penicillum amagaskiense*, *P. notatum*, and *P. vitale*	β-D-glucose enzyme from *A. niger* weakly acts on 2-deoxy-D-glucose, mannose, and galactose. *P. notatum* enzyme acts weakly on mannose and galactose (1% or less). Various structural modifications of β-D-glucose give activities ranging from 0–10% of that of β-D-glucose *per se*.	4.5–7.5; unstable at 8.1; assay usually done near 5.9.	30°C–60°C; assay near 30°C.	Inhibited by Cu^{2+} and -SH complexing agents.	Glucose oxidase is FAD-requiring enzyme that has a significant preference for β-glucose (160×) over α-glucose. The reaction catalyzed is β-D-glucopyranose → δ-D-gluconolactone (Gluc. Ox—FAD → Gluc. Ox–$FADH_2$; O_2 → H_2O_2, Enzyme catalyzed) δ-D-gluconolactone → (Nonenzymatic) D-gluconic acid
Lipoxygenase (EC 1.13.11.12) (lipoxidase)	Soy beans, alfalfa, peas, beans, peanuts, radishes, and potatos	Polyunsaturated fatty acids containing a *cis*, *cis*-1,4-pentadiene group (e.g., linoleic, linolenic and arachidonic acids). Also carotenes are degraded by lipoxygenase.	7–8	25°C	Ca^{2+} (activates one of the two enzymes in soybeans)	Only the *cis*, *cis* configuration is attacked by the enzyme. *Cis*, *trans* or *trans*, *trans* forms are not attacked or are competitive inhibitors. The reaction sequence for lipoxygenase and (*cis*,*cis*,*cis*) 8,11,14-eicosatrienoic acid is as follows: $CH_3—(CH_2)_4$—CH=CH—$C(H_R)(H_S)$—CH=CH—CH_2—CH=CH—$(CH_2)_6$—COOH ↓ $CH_3—(CH_2)_4$—CH=CH—Ċ(H_R)—CH=CH—CH_2—CH=CH—$(CH_2)_6$—COOH ↓

(*continued*)

Table 7.3. (*Continued*)

Enzyme	Sources	Substrates	pH optimum	Temp optimum	Metal ions involved	Mode of action and product(s)
						CH_3—$(CH_2)_4$—C(H)—CH=CH(H_R)—CH=CH—CH_2—CH=CH—$(CH_2)_6$—COOH ↓ O_2 CH_3—$(CH_2)_4$—C(H)(O—O)—CH=CH(H_R)—CH=CH—CH_2—CH=CH—$(CH_2)_6$—COOH ↓ CH_3—$(CH_2)_4$—C(H)(O—O—H)—CH=CH(H_R)—CH=CH—CH_2—CH=CH—$(CH_2)_6$—COOH
Lipase (EC 3.1.1.3)	Pancreas, milk, and Castor bean Microorganisms and cereal grains (wheat)	Triacylglycerol	Pancreas—8–9; milk—9; castor bean—6.3 (8.5–10.5 for unripe); microorganisms—5.6–8.5; cereal grains—7.4	30°–40°C	Most stimulated by Ca^{2+}; porcine pancreatic lipase is activated by NaCl; heavy metals inhibit.	Lipases are usually rather nonspecific mono-, di- and triglycerides from various sources and some primary alcohol esters are hydrolyzed. Pancreatic lipase hydrolyzes triglycerides more rapidly than simple alcohol esters. Shorter chain triglycerides are hydrolyzed more rapidly than longer. The α or α^1 positions are attacked more readily than the β (*sn* 1 and 3 *vs* 2 positions).

				Triglyceride → ① 1,2-diglyceride / ② 2,3-diglyceride → 2-monoglyceride; ↓ + $RCOO^-$ → glycerol + $RCOO^-$; + $R—COO^-$ Generally the rate of glyceride hydrolysis is Triglyceride > diglyceride > 1-monoglyceride > 2-monoglyceride Since the glycerides are not water soluble, an emulsion must be formed, then the lipolytic reaction occurs at the polar:apolar interface. Ordinary enzyme kinetics is not applicable with accuracy, e.g., to aqueous solutions.
Esterases (EC 3.1.1.__)	Widely distributed	See I.UB. Enzyme Nomenclature (1978), Academic Press.	—	

[a] Note: Where information was not found or was incomplete, the column was left blank.

active due to unfavorable hydrogen ion concentration. As tissues become more acid following death, the proteolytic enzymes are activated resulting in protein hydrolysis. During the aging of beef at approximately 4.4°C for 1–4 weeks, which has been a common commercial practice for many years, the action of proteolytic enzymes "tenderize" the meat.

Specific enzymes are involved in the synthesis of various flavor precursors, which in turn are converted into the flavor itself. For example, in celery, onions, tomatoes, bananas, oranges, and pineapples the development of flavor has been attributed to one enzyme activity or another.

Enzymes have been associated with changes in the color of foods. In some cases, color components of foods are broken down; for example, in the artificial ripening of fruits the green color is caused to disappear and the previously masked orange and yellow colors allowed to appear. Another example is the blanching of spinach at 77°C, which preserves the green color. This is probably related to the maximum temperature 77°C for chlorophyllase activity which converts chlorophyll into chorophyllin which possesses a green color.

Other examples of the changes caused by naturally occurring enzymes in foods are enzymatic browning catalyzed by phenolases, lipolysis catalyzed by lipases, and oxidations catalyzed lipoxidase. Phenolases are especially important in the development of brown coloration on fruits and vegetables after exposure of the interior to oxygen. Both monohydric and dihydric phenols act as substrates producing quinones; polymerization of these quinones is the well-known browning reaction, which is discussed later in this chapter. These enzymatic reactions can be controlled by blanching prior to processing and, in some cases, by inhibition of the enzyme through the use of chemicals such as sulfite, SO_2 gas, or $NaHSO_3$.

Rapid and extensive enzymatic changes may take place when the tissues of fruits and vegetables are damaged, during harvesting and transport, by freeze injury during storage, by insect and microbial attack, or during processing operations. These changes include darkening of peeled and cut fruits and vegetables, the development of off-flavors in bruised vegetables, etc.

Milk and Milk Products

A considerable number of enzymes occur in milk. Many of these undoubtedly find their way into the milk from the blood during the process of milk synthesis in the mammary gland. Most of the enzymes present in milk exert little effect on the milk components and, in some instances, have no substrate present to act upon. The presence of certain enzymes is used to indicate milk quality and proper pasteurization.

Lipases must be taken into account in the processing of dairy products, since they may lead to undesirable rancidity. Lipolytic activity varies with the pH of milk, the degree of homogenization, and the stage of lactation (greatest near the end of the period). Lipases are inactivated by pasteurization temperatures.

When milk is properly pasteurized, phosphatase inactivation is 99.9% complete. Because the conditions of inactivation are practically identical with those of pasteurization of milk, the absence of phosphatase activity is a valid indicator of the adequate pasteurization of milk and milk products.

Cereal Products

Cereals are grown for their edible seeds and include wheat, rye, barley, corn, oats, and rice. They are, in general, the cheapest sources of food energy and normally constitute about one-third or more of the caloric intake of humans. Cereals as harvested and stored are living materials and contain a number of active enzymes; a few of these play an important role in cereal technology.

Among the enzymes in cereal grains are α- and β-amylases. These enzymes are of great importance in flour for breadmaking. The α-amylase, for example, has two main functions. First, it provides for both yeast activity and gas production. Second, it affects dough properties such as viscosity and improves bread quality. Both the α- and β-amylases are used in converting starch to sugar. The β-amylase, because it does not attack damaged starch granules, cannot convert enough starch to the sugar maltose; however, α-amylase does attack these damaged granules of starch and converts it to dextrins. These are then hydrolyzed by β-amylase to maltose.

A loss of α-amylase activity results from mechanical harvesting of wheat, but the β-amylase activity is not affected by it. Thus, it is important to increase the α-amylase present in flour in order to adequately convert starches to sugars. To obtain increased amounts of α-amylase, malted flours which are usually produced from wheat or barley are added to the flour stream at the mill at a level of about 0.25–0.40%. This results in an addition of 10–15 amylase units per 100 g of flour.

Protease activity is very important in cracker production to hydrolyze some of the gluten; this hydrolysis is necessary to achieve the proper structure and texture in the finished cracker. Small amounts of proteases are introduced into flour when malted flour is added to increase β-amylase activity. Cracker dough can also be supplemented with added fungal proteases to improve its properties.

Flour also contains a lipoxidase enzyme, which can oxidize carotenoid pigments during oxidation of polyunsaturated fatty acids. Lipoxidase may be responsible for the bleaching of flour during natural aging, as well as for oxidative rancidity in cereals and cereal products.

Vegetables and Fruits

The changes that occur in fruit and vegetables as a result of enzymatic activity are usually disadvantageous. To assure reasonable keeping quality, special attention must be paid to the temperature, humidity, ventilation, and maturity during harvesting, packing, storing, and processing of vegetables and fruits.

Sweet corn, when freshly picked, contains appreciable sugar in the kernels. In a matter of a few hours, the sugar has disappeared with a resulting loss of sweetness and toughening of textures. Potatoes held in storage have a higher sugar content at low temperatures than at high temperatures. Potatoes to be used for chip manufacture are placed at a higher temperature for a period of time before chipping to reduce the sugar content to a minimum because potatoes with a high sugar content yield dark-colored chips. In ripening, enzymatic conversion of pectic substances occurs, resulting in a general softening and change in texture.

Enzymes do not accelerate reactions unless they are in direct contact with their substrate(s). For instance, fruits do not show browning unless they are bruised or cut, which permits contact of the enzyme polyphenol oxidase with oxygen and the polyphenols of the fruit. The proteases of fresh pineapple do not exert a proteolytic function because pineapple contains little protein. However, if fresh pineapple is added to a gelatin dessert, the pineapple proteases act on the gelatin proteins, causing liquefaction.

Enzymatic Browining

Probably the most common enzymatic reaction in fruits and vegetables is browning. Because browning reduces the quality of produce, several techniques are used to reduce or eliminate the reaction.

Browning of fruits such as apples, pears, peaches, and apricots, and of vegetables such as potatoes occurs when the tissue is exposed to oxygen. This exposure is commonly due to bruises, cuts, and other injury to the peel. The browning reaction occurs when the enzyme-catalyzed reaction of oxygen with certain phenolic compounds produces quinone structures, which polymerize to form brown-colored products. The enzyme that catalyzes these reactions is most frequently called polyphenol oxidase (*o*-diphenol:oxygen oxidoreductase, EC 1.10.3.1). This enzyme has also been referred to as phenolase, tyrosinase, catechol oxidase, and potato oxidase. These names are taken from the varied substrates of the enzyme or its source.

Polyphenol oxidase is a copper-containing enzyme that can undergo reversible oxidation and reduction in the process of hydroxylation and oxidation. In hydroxylation, Cu^{+} is oxidized to Cu^{2+} and in oxidation, Cu^{2+} is reduced to Cu^{+}. Since mushroom polyphenol oxidase can exist in multiple molecular forms, it can exhibit variable hydroxylation and oxidation activity depending on the form present.

Polyphenol oxidase catalyzes two basic types of reactions, hydroxylation and oxidation. Hydroxylation (cresolase activity) occurs only with phenols containing one hydroxyl group and yields *ortho*-diphenols, as follows:

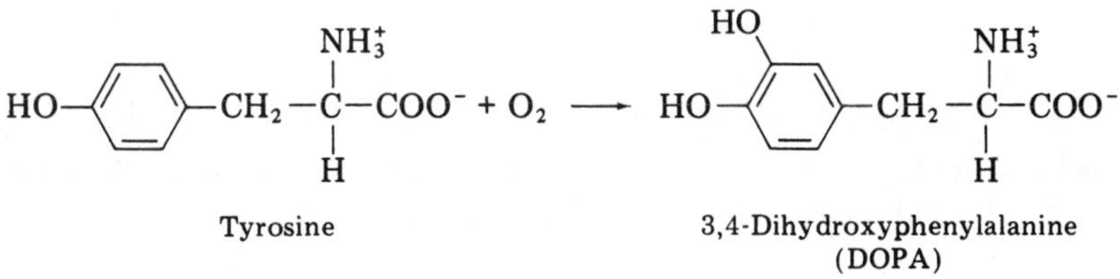

$$HO-C_6H_4-CH_2-\underset{H}{\overset{NH_3^+}{C}}-COO^- + O_2 \longrightarrow (HO)_2C_6H_3-CH_2-\underset{H}{\overset{NH_3^+}{C}}-COO^-$$

Tyrosine — 3,4-Dihydroxyphenylalanine (DOPA)

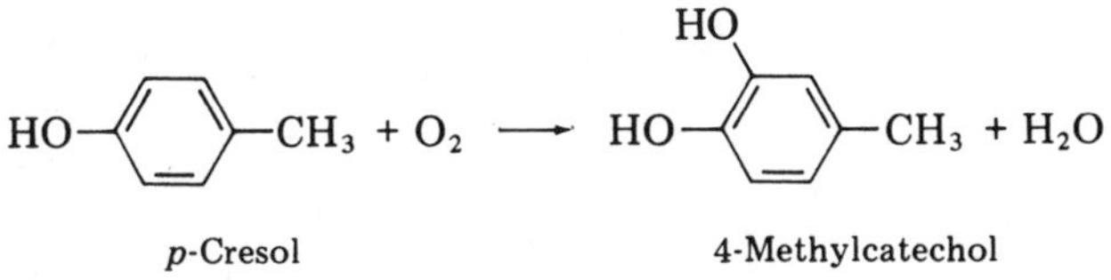

$$HO-C_6H_4-CH_3 + O_2 \longrightarrow (HO)_2C_6H_3-CH_3 + H_2O$$

p-Cresol — 4-Methylcatechol

The oxidation reaction (catecholase activity) occurs with *o*-diphenols, such as catechol, DOPA, or 4-methylcatechol, to form the corresponding benzoquinone.

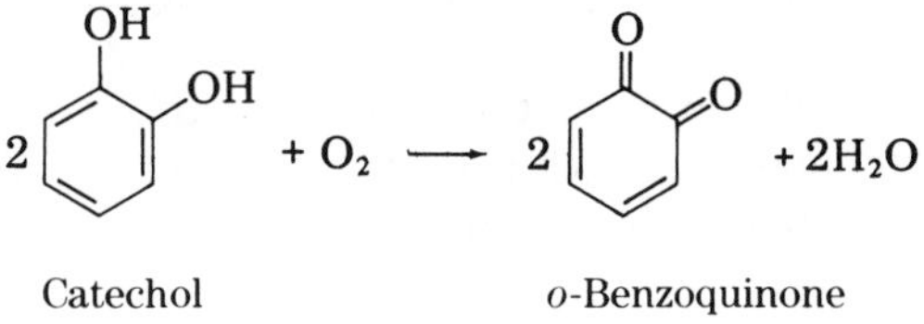

$$2\ C_6H_4(OH)_2 + O_2 \longrightarrow 2\ C_6H_4O_2 + 2H_2O$$

Catechol — *o*-Benzoquinone

Substrate specificity varies considerably among polyphenol oxidases from various sources. Potato polyphenol oxidase utilizes chlorogenic acid, caffeic acid, catechol, DOPA, and *p*-cresol. Chlorogenic acid is especially ubiquitous as a substrate for polyphenol oxidase in vegetables such as potatoes and in fruits such as apples and pears.

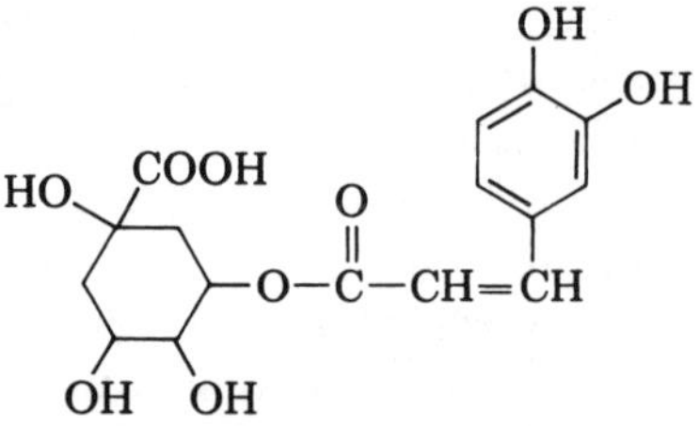

Chlorogenic acid

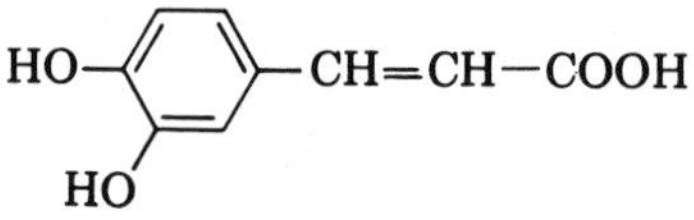

Caffeic Acid

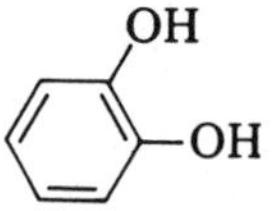

Catechol

Peach polyphenol oxidase uses 4-methylcatechol, gallic acid, protocatechuic acid, and dopamine as substrates, as well as the compounds used by the potato enzyme with the exception of *p*-cresol.

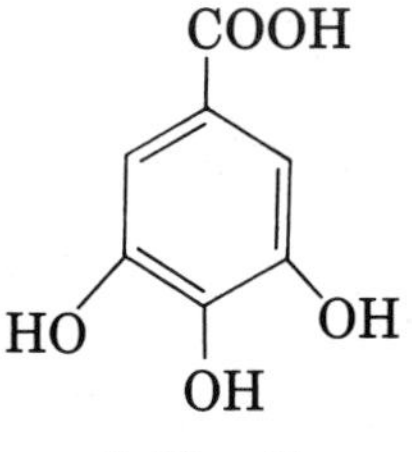

Gallic acid

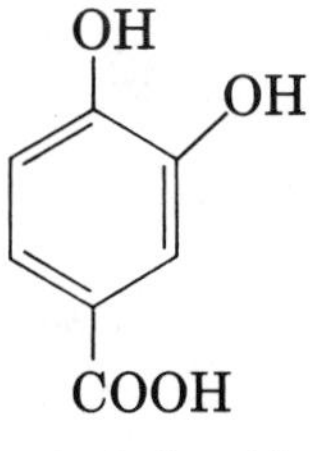

Protocatechuic acid

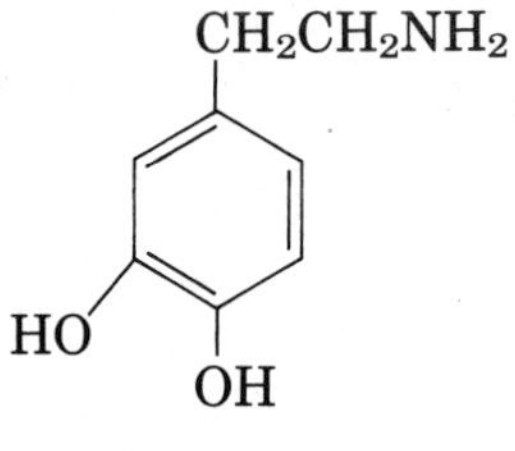

Dopamine

Prevention of Enzymatic Browning. Several methods have been used to control enzymatic browning. These include the addition of antioxidants or inhibitors, exclusion of oxygen, heating, and lowering of the pH.

Heating is one of the most convenient methods for prevention of browning. Polyphenol oxidases from various sources differ somewhat in their susceptibility to heat inactivation. However, most are inactivated almost completely upon heating to 90°–95°C for approximately 7 sec. This type of inactivation is complicated by the enzyme's pH dependence, which also varies considerably among polyphenol oxidases from different fruits and vegetables.

Sodium chloride, sulfur dioxide, sodium sulfite, and sodium hydrogen sulfite are used to inhibit polyphenol oxidase activity. The browning reaction is significantly inhibited by 1000 ppm NaCl or by 1 ppm SO_2. Ascorbic acid also has been used to prevent polyphenol oxidase-induced browning. Ascorbic acid acts by keeping the *o*-diphenol in the reduced form and by lowering the pH.

The optimum pH for polyphenol oxidases is 6–7; if the pH of the medium is lowered to 3 or below, negligible activity is observed. Citric acid, malic acid, ascorbic acid, and other organic acids have been used to lower the pH of fruits, thus inhibiting polyphenol oxidase and preventing enzymatic browning.

Meats

The enzyme-catalyzed metabolism in postmortem muscle results in the development of rigor mortis. Meat in the state of rigor is tough; however, if meat is allowed to age for a few days in a cooler, it gradually begins to soften. This softening or tenderization is thought to be due to the activity of proteolytic enzymes, which were inactive in the living animal.

Rigor mortis also occurs in fish and seafood products. However, proteolysis is undesirable in this case because in these foods it is usually associated with microbial action. As a consequence, processes such as freezing, canning, smoking, salting, and drying are used to prevent proteolysis in seafood products.

☐ COMMERCIAL APPLICATIONS

Enzymes were used in a variety of ways long before they were recognized as definite chemical substances that occurred in living cells. Processes for making bread, wine, alcohol, vinegar, sauerkraut, and pickles have been known from antiquity; these and many other food-related processes all depend on the action of enzymes.

Enzymes have several characteristics that make them desirable for use in industrial processes: (1) they accomplish a reaction more efficiently than other catalysts; (2) their rates of reaction can be readily controlled by adjusting temperature, pH, and reaction time; (3) their activity may be destroyed by heating to sufficiently high temperatures; (4) they are natural in origin and nontoxic, and therefore in most cases may remain in the finished product without any harmful consequences; and (5) they exhibit greater specificity than other catalysts and their activity can be readily standardized. In modern food technology, the commercial applications of enzymes are many and varied. Representative applications are listed in Table 7.4.

Carbohydrases

The carbohydrases hydrolyze polysaccharides and oligosaccharides. Some of the important enzymes in this group are listed in Table 7.3. In this group, the amylases have the most commercial applications.

There are three types of amylases: α-amylases, which hydrolyze α-(1→4) linkages in large starch molecules, in a random fashion, to form dextrins; β-amylases, which convert starch into maltose by splitting off disaccharide units progressively from the nonreducing end of the chain polymer; and glucoamylases, which produce glucose by progressive hydrolysis from the nonreducing end of the starch molecule. The latter enzyme is capable of converting starch to glucose without the formation of dextrins and maltose. (See Enzymatic Breakdown of Polysaccharides in Chapter 4.)

α-Amylases may be obtained from animals, higher plants, fungi, and bacteria, while β-amylases are produced only by higher plants, particularly cereals. These enzymes vary widely in sugar-forming ability, in thermal stability, and in the rate and extent of hydrolysis. For example, α-amylase is often referred to as the liquefying enzyme, whereas β-amylase is often referred to as the sugar-producing enzyme.

Invertases, another type of carbohydrase, may be isolated from yeast (*S. cerevisiae* or *S. carlsbergensis*) and from fungi (*A. oryzae, A. niger,* and *M. verrucaria*). They are primarily β-fructosidases and hydrolyze sucrose to glucose and fructose. These enzymes are used in the manufacture of invert sugar, which is more soluble than sucrose. Invert sugar is used in confections, candies, frozen desserts, and liqueurs.

Pectic Enzymes

Pectins are polymers made up of chains of galacturonic acid units linked by α-(1→4) glycosidic linkages (see Pectins in Chapter 4). Three pectic substances are known: protopectin, pectinic acids, and pectic acids. Protopectin, which occurs in the cell wall of plants, is the parent

Table 7.4. Commercial Applications of Enzymes in Food Production

Processing difficulty or requirement	Enzyme function	Enzyme used	Enzyme source
MILLING AND BAKING			
High dough viscosity	Reduces dough viscosity by hydrolyzing starch to smaller carbohydrates	Amylase	Fungal
Slow rate of fermentation	Accelerates process	Amylase	Fungal
Low level of sugars resulting in poor taste, poor crusts, and poor toasting characteristics	Increases sugar levels by converting starch to simple sugars	Amylase	Bacterial
Staling of bread	Enables bread to retain freshness and softness longer	Amylase	Bacterial
Mixing time too long for optimum gas retention of doughs	Reduces mixing time and makes doughs more pliable by hydrolyzing gluten	Protease	Fungal
Curling of sheeted dough for soda crackers as dough enters continuous cracker ovens	Prevents curling of sheeted dough	Protease	Fungal
Poor bread flavor	Aids in flavor development	Lipoxidase Protease	Soy flour Fungal
Off-color flour	Bleaches natural flour pigments and lightens white bread crumbs	Lipoxidase	Soy flour
Low loaf volume and coarse texture	Increases volume and reduces coarseness by hydrolyzing pentosans	Pentosanase	Fungal
MEATS			
High fat content	Aids in removal or reduction of fat content	Lipase	Fungal
Toughness	Tenderizes meat by hydrolyzing muscle protein and collagen	Protease	Ficin (figs) Papsin (papaya)
Serum separation of fat in meat and poultry products	Produces liquid meat products and prevents serum separation of fat	Protease	Fungal
High viscosity of condensed fish solubles	Reduces viscosity while permitting solids levels over 50% without gel formation of the condensed fish solubles	Protease	Fungal

(*Continued*)

Table 7.4. (*Continued*)

Processing difficulty or requirement	Enzyme function	Enzyme used	Enzyme source
Protein shortage	Prepare fish protein concentrate	Protease	Fungal and bacterial
DISTILLED BEVERAGES			
Thick mash	Thins mash and accelerates saccharification	Amylase	Bacterial and malt
Chill haze	Chillproofs beer	Protease	Fungal, bacterial, and papain
Low runoff of wort	Assists in physical disintegration of resin and improves runoff of wort	Amylase	Fungal and malt
FRUIT PRODUCTS AND WINES			
Apple juice haze	Clarifies apple juice	Pectinase	Fungal
High viscosity due to pectin	Reduces viscosity by hydrolyzing pectin	Pectinase	Fungal
Slow filtration rates of wines and juices	Accelerates rate of filtration	Pectinase	Fungal
Low juice yield	Facilitates separation of juice from the fruit, thus increasing yield	Pectinase	Fungal
Gelling of fruit concentrates	Prevents pectin gel formation and breaks gels	Pectinase	Fungal
Poor color of grape juice	Improves color extraction from grape skins	Pectinase	Fungal
Fruit wastes	Produces fermentable sugars from apple and grape pomace	Cellulase	Fungal
Sediment in finished product	Helps prevent precipitation and improves clarity	Pectinase	Fungal
SYRUPS AND CANDIES			
Controlled levels of dextrose, maltose, and higher saccharides	Controls ratios of dextrose, maltose, and higher saccharides	Amylase	Bacterial, fungal, and malt
High-viscosity syrups	Reduces viscosity	Amylase	Bacterial and fungal
Sugar loss in scrap candy	Facilitates sugar recovery from scrap candy by liquefaction of starch content	Amylase	Bacterial
Filterability of vanilla extracts	Improves filterability	Cellulase	Fungal

Table 7.4. (*Continued*)

Processing difficulty or requirement	Enzyme function	Enzyme used	Enzyme source
MISCELLANEOUS			
Poor flavors in cheese and milk	Improves characteristic flavors	Lipase	Fungal
Tough cooked vegetables and fruits	Tenderizes fruits and vegetables prior to cooking	Cellulase	Fungal
Inefficient degermination of corn	Produces efficient degermination of corn	Cellulase	Fungal
High set times in gelatins	Reduces set times of gelatin without significantly altering gel strength	Protease	Fungal
Starchy taste of sweet potato flakes	Increases conversion of sweet potato starch	Amylase	Fungal and bacterial
High viscosity of precooked cereals	Reduces viscosity and allows processing of precooked cereals at higher solid levels	Amylase	Fungal and bacterial

Source: Pulley (1969).

substance of this class of compounds. During the ripening of fruits, there is a decrease in protopectin and a proportional increase in pectinic acids. These contain more than 10% methyl ester groups, and it is these esterified substances that form semisolid gels with sugar and acid of the kind found in jellies. Acid hydrolysis of the methyl ester groups yields pectic acids, while partial hydrolysis yields substances known as low-methoxyl pectins. The latter substances will form gels with small amounts of calcium ions.

The enzymes that catalyze the breakdown of pectic substances may be classified into two groups: the pectases (pectin methylesterases) and the pectinases (polygalacturonases). The latter enzymes may be further divided into endo and exo enzymes. The endopolygalacturonases cleave α-(1→4) linkages within a chain, while the exo enzymes catalyze the progressive removal of galacturonic acid molecules from the nonreducing end of the chain. A group of enzymes that catalyze a nonhydrolytic-type of cleavage of the α-(1→4) linkage are referred to as transeliminases. These enzymes show the same effect as the polygalacturonases, i.e., a lowering in viscosity of pectin and the production of one mole of reducing sugar for each α-(1→4) linkage hydrolyzed.

Glucose Oxidases

A number of fungi produce glucose oxidases, which catalyze the oxidation of β-D-glucose to gluconic acid (**4.14**). Commercial preparations of glucose oxidase usually contain catalase and this has proven to be advantageous for most uses of this enzyme in foods. This system is used extensively for the removal of glucose from egg whites and whole eggs prior to drying to retard darkening and for reducing dissolved and headspace oxygen in packaged foods containing glucose.

Proteases

A protease is an enzyme that catalyzes the hydrolysis of proteins to yield proteoses, peptones, polypeptides, and small amounts of amino acids (see Proteolysis in Chapter 6). There are two types of proteases: exopeptidases and endopeptidases. The exopeptidases, which split off terminal amino acids by hydrolyzing the peptide linkage, may be further divided into two types: carboxypeptidases and aminopeptidases. The former enzymes act on the C-terminal peptide linkage, while the latter enzymes act on the N-terminal peptide bond. The exopeptidases find little use in the food processing industry. The important proteases are the endopeptidases, which hydrolyze internal peptide linkages to yield peptides.

A large number of commercial plant, animal, and microbial proteases are available for use in food processing. The commercial preparations are generally used on the basis of their specificity; however, factors such as heat stability, pH optimum, the presence of activators or inhibitors, availability, and price also may influence the choice of an enzyme preparation. For example, in cheesemaking rennin is used to catalyze the formation of milk curds, even though many proteases will clot milk, because rennin splits only a specific bond or bonds, whereas the other proteases catalyze a more extensive hydrolysis of casein. In contrast, the production of protein hydrolyzates requires proteases that lead to extensive hydrolysis of the protein, yielding peptides and amino acids.

The heat stability of papain is one of the main reasons for its extensive use in meat tenderizing. This heat stability permits papain-catalyzed reactions to continue for a time during the cooking, frying, or roasting of meat, and as a consequence more extensive hydrolysis can be obtained.

The optimum pH of a protease affects its use in foods. For example, in the chillproofing of beer, a protease is used to prevent the formation of "chill haze," which contains protein, tannin, and carbohydrates. The normal pH of beer is 4.5; therefore, an enzyme is required that is active at this pH. Papain alone or in combination with pepsin, bromelin, or fungal protease is used in the application.

Lipases

Lipases hydrolyze insoluble fats and fatty acid esters. Enzyme catalysis takes place at the oil–water interface of the emulsion, and as a consequence enzyme reaction velocity is dependent on the surface area. The lipases show considerable specificity with respect to chain length of the fatty acids, the degree of saturation, position of the fatty acids, and physical state of the substrate. In general, lipases hydrolyze triglycerides of fatty acids with 4–10 carbon atoms more rapidly than the longer-chained fatty acids. Lipases remove the long-chain saturated fatty acids (C_{12} to C_{18}) and the most common unsaturated fatty acids (C_{18}) at about the same rate. The products formed on hydrolysis are largely 1,2-diglycerides and 2-monoglycerides. For a fixed amount of enzyme, lipase activity increases with an increase in the surface area of the emulsified fat.

The action of lipolytic enzymes is important in the dairy industry. A high degree of lipolysis is necessary in the manufacture of certain Italian cheeses (e.g., Romano and Provolone). Traditionally, rennin paste is used in the manufacture of these cheeses because the paste contains lipases as well as proteolytic enzymes.

Butterfat-containing products are treated with pregastric lipase to produce a wide variety of food flavors. For example, in the chocolate industry, cultured butter or pregastric lipase-modified butter products are used to enhance the flavor of milk chocolate, butter creams, and caramels. At low levels of free fatty acids, the flavor of the confection is enhanced without the formation of a new flavor; at intermediate levels, a buttery flavor is created; and at high levels, a chessy flavor is produced. Lipase-modified butter products are also used to enhance the flavors of food products such as margarines, shortenings, popcorn oils, bakery products, and vegetable oils.

Insolubilized Enzymes

Because enzymes in their native state are water soluble, removing them or terminating their reactions in food processing operations presents some problems. It may be necessary to add inhibitors to terminate the reaction or to precipitate the protein-containing enzyme for removal. Purification of the product is often a problem when the enzyme is soluble. For these reasons insolubilized enzymes afford several advantages: easy separation of the enzyme from the reaction product(s); reuse of the enzyme; modification of enzyme properties (e.g., specificity, Michaelis constant, and pH optimum); and generally increased stability of the enzyme to both heat and pH.

Insolubilization can be achieved by attaching an enzyme to an insoluble carrier. Early attempts to insolubilize enzymes involved adsorption of enzymes on charcoal and similar adsorbants. Entrapment of enzymes in a copolyacrylamide matrix has also been investigated. These methods proved unsatisfactory due to desorption of the enzyme from the adsorbant or loss of enzyme protein from the matrix of the polyacrylamide gel. More recently techniques have been developed to covalently attach enzymes to insoluble carriers.

The techniques for immobilizing (or insolubilizing) enzymes have been grouped by Goldstein and Manecke (1976) into four basic categories:

1. Adsorption on inert supports or ion-exchange resins.
2. Entrapment by occlusion with cross-linked gels or by encapsulation with microcapsules, hollow fibers, liposomes, and fibers.
3. Cross-linking by bi- or multifunctional reagents, often following adsorption or entrapment with a structure of defined geometry.
4. Covalent binding to polymeric supports via functional groups nonessential for biological activity of the protein.

Formation of Covalently Bound Enzymes

Nowadays, covalent binding to a variety of carriers is the most widely used method for producing immobilized enzymes. The main reactions and carriers used in this technique are illustrated in this section.

- Protein coupling to polymeric acyl azides:

$$\vert\!-\overset{\displaystyle O}{\overset{\|}{C}}-NHNH_2 \xrightarrow[HCl]{NaNO_2} \vert\!-\overset{\displaystyle O}{\overset{\|}{C}}-N_3 \xrightarrow{H_2N\text{-protein}} \vert\!-\overset{\displaystyle O}{\overset{\|}{C}}-NH\text{-protein}$$

- Protein coupling to polymeric acid anhydrides:

$$\left|\begin{array}{l} -\overset{\displaystyle O}{\overset{\|}{C}} \\ \quad\;\; \diagdown \\ \qquad\quad O \\ \quad\;\; \diagup \\ -\underset{\displaystyle O}{\underset{\|}{C}} \end{array}\right. + H_2N\text{-protein} \rightarrow \left|\begin{array}{l} -\overset{\displaystyle O}{\overset{\|}{C}}\text{—NH-protein} \\ \\ \\ \\ -\underset{\displaystyle O}{\underset{\|}{C}}\text{—OH} \end{array}\right.$$

- Protein coupling to carbodiimide-activated[1] carboxylic polymers:

$$\left|\,-\overset{\displaystyle O}{\overset{\|}{C}}\text{—OH}\right. + \begin{array}{c} R \\ | \\ N \\ \| \\ C \\ \| \\ N \\ | \\ R' \end{array} \rightarrow$$

$$\left|\,-\overset{\displaystyle O}{\overset{\|}{C}}\text{—O—}\right.\begin{array}{c} R \\ | \\ NH \\ | \\ C \\ \| \\ N \\ | \\ R' \end{array} \begin{array}{l} \xrightarrow{H_2N\text{-protein}} \left|\,-\overset{\displaystyle O}{\overset{\|}{C}}\text{—NH-protein}\right. + R'\text{—NH—}\overset{\displaystyle O}{\overset{\|}{C}}\text{—NH—}\mathbf{R} \\ \\ \xrightarrow{\text{Rearrangement}} \left|\,-\overset{\displaystyle O}{\overset{\|}{C}}\text{—}\overset{\displaystyle R'}{\overset{|}{N}}\text{—}\overset{\displaystyle O}{\overset{\|}{C}}\text{—NH—R}\right. \end{array}$$

[1] The carbodiimide can be 1-cyclohexyl-3[2-(4-*N*-methylmorpholinium)ethyl] carbodiimide or 1-ethyl-3(3-dimethylaminopropyl) carbodiimide.

- Protein coupling to carboxylic polymers activated with *N*-alkyl-5-phenylisoxazolium salt:

$$\text{|}-C(=O)-O^- + \text{N-alkyl-5-phenylisoxazolium (R}-N^+\text{)} \longrightarrow \text{|}-C(=O)-O-C(C_6H_5)=CH-C(=O)-NH-R \xrightarrow{H_2N\text{-protein}}$$

$$\text{|}-C(=O)-NH\text{-protein} + C_6H_5-C(=O)-CH_2-C(=O)-NHR$$

- Protein coupling to carboxylic polymers activated with *N*-ethoxycarbonyl-α-ethoxy-1,2-dihydroquinoline:

$$\text{|}-C(=O)-O^- + \text{1,2-dihydroquinoline (N}-COOC_2H_5\text{, 2-}OC_2H_5\text{)} \longrightarrow \text{|}-C(=O)-O-C(=O)-O-C_2H_5 + \text{quinoline} + C_2H_5OH$$

$$\xrightarrow{H_2N\text{-protein}} \text{|}-C(=O)-NH\text{-protein} + C_2H_5OH + CO_2$$

- Protein coupling via 2,4-dintrofluorophenyl group:

- Protein coupling to cyanogen bromide-activated polysaccharide polymers:

- Protein coupling to polymers containing isothiocyanate functional groups:

$$|\!-N{=}C{=}S \xrightarrow{H_2N\text{-protein}} |\!-NH-\overset{\overset{\large S}{\|}}{C}-NH\text{-protein}$$

- Protein coupling to polymers containing imidoester function groups:

$$|\!-C{\equiv}N + R-OH + H^+ \rightarrow$$

$$|\!-\overset{\overset{NH_2^+}{\|}}{C}-O-R + H_2N\text{-protein} \rightarrow |\!-\overset{\overset{NH_2^+}{\|}}{C}-NH\text{-protein}$$

- Protein coupling to polymers containing aldehyde functional groups:

$$|\!-CHO \xrightleftharpoons{H_2O} |\!-CH(OH)(OH) \xrightleftharpoons{H_2N\text{-protein}} |\!-CH(OH)(NH\text{-protein})$$

$$\rightleftharpoons |\!-CH{=}N\text{-protein}$$

- Protein coupling to polymeric diazonium salts:

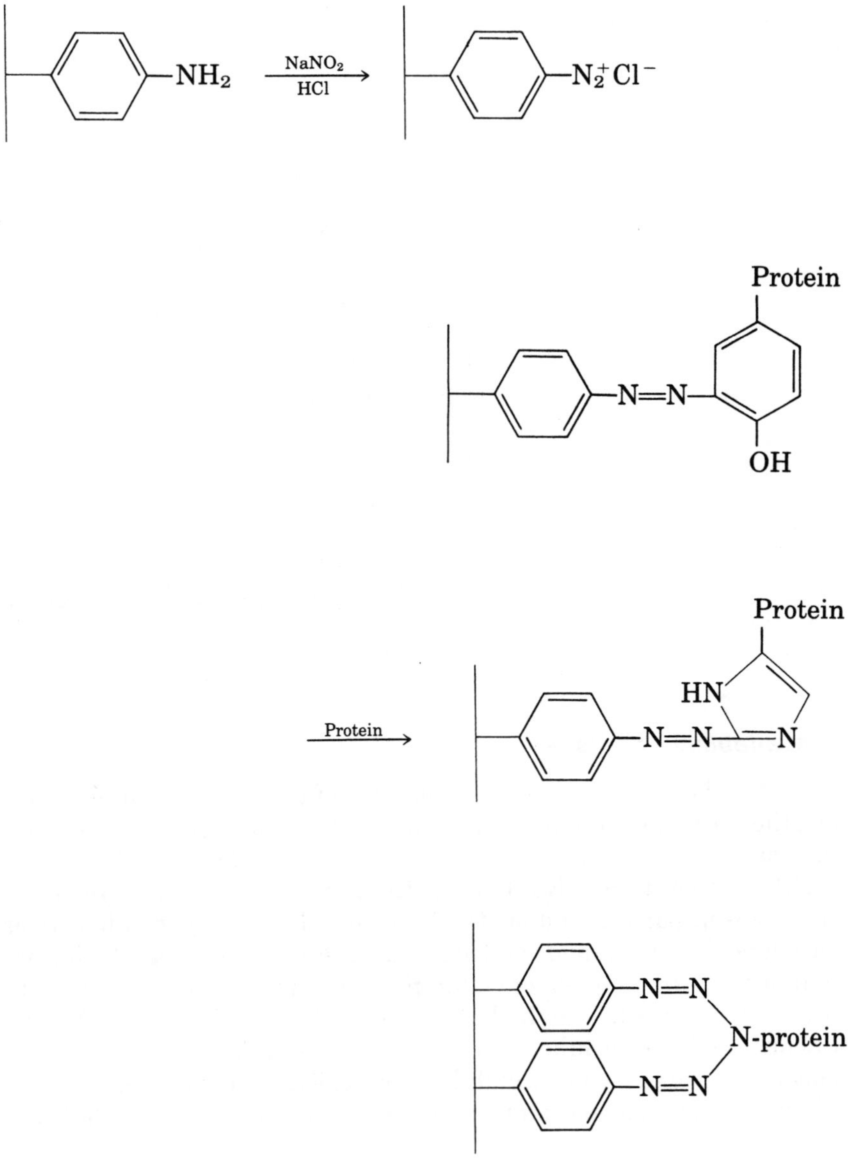

$$\text{P}-C_6H_4-NH_2 \xrightarrow[\text{HCl}]{\text{NaNO}_2} \text{P}-C_6H_4-N_2^+Cl^-$$

Protein

N=N

OH

Protein $\longrightarrow$

Protein

HN

N=N

N

N=N

N-protein

N=N

- Protein coupling to polymers containing sulfhydryl functional groups via thiol-disulfide interchange reactions:

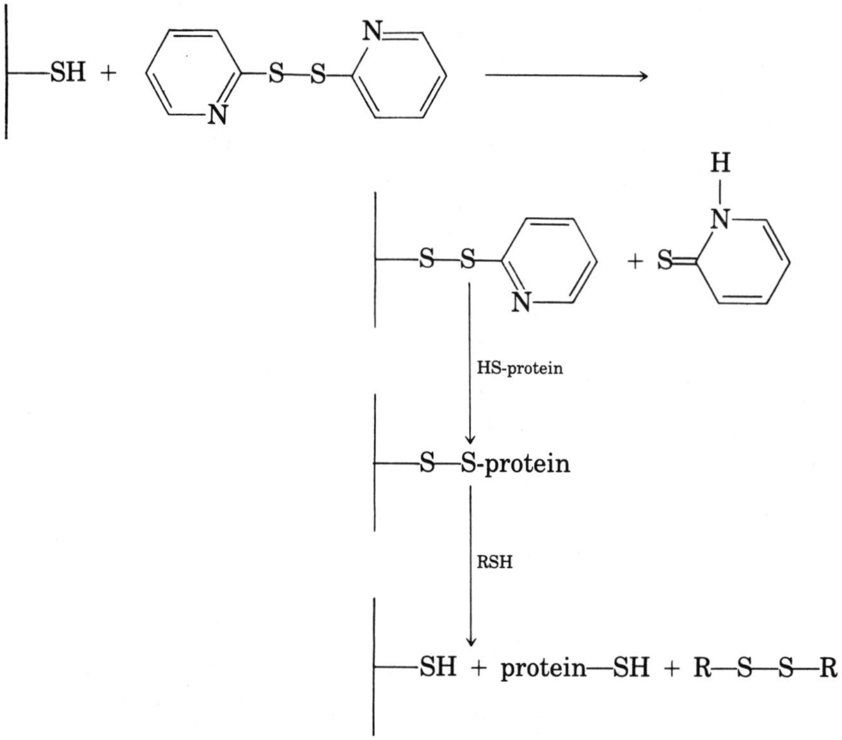

—SH + (2-pyridyl)—S—S—(2-pyridyl) ⟶

—S—S—(2-pyridyl) + S=(pyridine-2-thione, N—H)

↓ HS-protein

—S—S-protein

↓ RSH

—SH + protein—SH + R—S—S—R

Availability and Uses

Several biochemical supply firms now offer insolubilized enzymes, and the use of these products in commercial food processing is gradually increasing.

Miles Laboratories (Kankakee, IL) offers insolubilized chymotrypsin, trypsin, papain, and subtilisin attached to a polyanionic carrier (ethylene–maleic anhydride copolymer), to a water-insoluble diazonium salt of *p*-amino-DL-phenylalanine-L-leucine copolymer, or to a neutral synthetic resin (starch (dialdehyde) methylene-dianiline). Aldrich Chemical Co. (Milwaukee, WI) distributes a number of carriers that may be used for enzyme insolubilization. These are hydrophilic carriers of cross-linked polyacrylamides marketed as Enzacryl[2] and fall into

[2] Trademark of Koch-Light Ltd, Colnbrook Bucks, England.

five main categories: (1) Enzacryl AA has been used to insolubilize α-amylase, β-amylase, γ-amylase, and carboxypeptidase; (2) Enzacryl AH insolubilizes α-amylase; (3) Enzacryl Polyacetal insolubilizes trypsin, papain, α-amylase, dextranase, and urease; (4) Enzacryl Poly-thiol couples with compounds that contain disulfide bridges (e.g., lipoic acid and insulin); and (5) Enzacryl Polythiolactone binds proteins that contain side-chain amino groups (lysine) and hydroxyl groups (serine and tyrosine). Pharmacia Fine Chemicals Company (Piscataway, NJ) has immobilized a number of enzymes by coupling them with cyanogen bromide-activated Sepharose,[3] which is a beaded form of agarose, a polysaccharide polymer. Among the enzymes reported to be immobilized by Sepharose are chymotrypsin, rennin, trypsin, trypsinogen, protease, and ribonuclease. Sepharose-insolubilized enzymes offered by various companies include bromelin, ficin, glucose oxidase, peroxidase, alcohol dehydrogenase, asparagenase, hexokinase, glucose-6-phosphate dehydrogenase, and others.

Insolubilized enzymes so far have been used primarily in batch processes, although continuous methods would be preferable. Fluidized-bed reactors have shown some promise as an alternative to batch processes or less desirable continuous methods such as fixed-bed reactors. The insolubilized enzyme is suspended in the fluidized-bed reactor and the substrate solution flows out of the top; the suspended insoluble enzyme is kept in suspension by agitation from the substrate flow, which is counteracted by gravity. This procedure has been used to convert corn starch to glucose using insolubilized α-amylase. Insolubilized papain has been used to hydrolyze protein in beer, which in turn prevents the hazing of beer when chilled. Use of an insolubilized enzyme shortens the chillproofing process from days to minutes. Undoubtedly many other processes could be improved by use of insolubilized enzymes. The insolubilization of more than one enzyme on the same carrier offers exciting possibilities for coupled reactions.

☐ ENZYME ASSAYS

In food technology, enzymes are used for three major purposes. First, in some cases the activity of a particular enzyme in a food product is a measure of the previous history of the food; for example, a low level of phosphatase in milk indicates that proper pastuerization has occurred. Second, enzymes may be used to determine the concentration

[3] Trademark of Pharmacia Fine Chemicals.

of a food component; for example, glucose oxidase is used to determine glucose. And finally, as discussed already, enzymes are used to carry out various transformations of food components. For all these applications, accurate and rapid enzyme assays are important.

As discussed earlier (see Steady-State Kinetics section), the velocity of many enzyme-catalyzed reactions is directly proportional (first-order kinetics) to the concentration of enzyme when the substrate is present in excess. When the enzyme concentration is held constant, the reciprocal of the velocity is directly proportional to the reciprocal of the substrate concentration. Thus, the reaction conditions used in a particular enzyme assay will depend on whether enzyme concentration (measured as activity) or substrate concentration is being determined.

Determination of Enzyme Activity

When the purpose of an assay is to determine enzyme activity, the substrate concentration must be maintained high enough so that the reaction exhibits zero-order kinetics relative to the substrate and first-order kinetics relative to the enzyme. This allows the enzyme concentration (as activity) to be measured as a pseudo first-order reaction in which the velocity of the reaction is dependent upon the enzyme concentration. Under these conditions

$$v = k'[E] \text{ at constant } [S]$$

where k' = the first-order rate constant for the particular conditions of the reactions. At high levels of substrate, K_m is proportional to $k_2[E]_t$ and the Michaelis–Menten equation becomes

$$v = \frac{k_2[E]_t}{(1 + K_m/[S])}$$

where $[E]_t$ = total enzyme concentration. Thus, at a high and constant substrate concentration, the measured initial velocity is directly proportional to $[E]_t$, as shown in Fig. 7.1 (curve a).

In assays of this type, the reaction velocity may be measured either by a static (discontinuous) method or by a dynamic method. In the static method, the substrate and enzyme are mixed at zero time; aliquots of the assay mixture are withdrawn at specific time intervals and the reaction quickly stopped; then the concentration of remaining substrate or of product formed is determined. In the dynamic assay method, the substrate (or product) concentration is continuously monitored for a specified time period. With either method, the change in substrate (or product) concentration from zero time to time t is a measure of enzyme activity. The results are usually expressed as units of enzyme activity per milliliter or gram of enzyme under the specified assay conditions (e.g., temperature, pH). Because the substrate concentration is high relative to the enzyme concentration, the slight change in substrate concentration during the course of the assay does not affect the velocity of the reaction.

It is relatively easy to determine the activity of enzymes that catalyze reactions in which either the substrate or product (or a derivative of either) absorb ultraviolet or visible light. In such assays, the change in absorbance under appropriate conditions is a measure of the enzyme activity. The following assay methods for polyphenol oxidase, peroxidase, and catalase all involve measurement of absorbance to follow the course of the reaction.

Polyphenol Oxidase

As discussed in the section on enzymatic browning, polyphenol oxidase hydroxylates tyrosine to *o*-dihydroxyphenylalanine (DOPA), which is subsequently oxidized to *o*-quinone. In this assay, the rate of increase in the absorbance of *o*-quinone to 280 nm is proportional to enzyme concentration and is linear for approximately 5–10 min after an initial lag period.

Reagents

1. 1.0 m*M* L-tyrosine in aqueous solution (substrate).
2. 0.5 *M* phosphate buffer, pH 6.5.

Preparation of Enzyme Extract

To prepare enzyme extract, blend a 50-g portion of a vegetable or fruit (fresh, frozen, or dried) in a high-speed blender for 3 min, using enough water to make the blending process convenient. Remove larger particles by filtering. Larger particles and cellular debris may also be removed by centrifugation at 10,000 *g* for 10 min at 2°C.

Assay Procedure

Prepare assay mixtures, as indicated in Table 7.5, in 3-ml silica cuvettes with a 1-cm pathlength. Add all of the components except the enzyme. Prepare a nonenzymatic blank (Tube 1). After mixing, bubble oxygen into the mixture for 5 min. The reaction is initiated by adding the appropriate amount of enzyme extract followed by thorough mixing.

Immediately place the cuvette in an appropriate recording spectrophotometer with the wavelength set to 280 nm. The recorder should be adjusted so that an absorbance of 1.0 corresponds to the full scale.

Table 7.5. Assay Mixtures for Determining Polyphenol Oxidase Activity

	Volume per tube (ml)				
Component	1	2	3	4	5
Buffer	1.0	1.0	1.0	1.0	1.0
Substrate	1.0	1.0	1.0	1.1	1.2
Water	1.0	0.95	0.9	0.8	0.7
Enzyme	—	0.05	0.1	0.1	0.1

For most accurate results, the change in absorbance should be about 0.1/min. If the enzyme is too dilute to measure accurately, add more enzyme extract and decrease the water in the assay mixture proportionately. If the enzyme is too concentrated, dilute to a convenient concentration. For optimum results the enzyme activity should be measured when [S] is at V_{max}. V_{max} may be determined graphically from a Lineweaver–Burk plot (Fig. 7.3 on p. 288).

Calculations

The activity of polyphenol oxidase is commonly expressed as the change in absorbance per minute for the initial reaction. One unit of activity is defined as the amount of enzyme required to give an absorbance change of 0.001/min. From the observed change in absorbance per minute (correcting for blank if necessary), calculate the enzyme activity per milliter of enzyme extract. The results should be expressed for the reaction temperature employed.

Peroxidase

In this method, which is particularly suitable for determining peroxidase activity in milk, *p*-phenylenediamine is oxidized by the action of peroxidase on hydrogen peroxide.

Reagents

1. 0.3 *N* Hydrogen peroxide—Dilute 1.75 ml of 30% hydrogen peroxide to 100 ml with deionized-distilled water (keep cold).
2. 2.0% *p*-Phenylenediamine—Dissolve 1.0 g of *p*-phenylenediamine in 25 ml of hot distilled water; filter while hot. Make to final volume of 50 ml when cool.

Assay Procedure

Add 10 μl of milk and 50 μl of *p*-phenylenediamine to a tube containing 4.0 ml of distilled water using Eppendorf or similar pipettes.

Mix the contents; then add 10 μl of hydrogen peroxide and mix. Transfer the assay mixture to a 3-ml curvette with a 1-cm pathlength and measure the change in absorbance at 490 nm for 3 min with a recording spectrophotometer. The cuvette compartment should be thermostated at 20°C. Prepare a nonenzymatic oxidation blank by the same procedure omitting the hydrogen peroxide.

Samples are read against a *p*-phenylenediamine blank prepared by mixing 10 μl of hydrogen peroxide and 50 μl of *p*-phenylenediamine with 4.0 ml of water. These blanks should be prepared every 10 min or less.

Calculations

Peroxidase activity determined by this method is expressed in a unit called the purpurogallin number (PZ), which is the number of milligrams of purpurogallin formed per milligram of enzyme preparation when it acts upon 5 g of pyrogallol in 2 liter of water plus 10 ml of 0.5% hydrogen peroxide for 5 min at 20°C. To convert the observed change in absorbance (X) during the peroxidase assay to PZ, use the following equation:

$$PZ = \left(\frac{X}{0.0131} + 0.035\right) \times 10^{-3}$$

Catalase

The enzymatic breakdown of hydrogen peroxide to water and oxygen, catalyzed by catalase, is conveniently followed by the decrease in absorbance of hydrogen peroxide at 240 nm. If any components of the enzyme extract absorb appreciably at 240 nm, they should be removed. A nonenzymatic blank can be run to correct for any reaction between components of the extract and hydrogen peroxide.

Reagents

1. 0.059 M H_2O_2 prepared by diluting 0.3 ml of 30% H_2O_2 to 50 ml with 0.05 M (pH 7.0) phosphate buffer.
2. Commercial enzyme preparation (preferably lyophilized) to standardize procedure.

Preparation of Enzyme Extract

Blend a sample of meat or vegetable in a high-speed blender with enough water to give a workable slurry. Centrifuge in a refrigerated centrifuge for 15 min at 10,000 *g*. Carefully remove the supernatant, which contains the enzyme.

To prepare a nonenzymatic blank, heat the enzyme extract for 3 min at 100°C, then filter through a 0.45-μm Millipore filter. Often interfering substances can be removed by passing the enzyme extract through a short (10–15 cm) column of Sephadex G-100 or similar gel permeation media. This should be performed in 0.05 M (pH 7) phosphate buffer at 4°C. The emerging enzyme can be monitored at 280 nm in a flow cell or a column monitor.

Assay Procedure

Add 2 ml of the enzyme extract (properly diluted) to a 3-ml cuvette with a 1-cm pathlength. Prepare a control cuvette containing 2 ml of water. Pipette 1 ml of H_2O_2 solution into the enzyme solution, mix, and place immediately in spectrophotometer with wavelength setting at 240 nm. The absorbance should be recorded on a strip-chart recorder or determined at 10-sec intervals for 2 min. If necessary, a nonenzymatic blank should be assayed.

Calculations

In this assay, one unit of enzyme activity is defined as the amount of enzyme required to decompose 1 μmole of H_2O_2/min under the specified assay conditions at 25°C. The molar absorptivity of H_2O_2 in a 1-cm cuvette is 43.6 at 240 nm. Thus, the observed change in absorbance (ΔA) per minute can be converted to enzyme activity as follows:

$$\text{Units/ml} = \frac{(\Delta A/\text{min})(\text{dilution factor})\,(3)(1000)}{43.6 \times 2}$$

This measurement is most useful as a comparative method. Activities can be expressed as units of activity per milligram of protein if a protein determination is made. Although the protein in an unpurified sample is heterogeneous, some indication of relative specific activity can be obtained.

Determination of Substrate Concentration

If the purpose of an enzyme assay is to determine the concentration of substrate, then the reaction must be forced into pseudo first-order kinetics relative to the substrate by having a constant amount of enzyme present. Under these conditions, if assays are run with known concentrations of substrate, varying from about 0.01 K_m to 100 K_m, then a standard curve in the form of a Lineweaver–Burk plot of $1/v$ versus $1/[S]$ can be prepared. The concentration of a substrate in a test solution can be determined by measuring the reaction velocity with a known volume of the test solution and reading the corresponding substrate concentration from the standard curve.

Enzymatic methods for determining substrate concentration often are tedious and subject to considerable error if not performed carefully. However, they have the advantage of being quite specific, and therefore may permit determination of a compound in a mixture without prior separation from similar compounds. A good example of this is the enzymatic determination of glucose. Many of the carbohydrate tests described in Chapter 4 cannot distinguish between glucose and other simple sugars, and thus cannot be used to quantitatively determine glucose in a mixture. The glucose oxidase/peroxidase determination of glucose is specific and gives accurate results even in the presence of other simple sugars.

Glucose Oxidase/Peroxidase Determination of Glucose

In the glucose oxidase/peroxidase method for determining glucose, two enzyme reactions are "coupled" to allow a more convenient analysis. Glucose oxidase oxidizes glucose to produce H_2O_2, which serves as a substrate for peroxidase. A dye (*o*-dianisidine) reacts in the presence of H_2O_2 and peroxidase to give a stable product proportional to the original glucose concentration. The overall reaction catalyzed is as follows:

$$\text{Glucose} + H_2O + O_2 \xrightarrow{\text{Glucose oxidase}} H_2O_2 + \text{Gluconic acid}$$

$$H_2O_2 + o\text{-Anisidine} \xrightarrow{\text{Peroxidase}} H_2O + \text{Oxidized colored product}$$

The colored product absorbs at 540 nm. A glucose standard curve is used to determine the concentration of glucose in the original food product.

Reagents

1. Buffer—1.0 *M* phosphate, pH 6.0.
2. Purified horseradish peroxidase—aqueous solution.
3. 1% Aqueous solution of *o*-dianisidine.
4. Glucose substrate—18% aqueous solution.

Procedure

Add 0.1 ml of *o*-dianisidine dye solution to 12 ml of buffer. Follow the protocol listed in the following. Use 3 ml cuvettes (1-cm path) and thermostat the spectrophotometer at 25°C. The peroxidase activity in the coupled assay should be high enough to assure that this reaction is not the rate-limiting reaction. The activity of the peroxidase in the reaction must also be low enough to allow a sufficient period for analysis, i.e., 2–4 min.

	Assay mixture	*Control*
Glucose	0.3 ml	0.3 ml
Dye-buffer	2.5 ml	2.6 ml
Peroxidase	0.1 ml	0.1 ml
Glucose oxidase	0.1 ml	—

Using the control, zero the instrument against the control at 460 nm. Start the reaction by adding the glucose oxidase—containing solution to the cuvette and stir. Record the increase in absorbance for 2–4 min.

Calculations

$$\text{Units of glucose oxidase/mg} = \frac{A_{460}/\text{min}}{11.3 \times \text{mg enzyme/ml reaction mixture}}$$

One unit of activity is that amount of enzyme producing one micromole of H_2O_2 per minute at 25°C. This procedure can be modified to analyze, quantitatively, glucose-containing solutions. In the modification a standard curve for glucose is constructed with glucose oxidase and peroxidase at zero-order concentration and the working range of glucose at first-order kinetics concentrations. Various syrups can be analyzed by the glucose oxidase–peroxidase coupled method.

The concentration of glucose may be determined by modifying the preceding procedure used to determine the units of activity of glucose oxidase.

The amount of glucose oxidase must be increased to a level that the velocity (ΔA) is too rapid to conveniently measure. Then hold the glucose oxidase concentration at that level and begin reducing the concentration of known glucose to a conveniently measurable velocity. Prepare a standard curve of A_{460}/min vs. concentration of glucose. When a linear range of glucose is found, this may be used to determine glucose concentration in unknown solutions such as syrups. This procedure will essentially make the reaction zero order with respect to peroxidase and glucose oxidase and first order relative to glucose concentration.

☐ SELECTED REFERENCES

AMERICAN MEAT INSTITUTE FOUNDATION. 1960. The Science of Meat and Meat Products. W. H. Freeman and Co., San Francisco.

AOAC. 1980. Official Methods of Analysis. 13th ed. W. Horwitz (editor). Assoc. of Official Analytical Chemists, Arlington, VA.

AOAC. 1984. Official Methods of Analysis. 14th ed. S. Williams (editor). Assoc. of Official Analytical Chemists, Arlington, VA.

AYLWARD, F. and HAISMAN, D. R. 1969. Oxidation systems in fruits and vegetables—their relation to the quality of preserved products. Ad. Food Res. *17,* 2–77.

BECKHORN, C. J., LABBEE, M. D., and UNDERKOFLER, L. A. 1965. Production and use of microbial enzymes for food processing. J. Agric. Food Chem. *13*, 30–34.

BERNHARD, S. A. 1968. The Structure and Function of Enzymes. W. A. Benjamin, New York.

CLELAND, W. W. 1967. Enzyme kinetics. Ann. Rev. Biochem. *36*, 77.

CLELAND, W. W. 1970. Steady-state kinetics. *In* The Enzymes, 3rd ed., Vol. II. P. D. Boyer (editor). Academic Press, New York.

DESNUELLE, P. and SAVARY, P. 1963. Specificities of lipases. J. Lipid Res. *4*, 369–384.

DIXON, M. and WEBB, E. C. 1964. Enzymes, 2nd ed. Academic Pres, New York.

EPTON, R. and THOMAS, T. H. 1971. An Introduction to Water-Insoluble Enzymes. Koch-Light Labs. Ltd., Colnbrook Bucks, England.

EPTON, R. and THOMAS, T. H. 1971. Improving nature's catalysts. Aldrichimica Acta. *4*, 61–65.

FERDINAND, W. 1976. The Enzyme Molecule. John Wiley & Sons, New York.

FERSHT, A. 1977. Enzyme Structure and Mechanism. W. H. Freeman and Co., San Francisco.

FRUTON, J. S. 1971. Pepsin. *In* The Enzymes, 3rd ed., Vol. III. P. D. Boyer (editor). Academic Press, New York.

FURIA, T. E. (editor). 1972. Handbook of Food Additives. 2nd ed. Chemical Rubber Co., Cleveland, OH.

GERHART, J. C. and PARDEE, A. B. 1962. The enzymology of control by feedback inhibition. J. Biol. Chem. *237*, 891.

GLAZER, A. N. and SMITH, E. L. 1971. Papain and other plant sulfhydryl proteolytic enzymes. *In* The Enzymes, 3rd ed., Vol. III. P. D. Boyer (editor). Academic Press, New York.

GOLDSTEIN, L. and MANECKE, G. 1976. The chemistry of enzyme immobilization. *In* Applied Biochemistry and Bioengineering, Vol. I. L. B. Wingard, Jr., E. Katchalski-Katzir, and L. Goldstein (editors). Academic Press, New York.

HARTSUCK, J. A. and LIPSCOMB, W. L. 1971. Carboxypeptidase A. *In* The Enzymes, 3rd ed., Vol. III. P. D. Boyer (editor). Academic Press, New York.

HESS, G. P. 1971. Chymotrypsin-Chemical properties and catalysis. *In* The Enzymes, 3rd ed., Vol. III. P. D. Boyer (editor). Academic Press, New York.

HEWITT, E. J., MACKAY, D. A. M., KONIGSBACHER, K., and HASSELSTROM, T. 1956. The role of enzymes in food flavors. Food Technol. *10*, 487–489.

INTERNATIONAL UNION OF BIOCHEMISTRY. 1978. Enzyme Nomenclature. Academic Press, New York.

JOSLYN, M. A. and HEID, J. L. 1964. Food Processing Operations. Vol. II. AVI Publishing Co., Westport, CT.

JOSLYN, M. A. and PONTING, J. D. 1951. Enzyme-catalyzed oxidative browning of fruit products. Adv. Food Res. *3*, 1–46.

KEIL, B. 1971. Trypsin, *In* The Enzymes, 3rd ed., Vol. III. P. D. Boyer (editor). Academic Press, New York.

LEINER, I. E. and FRIEDENSON. 1970. Ficin. Methods In Enzymol. *19*, 261–273.

MAHLER, H. R. and CORDES, E. H. 1971. Biological Chemistry. 2nd ed. Harper and Row, New York.

METZLER, D. E. 1977. Biochemistry. Academic Press, New York.

MONOD, J., CHANGEUX, J. P. and JACOB, F. 1963. Allosteric proteins and cellular control systems. J. Mol. Biol. *6*, 306–329.

MORRISON, M. and SHONBAUM, G. R. 1976. Peroxidase-catalyzed halogenation. Ann. Rev. Biochem. *45*, 861–888.

PLOWMAN, K. 1972. Enzyme Kinetics. McGraw Hill, New York.

POMERANZ, Y. and MELOAN, C. E. 1971. Food Analysis: Theory and Practice. AVI Publishing Co., Westport, CT.

PULLEY, J. E. 1969. Enzymes simplify processing. Food Eng. *41*(2), 68–71.

REED, G. 1975, Enzymes in Food Processing. 2nd ed. Academic Press, New York.

REXOVÁ-BENKOVA, L. and MARKOVIČ, O. 1976. Pectic enzymes. Adv. Carbohydr. Chem. Biochem. *33*, 323–385.

SCHULTZ, H. W. (Editor). 1960. Food Enzymes. AVI Publishing Co., Westport, CT.

SCHWEIGERT, B. S. 1960. Food aspects of enzymes affecting proteins. *In* Food Enzymes. H. W. Schultz (editor). AVI Publishing Co., Westport, CT.

SEGEL, I. H. 1975. Enzyme Kinetics. John Wiley & Sons, New York.

SEGEL, I. H. 1976. Biochemical Calculations. 2nd ed. John Wiley & Sons, New York.

STRYER, L. 1981. Biochemistry. 2nd ed. W. H. Freeman and Co., San Francisco.

UNDERKOFLER, L. A. 1961. Production and application of plant and microbial proteinases. Soc. Chem. Ind. (London), Monograph *11*, 48–70.

VAN BUREN, J. P., MOYER, J. C. and ROBINSON, W. B. 1962. Pectin methylesterase in snapbeans. J. Food Sci. *27*, 291–294.

WALSH, C. 1979. Enzymatic Reaction Mechanisms. W. H. Freeman and Co., San Francisco.

WEILAND, H. 1972. Enzymes in Food Processing. Noyes Data, New Jersey.

WHITAKER, J. R. 1972. Principles of Enzymology for the Food Sciences. Marcel Dekker, New York.

WOLD, F. 1971. Macromolecules Structure and Function. Prentice-Hall, Englewood Cliffs, NJ.

ZEFFREN, E. and HALL, P. L. 1973. The Study of Enzyme Mechanisms. John Wiley & Sons, New York.

8

The Vitamins

☐ INTRODUCTION

An early landmark leading to the recognition of vitamins as a dietary essential occurred in 1881, when Lunin demonstrated that mice could not survive on purified diets consisting of proteins, carbohydrates, fats, and inorganic salts but would develop normally when milk was added. Lunin concluded that milk contained a substance essential for life and growth. Another early discovery, which demonstrated the specificity of the dietary essentials, was Eijkman's production of beriberi in chickens in 1897 by feeding them polished rice. This disease was prevented or cured by feeding either a diet of unpolished rice or a diet to which alcoholic extract of rice bran was added. In 1906, Hopkins confirmed the findings of Lunin and suggested that these essential nutrients be designated as *accessory food factors.*

The work of Eijkman was confirmed by Funk in 1911, who isolated from rice polishings a crystalline fraction that cured beriberi. This fraction was amine-like in chemical properties; hence, he suggested the substance be called *vitamine.* The terminal *e* was deleted in the early 1920s when it was realized that some substances effective in curing deficiency diseases were not amines.

As research progressed, findings indicated that many types of nutrients were necessary for growth and reproduction. Osborne and Mendel and McCollum and Davis discovered that growing rats require a fat-soluble factor for normal growth. This factor, found in certain animal fats such as butter, was designated fat-soluble A. The water-soluble factor, which cured beriberi, was designated water-soluble B.

Holst and Frolich, working in Norway, showed that guinea pigs restricted to a diet of cereals and hay developed scurvy while those fed fresh foods did not. The scurvy-preventing factor is now known as vitamin C. Similarly, the value of fish-liver oils in curing or preventing rickets in children was explained by the isolation and identification of vitamin D.

The obvious difference in solubility between the antiberiberi and antiscorbutic vitamins on the one hand and the antirachitic vitamin on the other led to the classification of vitamins as water soluble or fat soluble. Subsequent identification of the structures and metabolic functions of the vitamins have added much to our understanding of these dietary essentials. However, a biochemical explanation has not been discovered in all cases for the observed deficiency diseases caused by a lack of these essential nutrients.

Vitamins may be defined as complex organic compounds required in catalytic amounts for the proper functioning of living cells and not synthesized in amounts to meet physiological needs. The quantities required may range from micrograms per day (e.g., vitamin B_{12}) to milligrams per day (e.g., vitamin C).

Although it is convenient to list specific symptoms resulting from an inadequate intake of an individual vitamin, deficiencies due to a single vitamin are seldom encountered in practice. A diet that is deficient in one vitamin is usually deficient in several vitamins. As a result, the vitamin that is lowest in amount relative to its dietary requirements will probably determine the major signs of a deficiency but the symptoms of other deficiencies will also be present. Deficiency diseases, therefore, are most frequently clinical manifestations of several vitamin deficiencies. The recommended daily dietary allowances for the vitamins are listed in Table 8.1.

☐ WATER-SOLUBLE VITAMINS

The water-soluble vitamins have many properties in common. They have a similar distribution in foods, which explains why a deficiency of several factors is observed more frequently than a deficiency of a single vitamin. This group of vitamins must be supplied in the diet every day because dietary excesses are excreted in urine and not stored to an appreciable extent. The water-soluble vitamins function as coenzymes for specific enzyme systems essential for various metabolic processes of the cell.

Table 8.1. Recommended Daily Allowances for Vitamins—1980[a]

		Weight		Height			Fat-soluble vitamins			Water-soluble vitamins						
	Age (years)	(kg)	(lb)	(cm)	(in)	Protein (g)	Vitamin A (μg RE)[b]	Vitamin D (μg)[c]	Vitamin E (mg α-TE)[d]	Vitamin C (mg)	Thiamin (mg)	Riboflavin (mg)	Niacin (mg NE)[e]	Vitamin B-6 (mg)	Folacin[f] (μg)	Vitamin B-12 (μg)
Infants	0.0–0.5	6	13	60	24	kg × 2.2	420	10	3	35	0.3	0.4	6	0.3	30	0.5[g]
	0.5–1.0	9	20	71	28	kg × 2.0	400	10	4	35	0.5	0.6	8	0.6	45	1.5
Children	1–3	13	29	90	35	23	400	10	5	45	0.7	0.8	9	0.9	100	2.0
	4–6	20	44	112	44	30	500	10	6	45	0.9	1.0	11	1.3	200	2.5
	7–10	28	62	132	52	34	700	10	7	45	1.2	1.4	16	1.6	300	3.0
Males	11–14	45	99	157	62	45	1000	10	8	50	1.4	1.6	18	1.8	400	3.0
	15–18	66	145	176	69	56	1000	10	10	60	1.4	1.7	18	2.0	400	3.0
	19–22	70	154	177	70	56	1000	7.5	10	60	1.5	1.7	19	2.2	400	3.0
	23–50	70	154	178	70	56	1000	5	10	60	1.4	1.6	18	2.2	400	3.0
	51+	70	154	178	70	56	1000	5	10	60	1.2	1.4	16	2.2	400	3.0
Females	11–14	46	101	157	62	46	800	10	8	50	1.1	1.3	15	1.8	400	3.0
	15–18	55	120	163	64	46	800	10	8	60	1.1	1.3	14	2.0	400	3.0
	19–22	55	120	163	64	44	800	7.5	8	60	1.1	1.3	14	2.0	400	3.0
	23–50	55	120	163	64	44	800	5	8	60	1.0	1.2	13	2.0	400	3.0
	51+	55	120	163	64	44	800	5	8	60	1.0	1.2	13	2.0	400	3.0
Pregnant						+30	+200	+5	+2	+20	+0.4	+0.3	+2	+0.6	+400	+1.0
Lactating						+20	+400	+5	+3	+40	+0.5	+0.5	+5	+0.5	+100	+1.0

Source: Food and Nutrition Board (1980).

[a] The allowances are intended to provide for individual variations among most normal persons as they live in the United States under usual environmental stresses. Diets should be based on a variety of common foods in order to provide other nutrients for which human requirements have been less well defined.

[b] Retinol equivalents. 1 retinol equivalent = 1 μg retinol or 6 μg β-carotene.

[c] As cholecalciferol = 400 IU of vitamin D.

[d] α-tocopherol equivalents. 1 mg *d*-α tocopherol = 1 α-TE.

[e] 1 NE (niacin equivalent) is equal to 1 mg of niacin or 60 mg of dietary tryptophan.

[f] The folacin allowances refer to dietary sources as determined by *Lactobacillus casei* assay after treatment with enzymes (conjugases) to make polyglutamyl forms of the vitamin available to the test organism.

[g] The recommended dietary allowance for vitamin B_{12} in infants is based on average concentration of the vitamin in human milk. The allowances after weaning are based on energy intake (as recommended by the American Academy of Pediatrics) and consideration of other factors, such as intestinal absorption.

Thiamin (Vitamin B_1)

Thiamin is a crystalline compound, which is relatively heat stable in dry form but heat labile in solution, particularly when it is alkaline. It is oxidized in the presence of mild oxidizing agents to form thiochrome, a fluorescent pigment. This reaction is the basis for the chemical assay of thiamin.

Thiamin has a molecular weight of 337 and the following chemical structure:

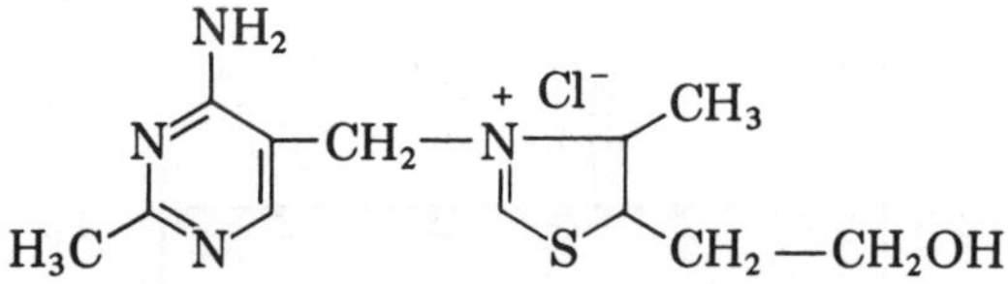

NH_2, Cl^-, CH_3, CH_2—N^+, H_3C, N, N, S, CH_2—CH_2OH

Thiamin chloride

The thiamin molecule is composed of a pyrimidine nucleus and a thiazole nucleus connected by a methylene bridge. The pyrimidine moiety is relatively stable, whereas the thiazole moiety opens readily in alkaline solution. Thiamin contains an amino group on the pyrimidine ring and a quaternary nitrogen on the thiazole ring. The thiazole ring also contains an ethoxy group, which is reactive in much the same way as other hydroxyl groups and is essential for vitamin activity.

Thiamin is absorbed from the small intestine and is phosphorylated in the intestinal mucosa. Though salts of the vitamin are not completely absorbed from the intestinal tract, modified derivatives (thiamin propyl disulfide, benzoylthiamin disulfide, and others) that are biologically active are absorbed to a greater degree in weak alkaline solution. These derivatives are relatively insoluble and are stable to heat; therefore, they may be of value in the enrichment of foods.

Biosynthesis

The details of the biosynthetic pathways for the pyrimidine and the thiazole moieties are unknown. However, enzymes in certain plant materials catalyze condensation of the preformed thiazole and pyrimidine moieties of thiamin, as shown in Fig. 8.1.

Thiamin pyrophosphate, the coenzyme form of thiamin, is synthesized by the first transfer of the pyrophosphate group from ATP, as follows:

$$\text{Thiamin} + \text{ATP} \rightarrow \text{Thiamin pyrophosphate (TPP)} + \text{AMP}$$

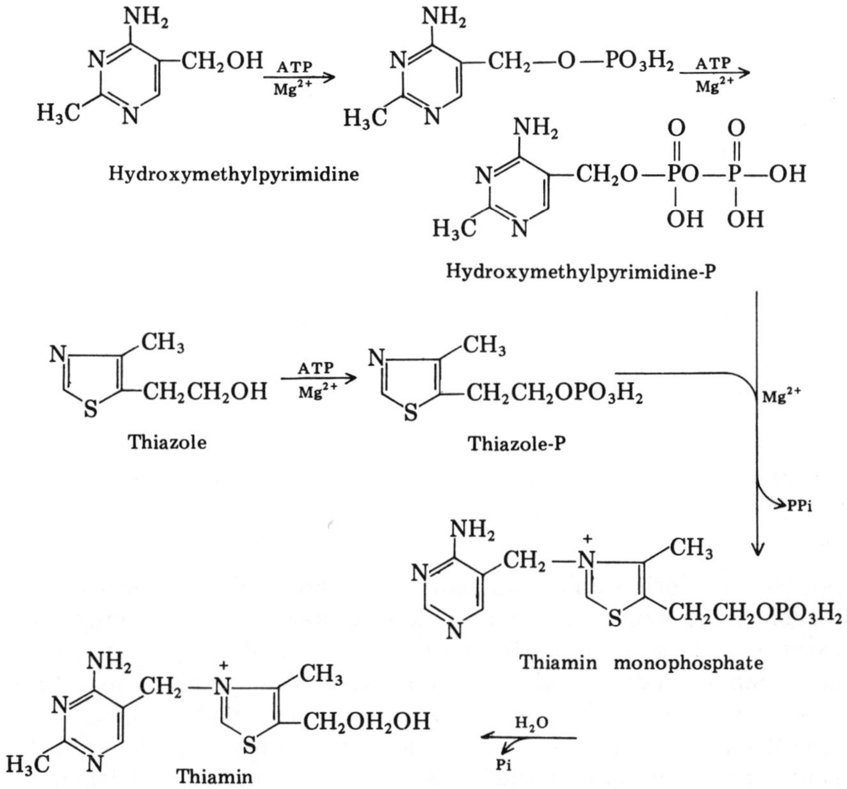

Hydroxymethylpyrimidine

Hydroxymethylpyrimidine-P

Thiazole

Thiazole-P

Thiamin monophosphate

Thiamin

Figure 8.1. Condensation of pyrimidine and thiazole moieties of thiamin (vitamin B_1), which is catalyzed by certain plant enzymes.

Function and Effects of Deficiency

Thiamin pyrophosphate functions in carbohydrate metabolism as a coenzyme (cocarboxylase) in the decarboxylation of α-keto acids and in the utilization of pentose in the hexose monophosphate shunt. A mechanism by which thiamin pyrophosphate exerts its action is outlined in Fig. 8.2. The active aldehyde complex can give rise to acetaldehyde, acetoin, or acetyl-SCoA. Thiamin and lipoic acid are essential components of the multienzyme pyruvate dehydrogenase complex and the glutamic dehydrogenase complex. The latter enzyme is required for oxidative decarboxylation of α-keto acids and their subsequent conversion to acyl-SCoA derivatives.

In view of the function of thiamin in carbohydrate metabolism, it generally has been assumed that requirements for this vitamin are

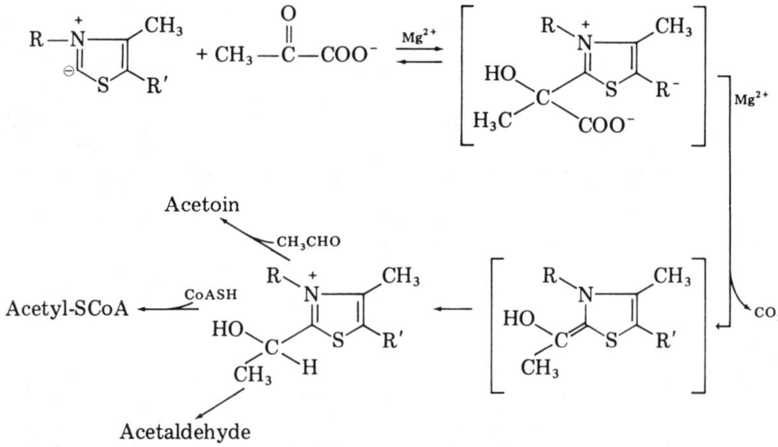

Figure 8.2. Proposed mechanism for thiamin coenzyme function in decarboxylation reactions.

related to caloric intake, particularly to those calories derived from carbohydrate. When protein and fat serve as a source of energy, they exert a sparing action on thiamin needs. Thiamin requirements are dependent on body size, physical activity, environmental temperature, and the physiological state of an individual. The most striking symptom of a thiamin deficiency is a disease of the nervous system, polyneuritis in lower animals and beriberi in man. Experimental deficiency symptoms can be produced by use of an analogue of thiamin, such as pyrithiamin or oxythiamin.

Recommended Dietary Allowances

All animals except ruminants require an exogenous source of thiamin. For the average man and woman, the recommended daily dietary allowance is 1.0–1.4 mg. Thiamin needs are increased when there is greatly increased metabolism such as occurs in febrile conditions, hyperthyroidism, or muscular activity. Additional allowances of 0.4 mg/day during pregnancy and 0.5 mg/day during lactation are recommended.

Distribution in Foods

Thiamin is distributed widely in nature. Among the richer sources, none of which is highly potent, are the whole cereal grains, nuts, pork, and eggs. Pork contains several times as much thiamin as beef (0.193

mg/100 gm and 0.009 mg/100 gm, respectively). Butter and cheese contain only negligible amounts, while milk is a better source. Potatoes furnish significant quantities of thiamin when used extensively in the diet; one medium potato, 2.5-in. diameter, contains 0.10 mg B_1. Legumes are a good source of this vitamin, but the usual methods of cooking them cause a considerable loss in the amount available for ingestion. Wheat flour is a poor source of thiamin, inasmuch as most of the vitamin is removed in the milling process.

Thiamin, one of the more labile of the vitamins, undergoes considerable loss during the preparation of food for human consumption. In aqueous solutions at pH 3.5 thiamin can be heated to 120°C for 20 min without deterioration, but if the pH of the solution is above 5.5, losses are marked. Thiamin is leached out of foods during wet processing, such as the blanching of vegetables. Cooking by boiling in an excessive amount of water and discarding the cooking water results in loss of thiamin. Washing before cooking also can cause a thiamin loss. For example, washing whole rice once can result in loss of 25% of the thiamin.

Baking has an adverse effect on the thiamin content of cereal products. The losses vary with the product, the duration of heating, the temperature, the amount of surface area exposed, and the amount of alkaline baking powder present. Baking of bread can result in losses of 15–25%; baking rolls to the usual degree of brownness causes losses of about 15% of the thiamin.

Thiamin losses from meat are related to cooking conditions, the size of the cut, and the fat content. Fluids from the thawing of frozen meats, from cooking of meats, and from slicing of meats contain significant amounts of thiamin. Canned meats undergo loss of thiamin during storage. For example, canned pork luncheon meat may lose as much as 30% of its thiamin content by the end of 6 months' storage at 21°C.

Cereals stored as whole grain can undergo losses, the extent of which depend on the moisture content. Similarly, fortified white flour loses thiamin even under favorable storage conditions.

Riboflavin (Vitamin B_2)

Riboflavin was first isolated from milk in 1879 as lactochrome, a natural yellow fluorescent pigment in milk whey. Not until 1933 was it shown that this essential dietary factor of milk whey was similar to the yellow fluorescent pigments isolated from eggs and liver. In the meantime, Warburg and Christian had isolated from yeast a flavin-protein complex, "yellow enzyme," necessary for cell respiration.

Chemical studies soon demonstrated that the fluorescent pigment associated with the "yellow enzyme" was the same as that of milk, eggs, and liver. Degradation studies of the pigment molecule showed that it consisted essentially of a cyclic ring structure (isoalloxazine) and a side chain of pentose sugar (ribose). The structure of riboflavin, confirmed by total synthesis, is as follows:

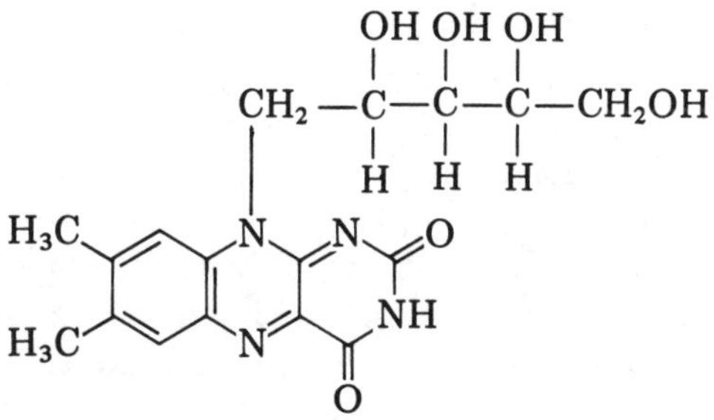

Riboflavin
6,7-Dimethyl-9-(1′-D-ribityl)-isoalloxazine

Riboflavin is sparingly soluble in water. It is stable to heat, air, and oxygen, but is light sensitive. Exposure of the vitamin to either ultraviolet irradiation or to visible light results in irreversible decomposition. Riboflavin is sensitive to alkalies but is quite stable in strong acid solutions.

Biosynthesis

Riboflavin is synthesized by all green plants and by most bacteria, molds, fungi, and yeast. The initial stages of biosynthesis are the same as those in purine biosynthesis. The first established intermediate precursor is 6,7-dimethyl-8-ribityllumazine.

Riboflavin synthesis is completed by the catalytic action of riboflavin synthetase in a reaction involving two molecules of the substituted lumazine. In this reaction the diazine ring of one molecule is ruptured and a four-carbon unit is added to the second lumazine molecule to form riboflavin and 4-ribitylamino-5-amino-2,6-dihydroxypyrimidine (Fig. 8.3).

Function and Effects of Deficiency

Riboflavin, an essential dietary constituent for mammals, is part of the structure of two flavin coenzymes, flavin mononucleotide (FMN) and flavin adenine dinucleotide (FAD). Flavin mononucleotide (ribo-

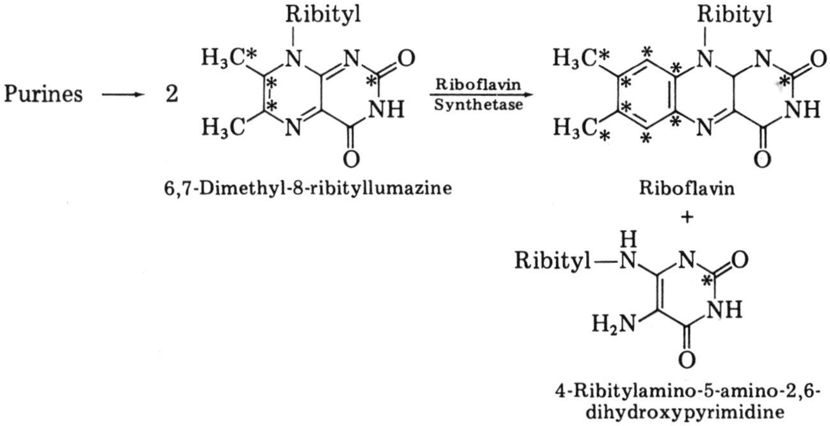

Figure 8.3. Final steps in biosynthesis of riboflavin (vitamin B_2).

flavin 5′-phosphate) is formed from riboflavin and ATP in a reaction catalyzed by flavokinase.

$$\text{Riboflavin} + \text{ATP} \rightarrow \text{FMN} + \text{ADP}$$

FAD is formed from FMN and ATP in a reaction catalyzed by flavin nucleotide pyrophosphorylase.

$$\text{FMN} + \text{ATP} \xrightarrow{Mg^{2+}} \text{FAD} + \text{PP}$$

Both coenzymes undergo reversible oxidation-reduction when participating in enzyme-catalyzed reactions. The oxidized and reduced forms are shown below.

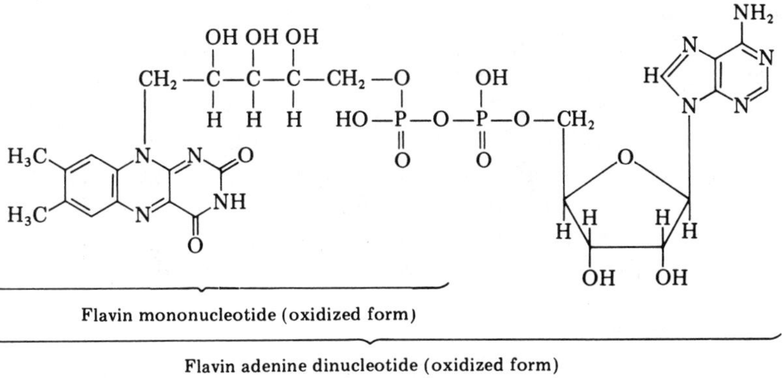

It is a common observation that humans with one vitamin deficiency are likely to have other associated vitamin deficiencies. This is man-

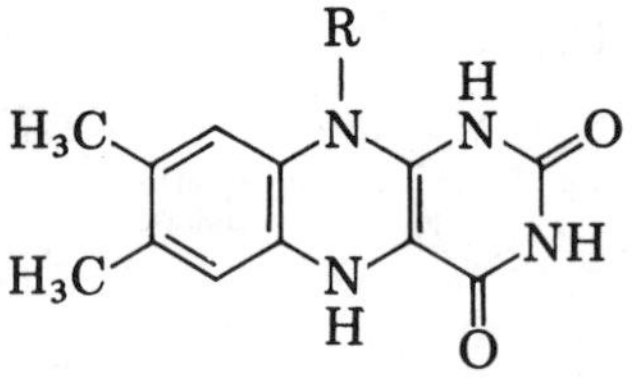

FMN or FAD (reduced form)

ifested especially with respect to thiamin, riboflavin, and niacin since these water-soluble vitamins often occur together in foods. Consequently, humans with either beriberi (thiamin deficiency) or pellagra (niacin deficiency) generally are also deficient in riboflavin.

Riboflavin deficiency in man is characterized by fissuring in the angles of the mouth (cheilosis), a localized seborrheic dermatitis of the face, a magenta-colored tongue, and organic disorders of the eye (corneal vascularization). Despite the fact that riboflavin functions as the prosthetic group of flavoproteins concerned with oxidative processes, the clinical changes attributed to its deficiency are minor. In general, both FMN and FAD decrease in concentration in riboflavin deficiencies and both decrease more rapidly in organs such as liver and kidney than in heart and brain. Inasmuch as riboflavin is essential for growth, there is a relationship between protein metabolism and riboflavin retention and excretion; animals on low-protein diets excrete increased amounts of riboflavin. The amount of riboflavin excreted in the urine tends to reflect dietary supply and body stores.

Recommended Dietary Allowances

Riboflavin requirements are related closely to energy expenditure. The Food and Nutrition Board of the National Research Council recommended, in 1980, an intake of 0.6 mg/1000 kcal. Thus, the daily allowance falls between 1.4 and 1.7 mg for men and 1.2 and 1.3 mg for women. The daily allowance is increased by 0.3 mg during pregnancy and by 0.5 mg during lactation. The requirements for children are approximately the same as for adults, i.e., 0.6 mg/1000 kcal. For elderly people and others whose caloric intake may be less than 2000 kcal, a minimum intake of 1.2 mg/day is recommended.

Distribution in Foods

Yeast is possibly the best source of riboflavin. Other good sources are milk, cheese, eggs, whole-grain and enriched cereals, green leafy vegetables, peas, lima beans, and most meats. Fish is a poorer source of the vitamin than beef, pork, or chicken. Fruits contain insignificant amounts.

Milk is an important source of riboflavin. Since this vitamin is stable to heat, pasteurization and spray-drying of milk have very little effect on riboflavin content. The ease with which riboflavin is destroyed by light can result in an extensive loss of riboflavin if milk is exposed to direct sunlight.

Riboflavin is one of the nutrients used in the enrichment of bread, alimentary paste products (macaroni, spaghetti, noodles, and vermicelli), corn meal and corn grits, rice, and farina. Enrichment standards, which set minimum and maximum levels, include not only riboflavin but also niacin, thiamin and iron.

Niacin (Nicotinic Acid)

Niacin is one of the most stable of the vitamins. It is stable to acids, bases, mild oxidizing agents, heat, and light. It is sparingly soluble in cold water but soluble in hot water and alcohol. Nicotinic acid and niacinamide are active biologically, but the amide is the form that exists in nature.

Biosynthesis

Tryptophan can be converted into niacin in the body. The pathways of the conversion of tryptophan into niacin involve several intermediary steps. One pathway involves the oxidation of tryptophan to kynurenine, which is converted by way of a number of intermediates to nicotinic acid mononucleotide. This substance is utilized in the synthesis of nicotinamide adenine dinucleotide (NAD^+), as shown in Fig. 8.4.

Function and Effects of Deficiency

Niacin in the form of NAD^+ acts as a coenzyme that functions as an acceptor of electrons from a substrate. The protein moiety of each enzyme utilizing NAD^+ confer substrate specificity. It will be noted that in addition to a hydride ion being transferred from the substrate to the nicotinamide moiety, a proton is liberated into the medium, as follows:

$$NAD^+ \text{ (or } NADP^+) + MH_2 \rightleftharpoons M + NADH \text{ (or NADPH)} + H^+$$

Less frequently, the coenzyme may be the phosphorylated derivative of NAD^+, nicotinamide adenine dinucleotide phosphate ($NADP^+$),

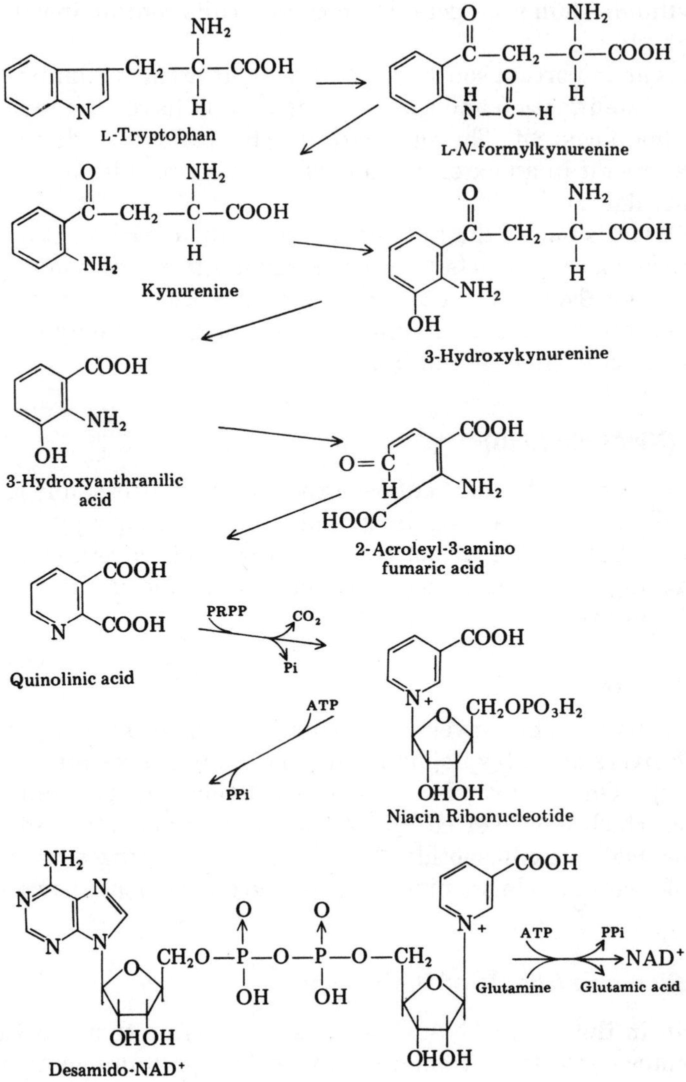

Figure 8.4. Biosynthesis of NAD^+ from L-tryptophan.

which functions in the same way as NAD^+. These coenzymes participate in many biochemical pathways including pyruvate metabolism, citric acid cycle, lipid metabolism, and synthesis of high-energy bonds. Reduction and oxidation of NAD^+ occur at the C-4 position. The oxidation and reduction process is stereospecific to yield either the H_R isomer or the H_S isomer.

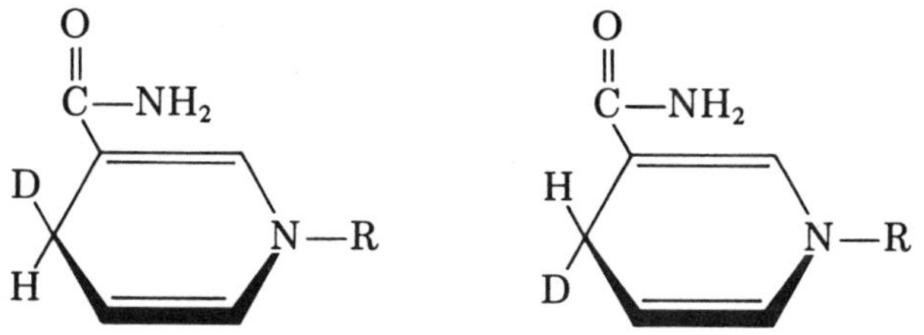

H_R isomer

H_S isomer

These isomers are based on the configuration of the monodeuterated reduced NAD^+ derivatives, i.e., above or below the plane of the ring. Alcohol dehydrogenase and lactic dehydrogenase utilize the α-isomer while α-glycerophosphate dehydrogenase and L-glutamic dehydrogenase utilize the β-isomer.

A deficiency of niacin leads to the disease known as pellagra. Since tryptophan is a precursor of niacin in man, pellagra must be considered as deficiencies of both the vitamin and the amount and the kinds of dietary protein available. In human dietaries, 60 mg of dietary tryptophan is equivalent to 1 mg of niacin. A characteristic glossitis and stomatitis usually appear early in pellagra. A dermatitis may appear on any part of the body, but generally it is confined to areas exposed to sunlight. Nonspecific mental symptoms include irritability, confusion, mental anxiety, and depression, which can develop to delirium and dementia. Severe persistent diarrhea occurs only in advanced cases.

Recommended Dietary Allowances

The term *niacin equivalent* (NE) is used to express the combined values of dietary sources of this vitamin and that converted from dietary tryptophan. One niacin equivalent (NE) is equal to 1 mg of niacin or 60 mg of tryptophan. The values recommended for men are 16–18 mg NE/day. For women the range is 13–15 mg/day. For children up to 10 years, the allowance varies from 8 to 15 mg/day.

Distribution in Foods

The richest sources of niacin are yeast, liver, lean meats, poultry, beans, peas, nuts, and whole-grain and enriched cereal products. Milk and eggs are low in this vitamin, but they are effective in preventing or curing pellagra probably because of their high-quality protein.

Pyridoxine Group (Vitamin B_6)

Vitamin B_6 does not denote a single substance but refers to a group of three related substances: pyridoxine, pyridoxal, and pyridoxamine.

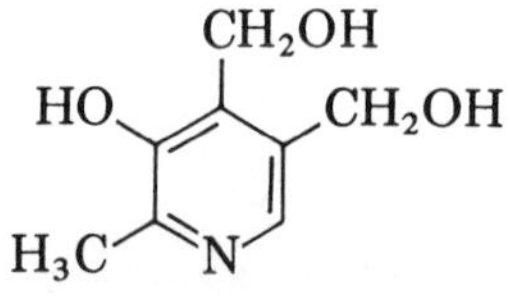

Pyridoxine 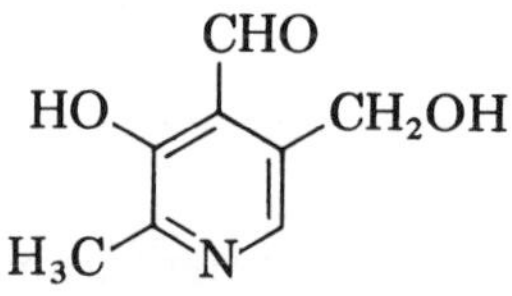 Pyridoxal 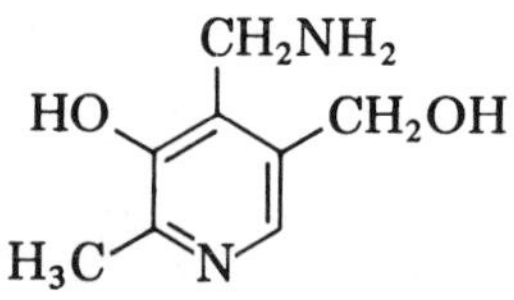 Pyridoxamine

Pyridoxine is found in plant products, while pyridoxal and pyridoxamine are found primarily in animal products. These substances are present in either the free or bound forms.

Pyridoxine is readily soluble in water, acetone, and alcohol, and slightly soluble in ether and chloroform. It is stable to heat in either acidic or basic solution. Pyridoxal and pyridoxamine decompose rapidly at high temperatures. Pyridoxine is destroyed on irradiation in neutral or alkaline solutions, but is stable in acid solution. All three substances are destroyed by oxidizing agents such as nitric acid, permanganate, and hydrogen peroxide.

Biosynthesis

Pyridoxine is synthesized by green plants and microorganisms but the biosynthetic pathway is not completely elucidated. Serine, glycine, glycolaldehyde, and tryptophan have been implicated as possible precursors of pyridoxine synthesis in *E. coli*.

Function and Effects of Deficiency

Vitamin B_6 occurs in tissues predominantly as the phosphates of pyridoxal and pyridoxamine. Pyridoxine is phosphorylated in the liver by a specific kinase and oxidized to pyridoxal phosphate (coenzyme form) by a specific flavoprotein. The latter form functions as a coenzyme in a number of chemical reactions involving amino acids. The interrelationships are as follows:

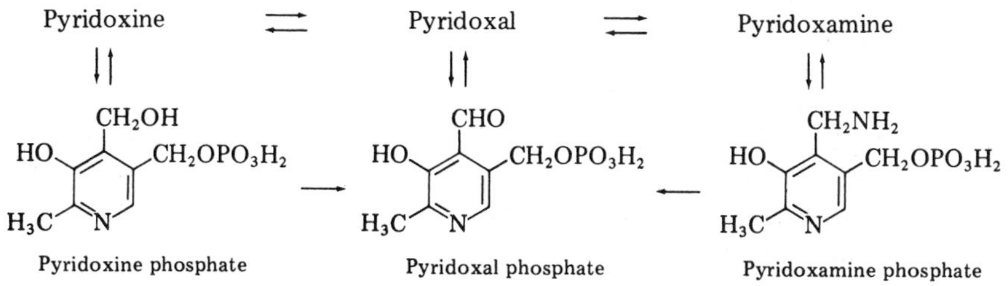

Pyridoxine ⇌ Pyridoxal ⇌ Pyridoxamine

Pyridoxine phosphate → Pyridoxal phosphate ← Pyridoxamine phosphate

Among the enzymes utilizing pyridoxal phosphate are amino acid decarboxylases, transaminases, enzymes involved in tryptophan metabolism (tryptophan to niacin), racemases (formation of equilibrium mixture of DL-alanine from D- or L-alanine), and cystathionase (conversion of cystathionine to serine and homocysteine). This involvement in many amino acid reactions explains the increased need for the vitamin when diets are high in protein.

The occurrence of vitamin B_6 is so widespread that a deficiency rarely occurs in humans. However, a vitamin deficiency can be induced by the administration of antagonists such as deoxypyridoxine and isonicotinic acid hydrazide, which are used in tuberculosis therapy. The symptoms observed following use of these compounds are skin lesions, anemia, and convulsive seizures.

The effect of pyridoxine deficiency is more dramatic in infants than in adults. A deficiency of this nutrient was the cause of an outbreak of convulsive seizures and nervous irritability in infants, less than 6 months old, fed a commercial canned liquid-milk formula. It is now known that the deficiency was caused by the destruction of vitamin B_6 during heat processing.

Recommended Dietary Allowances

The human requirement for vitamin B_6 is not known, but 1.8–2.0 mg daily is recommended as a reasonable allowance for adults. It is believed that the need for the vitamin increases during pregnancy, during lactation, with aging, and in such special situations as radiation exposure, cardiac failure, and isoniazid therapy for tuberculosis. The recommended allowance for women is 2 mg/day. However, during pregnancy or lactation, an additional 0.6 and 0.5 mg/day, respectively, are recommended. Since the B_6 group is widely distributed in both plant and animal foods, an average mixed diet will provide an adequate intake.

Distribution in Foods

The vitamin B_6 group is widely distributed in foods and occurs in both free and bound form. The best sources of this vitamin are liver, muscle meats, fish, whole-grain cereals, brewers' yeast, milk, egg yolk, lettuce, vegetables, bananas, lemons, oatmeal, and nuts. Generally, foods rich in the other members of the B-complex are also good sources of the pyridoxine group.

Pantothenic Acid

Pantothenic acid was partially purified from a variety of sources, but it was not recognized as the same factor in each case. Thus, it was referred to as the chick antidermatitis factor, the liver filtrate factor, the chick antipellagra factor, and the yeast factor. Finally, it was isolated from yeast by Williams and his colleagues. The significance of this nutrient in animal nutrition was established by Jukes, Woolley, and their associates. Its chemical structure is given below:

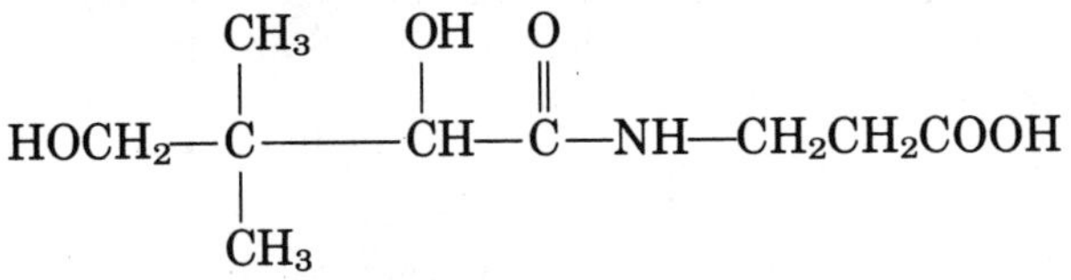

$$HOCH_2-\overset{\displaystyle CH_3}{\underset{\displaystyle CH_3}{\overset{|}{\underset{|}{C}}}}-\overset{\displaystyle OH}{\overset{|}{C}}H-\overset{\displaystyle O}{\overset{\|}{C}}-NH-CH_2CH_2COOH$$

Pantothenic acid
D(+)-*N*-(2,4-Dihydroxy-3,3-dimethylbutyryl)-β-alanine

Chemically, pantothenic acid is an unstable, viscous oil. It is customary to use the vitamin in the form of the calcium salt, a white solid, which dissolves freely in water, is bitter to the taste, and is readily hydrolyzed by acid or alkali. The vitamin is relatively stable to heat in neutral solution and long storage in foods.

Biosynthesis

Pantothenic acid can be synthesized by green plants and most microorganisms but not by mammalian tissues. The biosynthetic pathway in yeast, neurospora, and bacteria (e.g., *Escherichia coli, Neurospora crassa*) is shown in Fig. 8.5. Pantoic acid is formed from α-ketoisovaleric acid by the addition of a one-carbon unit to give ketopantoic acid, which is reduced to pantoic acid. The formation of pantothenic acid from pantoic acid and β-alanine is catalyzed by an ATP-dependent enzyme system. Pantothenic acid is then incorporated into coenzyme

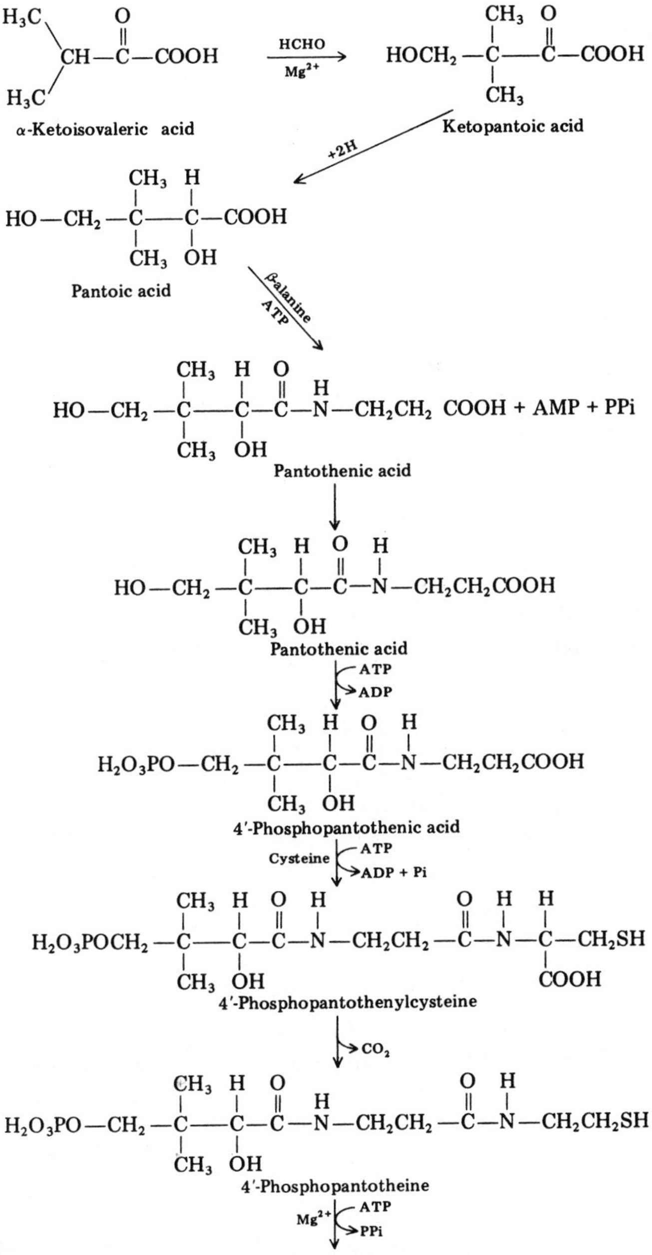

α-Ketoisovaleric acid

HCHO, Mg^{2+} →

Ketopantoic acid

+2H

Pantoic acid

β-alanine, ATP

$COOH + AMP + PPi$

Pantothenic acid

Pantothenic acid

ATP → ADP

4′-Phosphopantothenic acid

Cysteine, ATP → ADP + Pi

4′-Phosphopantothenylcysteine

→ CO_2

4′-Phosphopantotheine

Mg^{2+}, ATP → PPi

Figure 8.5. Biosynthesis of pantothenic acid and coenzyme A. Animals can form coenzyme A from pantothenic acid, but cannot synthesize pantothenic acid.

Continued.

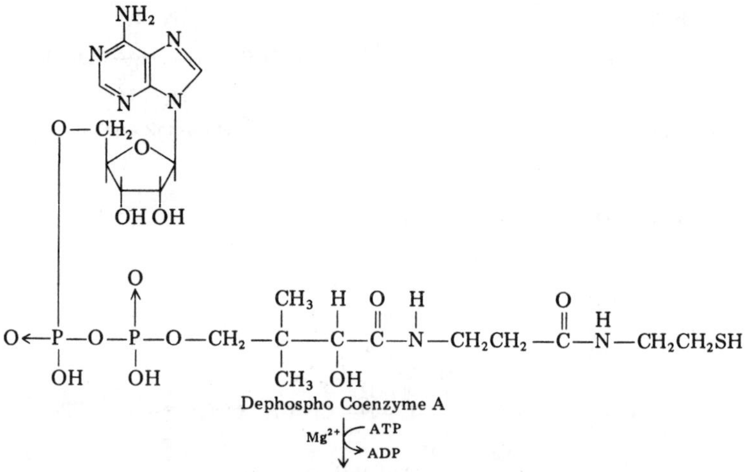

$$O \leftarrow \overset{\displaystyle OH}{\underset{}{P}}-O-\overset{O}{\uparrow}{P}(OH)-O-CH_2-C(CH_3)_2-CH(OH)-\overset{O}{\overset{\|}{C}}-NH-CH_2CH_2-\overset{O}{\overset{\|}{C}}-NH-CH_2CH_2SH$$

(adenosine: NH_2, N, N, N, N; O—CH_2; O; OH OH)

Dephospho Coenzyme A

Mg^{2+}, ATP → ADP

Coenzyme A (Phosphorylated in 3′ position of ribose)

Figure 8.5. *Continued.*

A (CoA) both in animals and microorganisms. The proposed pathway for biosynthesis of the coenzyme is also shown in Fig. 8.3.

Function and Effects of Deficiency

Pantothenic acid has a vital role in metabolism because it is a component of coenzyme A. This coenzyme is required for the metabolism of two-carbon compounds, notably acetyl groups. In general the $-SH$ group of CoA acts as an acyl acceptor, as follows:

$$R-\overset{O}{\overset{\|}{C}}-O^- + CoASH \rightarrow R-\overset{O}{\overset{\|}{C}}-SCoA$$

Coenzyme A is involved in the aerobic utilization of carbohydrates and in the breakdown and synthesis of fatty acids and lipids. Coenzyme A functions in the synthesis of sterols, steroid hormones, porphyrin, and acetylcholine.

Panthothenic acid is so widely distributed in foods that a deficiency disease due to a lack of the vitamin has not been observed in man on a natural diet. If, however, the pantothenic acid antimetabolite (antagonist) omega-methyl pantothenic acid is given with a deficient diet, an illness closely resembling the deficiency observed in animals develops. The syndrome consists of headaches, fatigue, abdominal distress, sleep disturbances, numbness and tingling of hands and feet, nausea, and malaise. The syndrome is not relieved by administration of pantothenic acid although improvement is noted on a good high-protein diet; consequently, it is believed that the deficiency effects observed are due to the toxicity of the antagonist rather than to a complete lack of pantothenic acid. An increase in susceptibility to infections is associated with pantothenic acid deficiency.

Recommended Dietary Allowances

The human requirement for pantothenic acid has not been established but is probably not above 5 mg daily. The Food and Nutrition Board of the National Research Council estimates that the average American diet furnishes about 7 mg of pantothenic acid daily, which apparently is adequate. Intakes vary between 5 and 20 mg/day. Presently, evidence indicates that an intake of 4–7 mg/day would be adequate for adults; a higher intake may be needed during pregnancy and lactation.

Distribution in Foods

Pantothenic acid is present in all plant and animal tissues. The greatest amounts of this nutrient occur in egg yolk, liver, kidney, fresh vegetables, and yeast. Skim milk, buttermilk, lean beef, grains, white potatoes, tomatoes, broccoli, cauliflower, fruits, peanuts, and molasses also are food sources. Egg white, beets, turnips, corn, rice, and apples are poor sources.

Biotin

Three separate lines of research led to the discovery of biotin. In 1916, Bateman showed that rats fed a diet rich in raw or uncooked egg white developed a peculiar skin disorder characterized by extreme losses of body hair and finally death. This condition was known as egg-white injury. The toxic factor in the egg white was identified as avidin. Györgyi, in 1931, isolated from yeast a protective factor, vitamin H, that was similar to the anti-egg-white injury factor. In 1933 a factor necessary for the growth and respiration of the nitrogen-fixing organism *Rhizobium* was discovered and designated as coenzyme R. In 1936, Kögl and Tönnis isolated from egg yolk a crystalline material, which they termed biotin. Finally, in 1940, Györgyi showed that vitamin H, coenzyme R, and biotin were identical. The structure of biotin is presented below:

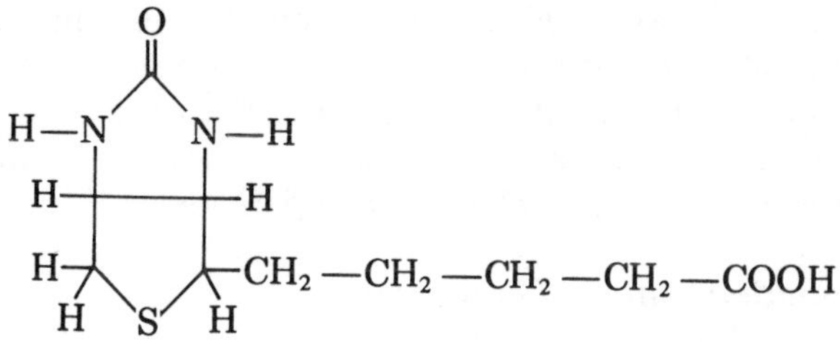

$$\text{Biotin: bicyclic ureido (}O{=}C\text{, two N–H) fused to tetrahydrothiophene ring (S), with side chain } -CH_2-CH_2-CH_2-CH_2-COOH$$

Biotin

Biotin is readily soluble in hot water, sparingly soluble in cold water, stable to heat, and labile to strong acids, alkalies, and oxidizing agents. Biotin combines with avidin, a glycoprotein found in raw egg white, to form an insoluble substance that cannot be absorbed from the intestinal tract. Avidin is readily inactivated by heating.

Biosynthesis

The pathway of biotin synthesis in plants and microorganisms is not known. The biosynthesis of biotin has been accomplished with *Achromobacter* IVS using isovaleric acid and [3-^{14}C]cysteine or $^{14}CO_2$. The postulated pathway is shown in Fig. 8.6.

Function and Effects of Deficiency

Biotin is essential for the activity of many enzyme systems in plants, microorganisms, and animals. It serves as the prosthetic group of a series of enzymes involved in synthesis and breakdown of fatty acids and amino acids through the addition of CO_2 and its removal from active compounds, and in deamination reactions of amino acids. Acetyl-SCoA carboxylase, a key enzyme in the synthesis of fatty acids in *E. coli*, catalyzes the following reaction:

$$CH_3-\overset{\overset{\displaystyle O}{\|}}{C}-SCoA + {}^{14}CO_2 + ATP \xrightarrow{Mg^{2+}}$$

$$HOO^{14}C-CH_2-\overset{\overset{\displaystyle O}{\|}}{C}-SCoA + ADP + Pi$$

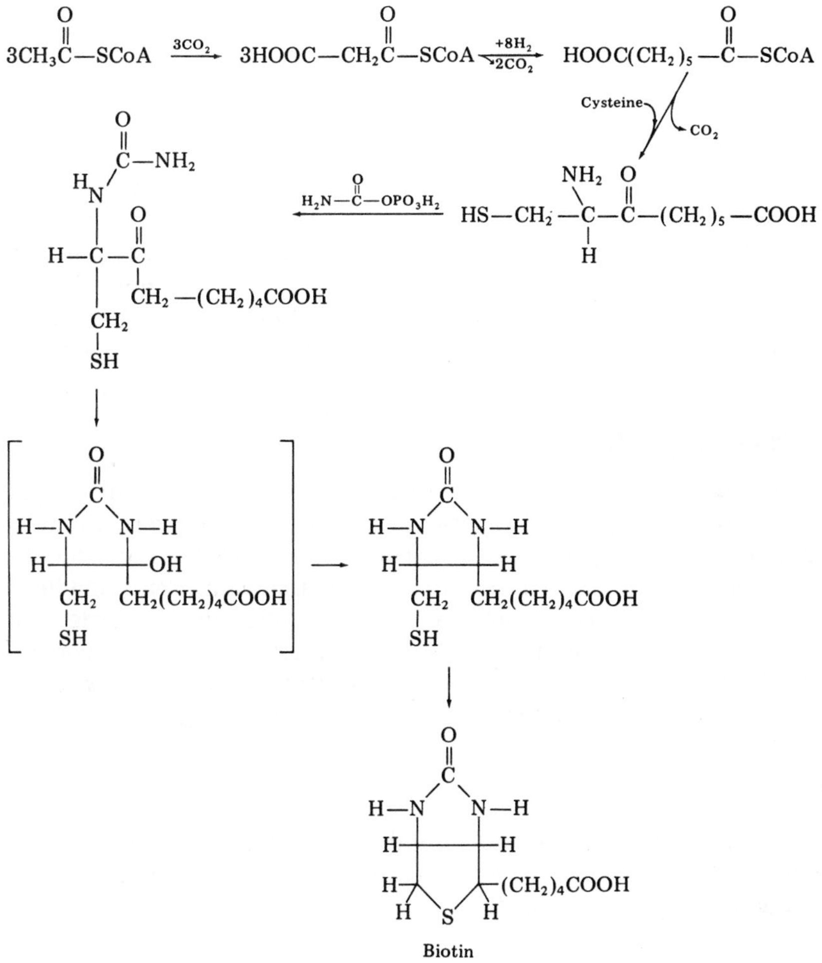

$$3CH_3C(=O)—SCoA \xrightarrow{3CO_2} 3HOOC—CH_2C(=O)—SCoA \xrightarrow[-2CO_2]{+8H_2} HOOC(CH_2)_5—C(=O)—SCoA$$

$$\xrightarrow[-CO_2]{\text{Cysteine}} HS—CH_2—CH(NH_2)—C(=O)—(CH_2)_5—COOH$$

$$\xrightarrow{H_2N—C(=O)—OPO_3H_2} H_2N—C(=O)—NH—CH(CH_2SH)—C(=O)—CH_2—(CH_2)_4COOH$$

$$\longrightarrow \left[\text{cyclic ureido: H—N, N—H; H, OH; } CH_2SH,\ CH_2(CH_2)_4COOH\right] \longrightarrow \text{cyclic ureido: H—N, N—H; H, H; } CH_2SH,\ CH_2(CH_2)_4COOH \longrightarrow \text{Biotin}$$

Biotin

Figure 8.6. Proposed pathway for biosynthesis of biotin in *Achromobacter* IVS.

In this reaction, biotin forms N-carboxybiotin-enzyme in which CO_2 attaches to the N-1 position of biotin. Labeled $^{14}CO_2$ appears first in the biotin as ^{14}C-N (carboxy) then later in malonyl-SCoA indicating that biotin is, in fact, mediating the transfer of CO_2.

Biotin deficiency in man does not occur naturally through a dietary shortage. Experimental deficiency, induced by a low-biotin diet containing large quantities of dried, uncooked egg white, results in scaly dermatitis, muscle pains, lassitude, anorexia, insomnia, and a slight anemia. All of these symptoms clear after feeding biotin. Spontaneous biotin deficiency in man seems unlikely because of intestinal bacterial synthesis of this compound. The total urinary and fecal excretions of biotin by man exceeds his dietary intake.

Recommended Dietary Allowances

Dietary requirements for biotin have not been established. The Food and Nutrition Board of the National Research Council suggest that diets providing a daily intake of 0.10–0.30 mg/day are adequate for adults. The lower level of intake would provide replacement for the upper level of excretion, which is approximately 0.046 mg/day. For children and infants it is suggested that biotin intake be 0.050 mg/1000 kcal.

Distribution in Foods

Good sources of biotin are peanuts, chocolate, egg yolk, milk, organ meats (liver, kidney), most vegetables (peas and cauliflower), and some fruits (bananas, grapefruit, and strawberries). Foods having moderate or low concentrations are meat, milk products, cereals, bread, and flour. For humans and some animals, synthesis by intestinal bacteria is an important source of biotin. Although royal jelly generally is not considered as a human food, it is the richest source of biotin, having approximately 410 μg/100 gm.

Folacin

The term folic acid refers to a number of derivatives of folacin (pteroylmonoglutamic acid), all of which are interconvertible. The structure of folacin is represented as follows:

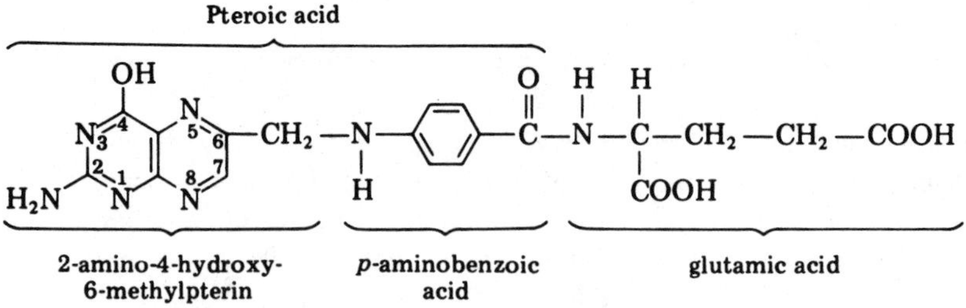

Pteroic acid

2-amino-4-hydroxy-6-methylpterin

p-aminobenzoic acid

glutamic acid

Pteroylmonoglutamic Acid

Folacin is a yellow crystalline compound, slightly soluble in cold water and moderately soluble in hot water. It is stable to heat in neutral or alkaline solutions but is inactivated when heated in acidic solutions. Either partial or complete inactivation is caused by sunlight, oxidation, reduction, esterification, and methylation. It is readily absorbed from the gastrointestinal tract and is stored primarily in the liver.

Biosynthesis

Guanosine is utilized directly for synthesis of 2-amino-4-hydroxy-6-hydroxymethyldehydropteridine, which can be converted to dihydrofolic acid via dihydropteroic acid with partially purified enzymes. The physiologically active form is the reduced tertahydrofolic acid, which is formed by the action of folacin reductase. The synthetic pathway is shown in Fig. 8.7.

Function and Effects of Deficiency

Folacin coenzymes function in the transfer and the utilization of one-carbon groups, such as formyl, hydroxymethyl, and methyl, particularly in the synthesis of purine and pyrimidine ribotides and deoxyribotides, and in amino acid interconversions. The oxidized folacin derivatives are enzymatically reduced to tetrahydrofolic acid (FH_4), which functions as the active coenzyme in transferring one-carbon fragments, e.g., serine → glycine + 1-C. These transfers can be mediated either by the attachment of the one-carbon unit to either the N-5 or N-10 position of pteroic acid or by forming a methylene bridge between N-5 and N-10, as shown below and on the next page:

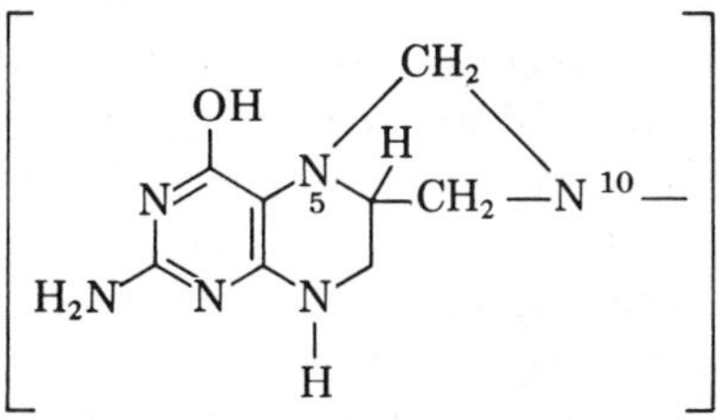

FH_4—N^5,N^{10}-Methylene

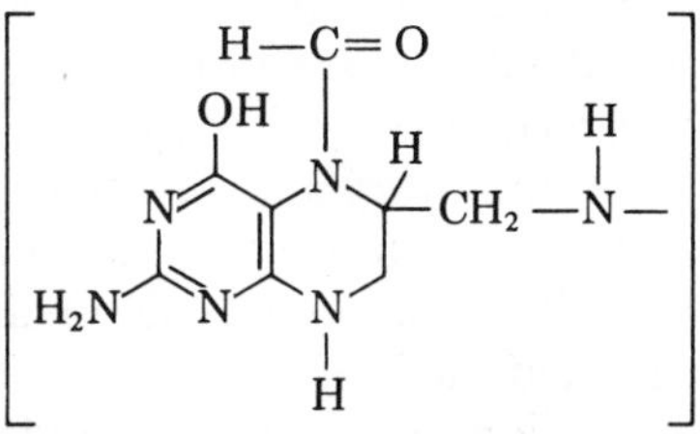

FH_4—N^5-Formyl

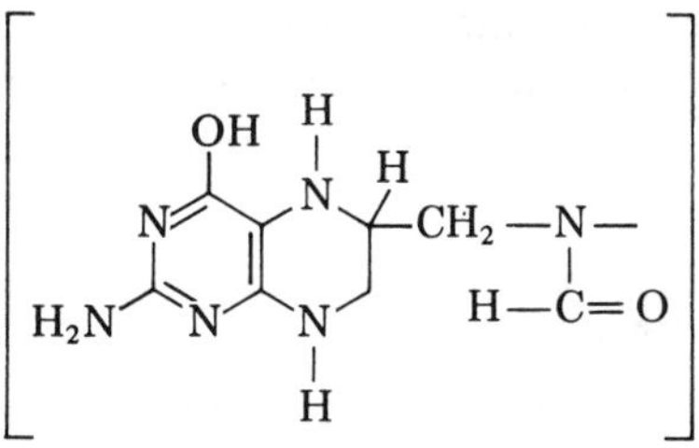

FH_4—N^{10}-Formyl

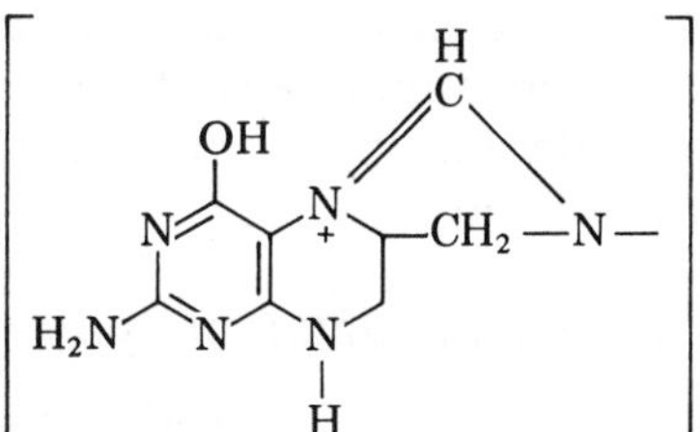

FH_4—N^5,N^{10}-Methenyl

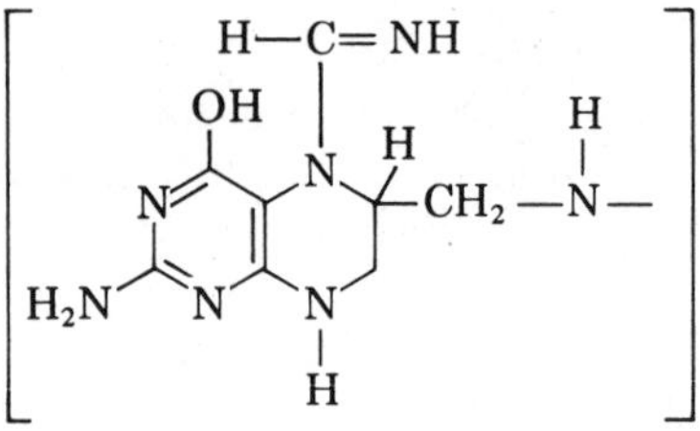

FH_4—N^5-Formimino

An example of the removal of a one-carbon unit is shown in the degradation of serine to glycine: FH_4 + serine → glycine + FH_4-N^5, N^{10} methylene.

The synthesis of N^5,N^{10}-methylenetetrahydrofolate is catalyzed by serine hydroxymethyltransferase as shown in the preceding reaction. Pyridoxal phosphate is also a coenzyme in the reaction. The N^5,N^{10}-methylenetetrahydrofolate formed in the reaction can subsequently be converted to tetrahydrofolate via the reaction:

$$CO_2 + NH_4^+ + N^5,N^{10}\text{-methylenetetrahydrofolate} + \text{NADH} \xrightarrow{\text{glycine synthase}} \text{glycine} + \text{tetrahydrofolate} + \text{NAD}^+$$

Thus, tetrahydrofolate derivatives serve as donors of one-carbon units in a variety of biosynthetic reactions.

A deficiency of folacin causes certain anemias, characterized by large erythrocytes in the peripheral blood and the accumulation in the bone marrow of immature red blood cells (megaloblasts). Dietary deficiencies of folacin are difficult to produce, since intestinal bacteria synthesize folacin. A deficiency can be induced experimentally through use of inhibitory analogs such as aminopterin and amethopterin. These analogs interfere with the biosynthesis of many tissue constituents; thus, these compounds are utilized in the treatment of some leukemias.

Folacin deficiency may arise from inadequate dietary intake, malabsorption, excessive demands by the tissues of the body (pregnancy and lactation), and metabolic derangements. Manifestations of a deficiency include glossitis (a sore, red, smooth tongue), diarrhea, gastrointestinal disturbances, and anemia. In man the anemias of pellagra, sprue, and pregnancy often improve when folacin is administered.

Recommended Dietary Allowances

The dietary allowance for the normal adult is 0.4 mg daily. A daily allowance of 0.8 mg is recommended during pregnancy and of 0.5 mg

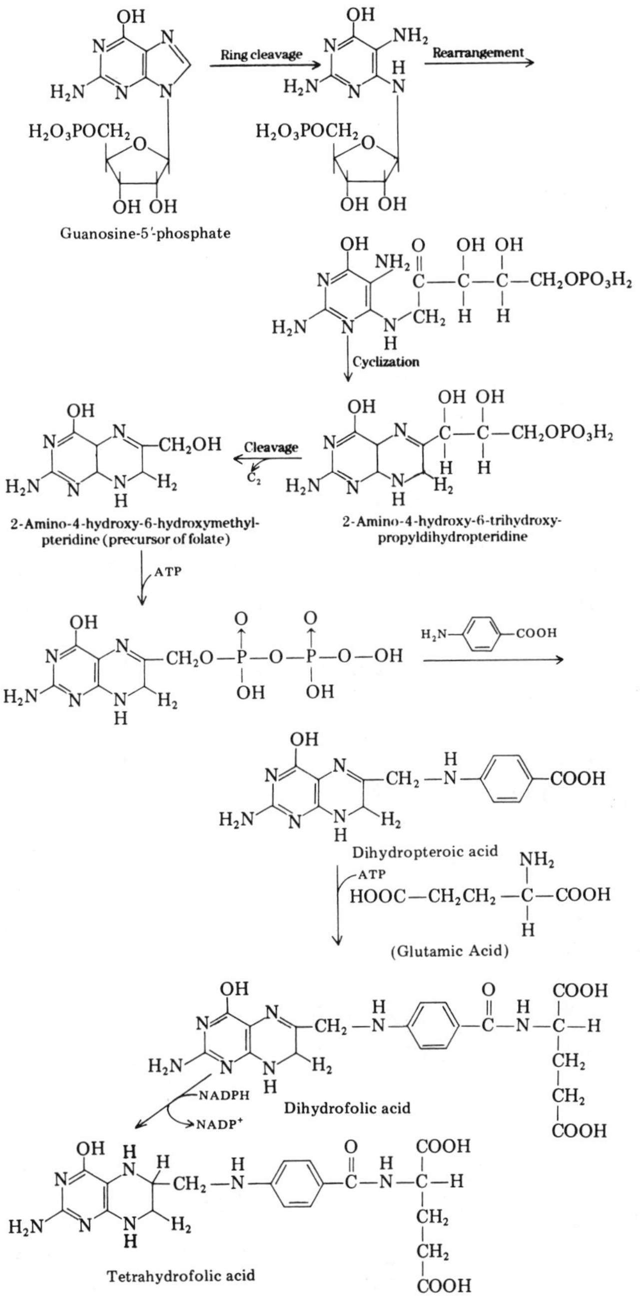

Guanosine-5′-phosphate

Ring cleavage

Rearrangement

Cyclization

Cleavage

2-Amino-4-hydroxy-6-hydroxymethylpteridine (precursor of folate)

2-Amino-4-hydroxy-6-trihydroxypropyldihydropteridine

ATP

Dihydropteroic acid

ATP

(Glutamic Acid)

NADPH

$NADP^+$

Dihydrofolic acid

Tetrahydrofolic acid

Figure 8.7. Biosynthesis of tetrahydrofolic acid, the physiologically active form of folic acid.

during lactation. Infants and children should receive 0.030–0.3 mg/day.

Distribution in Foods

Excellent sources of folacin include liver, dark green leafy vegetables, asparagus, broccoli, dry beans, cabbage, sweet corn, beet greens, bananas, strawberries, and whole wheat products. Meat, tubers, and cereals are fair sources.

Vitamin B_{12} (Cobalamin)

Pernicious anemia was first treated successfully in 1926 by Minot and Murphy through ingestion of large quantities of liver or of a specially prepared liver extract. In 1929, Castle suggested that two factors were necessary to restore normal cell count in a pernicious anemia patient: a factor in normal gastric juice called the intrinsic factor (a mucoprotein secreted in the stomach) and one in food called the extrinsic factor. Purification of the extrinsic factor was hampered because of the minute amounts present in foods and the necessity of using humans for assaying it. Shorb observed that a growth factor for *Lactobacillus lactis* in purified liver extracts could be correlated with antipernicious anemia activity. This observation led to the isolation of this factor (B_{12}) from liver extract.

Vitamin B_{12} is a slightly water soluble, dark red crystalline compound that is labile to strong acids and alkalies, and to light. This vitamin contains a porphyrin nucleus and cobalt, a heavy metal (Fig. 8.8). Cyanide is an artifact of isolation. The coenzyme form of vitamin B_{12} has a 5′-deoxyadenosyl group attached to the cobalt through the 5′-methylene group. This replaces the cyanide of the well-known cyanocobalamin. Vitamin B_{12} is not a single substance but consists of several closely related compounds with similar activity. In contrast to most of the B vitamins, vitamin B_{12} is not generally synthesized by higher plants; some microorganisms require it for growth.

Function and Effects of Deficiency

One specific reaction involving the B_{12} coenzyme is the isomerization of methylmalonyl CoA to succinyl CoA. this reaction is important in the utilization of odd-number fatty acids that ultimately yield propionyl-SCoA via β-oxidation. The synthesis of methionine from homocysteine is catalyzed by another of the coenzyme forms of B_{12}. It is essential also for the normal functioning of all cells but particularly

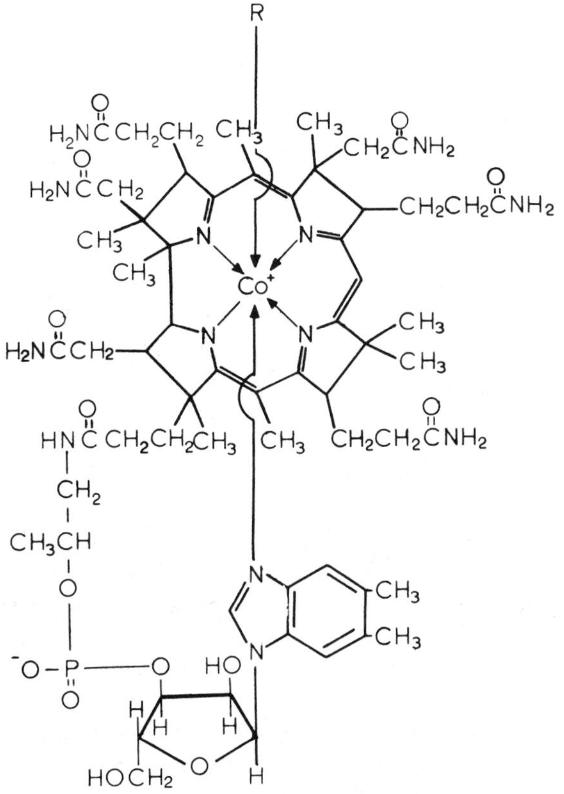

Figure 8.8. Structure of vitamin B_{12}. R = cyanide in cyanocobalamin. R = 5′-deoxyadenosyl (via 5′methylene) in coenzyme form of vitamin B_{12}.

those of the bone marrow, the nervous system, and the gastrointestinal tract.

The symptoms of a vitamin B_{12} deficiency are the same as those for pernicious anemia. Deficiency is characterized by macrocytic anemia, glossitis, degenerative lesions in the spinal cord and peripheral nerves, and achlorhydria (no HCl in gastric juice).

A deficiency is due primarily to a failure of absorption of the vitamin from the intestinal tract in the absence of the intrinsic factor in the gastric juice rather than a dietary inadequacy. Persons who subsist on a strict vegetarian diet may show low blood serum levels of vitamin B_{12} but anemia is uncommon. Pernicious anemia may occur in persons who have had surgical removal of the stomach, who are infected with fish tapeworm, and who are suffering from malabsorption syndromes, such as sprue.

Recommended Dietary Allowances

The recommended daily allowance for vitamin B_{12} for adults is 3 μg/day and for infants and children is 0.5–3.0 μg/day. Pregnant or lactating females should have an additional 1.0 μg in their diets. Average American diets are believed to meet these allowances.

Distribution in Foods

Vitamin B_{12} occurs in animal protein foods. The best sources are liver and kidney. Other sources include muscle meat, fish, oysters, milk, cheese, and eggs. Plant tissues are devoid of this vitamin.

Ascorbic Acid (Vitamin C)

Scurvy, a disease that has been known and described for several centuries, was at one time prevalent during famines and wars. Sailors on a long voyage were commonly affected with this disease. It is now known that insufficient fresh fruits and vegetables, which supply vitamin C, was the causal factor. Even with the modern knowledge of dietary essentials, there is evidence that subclinical scurvy is fairly prevalent in our society.

Ascorbic acid is a white, water-soluble, crystalline compound. It is reasonably stable in acid solution but sensitive to oxidation, which is accelerated by alkalies, iron and copper salts, heat, oxidative enzymes, air, and light. L-Ascorbic acid, the physiologically active compound, readily loses two of its hydrogen atoms to form L-dehydroascorbic acid. Reactions between these two forms are reversible, but further oxidation of the latter to L-ketogulonic acid results in an irreversible loss of physiological activity.

Biosynthesis

Plants and all mammals except man, monkey, and the guinea pig have the ability to synthesize ascorbic acid. Experiments have shown that the administration of uniformly labeled glucose will result in a uniformly labeled ascorbic acid. On the basis of these studies, the major steps in the synthesis of L-ascorbic acid in animals appear to be those indicated in Fig. 8.9. D-Galactose can also serve as a precursor in the synthesis of ascorbic acid. Interestingly, the L-hexose sugars, which closely resemble L-ascorbic acid, do not serve as precurosrs. Animals that do not have the enzyme necessary to convert L-gulonolactone to 3-keto-L-gulonolactone, need L-ascorbic acid as an essential dietary constituent.

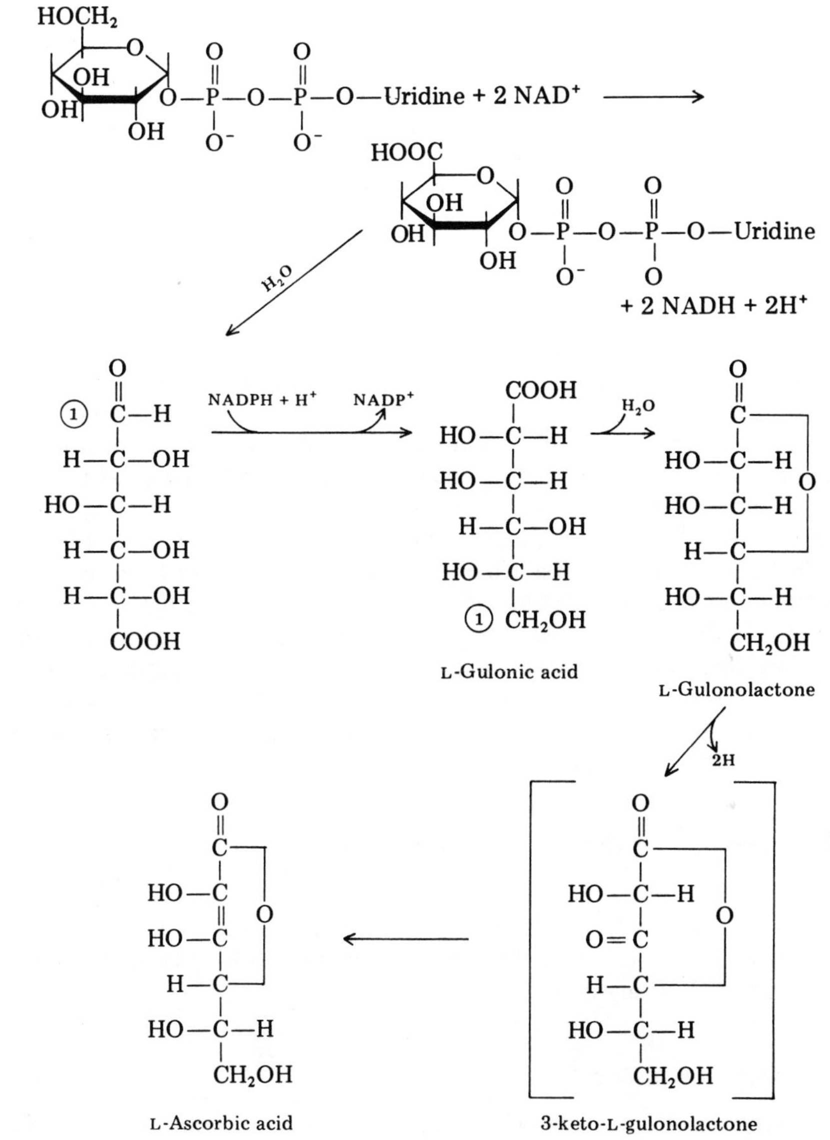

HOCH2 ... O—P—O—P—O—Uridine + 2 NAD+ ⟶

HOOC ... O—P—O—P—O—Uridine + 2 NADH + 2H+

H2O

NADPH + H+ NADP+

COOH, HO—C—H, HO—C—H, H—C—OH, HO—C—H, (1) CH2OH

L-Gulonic acid

H2O

L-Gulonolactone

2H

L-Ascorbic acid

3-keto-L-gulonolactone

Figure 8.9. Biosynthesis of L-ascorbic acid (vitamin C).

Function and Effects of Deficiency

All of the biochemical functions of vitamin C are not well-established. The reversible oxidation of L-ascorbic acid to L-dehydroascorbic acid suggests that the vitamin participates in the oxidative degradation of tyrosine (*p*-hydroxyphenylpyruvic acid to homogentisic acid) and in the conversion of folacin to the biologically active folinic acid; but ascorbic acid does not appear to be required as a specific cofactor in either process. However, the copper-containing enzyme dopamine hydroxylase, which converts 3,4-dihydroxyphenylamine to norepinephrine, requires ascorbate as a necessary co-substrate.

Ascorbic acid enhances the absorption of iron from the intestinal tract, participates in the mobilization of iron reserves and in electron transport, and promotes wound healing, which involves the formation of new connective tissue by conversion of proline to hydroxyproline in the synthesis of collagen and elastin. Ascorbic acid is essential for the formation and maintenance of the intercellular substance that binds together the cells of bone, teeth, blood capillary walls, and cartilage.

The disease that results from a deficiency of vitamin C in the diet is known as scurvy. In human beings the development of the disease is slow. In the early stages, the individual becomes lazy, loses appetite, and develops anemia. Later the gums become sore, bleed, and the teeth loosen. There may be pain in the muscles and joints. Subcutaneous hemorrhages are characteristic of the disease and death is sometimes due to internal bleeding.

Recommended Dietary Allowances

Recommendations for daily dietary allowances of vitamin C are 60 mg for adults, 45 mg for children, and 35 mg for infants. During pregnancy an additional 20 mg/day are recommended and during lactation, an additional 40 mg/day.

Distribution in Foods

The outstanding sources of vitamin C are citrus fruits (oranges, lemons, limes, and grapefruit), other fresh fruits, and green leafy vegetables. Tomatoes, cabbage, cauliflower, broccoli, and Brussels sprouts provide liberal quantities of ascorbic acid. Potatoes, green beans, peas, and apples are important sources of the vitamin because they are consumed in large quantities.

The vitamin C content of fruits and vegetables varies with conditions under which they are grown, stored, and processed. For example, the amount of sunlight available during ripening affects the ascorbic acid content. During processing there is considerable loss in the vitamin C content due to cutting, bruising, and prolonged storage at room temperatures. Freezing and frozen storage results in little loss of this vitamin. Because ascorbic acid is readily oxidized when exposed to air and is easily leached in water, losses during the cooking of food can be reduced by cooking in a minimum amount of water in a covered container.

Choline

There is some question as to whether or not choline should be classified as a vitamin because a deficiency cannot be produced in animals receiving a diet adequate in protein. Choline is synthesized in the body by transmethylation involving methionine, folic acid, vitamin B_{12}, betaine, and certain other compounds. The structure of choline (trimethylaminoethanol) is shown below:

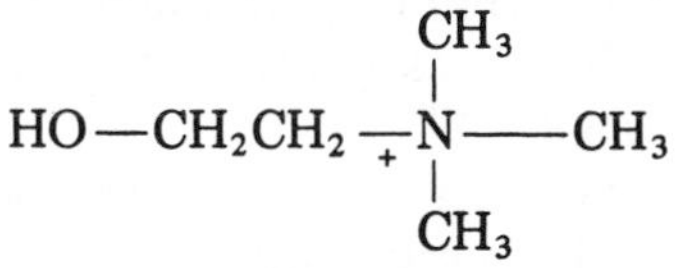

$$HO{-}CH_2CH_2{-}\overset{+}{N}(CH_3)_3$$

Choline (trimethylaminoethanol)

Choline functions in the body as a source of labile methyl groups and in the formation of phospholipids. Choline is an important constituent of acetylcholine, which functions in the transmission of nerve impulses.

Though choline deficiency has not been demonstrated in man, some evidence suggests that it may prevent fatty infiltration of the liver.

The richest known source of choline is egg yolk. Other good sources are liver, kidney, lean meat, yeast, skim milk, soybeans, beans, peas, and wheat germ. The average diet contains 250–600 mg/day, an amount known to be adequate when compared to animal requirements.

Glutathione

Glutathione is found in animal cells, bacteria, green plants, yeast, and other types of cells. It is a tripeptide with the following structure:

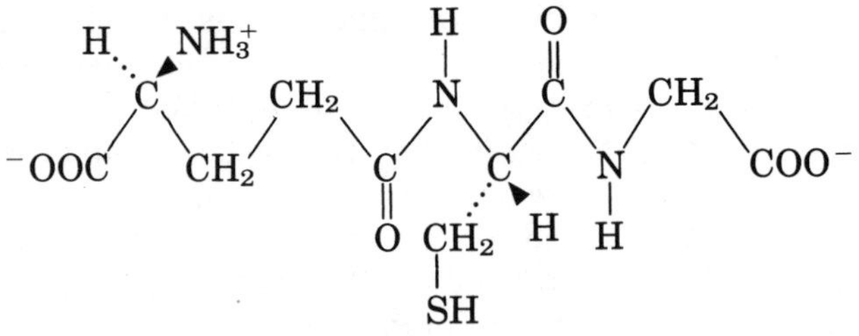

$$^{-}OOC-CH(NH_3^+)-CH_2-CH_2-C(=O)-NH-CH(CH_2SH)-C(=O)-NH-CH_2-COO^-$$

Glutathione
(γ-L-Glutamyl-L-cysteinylglycine)

Biosynthesis

Glutathione is a member of the γ-glutamyl cycle which functions in mammalian cells. The cycle makes use of the γ-carboxyl group of glutamate, the same carboxyl group that carries ammonia in the form of glutamine. In the amino acid transport process, glutathione supplies the activated γ-glutamyl group. The amino acid to be transported reacts by transpeptidation (Step 1), presumably in the cell membrane. The resulting γ-glutamylamino acid enters the cytoplasm and releases the free amino acid by an internal displacement by the free amino group (Step 2). The cyclic product, 5-oxoproline, is then available for an ATP-requiring reaction (Step 3).

Cysteinylglycine, formed in the initial transpeptidation, is hydrolyzed by a peptidase. Then, glutathione is regenerated in two steps which require ATP. Glutathione is typically found in animal cells in concentrations of 1.0–5.0 mM.

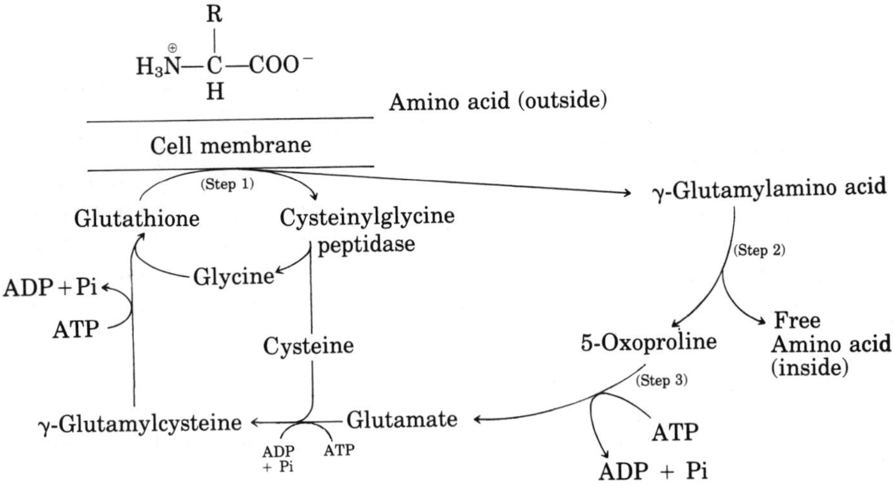

$$H_3\overset{\oplus}{N}-CH(R)-COO^-$$

Amino acid (outside)

Cell membrane

(Step 1)

Glutathione

Cysteinylglycine

γ-Glutamylamino acid

peptidase

(Step 2)

ADP + Pi

Glycine

ATP

Cysteine

5-Oxoproline

Free Amino acid (inside)

(Step 3)

γ-Glutamylcysteine

Glutamate

ATP

ADP + Pi

ADP + Pi

ATP

Glutathione is known to be an intracellar reducing agent that functions to protect -SH groups of proteins (particularly enzymes). Glutathione undergoes oxidation, as follows:

$$2\ \text{Glutathione-SH} \rightarrow \text{G-S-S-G} + 2_e^- + 2H^+$$

Glutathione functions in concert with other enzymes and cellular components to remove H_2O_2 formed by the reaction of autoxidizable drugs with O_2. The postulated sequence of these reactions is outlined below:

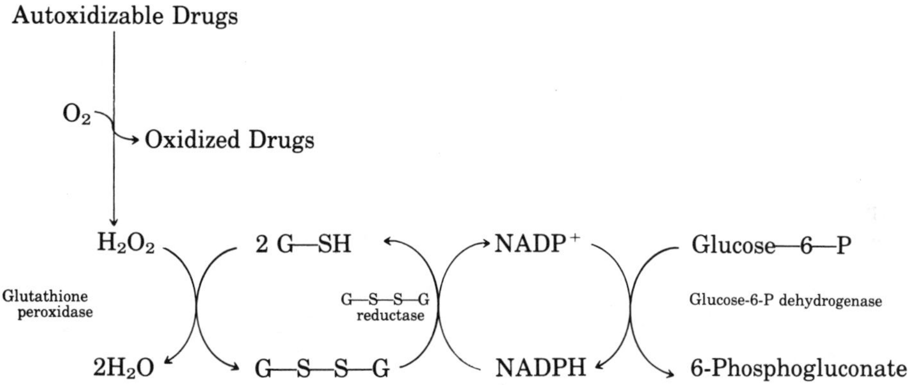

Autoxidizable Drugs

O_2 → Oxidized Drugs

H_2O_2 2 G—SH $NADP^+$ Glucose—6—P

Glutathione peroxidase G—S—S—G reductase Glucose-6-P dehydrogenase

$2H_2O$ G—S—S—G NADPH 6-Phosphogluconate

Glutathione peroxidase contains selenium and is thought to be the major H_2O_2-decomposing enzyme in red blood cells. Excessive concentrations of H_2O_2 will damage erythrocytes by causing excessive oxidation of hemoglobin to Fe^{3+}-containing methemoglobin; H_2O_2 also can attack double bonds of unsaturated fatty acids of the phospholipids in cell membranes. This latter reaction is thought to be the major cause of hemolytic anemia induced by drugs in certain individuals. Because of the involvement of glutathione in the removal of H_2O_2, sensitivity to certain drugs can result from a deficiency of glutathione or of glutathione reductase.

☐ FAT-SOLUBLE VITAMINS

The fat-soluble vitamins, which are soluble in fat and fat solvents, are found in nature closely associated with tissues that store fat. Therefore, intakes in excess of daily requirements are stored in the body and it is not necessary to have these accessory food factors in the diet every day. Deficiencies due to inadequate intake of fat-soluble vitamins are slow to develop.

Vitamin A

Vitamin A was the first fat-soluble vitamin to be recognized. In 1913, Osborne and Mendel and McCollum and Davis discovered that rats grew normally on diets containing either butterfat or an ether extract of egg yolk but failed to grow when lard was the only source of fat. By 1916, McCollum and his associates established that vitamin A was the growth-promoting factor, which is found only in certain animal tissues. The ultimate source of all vitamin A are carotenoids (α-, β-, and γ-carotenes), which are converted by animals into vitamin A. It occurs naturally in several different forms. See Table 8.2.

Structurally, vitamin A (retinol) consists of a β-ionone ring carrying two isoprene units with an all-trans configuration, as follows:

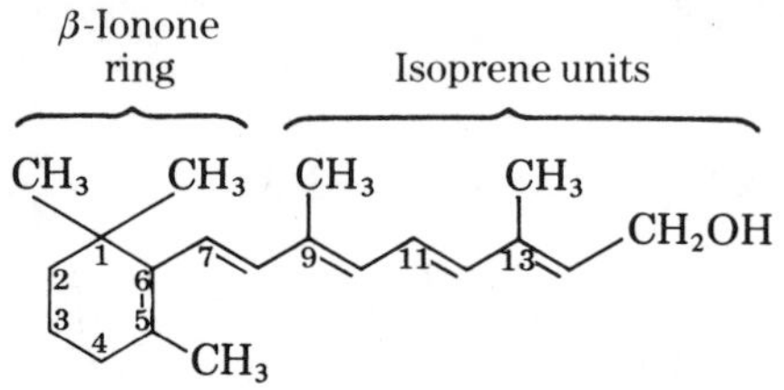

Retinol (vitamin A)
All-trans configuration

β-Carotene is structurally a double molecule of vitamin A, but this symmetrical molecule is only one-half as active as vitamin A. Thus, it is degraded *in vivo* with the formation of only one molecule of vitamin A.

Vitamin A crystals form pale yellow prisms that are alcohol and fat soluble. Vitamin A is susceptible to oxidation and autoxidizes very readily. It is heat stable in an inert atmosphere and readily stabilized by the addition of antioxidants such as α-tocopherol (vitamin E) and

Table 8.2. Some Carotenoids with Vitamin A Activity

Compound	Moiety structure[a]	Relative biological activity[b]
β-carotene	(Retinyl:)$_2$	100
α-carotene	Retinyl: α-retinyl	53
γ-carotene	Retinyl: geranyl-geranyl-4,8-diene	43
Cryptoxanthin	Retinyl: 3-hydroxyretinyl	57

[a] Colon (:) indicates a double bond joining two moieties head to head.
[b] In reference to activity of β-carotene (=100).

hydroquinone. The vitamin is destroyed by exposure to ultraviolet light.

Biosynthesis

The distribution of carbon atoms in vitamin A and carotene suggests that their synthesis involves the polymerization of isoprenoid units. In plants the biosynthesis of carotene begins with mevalonic acid, which is formed by the condensation and subsequent reduction of three acetyl groups. Mevalonic acid is converted to isopentenyl pyrophosphate (Fig. 8.10), which serves as the building unit for formation of carotenes via subsequent condensations and cyclization reactions. In the human, conversion of ingested carotene to vitamin A takes place primarily in the cells of the intestinal mucosa via the following reactions:

$$\beta\text{-Carotene} \xrightarrow{\text{Oxidation}} \text{Retinal} \xrightarrow{\text{Reduction}} \text{Retinol (vitamin A)}$$

The structures of these compounds are shown in Fig. 8.11.

Ingested vitamin A is usually in the form of retinyl esters, which are hydrolyzed in the intestine, reesterified to palmitate, transported with the chylomicra, and stored in the liver.

Function and Effects of Deficiency

The complete biochemical role of vitamin A in the body has not been elucidated. Its role as a precursor of visual pigments is best understood. A specific vitamin A aldehyde (11-*cis*-retinal) in combination with a specific protein (opsin) forms the visual pigment rhodopsin. The retinal forms a Schiff base with the ϵ-amino of a lysine residue in opsin. On exposure to light, rhodopsin is converted to *trans*-retinal and opsin, which is accompanied by bleaching. *Trans*-retinal is converted to 11-*cis*-retinal by a specific isomerase. In the dark, the enzyme equilibrium favors the trans form and, thus, some mechanism exists for accumulation of the 11-cis-isomer. Rhodopsin is regenerated by the reaction of 11-*cis*-retinal with opsin. The sequence of events in the visual cycle is summarized as follows:

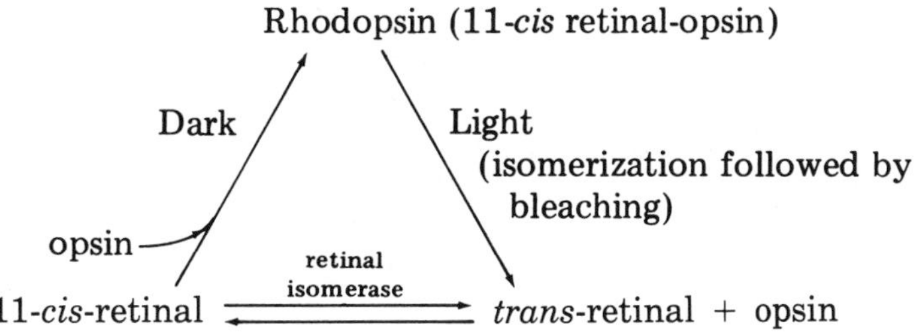

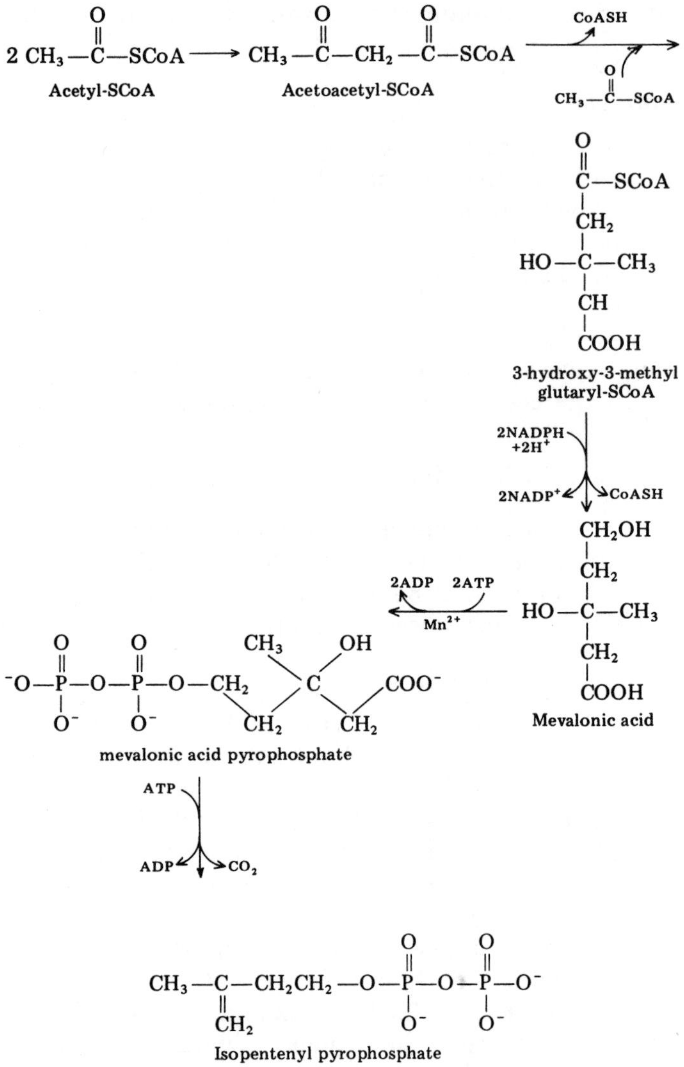

$$2\ CH_3-\overset{O}{\overset{\|}{C}}-SCoA \longrightarrow CH_3-\overset{O}{\overset{\|}{C}}-CH_2-\overset{O}{\overset{\|}{C}}-SCoA \xrightarrow[CH_3-\overset{O}{\overset{\|}{C}}-SCoA]{CoASH}$$

Acetyl-SCoA — Acetoacetyl-SCoA

$$HOOC-CH_2-\underset{CH_3}{\overset{OH}{C}}-CH_2-\overset{O}{\overset{\|}{C}}-SCoA$$

3-hydroxy-3-methyl glutaryl-SCoA

$$\xrightarrow[2NADP^+,\ CoASH]{2NADPH + 2H^+}$$

$$HOOC-CH_2-\underset{CH_3}{\overset{OH}{C}}-CH_2-CH_2OH$$

Mevalonic acid

$$\xrightarrow[Mn^{2+}]{2ATP \rightarrow 2ADP}$$

$$^-O-\underset{O^-}{\overset{O}{\overset{\|}{P}}}-O-\underset{O^-}{\overset{O}{\overset{\|}{P}}}-O-CH_2-CH_2-\underset{CH_3}{\overset{OH}{C}}-CH_2-COO^-$$

mevalonic acid pyrophosphate

$$\xrightarrow[ADP,\ CO_2]{ATP}$$

$$CH_3-\underset{\underset{CH_2}{\|}}{C}-CH_2CH_2-O-\underset{O^-}{\overset{O}{\overset{\|}{P}}}-O-\underset{O^-}{\overset{O}{\overset{\|}{P}}}-O^-$$

Isopentenyl pyrophosphate

Figure 8.10. Biosynthesis of isopentenyl pyrophosphate, a key intermediate in the synthesis of the fat-soluble vitamins.

A vitamin A-deficient diet leads to an impairment of adaptation to subdued light following exposure to bright light. This condition is known as night blindness, a measurement of which is a common method for evaluating vitamin A requirements in man. In later stages of a vitamin A deficiency the mucous membranes lose their power of

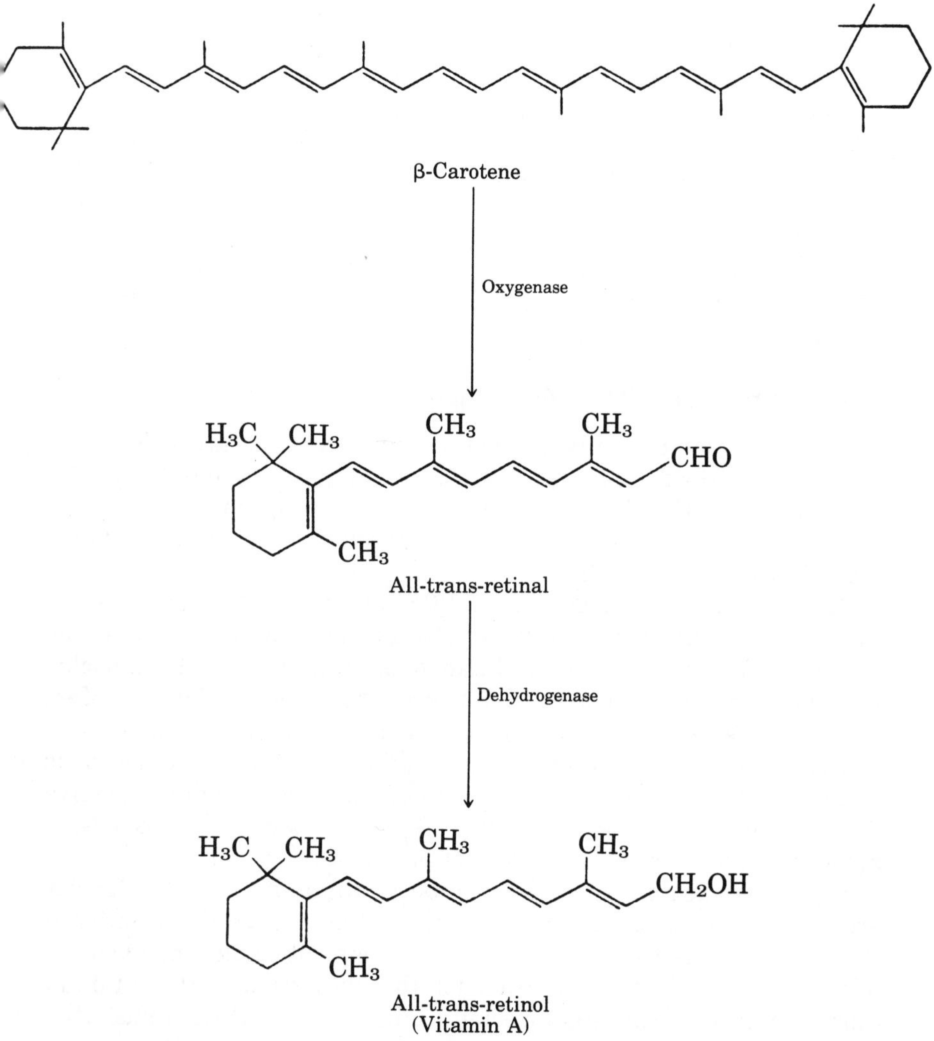

β-Carotene

Oxygenase

H_3C CH_3 CH_3 CH_3 CHO CH_3

All-trans-retinal

Dehydrogenase

H_3C CH_3 CH_3 CH_3 CH_2OH CH_3

All-trans-retinol
(Vitamin A)

Figure 8.11. Conversion of ingested carotene to vitamin A (retinol), which occurs in the intestinal mucosa.

secreting moisture, the eyelids become inflamed, and secondary infection usually follows. Finally blindness results. This is commonly referred to as xerophthalmia which means "dry eyes."

Vitamin A is essential for the integrity of the epithelial tissues of the body. Deficiency of the vitamin results in a stratified, keratinized epithelium (i.e., the epithelial cells become hard and scaly). These changes may occur in the alimentary, respiratory, and genitourinary tracts, as well as in the skin.

Vitamin A is necessary for normal growth and development. The vitamin is involved in bone growth, in particular in the activity of the osteoblasts. When vitamin A intake is inadequate, bones fail to grow in length, formation of the enamel of the teeth is impaired, with consequent degeneration of the teeth. This vitamin also functions in the maintenance of spermatogenesis in the male and fetal resorption in the female. It is involved in the synthesis of corticosterone from cholesterol in the adrenal cortex; thus, it is required for the synthesis of glycogen.

Recommended Dietary Allowances

The recommended daily allowances for vitamin A are 420 retinol equivalents (RE) for infants, 1000 RE for adult males and 800 RE for adult females. (One RE = 1.0 μg retinol or 0.6 μg pure β-carotene.) Recommended daily intakes during pregnancy are 1000 RE, and during lactation are 1200 RE.

Excessive intakes (10,000–20,000 RE) of vitamin A, over time, are likely to result in vitamin A toxicity. Resulting symptoms include bone decalcification, skeletal pain in infants, loss of hair, nausea, headache, and hyperirritability. The availability of vitamin A supplements often results in the use of excessive amounts of the vitamin in the diet. This, coupled with the increasing prevalence of adding vitamin A to various food products, increases the possibility of a person receiving excessive dietary amounts. The possibility of hypervitaminosis (excessive amounts) occurring in a normal diet is remote.

Among factors affecting the amounts of vitamin A needed in the diet are dietary fat, which is needed for absorption of vitamin A; vitamin E, which protects carotene and vitamin A against oxidation: adequate bile secretion, which is essential for the absorption of the vitamin; consumption of laxatives, especially mineral oils, which reduce the absorption of the vitamin; and protein malnutrition, which decreases intestinal absorption of vitamin A.

Distribution in Foods

Although plants do not contain vitamin A, foods of plant origin contain carotene, the precursor of vitamin A. Excellent sources of carotene include carrots, sweet potatoes, dark green leafy vegetables, winter squash, broccoli, apricots, and pumpkin. Yellow corn is the only cereal grain that contributes significant amounts of carotene to the diet. Foods that are excellent sources of vitamin A itself include whole milk, cheese made from whole milk, butter, enriched margarine, eggs, liver, glandular organs, and fish-liver oils (cod and halibut).

Vitamin D

Vitamin D refers to a group of similar sterol compounds. There are 11 sterols having vitamin D activity, but the two of greatest importance are vitamin D_2 (ergo-calciferol) and vitamin D_3 (cholecalciferol). Ergosterol (provitamin D_2) and 7-dehydrocholesterol (provitamin D_3), upon irradiation with ultraviolet light, give rise to the antirachitic vitamins D_2 and D_3 (Fig. 8.12). In plants, ergosterol is the predominant provitamin. Vitamin D_3, the form of the vitamin contained in most fish-liver oils, is produced by the action of sunlight on 7-dehydrocholesterol in the skin.

Vitamin D, being fat soluble, is absorbed from the intestines along with lipids. Consequently, conditions interfering with the absorption of fats will also reduce the absorption of vitamin D.

Vitamin D is relatively stable to heat and air oxidation, although prolonged exposure to light may cause some destruction.

Biosynthesis

The biosynthesis of vitamin D is quite similar to the biosynthesis of vitamin A. Mevalonic acid is the precursor for formation of isopentenyl pyrophosphate, which serves as the building unit for both vitamins.

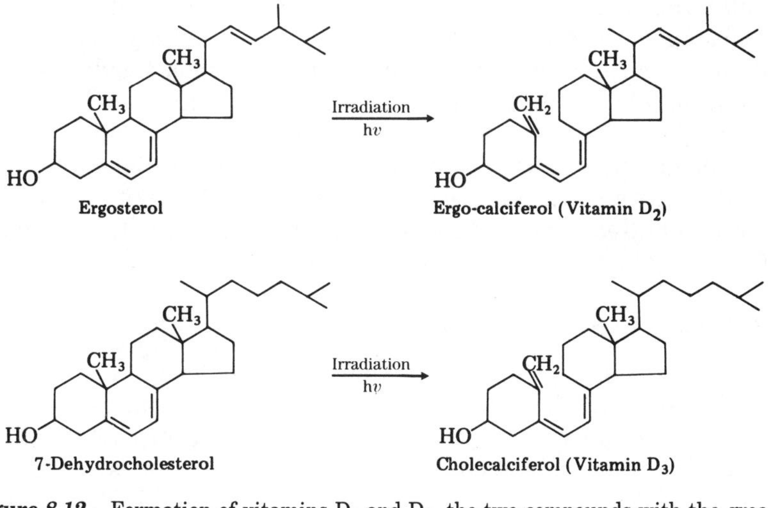

Figure 8.12. Formation of vitamins D_2 and D_3, the two compounds with the greatest vitamin D activity, from their provitamins.

In vitamin D synthesis, farnesyl pyrophosphate (C_{15}) condenses (head-to-head) to form squalene (C_{30}). The final steps of the biosynthetic process involve cyclization of squalene to form lanosterol (C_{30}), which is then converted to 7-dehydrocholesterol (Fig. 8.13).

Function and Effects of Deficiency

Vitamin D is essential at all ages to maintain calcium homeostasis and skeletal integrity. It is necessary for the absorption and utilization of calcium and phosphorous. A deficiency of this vitamin causes decreased intestinal absorption of calcium and phosphate; increased excretion of phosphate by the kidney; reduced blood calcium levels (hypocalcemia) and a failure to mobilize bone calcium; increased levels of alkaline phosphatase in the blood; and decreased concentrations of citrate and other Krebs' cycle components in body fluids.

The fact that there is a delay between the administration of vitamin D and a reversal of the physiologic changes that occur in a vitamin D deficiency suggested that the vitamin was metabolically converted to more active forms. Recent research has led to the isolation and identification of a biologically active metabolite, 25-hydrocholecalciferol (DeLuca 1969).

The principal manifestation of vitamin D deficiency in the young is rickets. In adults, the deficiency disease is known as osteomalacia and occurs especially in pregnant and lactating women.

Rickets is essentially a disturbance of calcium–phosphorus metabolism that results in deficient calcification of bones. Dental structures are also affected, but to a lesser degree than bone. The disease may be manifested by bowed legs, knock-knees, or enlarged joints. Malformation of the rib cage is another clinical symptom of the disease. The amounts of calcium and phosphorus in the diet, the availability of these elements, and their ratio to each other are additional factors involved in the incidence and severity of rickets. Since ultraviolet radiation in sunlight is involved in the production of vitamin D activity in the skin, environmental factors such as climate, season, mode of living, smoke, and fog may be correlated with the incidence of rickets.

Recommended Dietary Allowances

Although vitamin D can be synthesized in the body, it is a dietary essential and must be supplied in the diet. The recommended daily allowance for the vitamin is 400 IU/day or 10 μg of cholecalciferol. (One IU of vitamin D = 0.025 μg of cholecalciferol.) Males and females older than 18 years need less, e.g. 7.5 μg/day to age 22, then 5 μg.

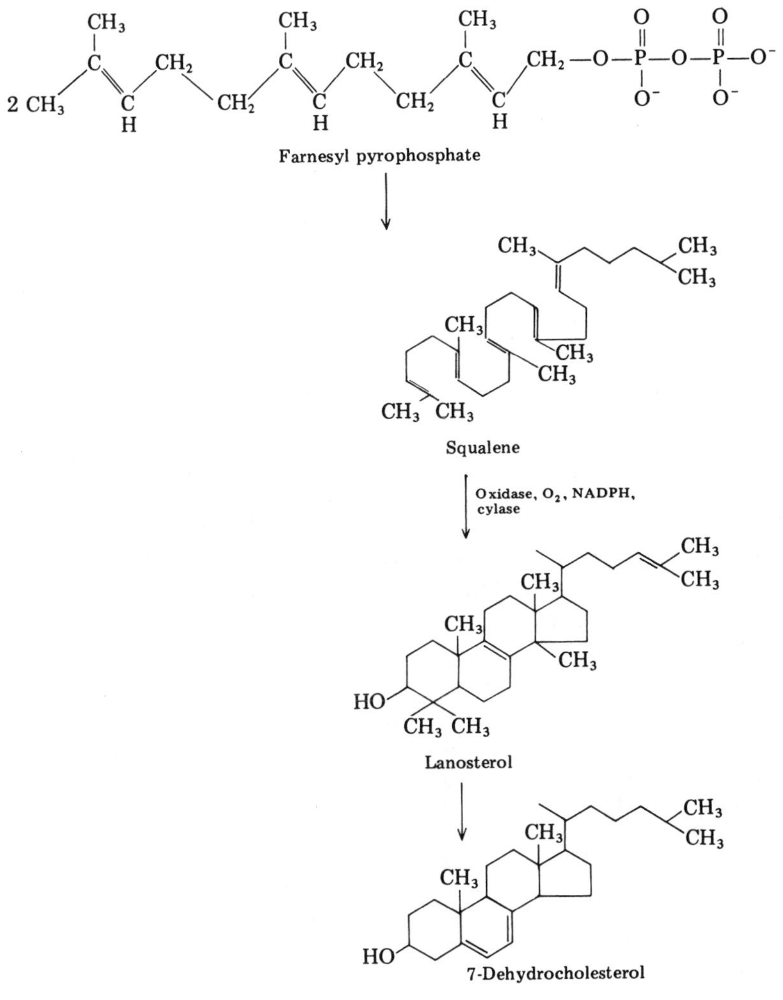

2 CH_3—C(CH_3)=CH—CH_2—CH_2—C(CH_3)=CH—CH_2—CH_2—C(CH_3)=CH—CH_2—O—P(O)(O^-)—O—P(O)(O^-)—O^-

Farnesyl pyrophosphate

↓

Squalene

↓ Oxidase, O_2, NADPH, cylase

Lanosterol

↓

7-Dehydrocholesterol

Figure 8.13. Final steps in biosynthesis of 7-dehydrocholesterol (provitamin D_3). Isopentenyl pyrophosphate is primary precursor for farnesyl pyrophosphate.

Excessive amounts of vitamin D (1000–2000 IU/kg/day) are usually toxic to children and adults, and may lead to the development of hypercalcemia. Adults receiving 50,000 IU/day will show symptoms of toxicity, including anorexia, vomiting, diarrhea, and loss of weight. Serum calcium and phosphorus levels are raised and various organs of the body are susceptible to calcium deposits.

Distribution in Foods

Most natural foods contain either little or negligible amounts of vitamin D. The best food sources are saltwater fishes (sardines, herring, mackerel, and tuna), which are high in body oils. Fish-liver oils are rich sources, but they are used primarily in medicinal preparations. Egg yolk, vitamin D-fortified foods such as milk, and some breads and breakfast cereals are good sources of vitamin D.

Vitamin E (α-Tocopherol)

Vitamin E, established as a dietary essential in 1922 by Evans and Bishop, was found to be necessary for reproduction in rats. Thus, the vitamin became known as the *antisterility factor*. There is, however, no evidence that vitamin E has any effect on human reproduction.

The tocopherols occur as light-yellow viscous oily liquids. The naturally occurring isomers are the α forms, while the synthetic product is a racemic mixture. They are insoluble in water, but soluble in fats and fat solvents. Vitamin E is not destroyed by acid, alkali, the process of hydrogenation, or by high temperatures, but is oxidized slowly by air and rapidly in the presence of rancid fats or iron salts. Decomposition occurs in the presence of ultraviolet light.

Vitamin E exists in several forms—α-, β-, γ-, and δ-tocopherols—all of which have been isolated from natural sources. The α form is the most active and the only one shown to be essential in the human diet. The tocopherols contain a substituted chroman ring and have the following general formula:

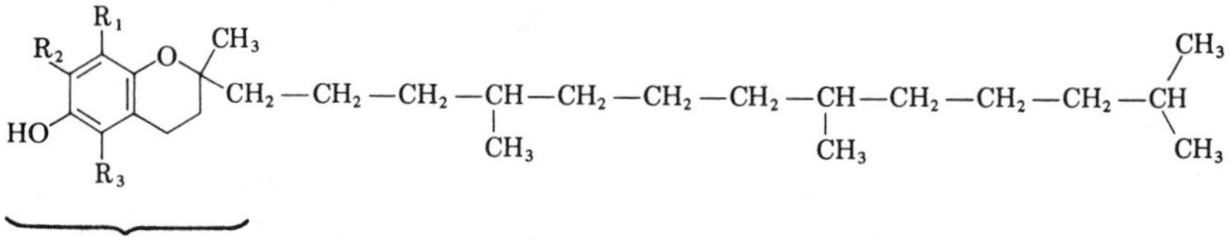

$$\text{(chroman ring: } R_1, R_2, HO, R_3, O, CH_3\text{)}-CH_2-CH_2-CH_2-\underset{CH_3}{CH}-CH_2-CH_2-CH_2-\underset{CH_3}{CH}-CH_2-CH_2-CH_2-CH(CH_3)_2$$

Chroman ring

The different forms differ in their R_1, R_2, and R_3 groups as follows:

α-tocopherol—R_1, R_2, R_3=CH_3

β-tocopherol—R_1, R_3=CH_3 and R_2=H

γ-tocopherol—R_1, R_2=CH_3 and R_3=H

δ-tocopherol—R_1=CH_3 and R_2, R_3=H

α-Tocopherol, with three methyl groups on the chroman ring, exhibits considerably greater biological activity than its β, γ, and δ homologs. When compared to α-tocopherol (100), the relative biological activity of β- and γ-tocopherols is 25 and 19, respectively; δ-tocopherol has little biological activity. Thus, it appears that optimal activity occurs when there are three methyl groups on the ring; however, the hydroxyl group on the chroman ring must be present for vitamin E activity.

Biosynthesis

Although the complete details of tocopherol biosynthesis is yet to be completely elucidated, the radioactive carbon of [^{14}C] shikimate is incorporated into both quinones and tocopherols by microorganisms and chloroplasts. Shikimate serves as a precursor of chorismate which, through a series of intermediates, ultimately forms tocopherols. The R_1 methyl group of the chroman ring originated from the methenyl carbon of chorismate.

Function and Effects of Deficiency

A specific function of vitamin E in human metabolism has not been clearly established. The most important recognized role of vitamin E is its antioxidant effect on dietary unsaturated fats and vitamin A. Tests with oxidizing agents such as hydrogen peroxide have indicated that vitamin E protects red blood cells against hemolysis. Vitamin E appears to be involved with the metabolism of polyunsaturated fatty acids, especially linoleic acid, in the body. For example, one study indicated that the vitamin E requirement could be directly correlated with the amount of polyunsaturated fatty acids in the diet (Horwit 1960). Consequently, as the intake of polyunsaturated fats increase, the need for vitamin E may increase. Selenium, chromenols, and certain antioxidants function as a partial substitute for the vitamin. Additional research is needed to determine whether or not vitamin E requirements should be increased.

Recommended Dietary Allowances

Although the biological role of vitamin E in animals is still unsolved, it is clearly an essential dietary nutrient. The recommended daily dietary allowances range from 3 α-tocopherol equivalents (α-TE) for infants to 10 α-TE for adult males. (One α-TE = 1 mg α-tocopherol.) For pregnant and lactating females the levels are 10 and 11 α-TE, respectively. These allowances are based on size; that is, the recommended allowance is the body weight in kilograms times approximately 0.15. Hence, there is a gradual increase in the vitamin E requirement throughout the period of growth to maturity.

Toxicity symptoms have not been reported for high intakes of vitamin E.

Distribution in Foods

Tocopherols occur in greatest concentrations in vegetable oils of which soybean oil is the richest. Dark green leaves, nuts, and the oil found in the germs of cereal grains, especially wheat germ oil, also are excellent sources. Eggs, margarine, liver, and dry navy beans are good sources.

Vitamin K

Vitamin K is known as the antihemorrhagic vitamin because it promotes blood coagulation. In 1929, Dam observed that newly hatched chicks developed a fatal hemorrhagic disease when raised on a ration adequate in all known vitamins and dietary essentials. The factor missing from the diet was present in the unsaponifiable nonsterol fraction of hog liver and of alfalfa. In 1935, the same investigator showed that the antihemorrhagic factor was associated with a decrease in the prothrombin concentration in the blood.

Subsequent research led to the isolation and identification of vitamin K in 1939. There are at least two naturally occurring substances, vitamin K_1 and K_2, capable of preventing hemorrhagic diathesis due to lowered prothrombin levels. Vitamin K_1 was isolated from alfalfa and K_2 was isolated from fish meal. Both substances are naphthoquinone derivatives with the following structures:

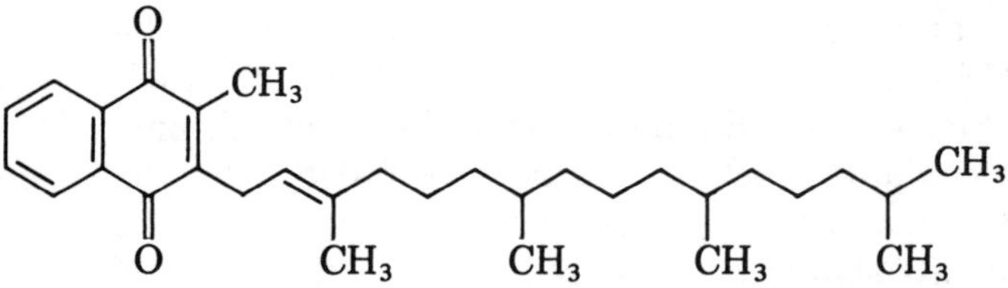

Vitamin K_1 (2-methyl-3-phytyl-1,4-naphthoquinone)

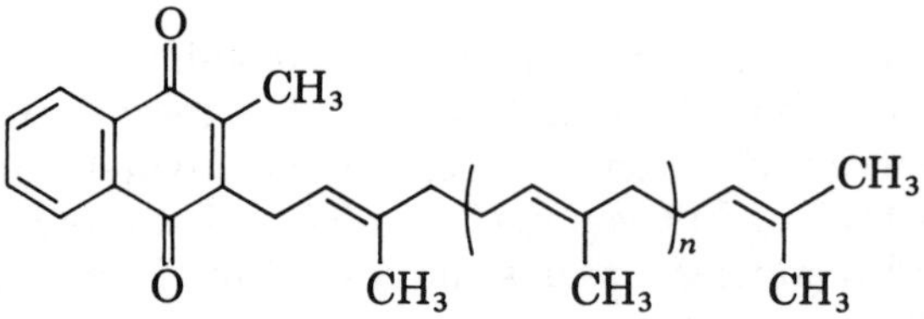

Vitamin K_2 (n = 4, 6, 7 or 8)

Many related compounds also have vitamin K activity. A synthetic compound, 2-methyl-1,4-naphthoquinone (menadione), is more potent than the natural vitamin K. Menadione, the principal form of the vitamin used for clinical purposes, is used as the reference standard for measurement of vitamin K activity.

Vitamin K is fat soluble and relatively stable to heat, oxygen, and moisture, but is destroyed by sunlight and alkalies. There is little or no destruction of the vitamin during food processing.

Biosynthesis

The metabolic pathways for the synthesis of vitamin K are not known. However, it appears the vitamin is synthesized in a manner similar to vitamins A and D, i.e., mevalonic acid → isopentenyl pyrophosphate → C_{20} unit, with subsequent condensations and cyclization.

Function and Effects of Deficiency

The exact biochemical function of vitamin K is not known. However, a well-established symptom of vitamin K deficiency is defective blood coagulation; more specifically, vitamin K is required to maintain normal plasma levels of prothrombin and three other clotting factors (VII, IX, and X). A reduction of any of these four factors, presumably all proteins, may be used to measure the action of vitamin K.

Recent findings indicate that vitamin K also may participate in oxidative phosphorylation and in mitochondrial electron transport.

Absorption of vitamin K is dependent upon the presence of bile salts in the upper intestinal tract. After absorption, it is transported to the liver where it catalyzes the synthesis of prothrombin (factor II). Consequently, any disease or injury that obstructs the flow of bile, or damages the liver in such a manner that synthesis of prothrombin is inhibited, will cause a reduction in prothrombin. Since the blood-clotting process depends on the conversion of fibrinogen to fibrin by the action of thrombin, the active form of prothrombin, decreased prothrombin formation will increase clotting time.

Recommended Dietary Allowances

The average balanced diet apparently contains adequate amounts of vitamin K. Therefore, no requirement is stated because a deficiency of vitamin K is unlikely. Therapeutic dosages of 2 mg of menadione injected intravenously will correct a vitamin K deficiency. Microgram amounts are required to prevent a decrease in blood-clotting factors that usually takes place in the blood of the newborn infant; this can be accomplished by administering a prophylactic dose of vitamin K to the infant soon after delivery. Excessive doses of menadione causes toxic symptoms (kernicterus) in the infant.

Vitamin K and dicoumarol are antagonistic. The latter substance, therefore, is used as an anticoagulant in the treatment of thrombosis.

Distribution in Foods

Vitamin K is widely distributed in nature. It is found in green, leafy vegetables such as kale, spinach, cabbage, and collards. Pork liver is a very rich source, while eggs and milk contain smaller amounts.

Intestinal bacteria are able to synthesize vitamin K_2, and this is likely the most important source of the vitamin.

☐ ANALYSIS

Vitamins are present in foods in concentrations ranging from a few micrograms to 10 milligrams per 100 grams of food. This wide range necessitates a variety of analytical techniques.

Many of the vitamins (e.g., riboflavin, tocopherols) fluoresce; others (e.g., thiamin) can be converted to compounds that fluoresce. This fluorescence is the basis for very sensitive assays that are reasonably free of interfering substances. A number of other vitamins (e.g., vitamins A, D, K) react with specific reagents to form colored products whose absorbance is proportional to the concentration of the vitamin. Assays of this type are often not very sensitive and may be hampered by interfering substances. In microbiological assays, growth of a test organism in the presence of a vitamin-containing sample is compared with growth in the presence of known amounts of the vitamin. Although microbiological assays are quite sensitive and specific, strict adherence to the established assay conditions are critical for accurate results. Vitamin B_{12}, folic acid, biotin, pantothenic acid, and vitamin B_6 are commonly assayed by microbiological methods.

The most common methods of vitamin analysis are listed in Table 8.3. Analytical procedures for five of the vitamins (vitamin A, riboflavin, thiamin, vitamin C, and niacin) are described in detail in the remainder of this chapter.

Determination of Vitamin A by Carr–Price Method

In the Carr–Price method, vitamin A is reacted with antimony trichloride (Sb_2Cl_3) in chloroform to give a blue-colored complex. This reaction is very rapid and sensitive. Maximum color develops within 3–5 sec, after which fading begins at a rate that is dependent on the nature of the sample, temperature, light intensity and other factors. The color intensity is determined spectrophotometrically at 620 nm and is proportional to the concentration of vitamin A. Discussion in this section is based on material supplied by Hoffman-LaRoche Inc., Nutley, NJ.

Although vitamin A absorbs in the ultraviolet (λ_{max} = 325 nm), direct spectrophotometric determination is not suitable for analysis of samples containing appreciable interfering absorption from substances that cannot be easily separated from vitamin A. Today, however, spectrophotometric analysis is commonly used for high-potency products containing pure, synthetic vitamin A (e.g., concentrates and pharmaceutical preparations). The Carr–Price method, despite its problems and limitations, is most useful for analyzing low-potency products, such as foods or feeds, that contain UV-absorbing interfering substances; this method also can be used with high-potency samples if a UV spectrophotometer is not available. In most cases, the Carr–Price method gives a close approximation of the true vitamin A potency, especially if the predominate form of the vitamin present is the all-trans isomer and the precautions discussed in the next section are observed.

Problems and Limitations

The determination of vitamin A by the Carr–Price reaction is beset by a number of problems, some of which are common to any assay method and due to the nature of vitamin A. Other problems arise from the nature of the Carr–Price reagent or are inherent in the handling of relatively low-potency products.

The presence of fats and emulsifiers in samples can lead to difficulties in extraction of vitamin A due to the formation of emulsions. Complete saponification of fats is necessary to prevent their appearance in the final solution used for colorimetry, thus avoiding the resultant turbidity in the Carr–Price reaction.

Moisture must be excluded from the Carr–Price reaction to avoid turbidity. Addition of several drops of acetic anhydride to the final aliquot taken for colorimetry or to the Carr–Price reagent will eliminate the effect of traces of water.

Vitamin A is sensitive to light; prevention of exposure is particularly important at very low concentrations in solution. Using low-actinic glassware or working in a dark room helps to prevent vitamin A losses. Examination of chromatographic columns with a UV light to locate

Table 8.3. Common Methods of Vitamin Analysis

Vitamin	UV or visible absorption λ_{max} (nm)	Fluorescence: Excitation λ_{max} (nm)	Fluorescence: Emission λ_{max} (nm)	Colorimetry or fluorimetry: Reagent	Colorimetry or fluorimetry: λ_{max} (nm)	Interferences	General comments
A	325	330–360	480	(1) Sb_2Cl_3 in $CHCl_3$ (2) Cl_3CCOOH or F_3CCOOH in CH_2Cl_2	610	Carotenoids and sterols	Color of complex fades rapidly (10–30 sec). Tri-fluoroacetic acid in CH_2Cl_2 appears to be better color developer. 1,3-Dichloro-2-propanol can be used for a pink color (λ_{max} at 555 nm) which is stable from 2–10 min. The blue color complex method is about 2.5 times more sensitive than ultraviolet methods for retinol. Direct sunlight and acid conditions must be avoided and antioxidants must be present during alkaline hydrolysis.
D	265	—	—	Sb_2Cl_3 in 1,2-dichloroethane	500	Sterols, carotenoids, vitamin A	Pink-colored complex fades with time. Ergocalciferol cannot be distinguished from cholecalciferol. Interferences by sterols, carotenoids, and vitamin A require that these be removed.
E	292 (alch.)	295	340	Fe^{3+} in 4,7-diphenyl-1,10-phenanthroline	534	Carotenoids, BHA, BHT	Fluorescence measures total free tocopherols. Tocopherol esters do not fluoresce. BHA interferes with both fluorescence and colorimetry. Tocopherols are sensitive to oxidation in alkaline media.
K	249 (phylloquinone) 248 (menaquinone-7) 244 (menadione)	—	—	(1) Sodium diethyl dithiocarbamate with sodium ethoxide (2) Ethyl acetonate with NH_3 or alkali (for menadione)	575 440–450	—	The blue color formed in the Irreverre-Sullivan reaction (sodium diethyl dithiocarbamate with sodium ethoxide) fades to a reddish-orange. This method may be used for both phylloquinone or menaquinones and is quite sensitive (4 μg of phylloquinone). The Irreverre-Sullivan reaction is unsuitable for menadione, which forms a red complex with ethyl acetonate and NH_3 or alkali. Vitamin K is sensitive to alkali and is destroyed by UV radiation.
C (ascorbic acid)	245 (acid) 265 (neutral)	—	—	(1) 2,6-dichlorophenol-indophenol (DI)	520	Reducing agents	Oxidation of ascorbic acid by DI can be followed by direct visual titration of photometrically if intense natural color are present. The DI method is not selective and is most applicable to high-potency products. In the DNH method, L-ascorbic acid and dehydroascorbic acid are oxidized to L-gulonic acid, which then reacts with DNH reagent to form a colored dinitrophenyl hydrazone. Dehydroascorbic acid reacts with *o*-phenanthroline to give a fluorescing derivative.
				(2) 2,4-dinitrophenyl-hydrazine (DNH)	520 (540 in presence of reductones)	Reductones in canned food	

		350 (fluorophor)	430 (fluorophor)	(3) *o*-phenanthroline	430		
B_6 (pyridoxine, pyridoxal, and pyridoxamine)	325 (pH 6.75)	352	437	Indophenol and 2,6-dichloroquinone chloriomide	—	—	Microbiological assay using an organism such as *Saccharomyces carlsbergensis* is the preferred method for foods or other materials where chemical interferences are present.
B_1 (thiamin)	235 (pH 10.6) 232 and 268 (pH 6.6) 245 (pH 2.9)	358 (375)	435	$K_3Fe(CN)_6$ in NaOH	ca 435	Other fluorescing substances	Thiamin is most conveniently measured by converting thiamin to thiochrome, which fluoresces at about 435 nm. Thiamin pyrophosphate must first be hydrolyzed to thiamin. In neutral or alkaline solutions, thiamin is rapidly destroyed probably due to the destruction of the thiazole moiety.
B_2	266, 371, and 475	ca 440, several reported	ca 565, several reported	—	—	High concentrations of iron and pigments. Caramelization also causes interferences.	Riboflavin fluoresces without chemical conversion. Riboflavin is extremely sensitive to light. The higher the pH of the solution, the greater will be the destruction of riboflavin in the presence of light.
Niacin and niacinamide	385, 262, (212, amide)	—	—	Cyanogen bromide	430–450 for pharmaceuticals 470 for cereals 550 for multivitamin preparations	High concentrations of pigments	Chemical methods for niacin first convert niacin derivatives such as NAD, NADP, and niacinamide into free niacin. Free niacin is then reacted (after separation from impurities) with cyanogen bromide to form a pyridinium compound, which undergoes rearrangement yielding derivatives that will couple with aromatic amines, giving colored compounds. Microbiological assay is also common.
B_{12} (cobalamin)	278, 361, 550 (H_2O)	275	305	—	—	—	Microbiological assay is method of choice because chemical methods are usually insensitive to levels normally in foods.
Folic acid (pteroylglutamic acid)	282, 350 (pH 7.0)	—	—	—	—	—	Microbiological assay is the only official method.
Pantothenic acid	358	—	—	Chlorinating solution, acidified phenol solution, potassium iodide			The chemical method uses acid to hydrolyze pantothenate to give β-alanine. The β-alanine is treated with a chlorinating solution and then with KI. Free iodine is measured spectrophotometrically. Microbiological assay also is possible.
Biotin	234	—	—	—	—	—	Microbiological assay is the method of choice because chemical methods lack sensitivity to the extremely low levels in foods.

positions of the vitamin A and other bands should be done with the least possible use of the UV light.

Because vitamin A is sensitive to oxidation by air, maintaining an atmosphere of inert gas over solutions of vitamin A is necessary to avoid oxidative losses. Nitrogen is preferred since carbon dioxide occasionally contains acidic contaminants that can cause significant losses of vitamin A in dilute solutions.

Purification of vitamin A extracts by column chromatography requires careful attention to detail in preparation of the adsorbent, packing of the column and carrying out of the chromatographic separation in order to assure satisfactory separation and recovery of vitamin A.

Carotenes, carotenoids, and other substances yield a blue color with the Carr–Price reagent. By estimating the concentration of β-carotene from the absorbance to 450 nm, a blue color correction factor can be calculated from a calibration curve for the Carr–Price reaction of β-carotene.

The Carr–Price assay overestimates vitamin A potency in the presence of isomers other than all-trans vitamin A or of breakdown products. All-trans, 2-cis, 6-cis, and 2,6-di-cis isomers yield the same color intensity in the Carr–Price reaction, whereas their relative biological potencies are 100%, 57%, 21%, and 24%, respectively. Anhydro-vitamin A and retro-vitamin A, which have practically no biological activity as vitamin A, also yield an intense blue color in the Carr–Price reaction. The maleic anhydride reaction of Robeson and Baxter (1947) can be used to determine the relative proportions of 2-cis + 2,6-di-cis versus all-trans + 6-mono-cis isomers, but there is no procedure suited to routine assays for separation and measurement of the individual isomers or breakdown products of vitamin A.

Extracts of low-potency products may contain substances that partially inhibit the development of the blue color in the Carr–Price reaction. A correction for such inhibition can be obtained by carrying out the colorimetric reaction with and without an added amount of a standard solution of vitamin A. This requires careful, replicate measurements due to the inherently lower precision of such an internal standard method.

The concentration, purity, and age of the Carr–Price reagent influence the amount of color produced with vitamin A. Careful, periodic calibration is required to insure satisfactory performance of the assay. The corrosive nature of this reagent necessitates careful handling.

The limits of error in vitamin A assays by the Carr–Price method depend on the nature and potency of the sample as well as on the care exercised in carrying out the procedure. For high-potency products, which can be handled by a rapid, single-extraction method, the assay is usually reproducible to within ±3%. For low-potency products, which require additional manipulations and chromatographic purification, the limits of error are more likely to fall in the range of ±5–20%. In addition, in the analysis of a dry mixture of a high-potency vitamin A product, uniformity of distribution of the vitamin A concentrate particles becomes a significant factor.

Special Equipment

1. Photoelectric colorimeter or spectrophotometer—Short-path cuvettes may be necessary when using a spectrophotometer due to more intense color development on occasions.
2. Low-actinic volumetric and Erlenmeyer flasks and separatory funnels.
3. Glass-stoppered centrifuge tubes of about 40-ml capacity.
4. Metal box to hold test tubes and provide protection from light during evaporation of solvents. It should have holes in the bottom to allow for immersion of tubes in warm water to facilitate evaporation.
5. Rapid-delivery, 10-ml pipette with stopcock.

Reagents

1. Chloroform (reagent grade)—Purify by redistillation if necessary to prevent color interference.
2. Carr–Price reagent—Dissolve the entire contents of a 1-lb bottle of CP antimony trichloride in about 1500 ml (5 lb) of chloroform with gentle warming and intermittent stirring. Cover the solid antimony trichloride with solvent as rapidly as possible to avoid picking up moisture from the air. Allow reagent to cool, filter through Whatman No. 12 fluted paper into a brown bottle and store in a cool, dark place.
3. Diethyl ether (U.S.P.) free of peroxides.
4. Hexane (Skellysolve B) with b.p. 60°–70°C—Redistill from all-glass apparatus and use only the 64°–68°C fraction.
5. U.S.P. vitamin A reference standard.
6. All-trans β-carotene.

Extraction of Vitamin A from Various Products

The procedure used for extraction of vitamin A is dependent on the nature of the product to be assayed. Such variables as the vitamin A potency, fat content, sugar content, and nature of the vitamin A source influence the decisions on sample size, amount of caustic required for saponification, use of water in saponification mixture, extraction procedure, etc. Details of the extraction procedures for vitamin A in particular products are described in AOAC (1984). Modifications for higher or lower potencies than those described can usually be made by suitable alteration of sample size, dilution, and/or size of aliquot taken for colorimetry.

Vitamin A Standard Calibration Curve

Accurately weigh 0.5–1.0 g of U.S.P. vitamin A reference standard (capsules containing an oil solution with a claimed potency of 100,000 U.S.P. units of vitamin A/g). Dissolve in sufficient chloroform to give a concentration (e.g., 15–25 U.S.P. units/ml) that will yield an absorbancy of about 0.5 when 2 ml are reacted with 10 ml of Carr–Price reagent. From this solution, make a series of dilutions to give solutions that are 80%, 60%, 40%, and 20% as concentrated.

Carry out the Carr–Price reaction and determine the absorbance of the solutions according to the procedure under *Carr–Price Colorimetry*. Using rectangular coordinate paper, draw the best fitting smooth curve through the origin and the five points obtained by plotting the absorbancies versus the corresponding known quantities of vitamin A. The calibration curve should be checked at frequent intervals, particularly when reagent solutions are renewed. Determine the slope of the plot of absorbance versus vitamin A concentration; this is the calibration constant (K). Thus, units of vitamin A = K × absorbance.

β-Carotene Standard Calibration Curve

Weigh exactly a quantity of about 50 mg of pure all-trans β-carotene from a sealed container into a 500-ml volumetric flask. (If remainder of the β-carotene is to be used again, ampule or container must be evacuated and flushed with CO_2 gas twice, and the container closed. Store container in cool place.) Dissolve in Skellysolve B and make to mark. Subdilute 10 ml to 100 ml with Skellysolve B to obtain approximately 10 μg/ml and make further dilutions to contain about 3, 2, 1, 0.5, 0.2 and 0.1 μg of carotene/ml. Add 5 mg of free tocopherol per 100 ml of Skellysolve B to stabilize the dilute standard solutions of carotene. (The addition of 10 to 50 times as much tocopherol as there is carotene present is reported to stabilize standard solutions of carotene completely for as long as 12 weeks at 20°C in diffuse daylight. The use of more than 50 μg/ml of tocopherols may produce measurable color.) Zero the spectrophotometer or colorimeter set at 455 nm with Skellysolve B as blank. Read each of the dilutions. Plot the absorbancies against concentrations to obtain a linear curve.

Calibration Curve for β-Carotene in the Carr–Price Reaction

When vitamin A is to be determined in extracts containing β-carotene, a correction must be applied for the absorbance due to reaction of carotene with the Carr–Price reagent. This correction is determined from a calibration curve prepared as follows: Evaporate aliquots of the β-carotene solution containing graded levels from 5 to 50 μg of β-carotene under nitrogen in a warm-water bath. Remove the last traces of solvent without benefit of heat. Add 2 ml of chloroform to each tube. Carry out the Carr–Price reaction as described in the next section. Take the steady absorbance reading within 4 sec as for vitamin A. Plot the observed absorbancies against μg of β-carotene/ml. Be sure that the curve is linear over a reasonable range of concentrations.

Carr–Price Colorimetry

Zero the spectrophotometer (620 nm) with a solution of 2 ml of chloroform + 10 ml of Carr–Price reagent.

Place the sample tube containing 2 ml of chloroform solution (containing unknown sample, known vitamin A standard, or β-carotene standard) in the colorimeter and add 10 ml of Carr–Price reagent from the rapid-delivery pipette or use a small-volume cuvette with 0.2 ml of chloroform solution and 1 ml of Carr–Price reagent. Take the maximum stable reading, which occurs within 3–5 sec and then begins to fade. Disregard the initial overswing due to bubbles caused by mixing. Examine the tube immediately after reading to see that the color is blue and without turbidity. Very slight traces of moisture can be taken care of by adding one or two drops of acetic anhydride before adding the Carr–Price reagent.

The observed absorbance of unknown vitamin A samples may need to be corrected for the presence of carotene or of color-inhibiting substances, as described in the next sections.

Correction for Color Due to Carotene

When the extraction of vitamin A from a test material yields a final extract (in ethyl ether or petroleum ether) that is yellow due to carotene, a correction should be made for the blue color formed by reaction of carotene with the Carr–Price reagent.

Read the absorbance of the yellow color of a 10-ml aliquot of extract in the spectrophotometer at 445 nm. Evaporate the solvent under a stream of nitrogen with gentle warming and immediately at dryness take up the residue with 2 ml of chloroform and carry out the Carr–Price colorimetry. Determine the carotene concentration in the extract from the β-carotene standard calibration curve. Using the calibration curve for β-carotene in the Carr–Price reaction, determine the blue color absorbance that this amount of β-carotene would contribute during Carr–Price colorimetry of the extract. This represents the correction factor for carotene (A_{car}).

Correction for Color Inhibition

Extracts of low-potency products may contain substances that partially inhibit the amount of blue color developed in the Carr–Price reaction. A correction for such inhibition may be made by colorimetry of an identical test sample aliquot to which has been added a known amount of vitamin A standard. One aliquot of the ether extract of the test sample containing about 10 U.S.P. units of vitamin A is evaporated and taken up in 2 ml of chloroform for colorimetry. An equal aliquot of the sample extract is evaporated and taken up in 1 ml of chloroform plus 1 ml of a chloroform solution of the U.S.P. vitamin A reference standard containing a known amount (about 10 U.S.P. units) of vitamin A. A third tube is prepared containing 1 ml of the standard vitamin A solution and 1 ml of chloroform.

Carry out Carr–Price colorimetry on the unknown, unknown + standard increment, and standard increment. Determinations should be done in duplicate because of the inherently low precision of the standard increment technique.

Calculations

The observed absorbance of an unknown test sample in the Carr–Price reaction must be corrected both for the presence of carotene and of color-inhibiting substances. First, subtract the carotene correction factor (A_{car}) from the absorbance values of the unknown and unknown + standard increment samples to give corrected values:

$$A_u = \text{obs. absorbance of unknown} - A_{car}$$

$$A_s = (\text{obs. absorbance of unknown} + \text{standard increment}) - A_{car}$$

Then, if A_i is the absorbance of the standard increment alone and K is calibration constant for vitamin A (determined from slope of calibration curve), the units of vitamin A per tube for an unknown sample can be calculated as follows:

$$\text{Vitamin A (U.S.P. units/tube)} = \text{K} \times A_u \times \frac{A_i}{A_s - A_u}$$

The absorbance readings of test samples should fall within the linear range of the vitamin A calibration curve.

Determination of Riboflavin by Fluorimetry

Riboflavin (vitamin B_2), in food products, can be determined by comparing the fluorescence (excitation at 440 nm; emission at 565 nm) of test samples with that of standard riboflavin solutions. With low-potency samples, an internal standard (riboflavin) is added to improve the detection limits. Regardless of the procedure used, special care should be taken to protect all stages of the assay from light. The pH should be maintained near 7.

Reagents

1. 0.1 *M* HCL—Dilute 8.2 ml concentrated HCL to 1 liter with water.
2. 0.1 *M* NaOH—Dissolve 4.0 g NaOH pellets (CP) per liter (aqueous).
3. 3% H_2O_2—Prepare fresh daily from 30% H_2O_2.
4. 4% $KMnO_4$—4.0 g $KMnO_4$ per 100 ml of aqueous solution.
5. *Sodium hydrosulfite*—Reagent grade ($Na_2S_2O_4$).
6. *Glacial acetic acid* (reagent grade).
7. 0.02 *M* acetic acid—Dilute 1.17 ml glacial acetic acid to 1 liter with water.
8. *Standard riboflavin stock solution (100 μg/ml)*—To 50 mg of U.S.P. riboflavin reference standard, previously dried at 105°C for 2 hr and stored, protected from light, in a desiccator over phosphorus pentoxide, add approximately 300 ml of 0.02 *M* acetic acid; heat the mixture on a steam bath, with frequent agitation, until the riboflavin has dissolved. Cool and add 0.02 *M* acetic acid to make 500 ml. Add a few drops of toluene and store at 4°C.
9. *Standard riboflavin internal standard solution (1 μg/ml)*—Dilute 10 ml of 100 μg/ml standard solution to 1 liter using 0.02 *M* acetic acid.

Apparatus

Any reliable spectrofluorometer equipped with an input filter of narrow transmittance range with a maximum at approximately 440 nm and an output filter of narrow transmittance range at approximately 565 nm. Modern spectrofluorometers equipped with microprocessors and/or computers with appropriate software are particularly convenient. Quantitative software can be used to make the report easily obtainable and rapid.

Procedures

1. Extraction procedure for whole wheat flour —Weigh 10 g of unenriched whole wheat flour and mix with 100 ml of 0.1 *M* HCl. Mix thoroughly with agitation. Autoclave at 122°C for 30 min. Mix thoroughly until consistent. Adjust the pH to 6.0 with 0.1 *M* NaOH. Immediately precipitate all remaining proteins by adding 0.1 *M* HCl to pH 4.5. This solution should contain no more than 0.1 mg of riboflavin per milliliter. Dilute the mixture to near 0.15 μg of riboflavin per milliliter and centrifuge until clear. Recheck the pH and add a small amount of 0.1 *M* HCl to an aliquot to determine if additional protein precipitation will occur. If precipitation is complete, adjust pH to 6.8 with 0.1 *M* NaOH. Recentrifuge if precipitation occurs. This is the sample solution.
2. Determination of riboflavin —Add 10 ml of the sample solution to each of 4 tubes. To two of the tubes, add 1.0 ml of Riboflavin Internal Standard (1.0 μg/ml). To the other two tubes add 1.0 ml of water and mix. Add, with mixing, 1.0 ml of conc. acetic acid and 1.0 ml of 4% $KMnO_4$. Allow oxidation to occur for approximately 2–3 min, then add, with mixing, 1.0 ml of 3% H_2O_2. The purple permanganate color must disappear within 10 sec. Agitate to allow O_2 bubbles to be expelled. Be sure that O_2 bubbles that cling to the tubes are removed. Measure the fluorescence of the tubes containing Sample and Internal Standard. This is recorded as *unknown* (*U*). Measure the fluorescence of the tubes containing Sample and Water. This is recorded as *blank* (*B*). To the *blank* (*B*) add 20 mg of $Na_2S_2O_4$, mix and measure the *minimum fluorescence* (*MF*) observed within 5 sec.

Calculation

$$\text{mg of riboflavin/ml of sample solution} = \frac{(B - MF)}{(U - B)} \times 10^{-4}$$

$(B - MF)/(U - B)$ ratio must be between 0.66 and 1.5.

Determination of Free Thiamin by Thiochrome Method

Reagents

1. 5 *N* HCl—410 ml concentrated HCl per liter.
2. Neutral potassium chloride solution—Dissolve 250 g KCl (A. R. grade) in H_2O to make 1 liter.

3. Acid Potassium chloride solution—Add 8.5 ml of concentrated HCl to 1 liter of the neutral KCl solution.
4. Sodium hydroxide solution 15%—Dissolve 15 g NaOH in H_2O to make 100 ml.
5. Potassium ferricyanide solution 1%—Dissolve 1 g $K_3Fe(CN)_6$ in H_2O to make 100 ml. Prepare solution on day it is used.
6. Oxidizing Reagent—Mix 4 ml of 1% $K_3Fe(CN)_6$ solution with the 15% NaOH solution to make 100 ml. Use within 4 hours.
7. Isobutyl alcohol—(A. R. Grade)—Redistill in all glass apparatus.
8. Thiamin hydrochloride stock standard solution—100 μg/ml. Transfer 100 mg of U.S.P. thiamin hydrochloride reference standard, previously dried at 105°C for 2 hours into a 1-liter flask, dissolve in 0.1 *N* HCl. (Thiamin is hygroscopic, avoid absorption of moisture during the weighing.)
9. Intermediate standard, 4 μg/ml—To 40 ml of stock standard add 19.2 ml of 5 *N* HCl and distilled water to make 1 liter.
10. Working standard, 0.2 μg/ml—Transfer 5.00 ml intermediate standard into a 100-ml volumetric flask and add 80 ml of acid potassium chloride solution plus water to make 100 ml.
11. Quinine sulfate stock solution—Use quinine sulfate solution or standard fluorescent glass block to govern reproducibility of the fluorometer. Dissolve 10 mg of quinine sulfate in 0.1 *N* H_2SO_4 to make 1 liter. Store in light-resistant containers.
12. Quinine sulfate working solution—As needed, dilute 15 ml of stock solution with 0.1 *N* H_2SO_4 to 500 ml. The fluorescence of this solution is approximately equivalent to that of the thiochrome formed from 1 μg of thiamin under the conditions described in this procedure. Store at about 10°C in glass-stoppered, light-resistant bottle.
13. Base exchange silicate—Purify artificially prepared silicate of base exchange type, in form of granular powder of "50–80 mesh" size, as follows: Place convenient quantity (100–500 g) base exchange silicate in suitable beaker, add enough hot 3% HOAc to cover material, and boil 10–15 min, stirring constantly. Let mixture settle and decant supernatant. Repeat washing 3 times, then wash similarly 3 times with hot KCl solution (part by weight KCl/4 volumes solution), and finally wash with boiling H_2O (*distilled* H_2O must be used) until last washing gives no reaction for Cl. Dry material at about 100°C and store in well-closed container. (Purified base exchange silicate may be purchased as Thiochrome Decalso, Fisher Scientific Co., No. T-97.)

14. 2.5 *M* Sodium Acetate Solution—Dissolve 205 g of anhydrous sodium acetate or 345 g of $NaC_2H_3O_2$ trihydrate in water and dilute to 1 liter.

Apparatus

1. Fluorophotometer, such as Pfaltz and Bauer (Model B), Farrand (Model A-3), Turner, or Coleman (Model 12 or 12-A), with filters for thiochrome assay.
2. Thiochrome reaction vessels, 30-ml capacity, with stopcocks.
3. Glass-stoppered centrifuge tubes, 50 or 60 ml capacity.
4. Rapid delivery (0.5 ml/sec) 3-ml pipette.
5. Automatic pipette (to deliver 15 ml of isobutanol).
6. Shaking machine.
7. Centrifuge.
8. Standard fluorescent glass block for B_1.
9. Chromatographic columns—Use glass chromatographic tubes (about 275-mm overall length, with reservoir capacity about 60 ml) consisting of 3 parts fused together with following approximate id (1) reservoir at top, 95-mm long, 30-mm diameter, converging into (2) adsorption tube, 145-mm long, 6-mm diameter, and at lower end (3) tube is drawn into capillary 35-mm long and of such diameter that when tube is charged, rate of flow will be $\leq$1 ml/min. Prepare tubes for use as follows: Place over upper end of capillary, with aid of glass rod, a plug of fine glass wool. Add to adsorption tube H_2O suspension of 1.0–2.0-g purified base exchange silicate, taking care to wash down all silicate from walls of reservoir. To keep air out of adsorption column, keep layer of liquid above surface of silicate during adsorption process. (Prevent tube from draining by placing rubber cap, filled with H_2O to avoid inclusion of air, over lower end of capillary.)

Extraction and Purification of Thiamin in Whole Wheat Flour

Weigh 10 g of unenriched whole wheat flour and mix with 100 ml of 0.1 *N* HCl. Mix thoroughly with agitation. Autoclave at 122°C for 30 min. Mix until consistency is achieved. Dilute this solution to contain approximately 0.2 μg of thiamin/ml. This is the *assay sample solution.*

An aliquot of the assay sample solution, which contains approximately 5 μg of thiamin, is passed through a base-exchange silicate chromatography column. The column is washed with three 5-ml portions of 95°C water. Do not allow the level of the liquid to drain below the top of the column bed. Elute the thiamin from the base-exchange silicate by washing with five 4.0–4.5-ml portions of near boiling acid–KCl solution. Collect the eluate in 25-ml volumetric flasks. Cool and dilute to 25 ml with acid–KCl solution.

Oxidation of Thiamin to Thiochrome

Five milliliters of the assay solution are transferred into a glass-stoppered centrifuge tube, 3 ml of alkaline ferricyanide are added with constant swirling, holding the tip of the pipette so that the stream of solution does not hit the side of the tube (use pipette that delivers 3 ml in 1–2 sec); 15 ml of isobutanol are added immediately and the tube stoppered and shaken briefly. A thiamin standard (5 ml of working standard) is run with each one or more unknown samples. After isobutanol has been added to all tubes, shake all tubes for 2 min in a shaking machine. The tubes are centrifuged for 1 min and the isobutanol poured or pipetted into the cuvettes, taking care, if pouring, to adjust the meniscus in all cuvettes to the same level. Duplicate runs are recommended.

Fluorophotometry

Set the fluorometer, using "standard glass block" B_1 or quinine working solution, at such a reference point so that 1 μg of oxidized thiamine in 15 ml of isobutanol will give a reading in the middle of the scale. The thiamin content of the oxidized sample is determined by comparing the intensity of fluorescence of the sample (U) with that of oxidized standard (S), correcting for blank fluorescence (b = blank of unknown; bs = blank of standard) of each of these solutions. Blank readings are taken after stirring with two drops of a solution of methanol : 5 N HCl (1 : 1 vol/vol).

Calculation

$$\mu\text{g Thiamin/5 ml assay solution} = \frac{(U - b)}{(S - bs)}$$

Remarks

Pipetting—Arrange the dilutions so that the 5 ml of final assay solution used for the oxidation step contains about 1 μg of thiamin. If the sample potency is very low, the assay solution may be diluted to contain about 0.5 μg per 5 ml. In this case, the working standard solution should also be diluted to contain 0.1 μg per ml.

Cuvettes—All cuvettes are numbered and checked, one against the other, with the working quinine solution. Correction values, if any, are noted and used.

Since the reading obtained on the thiamin standard may vary when different bottles of isobutanol are used, it is essential to use isobutanol from the same bottle for the given set of assays.

Photofluorometers with an input filter of narrow transmittance range with a maximum at about 365 nm and an output filter of narrow transmittance range with a maximum at 435 nm are satisfactory for thiamin fluorescence measurement. Take readings promptly after oxidation step and avoid excessive exposure of thiochrome solutions to the ultraviolet light.

Determination of Vitamin C by DI Titration Method

Most practical methods that have been proposed for the determination of L-ascorbic acid in foods have been based upon the reversible oxidation of L-ascorbic acid to dehydroascorbic acid. Iodine and 2,6-dichlorobenzenoneindophenol (DI) are the oxidants most commonly used for this reaction. The usefulness of iodine, however, is limited to fairly pure solutions of L-ascorbic acid, leaving DI as the agent of choice for food products. Discussion in this section is based on material supplied by Hoffman-LaRoche Inc., Nutley, NJ.

For many food products, direct visual titration with DI in a pH range of 1–3.5 gives a reasonably accurate and satisfactory measure of the vitamin C content. Directions for a simplified visual titration are given below. However, in some foods (e.g., strawberries, beets, prunes) the presence of intense natural colors makes it difficult, and often impossible, to carry out visual titration. Such products may be assayed by a photoelectric method, which effectively eliminates interference by pigments or by turbidity resulting from colloidal suspensions.

L-Ascorbic acid in solution slowly oxidizes; the rate is accelerated by the presence of oxygen, certain enzymes, and minute quantities of dissolved copper. Dehydroascorbic acid is the first product formed in the oxidation process. Both L-ascorbic acid and the dehydro form are biologically active vitamin C compounds, but dehydroascorbic acid does not react with DI. Therefore, to determine total ascorbic acid, the dehydroascorbic acid is reduced with hydrogen sulfide gas or other suitable reducing agents before titration with DI.

The formation of dehydroascorbic acid arises from the direct and/or enzymatic oxidation of L-ascorbic acid. For a critical evaluation of the vitamin C content of unpasteurized beverages, aged beverages (both pasteurized and unpasteurized) and aged sugar products, it is important to assay for total ascorbic acid (dehydroascorbic acid plus L-ascorbic acid), which corresponds to the total vitamin C content of the sample. This information becomes important when a maximum vitamin C claim is to be made for a product. With freshly prepared products it is usually sufficient to determine L-ascorbic acid only since little of the dehydro form is present or else the small amount that is present is only stable for a short period of time.

In general, however, it should be remembered that not only may dehydroascorbic acid be present in a product under certain conditions (dehydration, storage under conditions permitting oxidation, etc.). It may also arise as a result of oxidation during analysis if proper precautions are not taken.

Visual Titration for L-Ascorbic Acid

The vitamin C content (L-ascorbic acid only) of fruit juices and various canned, fresh, and frozen fruits and vegetables can be determined by simple titration with a standardized DI solution.

Reagents

1. 20% Metaphosphoric acid stock solution—Dissolve with shaking 100 g (3.5 oz) of solid reagent-grade metaphosphoric acid in about 400 ml of distilled water and dilute to 500 ml. Filter rapidly through a fluted filter paper into a glass-stoppered bottle. This stock solution can be used for about a month if it is refrigerated at 40°F.
2. 4% Metaphosphoric acid working solution—Dilute 1 part stock solution with 4 parts of distilled water. This working solution can be used for several days if it is kept refrigerated.
3. Sodium 2,6-dichlorobenzenoneindophenol (DI) solution—Dissolve 0.1 g of powdered DI in 100 ml of warm distilled water, cool, dilute to 500 ml, and filter through a fluted paper into an amber, glass-stoppered bottle. This solution can be used for 10–14 days if stored in the refrigerator. DI is available from the Eastman Kodak Company, Organic Chemical Division, Rochester, NY.
4. L-Ascorbic acid—Available from Hoffman-La Roche Inc., Nutley, NJ.
5. 3% Hydrogen peroxide solution—Prepare by dilution of 30% H_2O_2 and store in refrigerator.

Standardization of DI With L-Ascorbic Acid

The DI solution should be standardized daily in careful assay work. This requires the preparation of a fresh standard solution of L-ascorbic acid every day.

Weigh accurately (±0.01 mg) about 0.1 g of pure L-ascorbic acid into a 100-ml volumetric flask and dilute to mark with 2% metaphosphoric acid. Promptly titrate 2.0-ml aliquots of this standard L-ascorbic acid solution with the DI solution to the first pink color that persists for at least 10 sec (about 15–20 ml should be required). Express the strength of the DI as mg of L-ascorbic acid per ml of DI solution.

Assay Procedure (#1)

Carry out the DI titration with canned, fresh, frozen or thawed foods containing natural L-ascorbic acid or treated with crystalline L-ascorbic acid during processing as follows:

Place a representative weighed or measured sample of the food in a Waring blender, add 400–500 ml of 4% metaphosphoric acid, and blend for 1–3 min. Transfer the entire contents quantitatively to a 1-liter volumetric flask or glass-stoppered graduate and dilute to 1 liter with distilled water. After mixing thoroughly, filter the extract through a coarse fluted paper; store extract in refrigerator.

Pipette three 10-ml aliquots of the filtered extract into 125-ml Erlenmeyer flasks and titrate rapidly with the standardized DI solution until the first pink color that persists for at least 10 sec. Average the titration values and calculate the L-ascorbic acid content per sample from the following formula:

$$\text{L-ascorbic acid (mg)/sample} = V \times E \times 100$$

where V = ml of DI solution required to titrate 10 ml of sample extract; E = mg L-ascorbic acid/ml DI solution; and 100 = dilution factor.

Assay Procedure (#2)

To assay fruit juices or other liquids with no interfering coloration, dilute 2 fluid oz (59.2 ml) of the liquids to 100 ml with 4% metaphosphoric acid and mix well; filter if the mixture is not clear. Titrate three 10-ml aliquots of the extract in the manner previously described.

Average the titration values and calculate the L-ascorbic acid per ounce of juice as follows:

$$\text{L-ascorbic acid (mg)/oz} = V \times E \times 5$$

where 5 = dilution factor and V and E are the same in procedure #1.

Eliminating Interference

Sulfite, iron, and tin interfere in the determination of L-ascorbic acid by reacting with DI, causing erroneously high values. Interference by any of these substances can be detected by conducting the titration with and without treatment with H_2O_2. These interferences may be overcome by adding an equal volume of 3% H_2O_2 to the extract just before titration with DI. The peroxide–extract mixture should be titrated promptly with the DI solution, otherwise some L-ascorbic acid may be converted to dehydroascorbic acid. The presence of thiourea does not interfere in the determination of L-ascorbic acid by the DI method.

Visual Titration for Total Ascorbic Acid

The DI titration method can be used to determine the sum of L-ascorbic acid and dehydroascorbic acid if the sample is treated with a reducing agent before titration.

Reagents

1. Standardized DI solution as described in method for L-ascorbic acid.
2. Metaphosphoric acid sample extract as described under *Assay Procedure (#1 and #2)* in method for L-ascorbic acid.
3. Acetate buffer—Dissolve 250 grams of crystalline sodium acetate (trihydrate) in 250 ml of water and filter. Add 500 ml of glacial acetic acid and dilute to 1000 ml.
4. 5% of NaOH solution.
5. Alkaline pyrogallol solution—Dissolve approximately 25 g sodium hydroxide and 5 g pyrogallol in 250 ml of water. Prepare fresh for each use.
6. Caprylic alcohol.
7. Lead acetate paper.
8. Cylinder of hydrogen sulfide (H_2S) gas.
9. Cylinder of nitrogen gas.

Apparatus

Glass equipment is needed for bubbling the gases through the extracts. For the injection of H_2S, the gas train consists of $H_2S \rightarrow$ water $\rightarrow$ sample extract $\rightarrow$ NaOH; for sweeping out the H_2S with N_2 the gas train is $N_2 \rightarrow$ alkaline pyrogallol $\rightarrow$ extracts $\rightarrow$ NaOH.

Reduction of Dehydroascorbic Acid

Measure 40 ml of metaphosphoric acid sample extract into a 100-ml glass-stoppered graduate, add 4 ml of the acetate buffer, and dilute to 60 ml with water. The extract is now at a pH of about 3.5. Place 45–50 ml of this buffered extract in an amber test tube or cylinder fitted with a gas injection tube. (The gas injection tubes are fitted to the cylinders with standard tapered joints. However, if these are not available, rubber stoppers may be used; the latter should be boiled in dilute alkali before use.) Add 2–6 drops of caprylic alcohol to minimize foaming. Between the H_2S cylinder and the extracts, connect a gas-washing bottle containing water in order to saturate the gas with moisture. After the last extract connect another gas-washing bottle containing 5% NaOH to absorb the excess gas. Connect the train with rubber tubing and allow H_2S to bubble through at a moderate rate for 15–20 min, then close the H_2S cylinder valve, disconnect the tubing from the cylinder, and seal both ends of the system with pinch clamps. Allow to stand (away from direct sunlight) for about 2 hr. Then pass wet nitrogen (freed of oxygen by passage through alkaline pyrogallol) through the train until all H_2S is removed (2–3 hr required). Test for the presence of H_2S with lead acetate paper, which turns black or brown in the presence of H_2S.

Assay Procedure

Carry out titration with standardized DI solution (as described in previous method) on three 15-ml aliquots of the H_2S-treated extract. Calculate total ascorbic acid as shown under *Assay Procedure (#1)* in previous method. Carry out similar titrations on 15-ml aliquots of a freshly buffered and diluted extract that is not subjected to H_2S reduction; these samples determine L-ascorbic acid only. The amount of dehydroascorbic acid can be obtained by subtracting the value for L-ascorbic acid from that for total ascorbic acid.

Determination of Niacin by Microbiological Method

Microbiological assay is the method of choice for determining niacin in food products because chemical methods are not sensitive enough. This assay involves use of a test microorganism that requires niacin to grow. The test organism is inoculated into an assay medium that does not contain niacin and aliquots of the sample being assayed are added. Standard assays are run under the same conditions using the test organism and known amounts of niacin. After incubation for a period of time, growth of the organism (measured by the turbidity of the assay tubes at 660 nm) in the sample assays and standard assays is compared. The concentration of niacin in the sample can be determined from the standard curve of growth (turbidity reading) versus niacin concentration.

The method described in this section involves the determination of niacin in fortified pudding that contain about 2.0 μg niacin/5 oz. This method can be adapted to determination of niacin in other materials by suitable changes in the sample preparation procedure.

Reagents

1. 2 N H_2SO_4.
2. 1 N NaOH.
3. 1 N HCl.
4. Niacin standard stock solution—Accurately weigh 100 mg U.S.P. reference standard niacin, transfer to 1-liter volumetric flask, bring to volume with water, and mix. This standard stock solution should be kept under toluene at about 4°C.
5. Niacin standard working solution—On the day of assay bring the niacin stock solution to room temperature and dilute 5 ml to 500 ml with water; then take 5 ml of this solution and dilute again to 100 ml. This gives the standard working solution containing 0.5 μg/ml.
6. 10% Hydrolyzed casein—Reflux 100 g vitamin-free casein plus 500 ml constant boiling HCl (50% concentrated HCl: 50% H_2O) for 8–12 hr on hot plate (glass beads added to flask). Concentrate in vacuum to remove HCl. Add 500 ml distilled water and reconcentrate. Take down to thick syrup each time. Then take pH up to 3.0 with concentrated NaOH; make volume to 1000 ml. Add 20 g Darco G-60 charcoal. Stir for 1 hr. Filter. Repeat treatment if casein is not straw colored. Store in refrigerator under toluene.
7. Amino acids and purine solution—Dissolve 50 mg each of adenine, guanine, uracil, and xanthine in 10 ml concentrated HCl and 10 ml water with heat and add about 900 ml hot water. Add 1 g L-cystine and 2 g DL-tryptophan and stir. When all ingredients are in solution, cool and bring the volume to 1 liter. Keep under toluene at room temperature.
8. Vitamin solution—Place the following in 100-ml volumetric flask: 10 mg vitamin B_1; 10 mg vitamin B_2; 10 mg calcium pantothenate; 20 mg vitamin B_6; 1 mg *para*-aminobenzoic acid; and 25 μg biotin. Make up to 100 ml with distilled water and store under toluene in the refrigerator.
9. Salts A solution—50 g K_2HPO_4 and 50 g KH_2PO_4 made up to 500 ml with water. Add 5 drops concentrated HCl and store under toluene.
10. Salts C solution—In 500 ml water dissolve 20 g $MgSO_4 \cdot 7H_2O$, 1 g NaCl, 1 g $FeSO_4 \cdot 7H_2O$, and 3 g $MnSO_4 \cdot H_2O$. Add 1 ml concentrated HCl and store under toluene.

11. Liver–tryptone agar—To 400 ml hot water add the following: 4.0 g liver paste (Liver fraction 1, NF XI); 4.0 g tryptone; 4.0 g glucose; 0.8 g K_2HPO_4; 6.0 g agar; 2.0 ml Salts A solution; and 2.0 ml Salts C solution. After these are in solution, add 1.2 g $CaCO_3$.
12. Niacin assay medium (double strength)—Combine the following: 50 ml 10% hydrolyzed casein; 100 ml amino acids and purine solution; 10 ml vitamin solution; 5 ml Salts A solution; 5 ml Salts C solution; 20 g anhydrous dextrose; and 10 g anhydrous sodium acetate. Make up to 500 ml with water and adjust pH to 6.6–6.8. Complete media are available commercially. If one is used, an adjustment in the standard and sample potency in the final dilution may be necessary to achieve a suitable range of growth response.

Stock Cultures and Inoculum

The test organism, *Lactobacillus plantarum* ATCC 8014, is carried on liver–tryptone agar. Add 10 ml of the prepared liver–tryptone agar to each culture tube, plug, and sterilize 15 min at 15 lb; do not char. The culture is carried by daily transfer from stab to stab.

Broth inoculum medium is prepared by adding 50 μg niacin to 250 ml double-strength niacin assay medium, diluting to a volume of 500 ml with distilled water, and mixing thoroughly. Ten-ml aliquots are dispensed into each of 40 test tubes; the tubes are plugged with cotton and sterilized for 15 min at 121°C. They are stored in the refrigerator until used.

The day before the assay is set up, some of the organisms are transferred from the previous day's stab to broth inoculum with a sterile needle, and this inoculum is incubated for 16–24 hr at 37°C. After incubation, the inoculum is centrifuged for 8 min at 1500–2000 rpm, the supernatant is decanted, and 10 ml of sterile 0.9% saline are added. The organisms are suspended in the saline, and a subdilution of 1 ml of suspension in 100 ml sterile 0.9% saline is made. One drop of this diluted inoculum is added to all assay tubes except the uninoculated blanks and the sample color blanks.

Sample Assay Tubes

Place 50 g of pudding in a 600-ml beaker, add about 60 ml of water and 100 ml of 2 *N* H_2SO_4, and disperse pudding thoroughly with a stirring rod. Autoclave 30 min at 121°C. Cool, adjust pH to 6.0–6.5 by adding 1 *N* NaOH and then to 4.5 with 1 *N* HCl or H_2SO_4. Transfer to a 1-liter volumetric flask, bring to volume with distilled water, mix well and filter (use Celite Analytical Filter Aid if necessary). Pipette 10 ml of clear filtrate into a 50-ml beaker, adjust pH to 6.8, transfer quantitatively to a 200-ml volumetric flask, bring to volume with distilled water, and mix. This is the sample solution for assay. If turbid, filter before use and assay the clear filtrate.

Pipette sample solution for assay into duplicate tubes as follows:

ml Solution	*ml H_2O*
1	4
2	3
3	2
4	1
5	0
0	5 (uninoculated color blank)

Niacin Standard Assay Tubes

Pipette niacin standard working solution for assay into duplicate tubes as follows:

ml Niacin Solution	*μg Niacin*	*ml H_2O*
0	0	5.0 (uninoculated blank)
0	0	5.0 (inoculated blank
0.5	0.025	4.5
1.0	0.050	4.0
1.5	0.075	3.5
2.0	0.100	3.0
2.5	0.125	2.5
3.0	0.150	2.0
4.0	0.200	1.0
5.0	0.250	0

Assay Procedure

To all assay tubes and blanks add 5 ml of double-strength niacin assay medium, cap with aluminum caps, and sterilize 6 min at 121°C. Cool in cold water bath to uniform temperature, add one drop inoculum to all tubes except uninoculated blanks and color blanks, and incubate for 18 hr at 37°C. At the end of the incubation period, steam for 10 min at 100°C. Cool to room temperature and read in photoelectric colorimeter at 660 nm.

Calculations

Plot colorimeter readings for standard niacin tubes verus niacin concentration on two-cycle by one-cycle logarithmic paper using the single cycle for the colorimeter readings. The uninoculated sample tube is used as a color blank and, if necessary, a proportionate correction is made at each level before determining the niacin content of each tube. Determine the concentration of vitamin in each sample tube from the standard curve. Calculate the average niacin content in μg per ml, discarding any value exceeding ±10% of the average. The niacinamide content of the original pudding is calculated as follows:

$$\text{Niacinamide } (\mu\text{g})/5 \text{ oz pudding} = \text{Niacin } (\mu\text{g/ml}) \text{ in sample assay solution} \times 56.2$$

SELECTED REFERENCES

AOAC. 1980. Official Methods of Analysis. 13th ed. W. Horwitz (editor). Assoc. of Official Analytical Chemists, Arlington, VA.

AOAC. 1984. Official Methods of Analysis. 14th ed. S. Williams (editor). Assoc. of Official Analytical Chemists, Arlington, VA.

AUGUSTIN, J. A., KLEIN, B. P., BECKER, D. B. and VENUGOPAL, P. B. 1985. Methods of Vitamin Assay. 4th ed. John Wiley & Sons, New York.

BROCK, J. F. 1961. Recent Advances in Human Nutrition. Little, Brown and Co., Boston.

BURTON, B. T. (editor). 1965. Heinz Handbook of Nutrition. 2nd ed. McGraw-Hill, New York.

COBURN, S. P. 1981. The Chemistry and Metabolism of 4′-Deoxypyridoxine. CRC Press, Boca Raton, FL.

COURSIN, D. B. 1954. Convulsive seizures in infants with pyridoxine-deficient diets. J. Am. Med. Assoc. *1954*, 406–408.

DELUCA, H. F. 1969. Metabolism and function of vitamin D. *In* The Fat-Soluble Vitamins. H. F. DeLuca, and J. W. Suttie (editors). Univ. of Wisconsin Press, Madison, WI.

DELUCA H. F. and SUTTIE, J. W. 1970. The Fat Soluble Vitamins. The Univ. of Wisconsin Press, Madison, WI.

DOLPHIN, D. 1982. B_{12}. Vols. I and II. John Wiley & Sons, Canada.

FREED, M. 1966. Methods of Vitamin Assay. 3rd ed. Interscience Publishers, New York.

FOOD AND NUTRITION BOARD. 1980. Recommended Dietary Allowances. National Research Council, National Academy of Science, Washington, DC.

GLOVER, J. 1970. Biosynthesis of the fat-soluble vitamins. *In* Fat-Soluble Vitamins. R. A. Morton (editor). Pergamon Press, London.

GOODWIN, T. W. 1963. The Biosynthesis of Vitamins and Related Compounds. Academic Press, New York.

GYÖRGY, P. and BERNHARDT, W. L., JR. 1968. Biotin: V and VI. *In* The Vitamins, 2nd ed., Vol. II. W. H. Sebrell, Jr., and R. S. Harris (editors). Academic Press, New York.

GYÖRGY, P. and PEARSON, W. N. (editors). 1967. The Vitamins. Vols. VI and VII. Academic Press, New York.

HARDINGE, M. G. and CROOKS, H. 1961. Lesser known vitamins in foods. J. Am. Dietet, Assoc., *38*, 240–245.
HORWIT, M. K. 1960. Vitamin E and lipid metabolism in man. Am. J. Clin. Nutr. *8*, 451–461.
INTERDEPARTMENTAL COMMITTEE ON NUTRITION FOR NATIONAL DEFENSE. 1963. Manual for Nutrition Surveys. 2nd ed. National Institutes of Health, Bethesda, MD.
KING, R. D. (editor). 1978. Developments in Food Analysis Techniques—1. Applied Science Publishers, London.
KRAUSE, M. V. 1969. Food Nutrition and Diet Therapy. 4th ed. W. B. Saunders, Philadelphia.
KUTSKY, R. J. 1973. Handbook of Vitamins and Hormones. Van Nostrand Reinhold, New York.
LEWIN, S. 1976. Vitamin C: Its Molecular Biology and Medical Potential. Academic Press, London.
MEDER, H. and WISS, O. 1968. Vitamin B_6 groups: V. Occurrence in foods. *In* The Vitamins, 2nd ed., Vol. II. W. H. Sebrell, Jr., and R. S. Harris (editors). Academic Press, New York.
METZLER, D. E. 1977. Biochemistry, The Chemical Reactions of Living Cells. Academic Press, New York.
NORMAN, A. W. 1979. Vitamin D The Calcium Homeostatic Steroid Hormone. Academic Press, New York.
ROBESON, C. D. and BAXTER, J. G. 1947. Neovitamin A. J. Chem. Soc. *69*, 136–141.
ROBINSON, F. A. 1966. The Vitamin Co-Factors of Enzyme Systems. Pergamon Press, New York.
ROSENTHAL, H. L. 1968. Vitamin B_{12} group: V. Occurrence in foods. *In* The Vitamins, 2nd ed., Vol. II. W. H. Sebrell, Jr., and R. S. Harris (editors). Academic Press, New York.
SAUBERLICH, H. E. 1968. Vitamin B_6 group: VII. Biosynthesis of vitamin B_6. *In* The Vitamins, 2nd ed., Vol. II. W. H. Sebrell, Jr., and R. S. Harris (editors). Academic Press, New York.
SEBRELL, W. H., JR. and HARRIS, R. S. (editors). 1967. 1968. 1971. The Vitamins, Vols. I, II, and III. Academic Press, New York.
SEIB, P. A. and TOLBERT, B. M. 1982. Ascorbic Acid: Chemistry, Metabolism, and Uses. American Chemical Society, Washington, DC.
SELBY, M. A. and GREEN, J. 1970. The fortification of human and animal foods with fat-soluble vitamins. *In* Fat-Soluble Vitamins. R. A. Morton (editor). Pergamon Press, London.
U.S. DEPARTMENT OF AGRICULTURE. Agriculture Handbook *97*. Government Printing Office, Washington, DC.
VERGROESEN, A. J. 1975. The Role of Fats in Human Nutrition. Academic Press, London.
WALD, G. and HUBBARD, R. 1970. The chemistry of vision. *In* Fat-Soluble Vitamins. R. A. Morton (editor). Pergamon Press, London.
WHITE, A., HANDLER, P. and SMITH, E. L. 1968. Principles of Biochemistry. 4th ed. McGraw-Hill, New York.
WINESTOCK, C. H. and PLANT, G. W. E. 1965. The biosynthesis of coenzymes. *In* Plant Biochemistry. J. Bonner and J. E. Varner (editors). Academic Press, New York.
WOHL, M. G. and GOODHART, R. S. (editors). 1960. Modern Nutrition in Health and Disease. Lea & Febiger, Philadelphia.

9

Flavoring Agents

☐ INTRODUCTION

Foods may be evaluated either on the basis of their nutrient content or their appeal to the consumer. Although the nutrient content of a food is an important factor, it is nevertheless true that a food will not be chosen freely and consumed unless it appeals to the consumer. The willingness to use or eat a food is psychological and depends upon one's reactions to its sight, odor, feel, and even sound. Expressed another way, a food must appear tasty, it must taste right, and it must possess the texture one wants.

In many cases foods can be made more appetizing by the addition of *flavoring agents* such as spices, herbs, and flavorings. Of further help in this respect is the use of *flavor enhancers* (glutamate, 5′-nucleotides, maltol) and salt.

Flavoring agents may be natural or synthetic. In the past most flavoring agents were derived from natural sources. Among the most important were spices, aromatic seeds, and herbs. Generally speaking, when the natural product was from plants of tropical origin, it was considered a spice; when from plants of temperate climates, it was considered a herb. This distinction is somewhat tenuous, and it is sometimes difficult to classify a product as a spice or herb. All of these products have one thing in common: they contain an aromatic flavoring component, usually an essential oil, that enriches or alters the taste of a food. With few exceptions the particular plant product is used whole or ground after being dried.

☐ COMMON SPICES AND HERBS

Allspice (Pimento)

Allspice, *Pimenta dioica* (L.), is the dried fruit of an evergreen tree belonging to the Myrtle family. The tree is native to the West Indies, but is found in Mexico, Central America, and northern South America. The berries are harvested just before they ripen because the ripened fruits lose much of their quality. The berries are dried by exposing them to the sun for a period of 7–12 days. The common name, allspice, originates from the fact that the berries have a flavor similar to the combined flavors of nutmeg, cloves, and cinnamon. Most commercial allspice comes from Kingston, Jamaica.

Allspice contains a volatile oil, a fixed oil, resin, cellulose, pentosans, starch, pigments, etc. The starch granules are small, nearly circular, and uniform in size with a central dotted hilum. Allspice contains lumps of yellow, brown, or red resin, which is characteristic of this spice. The volatile oil constitutes 3–4.5% of allspice. The principal flavor component of the oil is eugenol, which makes up 60–75% of the oil.

Allspice is available both whole and ground. It is used for flavoring meats, gravies, sauces, relishes, pickles, preserves, puddings, cakes, and beverages. Ground allspice is a constituent of spice mixtures such as curry powder and pastry spice.

Adulterants reported in ground allspice include clove stems, nutshells, fruit stones, cereals, and dried fruit products.

U.S. Standards

Allspice shall contain not less than 8% quercitannic acid (calculated from the total oxygen absorbed by the aqueous extract), not more than 25% of crude fiber, not more than 6% of total ash, and not more than 0.4% of ash insoluble in hydrochloric acid.

Anise

Anise is the dried fruit of an annual herbaceous plant, *Pimpinella anisum* (L.), belonging to the Umbelliferae or Carrot family. *Pimpinella anisum* (L.) is indigenous to the eastern Mediterranean region, but is cultivated in many parts of the world including Spain, Syria, Turkey, India, China, Mexico, and Argentina.

Anise is used to flavor soups, cakes, cookies, and rolls; in confectionary to flavor candies and syrups.

Crushed anise seeds yield a steam volatile oil amounting to 1.5–3.5%. The principal flavor component of the oil is anethole (80–90%) together with other organic compounds such as methyl chavicol, anisketone, and anisaldehyde. Anise seed and anise oil are characterized by a licorice-like flavor and odor.

U.S. Standards

Star anise seed, the dried fruit of *Illicium verum* (Hook), contains an essential oil of similar chemical composition to that from anise even though it is different from anise botanically. It contains not more than 5% of total ash.

The dried fruit shall contain not more than 9% of total ash, and not more than 1.5% of ash insoluble in hydrochloric acid.

Capsicum Peppers

Several condiments are prepared from the dried fruits of a plant belonging to the Solanaceae or Potato family. It is native to tropical America, but is cultivated throughout the world. Capsicum fruits are the source of paprika, cayenne pepper, red pepper, and chili powder.

Paprika is produced from the fruit of *Capsicum annum*. The color of the fruits may range from a bright-red to a brick-red. The flavor is characterized by having little or no pungency. *Capsicum annum* grows in temperate climates, in countries such as Spain, Yugoslavia, Turkey, Mexico, Chile, and the United States. In the home paprika is used for flavoring and garnishing meat and fish dishes, poached and deviled eggs, salads, canapes, etc. In the food industry, it is used in the preparation of catsup, sauces, and other foods.

Cayenne pepper is made from the dried, ripe fruits of *C. frutescens* (L.), *C. boccatum* (L.), or some other smaller more pungent capsicum. These plants are grown in Africa, Japan, Louisiana, and other areas of the world. The pungency of this spice is due to a compound known as capsaicin; the spice must be used with discretion for a small amount will add considerable flavor to foods.

Red pepper is the ground product from large red peppers. It is milder than cayenne pepper and is used primarily in spaghetti, soups, stews, and Mexican-style dishes.

Chili powder is a combination of spices that includes a bland capsicum pepper. This condiment is used in the preparation of foods such as chili con carne and hot tamales. It is also used as a seasoning for eggs, stews, pork and beans, sausage products, etc.

The principal flavoring constituent of the capsicums is a nonvolatile crystalline substance known as capsaicin. Paprika owes its color to several carotenoids (oleoresin of paprika).

U.S. Standards

Paprika is the dried, ripe fruit of *Capsicum annum* (L.), contains not more than 8.5% of total ash, and not more than 1% of ash insoluble in hydrochloric acid. The iodine number of its extracted oils is not less than 125 nor more than 136.

Cayenne pepper is the dried, ripe fruit of *Capsicum frutescens* (L.), or some other small-fruited species of *Capsicum*. It contains not less than 15% of nonvolatile ether extract, not more than 1.5% of starch, not more than 28% of crude fiber, not more than 8% of total ash, and not more than 1.25% of ash insoluble in hydrochloric acid.

Caraway and Caraway Seed

Caraway is the dried fruit of a biennial herb belonging to the Carrot family, which is native to Europe. Imports of caraway into the United States are mainly from the Netherlands.

Caraway is used to flavor rye bread, cakes, cookies, cheese, and sauerkraut; it is available whole or ground.

The principal flavor constituents of the volatile oil of caraway are D-carvone (50–60%) and D-limonene, with smaller amounts of other related compounds. The volatile oil constitutes 3–5% of caraway.

U.S. Standards

The dried fruit of *Carum carvi* (L.) contains not more than 8% of total ash, and not more than 1.5% of ash insoluble in hydrochloric acid.

Cardamom and Cardamom Seed

Cardamom is the dried nearly ripe fruit of a herbaceous perennial plant belonging to the Ginger family. It is native to India and cultivated in India, Ceylon, and Central America (Guatemala).

Cardamom is sweet, spicy, and highly aromatic. It is available in the pod, as decorticated seeds (i.e., capsule removed), or ground. It is used as a flavoring for breads, buns, pastries, cakes and cookies. It is an essential ingredient of curry powder and is used sparingly in sausage and hamburger seasonings.

Cardamom contains from 2 to 10% volatile oil. The principal flavor constituents are cineole and D-terpinyl acetate.

U.S. Standards

The dried seed of cardamom, *Elettaria cardamomum* (Maton), shall contain not more than 8% of total ash, nor more than 3% of ash insoluble in hydrochloric acid.

Celery Seed

Celery seed is the dried ripe fruit of a biennial herb belonging to the Parsley family (Umbelliferae). The plant is cultivated in many parts of the world including France, India, Italy, Japan, and the United States.

Celery seed is used as a flavoring for tomato juice, sauces, soups, pickles, salads, and vegetable juices.

Celery seed contains 2–3% of a steam volatile oil of which the chief constituents are D-limonene (60%) and selinene (10–20%). The characteristic flavor is believed to be due to small amounts of a number of alcohols, anhydrides, acids, and lactones.

U.S. Standards

The dried fruit of *Celeri graveolens* (L.) shall contain not more than 10% of total ash, nor more than 2% of ash insoluble in hydrochloric acid.

Ceylon and Cassia Cinnamon

Ceylon cinnamon, the true cinnamon of commerce, is the dried inner bark of *Cinnamomum zeylanicum*, an evergreen tree belonging to the laurel family and native to Ceylon. This spice is not imported in large amounts into the United States but is in demand in other parts of the world.

The dried bark assumes the form of a cylinder, which is cut into short lengths and marketed as Ceylon cinnamon sticks. Ground cinnamon is used as a baking spice for cakes, breads, buns, cookies, and pies. Ceylon cinnamon has a fragrant odor and a warm, sweet, aromatic taste.

Ground Ceylon cinnamon contains starch grains about the size of allspice. Bark fibers are more numerous than in cassia cinnamon but no cork is present. Stone cells are somewhat similar to those in allspice.

The bark of Ceylon cinnamon contains 0.5–1.0% of steam volatile essential oil. The principal flavor constituent is cinnamic aldehyde (55–75%) and small amounts (4–8%) of eugenol.

Cassia cinnamon is prepared from the dried bark of *Cinnamomum cassia*, an evergreen tree native to South Vietnam and the eastern Himalayas. Cassia bark has a more intense aroma, a higher essential oil content, and is less delicately flavored than Ceylon cinnamon. These two species are obtained from two different species of plants grown in different areas. However, their nomenclature has been used so interchangeably that they have come to mean the same product.

More cassia cinnamon than Ceylon cinnamon is used in the United States. The primary sources for cassia cinnamon are Saigon cassia, *C. loureirii* (Nees); Korintji "thick quill," *C. burmannii* (Blume), grown in Sumatra; Padang "thin quill"; and *C. sintok* (Blume), produced in Malaysia.

Cassia is used by bakers, confectioners, fruit canners, and other food processors in the same manner as Ceylon cinnamon.

The volatile oil of cassia contains from 75 to 90% of cinnamic aldehyde together with smaller amounts of related aldehydes. Adulterants reported in the ground spice include exhausted cassia, ground bark, cassia buds, nutshells, and other cheaper material.

U.S. Standards

Cinnamon is the dried bark of cultivated varieties of *Cinnamomum zeylanicum* (Nees) or of *C. cassia* (L.) (Blume), from which the outer layers may or may not have been removed. Ceylon cinnamon is the dried inner bark of cultivated varieties of *Cinnamomum zeylanicum* (Nees). Saigon cinnamon and Saigon cassia are from the dried bark of cultivated varieties of *C. cassia* (L.) (Blume). Ground cinnamon and ground cassia are from the powder made from cinnamon. It shall contain not more than 5% of total ash, nor more than 2% of ash insoluble in hydrochloric acid.

Cloves

Cloves are the dried flower buds of *Eugenia caryophylata*, an evergreen tree of the Myrtle family. It is indigenous to the Moluccas or Spice Islands but is cultivated in many of the East Indian Islands, Madagascar, Zanzibar, Pemba, Penang (Malaysia), India, and Ceylon. When the green buds change to a reddish color, they are removed from the tree and dried. The dried flower buds must be handled with care. They range from $\frac{1}{4}$ to $\frac{1}{2}$ in. in length and are reddish brown.

Cloves are characterized by their strong, pungent flavor. The spice is available either whole or ground and is used as a flavoring for hams, roasts, stews, pickled fruits, preserves, desserts, cakes, puddings, and spicy sweet syrups.

Ground cloves contain tissue cells found in flower buds and broad bast fibers, but starch grains are absent. Cloves contain 12–21% of volatile oil; the principal flavor constituents are eugenol (70–90%), caryophyllene, vanillin, and small amounts of several other substances. Adulterants include exhausted cloves, clove stems, allspice, nutshells, and cereals.

U.S. Standards

Cloves, whole or ground, shall be the dried flower buds of *Eugenia caryophyllata* (Thunb.). They contain not more than 5% of clove stems, not less than 15% of volatile ether extract, not less than 12% quercitannic acid (calculated from the total oxygen absorbed by the aqueous extract), not more than 7% of total ash, nor more than 0.5% of ash insoluble in hydrochloric acid.

Coriander

Coriander is the dried fruit of an annual belonging to the Parsley family, which is native to the Mediterranean region. This spice is cultivated principally in northern Africa, Romania, parts of the Soviet Union, and the United States.

Coriander has an aromatic pungent taste and a very distinctive odor. It is available whole or ground. The whole seed is used in pickling spices, while the ground spice is a component of curry powder and a flavoring for pastries, buns, cookies, and processed meats.

The dried fruit contains 0.1–1% of volatile oil. The chief flavor constituent is D-linalool, an isomer of geraniol.

U.S. Standards

Coriander is the dried fruit of *Coriandrum sativum* (L.). It shall contain not more than 7% of total ash, nor more than 1.5% of ash insoluble in hydrochloric acid.

Cumin Seed

Cumin seed is the dried fruit of *Cuminum cyminum*, a plant belonging to the Parsley family (Umbelliferae). It, like coriander, is a native of the Mediterranean region but is also found in Central America. Cumin seed is similar in appearance to caraway seed but longer.

Cumin has a distinctive odor and an aromatic, hot and bitter taste. It is available whole or ground. The spice is used extensively in Mexican foods and also in the preparation of processed meats, pickles, cheese, and sausages. At home it is used to flavor soups and stew. Cumin seed is an essential ingredient of curry powder and chili powder.

Cumin seed contains about 2–4% of volatile oil. The principal flavoring constituent of the oil is cumaldehyde.

U.S. Standards

Cumin seed shall contain not more than 9.5% of total ash, not more than 1.5% of ash insoluble in hydrochloric acid, nor more than 5% of harmless foreign matter.

Dill

Dill seed is the dried fruit of an annual belonging to the parsley family (Umbelliferae), which is grown extensively in temperate areas such as the United States, Canada, Central Europe, and India. In the United States dill is grown primrily in the North Central states and in the Pacific Northwest. In addition to seeds, the leafy tops are also used and are referred to as dill weed.

Dill seed and dill weed are available whole or ground but are usually sold in the whole form. It has a pungent and faintly aromatic flavor. Dill seeds yield an essential oil that is an important flavoring in the pickle industry. The spice is used as a flavoring for meat, fish, sauces, and pickles; it is a component of meat seasonings.

The principal flavor constituents of dill are carvone (40–50%), D-limonene, and terpenes.

U.S. Standards

Dill seed is the dried fruit of *Anethum graveolens* (L.). It shall contain not more than 10% of total ash, nor more than 3% of ash insoluble in hydrochloric acid.

Fennel

Fennel, an aromatic perennial of the Parsley family, is native to southern Europe and the eastern Mediterranean area. Dried fennel seeds have an odor and taste somewhat similar to those of anise. The seeds are available whole or ground. They are used to flavor soups, fish dishes, bread, rolls, pastries, confections, and pickles.

The dried fennel fruit contains about 3–6% of volatile oil. The principal flavor constituents are anethole (50–60%), fenchone (20%), and various terpenes.

U.S. Standards

Fennel seed is the dried fruit of *Foeniculum vulgare* (Miller) and related species. It contains not more than 9% of total ash, nor more than 2% of ash insoluble in hydrochloric acid.

Garlic

Garlic is a bulbous herb, *Allium sativum* (L.), a member of the Lily family (Liliaceae), which is cultivated throughout the world. The major producing countries are the United States, Hungary, Taiwan, and Egypt.

Garlic is used extensively for flavoring all types of food products—soups, meats, gravies, sauces, salad dressings, bread, and many other products. Garlic is available either as powder or as salt. Garlic powder is the ground product of the dehydrated garlic cloves. Garlic salt is dehydrated garlic combined with table salt.

The volatile oil content of garlic is rather small, about 0.1%. The principal flavor constituents are diallyl disulfide, allyl propyl disulfide, and diallyl trisulfide.

Ginger

The source of ginger is a perennial herb belonging to the Ginger family. It is native to Southeast Asia and now is cultivated in Jamaica, India (Malabar Coast), Nigeria, Southern China, Japan, and Australia.

Ginger is obtained from the whole or partially peeled rhizomes (underground stems) and is referred to in the trade as "hands." The varieties of ginger are named for the area of production—Jamaica ginger, African ginger, Indian ginger, etc. Jamaica ginger is considered the finest, possessing the most delicate aroma and flavor. African ginger possesses the greatest pungency, a harsher flavor, and a higher yield of essential oil; it is used in the meat industry.

The aroma of ginger is spicy and pungent, while the taste is aromatic and hot. It is available whole, cracked, or ground. Oleoresin of ginger, which is obtained by extraction from ground ginger, is used as a flavoring in soft drinks. The spice is used as a flavoring in pies, cookies, cakes, gingerbread, puddings, pickles, and beverages. Cracked ginger is an ingredient of whole mixed pickling spice. Ground ginger is a component of curry powder, pastry spice, mincemeat spice, and other seasoning mixtures.

Ginger contains 1–3% of a volatile oil containing a sesquiterpene (zingiberene), the terpenes (α- and β-phellandrene, α-camphene), an alcohol (zingiberol), and small amounts of other substances.

U.S. Standards

Ginger is the washed and dried, or decorticated and dried, rhizome of *Zingiber officinale* (Roscoe). It must contain not less than 42% of starch, not more than 8% of crude fiber, not more than 1% of lime (CaO), not less than 12% of cold-water extract, not more than 7% of total ash, not more than 2% of ash insoluble in hydrochloric acid, nor less than 2% of ash insoluble in cold water.

Jamaica ginger contains not less than 15% of cold-water extract, and conforms in other respects to the standards of ginger. Limed ginger or bleached ginger is whole ginger coated with carbonate of calcium. It contains not more than 10% of total ash, and conforms in other respects to the standards for ginger.

Laurel Leaves (Bay Leaves)

Laurel leaves are obtained from an evergreen shrub, *Laurus nobilis* (L.), belonging to the Laurel family. The shrub is cultivated primarily in the countries bordering the Mediterranean. The dried leaves should not be confused with "Bay" from *Pimenta acres* (L.), or with California "Bay" from *Umbellularia california* (Nutt), which impart different flavors.

Laurel leaves are used to flavor meats, poultry, fish, soups, vegetables, and sauces. Oil of bay, which can be isolated from the dried leaves by steam distillation, is used in the preparation of pickling spices and in flavoring of vinegars. The principal flavoring constituent of the volatile oil is cineole.

Marjoram (Sweet Marjoram)

Marjoram is obtained from an annual herb belonging to the Mint family (Laviatae), which is cultivated in France, Germany, Italy, Chile, Mexico, and the United States. The spice consists of the dried leaves, with or without a small proportion of the flowering tops.

Dried marjoram leaves are available whole or ground. The spice has a fragrant odor and a taste that is aromatic, warm, and pleasantly bitter. It is used as a flavoring in soups, stews, salads, vegetables, chicken, and turkey stuffing, sausage, lamb, and mutton. The volatile oil (about 1%) contins terpenes and terpene alcohols.

U.S. Standards

The dried leaves, with or without a small proportion of the flowering tops of *Majorana hortensis* (M.), shall contain not more than 16% of total ash, not more than 4.5% of ash insoluble in hydrochloric acid, nor more than 10% of stems and harmless foreign material.

Mustard

Mustard is the dried ripe seed of an annual herb belonging to the family Cruciferae (Mustard family) and the genus *Brassica*. The term *mustard* refers to the seeds of different annuals: white or yellow mustard is obtained from *Sinapis alba* (L.) and *Brassica hirta* (Moench); black or brown mustard is obtained from *Brassica nigra* (L.) (Koch). The plants are cultivvated in many parts of the world including Argentina, Australia, Canada, China, Denmark, England, France, India, and the United States. Most of the mustard seed imported into the United States is from Canada.

Mustard is available as ground mustard and whole mustard seeds, which are used in pickling. Probably the most common products is prepared mustard—a mixture of ground mustard, salt, white vinegar, and spices, which is used on a variety of sandwiches, eggs, meats, and salad dressings. The yellow color of prepared mustard is due to the addition of ground Alleppey turmeric during its manufacturing.

Unlike most spices neither the dried, whole mustard seeds nor the ground mustard have a perceptible odor. Only after ground mustard is moistened and is allowed to stand for a few minutes can one perceive the characteristic odor of mustard. Once the moistened powder has developed its flavor, the flavor begins to deteriorate within an hour. The flavor can be stabilized by acidification (e.g., with vinegar) and by refrigerating moistened mustard powder.

White mustard seed contains the glucoside sinalbin, a flavor precursor that upon enzymatic hydrolysis (catalyzed by myrosin) yields the volatile oil known as sinalbin mustard oil (*p*-hydroxybenzylisothiocyanate). Black mustard contains the glucoside synigrin, which on enzymatic hydrolysis yields allylisothiocyanate.

U.S. Standards

White and black mustard seed must not contain more than 5% of total ash, nor more than 1.5% of ash insoluble in hydrochloric acid. White mustard seed must contain no appreciable amount of volatile oil. Black mustard seed (*Brassica nigra*) must yield 0.6% of volatile mustard oil (calculated as allylisothiocyanate).

Nutmeg and Mace

Two spices are obtained from the nutmeg tree, *Myristica fragrans* (Houtt); nutmeg is the dried ripe seed and mace is the dried arillode, which envelopes the shell containing the seed or nutmeg. Nutmeg is native to the Moluccas but is cultivated in Indonesia and the West Indies.

The flavor of both nutmeg and mace is warm and spicy. Nutmeg however, tends to be more delicate in aroma than mace. The two spices are available whole or ground, although whole mace is rarely found in grocery stores. These spices are used to flavor cakes, cookies, soups, preserves, and meat products such as frankfurters and bologna. Nutmeg is also frequently used as a condiment in dairy products such as egg nog, junket, and puddings.

The volatile oil of nutmeg contains a number of terpenes (e.g., pinene, camphene, and cymene) and myristicin. The oil of mace is similar to that of nutmeg.

U.S. Standards

Nutmeg is the dried seed of *Myristica fragrans* (Houtt), deprived of its testa, with or without a thin coating of lime (CaO). It contains not less than 25% of nonvolatile ether extract, not more than 10% of crude fiber, not more than 5% total ash, and not more than 0.5% of ash insoluble in hydrochloric acid. Mace is the dried arillus of *Myristica fragrans* (Houtt). It contains not less than 20%, not more than 30% nonvolatile ether extract, not more than 10% crude fiber, not more than 3% total ash, and not more than 0.5% of ash insoluble in hydrochloric acid.

Macassar nutmeg, Papua nutmeg, Male nutmeg, and Long nutmeg are from the dried seed of *Myristica argentiea* (Warb), deprived of its testa.

Macassar mace and Papua mace are the dried arillus of *Myristica argentiea*.

Origanum

Two types of origanum are recognized, Mexican and European. The former (*Lippia graveolens*) is more pungent than the latter (*Origanum vulgare*), which is sometimes called oregano. Origanum (or oregans) is composed of the dried leaves and flowering tops of a plant belonging to the Mint family. Origanum is produced principally in Spain and Greece.

Origanum, available either whole or ground, has a strong aromatic odor and a pungent taste. It is used to flavor soups, pizza, spaghetti, pasta sauce, omelets, beef stew, chili con carne, and other dishes. It is an ingredient of Mexican chili powder.

The principal flavor constituent of the volatile oil is carvacrol. Varying percentages of thymol are present together with terpenes, alcohols, and esters.

Parsley

Parsley is the dried leaf of *Petroselinum sativum* (Hoffin). It is used extensively in the flavoring of soups, meat and fish dishes, sauces, and vegetables. In a fresh, green condition it is used as a garnish. The principal flavor constituent of the volatile oil from parsley leaves is apiol, a phenol ether.

Pepper

Pepper is prepared from the dried fruit of *Piper nigrum* (L.) (Piperaceae), a vinelike shrub native to southern India. There are two types of pepper, both of which are derived from the same plant. Black pepper is the dried, whole, unripe fruit or peppercorn; white pepper is derived from the ripe berries of which the dark outer hull has been removed.

Pepper is cultivated in India, East Indies, Malay States, West Indies, and Brazil. The many types and grades of pepper available are usually identified by the districts in which they are produced or by the exporting ports.

Both black and white pepper are available whole, cracked, and ground. The flavor can be characterized as possessing a penetrating aroma and a hot, biting, pungent flavor. This spice is used as a seasoning and flavoring for meats, soups, fish, eggs, vegetables, gravies, and sauces. In the food industry it is used in the preparation of processed meats, pickles, soups, and sauces.

Pepper contains 1–3% of a volatile oil containing l-phellandrene, the alkaloids piperine and piperdine, and chavicine, a pungent substance.

U.S. Standards

Black pepper shall contain not less than 6.75% of nonvolatile ether extract, not less than 30% of starch, not more than 7% of total ash, nor more than 1.5% of ash insoluble in hydrochloric acid.

White pepper shall contain not less than 7% of nonvolatile ether extract, not less than 52% of starch, not more than 5% of crude fiber, not more than 3.5% of total ash, nor more than 0.3% of ash insoluble in hydrochloric acid.

Rosemary

Rosemary is from the dried leaves of a small evergreen shrub, *Rosmarinus officianalis* (L.), which is native to the countries bordering the Mediterranean. The spice has a pleasant odor and taste. It is available either whole or ground and may be used as a flavoring in soups, stews, vegetables, and salads.

Rosemary contains 1–2% of volatile oil whose principal flavor constituent is borneol.

Saffron

Saffron is the dried stigma of the flower of a plant belonging to the Iris family and closely related to the common garden crocus. The plant is cultivated in Spain, Portugal, France, and Italy.

Saffron has a sharp, penetrating odor and a pungent bitter taste. Because it imparts a yellow color to water, saffron is used in coloring and flavoring foods such as pastry, confections, breads, cheeses, and mixed dishes such as chicken and rice. Saffron is extremely expensive and a small amount is sufficient to flavor a food.

Saffron contains small amounts (0.5–1%) of a volatile oil that contains a glucoside (picrocrocin) and a coloring principle (crocin and crocetin).

U.S. Standards

Saffron is the dried stigma of *Crocus sativus* (L.). It shall contain not more than 10% of yellow styles and other foreign matter, not more than 14% of volatile matter when dried at 100°C, and not more than 7.5% of ash insoluble in hydrochloric acid.

Sage

Sage is prepared from the dried leaves of *Salva officinalis* (L.) (Laviatae). The plant is a member of the Mint family and is indigenous to areas surrounding the Adriatic Sea. Dalmation sage from Yugoslavia is highly regarded, but there are many other species of sage available for culinary use.

The odor of dried sage is fragrant and aromatic, while the taste is warm, spicy, and slightly bitter. It is available whole, crushed, or ground. Sage is used as a seasoning to flavor meats, sausages, baked fish, poultry, and a variety of stuffings.

This spice contains up to 2.5% of a volatile oil having as the characteristic component the ketone thujone and borneol.

Tarragon

Tarragon is composed of the dried leaves of *Artemisia dracunculus* (L.), a small perennial plant belonging to the Sunflower family, which is native to Russia and western Asia. There are two kinds of tarragon known to the spice trade: Russian tarragon (*Artemisia dracunculoides* (Pursh) and French tarragon (*Artemisia dracucunloides*). The latter is more pungent and aromatic and is the variety grown in the United States.

Tarragon in flavor is mildly aromatic, while in aroma it is licorice-like. It is used as a flavoring in vinegar, pickles, prepared mustard, soups, and stews. The principal flavoring component in the volatile oil of tarragon is methyl chavicol.

Thyme

Thyme consists of the dried leaves and flowering tops of *Thymus vulgaris* (L.), an annual that belongs to the Mint family. This plant grown primarily in the Mediterranean area.

Thyme has a fragrant aromatic odor and a pungent flavor. It is available whole or ground and is used as a flavoring for soups, various meat and fish dishes, vegetables, and poultry dressing.

Steam distillation of dried thyme leaves yields 1–2% of oil, of which the chief components are thymol and carvacrol.

Thyme, the dried leaves of *Thymus vulgaris*, must contain not more than 14% of total ash, nor more than 4% of ash insoluble in hydrochloric acid.

Turmeric

Turmeric is the dried rhizome of a perennial belonging to the Zingiberaceae or Ginger family. It is native to southern Asia and cultivated in China, Indonesia, India, Haiti, Jamaica, and other tropical lands. The quality, appearance, and color of whole turmeric varies according to its source. The rhizomes have a dark yellow-orange color internally.

Turmeric is available whole or ground. It has a spicy, peppery-like odor and a flavor reminiscent of mustard. In its ground form it is used extensively in curry-type flavors. In the home it is used to add flavor and color to pickles and relishes.

Turmeric contains about 5% of a volatile oil containing ketones (e.g., turmerone), alcohols, and a yellow coloring agent (curcumin).

☐ OLEORESINS

Oleoresins are prepared by percolating a volatile organic solvent through a ground spice or herb. Consequently, these extracts contain nonvolatile constituents as well as the steam volatile constituents (essential oils). The extracting solvent must be selected with care in order to produce the kind of extractive desired for a particular flavoring purpose. Extraction solvents include hydrocarbons, chlorinated hydrocarbons, acetone, methyl alcohol, and isopropyl alcohol. After extraction is completed, the organic solvent must be completely removed to assure compliance with the regulations of the Food, Drug, and Cosmetic Act dealing with solvent residues in oleoresins. The residue remaining after removal of the solvent is the oleoresin. This material consists of essential oils, soluble resins, nonvolatile oils, and other solvent-soluble materials that were present in the ground spice or herb.

An oleoresin is not identical to the herb or spice from which it is derived. For example, it has been demonstrated that the composition of the essential oil component in an oleoresin differs from the composition of the essential oil produced by steam distillation.

Certain spices are extracted for their oleoresins because it is the nonvolatile fraction that is necessary for a full-flavoring effect. For example, the volatile oil of either black pepper or ginger would have little of the pungent effect characteristic of these spices because the constituents that are responsible for this effect are not volatile and are present only in the oleoresin. Other spices are used as oleoresins because little or no volatile essential oil is present in the spice per se (e.g., turmeric, paprika, and capsicum).

Oleoresins are normally viscous and as a consequence may be difficult to handle. Several techniques are available to improve the handling characteristics of oleoresins. For example, a selective diluent may be added, such as propylene glycol or a "fatty" oil, or the oleoresin may be distributed on the surface of a dry edible carrier (e.g., dextrose, starch, soy protein, and salt). The oleoresin also may be treated to form a solution or an emulsion. For example, when a surfactant such as Polysorbate 80 is added to an oleoresin, a water-dispersible mixture results, and gum tragacanth may be used as an emulsifying agent (o/w). Finally, oleoresins may be encapsulated in substances such as food starch.

Oleoresins have several advantages over the natural herb or spice: uniformity of flavor stability, better storability, and the absence of viable bacteria.

☐ FLAVORING EXTRACTS

A flavoring extract is defined as a solution in ethyl alcohol of proper strength of the sapid and odorous principles derived from an aromatic plant, or parts of the plant, with or without its coloring matter, conforming in name to the plant used in its preparation. The flavoring extract may be reinforced with a synthetically produced flavor principle; however, in this case the resulting product must be labelled as containing artificial or imitation flavor. This labeling requirement holds even though the synthetic ingredient is naturally present in the extract.

Vanilla extract is perhaps the most widely used flavoring extract. Other flavoring extracts include lemon, lime, orange, almond, cinnamon, clove, and berries.

Vanilla Extract

Vanilla extract is obtained from the cured vanilla bean by water–alcohol extraction in a manner similar to the way coffee is percolated.

The FDA standards require that each 100 ml of vanilla extract contain the solubles extracted by 35% alcohol (by volume) from 10 g of vanilla beans. Standards of identity (see Chapter 1) now exist for various vanilla products including (a) concentrated extracts equivalent to *x*-fold, *x* being a whole number; (b) vanilla flavoring and concentrated vanilla flavoring; (c) vanilla oleoresins; and (d) vanilla–vanillin products.

The vanilla bean is the cured, full-grown, unripe fruit of *Vanilla planifolia* (Andrews) or *Vanilla tahitensis* (Moore). *Vanilla planifolia* is the variety of the vanilla bean that is most important in flavoring. It is native to Mexico and is now grown in Madagascar, Reunion Island, the Seychelles, French Polynesia, and Indonesia. *Vanilla tahitensis*, grown in Tahiti and Hawaii, produces beans of inferior quality, which are used chiefly for blending and giving body. The latter variety is easily distinguished from *V. planifolia* due to the presence of heliotropine.

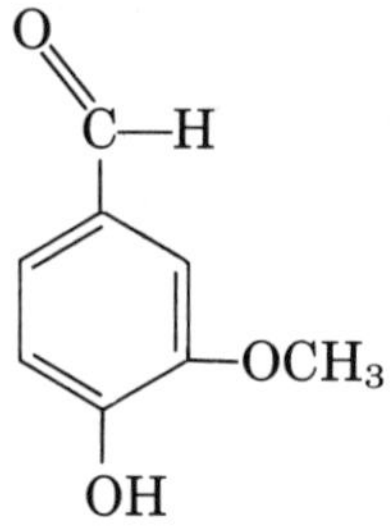

O=C–H, OCH_3, OH

vanillin

The principal flavoring constituent of the cured bean is vanillin (vanillin does not occur in the growing plant). This aromatic compound is found in a number of plants (e.g., cloves) and can be readily synthesized from coniferin and eugenol, from lignin and waste paper pulp, and from coal tar. In addition to vanillin, there are small amounts of other flavoring substances that contribute to the distinctive flavor of the extract. Coumarin, a compound with a vanilla-like flavor found in the tonka bean, *Depteryx odorata* (Aublet), has been used as an alternative for vanilla. In 1953, it was learned that coumarin apparently possessed a high level of toxicity; as a consequence its use in foods or food additives has been barred by regulations declared under the Food, Drug, and Cosmetic Act.

Vanillin, or imitation vanilla, is commercially available. However, because it is prepared from pure vanillin and vanillin acetated, it lacks certain secondary scents and minor flavor components (gums, resins, etc.) found in the natural product.

☐ ANALYSIS

Many of the common analytical determinations performed on spices are forms of proximate analysis, which was discussed in Chapter 2. In general, the addition of an adulterant to a spice is normally reflected in the gross composition of the spice. Thus, estimations of moisture, ash, nitrogen, fiber, reducing matter, starch, etc., are often sufficient to establish the purity of a spice and in many instances its conformity to the standard of identity established for it. In fact, the components most useful in detecting such adulteration are usually defined in the standard of identity of a spice.

Sample Preparation

Spice samples for analysis should be finely ground. If the material is not already finely ground, then grind a sample in a Wiley mill to pass through a 20-mesh sieve or through a sieve with 1-mm circular openings. Mix the ground material thoroughly, then store it in an air-tight container in a refrigerator. Before removing a sample for analysis, mix the ground spice thoroughly by stirring or rotating the container, since ground spices have a tendency to stratify. Remove approximately 2 g from the center of the material. The entire sample should be removed in one operation; adding or removing a portion from the balance should be avoided. When determining starch by the diastase method, the sample should be ground to an impalpable powder.

General Methods

Moisture

If normal drying procedures are applied to spices, the ether-extractable matter will volatilize and, thus, incorrect moisture values will be obtained. As a consequence, the moisture content of spices is usually determined by distillation with toluene. In this procedure, a weighed sample is covered with toluene and heated in a distilling flask. The boiling point of toluene is slightly higher than that of water. When boiling occurs, the toluene and water distill over, are condensed by a West condenser, and drop into a graduated collection tube (Bidwell–Sterling trap). The water being heavier separates to the bottom of the trap, while the toluene continuously flows back into the flask. When the distillation is completed, the volume of water in the collection tube is measured.

Procedure

Transfer a 40-g sample of the well-mixed ground spice into a 250-ml Erlenmeyer flask with standard taper joints (24/40). Add sufficient toluene, about 75–100 ml, to cover the sample completely. Insert a magnetic stirring bar and connect the flask to the side arm of a Bidwell-Sterling trap (24/40) and West condenser (24/40). Pour toluene through the condenser until the trap is filled. Place the assembled apparatus on an electric hot plate equipped with a magnetic stirrer. Heat the flask to boiling and distill at the rate of 1 or 2 drops/sec until no additional water is distilled over. Wash down the condenser with toluene by pouring toluene in the top. Dislodge any water droplets adhering to the walls of the condenser by brushing with a burette brush saturated with toluene, at the same time pouring toluene through the top of the condenser. Continue the distillation about 5 min longer, cool the trap to room temperature, and read the volume of water distilled.

Calculations

If the density of water is assumed to be 1.0, the percentage of water in the sample can be calculated as follows:

$$\% \text{ Water} = \frac{\text{Volume water (ml)} \times 1.0 \text{ mg/ml} \times 100}{\text{Wt sample (g)}}$$

Comments

Clean all glass apparatus with dichromate–sulfuric acid cleaning solution, rinse thoroughly with distilled water, then with alcohol and dry before use. The American Spice Trade Association (ASTA 1985) recommends using benzene instead of toluene for spices containing large proportions of sugars and other materials that may decompose at the temperature of boiling toluene (110°C). The boiling point of benzene is 80°C.

Total Ash

The percentage of total ash is specified in the standards of identity for many spices. The procedure given here is suitable for most spices, except nutmeg, mace, ginger, and cloves; these should be heated at 600°C rather than 550°C as specified in the procedure. For a fuller discussion of ashing, see Ash section in Chapter 2, p. 33.

Procedure

Weigh accurately an approximate 2-g sample of the well-mixed spice into a crucible. Set the crucible in the entrance of an open muffle furnace set at 550° ± 20°C so that the sample will smoke but not ignite. Then place the crucible in the muffle furnace for 30 min. Remove the crucible from the muffle furnace, cool, and add several drops of water to the ash. Carefully evaporate to dryness and again heat in the muffle furnace for 30 min. If the first wetting showed no visible carbon particles, transfer the crucible to a desiccator, cool to room temperature, and weigh immediately. Calculate the percentage of ash as follows:

$$\% \text{ Ash} = \frac{\text{Wt after ashing} \times 100}{\text{Wt before ashing}}$$

If the first wetting showed visible carbon particles repeat the wetting, evaporation, and heating steps until carbon particles are no longer visible. If carbon-free ash cannot be obtained in this manner, leach the ash with hot water, filter through a quantitative filter paper, and wash the paper thoroughly with hot water. Transfer the paper and contents to the crucible, dry and heat in the muffle furnace until the ash is white. Remove the crucible from the furnace and allow to cool. Transfer the filtrate obtained by leaching into the crucible and evaporate to dryness on a steam bath. Place the crucible in the muffle furnace for 30 min, remove, cool, and weigh as above.

Acid–Insoluble Ash

The percentage of acid–insoluble ash, like that of total ash, is specified in the standards of identity for many spices. The determination of acid–insoluble ash is performed on a sample already analyzed for total ash, as described in the previous procedure.

Procedure

Treat the total ash with 25 ml of diluted hydrochloric acid (prepared by mixing 1 part of concentrated hydrochloric acid with 2.5 parts distilled water). Cover the dish with a watch glass to prevent spattering and boil for 5 min. Filter through a Gooch crucible and wash the residue with hot water until the washings are acid free. Transfer the Gooch crucible and contents to the furnace and ignite at 550°C until it is free from carbon, cool in a desiccator, and weigh. Calculate the percentage of ash insoluble in hydrochloric acid.

Nitrogen

The total nitrogen content of spices generally is determined by the Kjeldahl procedure described in Chapter 2. Usually, a 5-g sample of well-mixed spice gives analytical values in the appropriate range.

However, because pepper and the capsicums (cayenne, red pepper, and paprika) are comparatively high in total nitrogen (protein), only 1-g samples of these spices are required.

Volatile and Nonvolatile Oils

The ether extraction method for determining volatile and nonvolatile oils is not suitable for spices high in volatile oils, such as nutmeg, mace, and the capsicums. This determination is important in evaluating the purity of spices because the taste of many spices is due to its ether-soluble components. The standards of identity for cayenne pepper, cloves, and pepper specify percentages of volatile or nonvolatile ether extract.

Procedure

Weigh a 2-g sample of the well-mixed ground spice into a paper extraction thimble. Place the thimble in a continuous extraction apparatus and extract with anhydrous ether for 20 hr. Transfer the ether solution of the extract to a tared dish and allow the ether to evaporate at room temperature until no appreciable ether odor is apparent. Place in a desiccator over concentrated sulfuric acid for 18 hr; then weigh the dish and contents to obtain the total ether extract. Heat the extract gradually and then to a minimum constant weight at 110° ± 2°C in a drying oven. The residue is the nonvolatile ether extract, and the difference between the total ether extract and the nonvolatile ether extract is the volatile ether extract.

The extracted spice residue remaining in the thimble may be used for the determination of crude fiber.

Volatile Oil

The oil obtained by steam distillation of a spice is defined as the volatile oil. The procedure described here is based on AOAC method 30.020 and the ASTA Clevenger trap method.

Apparatus

1. Volatile oil traps—Clevenger traps with T joints: (1) for oils with densities near or less than water and (2) for oils with densities greater than water (see AOAC 30.01, p. 565). These traps are available from Scientific Glass Apparatus Co., Bloomfield, NJ.
2. 1-Liter round-bottomed, short-necked flask. (29/42 T joint).
3. Heavy-duty magnetic stirrer and teflon-coated stirring bar.
4. West condenser-400 mm in length, with 24/40 T joint.
5. Electric heating mantle.

Procedure

Transfer a weighed sample of a well-mixed ground spice into a round-bottomed flask in an amount sufficient to yield 2–5 ml of volatile oil. Add water to the flask until it is slightly less than half full and mix the contents by swirling. Add a pea-sized pellet of an antifoam agent such as Dow Corning Antifoam A (or a piece of carnauba wax approximately 12 mm in diameter). Insert a stirring bar (or add some boiling chips), enclose the flask in the electric heating mantle, connect the appropriate Clevenger trap to the flask, attach the condenser to the top of the trap and place on a magnetic stirrer. Start stirrer and heating mantle. If a heating mantle is unavailable, an oil bath (hydrogenated cottonseed oil is satisfactory) can be used. Distill for at least 1 hr, stopping when no further oil is distilled for a 1-hr period. Allow flask to cool to room temperature. Water droplets adhering either to the oil layer or to the wall of the condenser may be dislodged by using a thin wire. Read the volume of oil to the nearest 0.02 ml and calculate the percentage of oil by volume. Concentration is expressed as ml of oil/100 g of spice.

Comments

Both the trap and condenser should be thoroughly cleaned with cleaning solution just before they are used. In the case of spices, (e.g., nutmeg, allspice, cloves, cinnamon) containing volatile oils lighter than water and fixed oils heavier than water, the distillation is stopped when the oil fraction obtained during 1 hr intervals is heavier than water.

If the volatile oil is to be analyzed further, transfer the oil to a glass-stoppered tube, separating it as completely as possible from water. Add a minimum quantity of anhydrous sodium sulfate to dry the oil; allow the sodium sulfate to settle before removing an oil sample for measurements of optical rotation or refractive index, or other procedures.

Alcohol Extract

The percentage alcohol (or water) extract is valuable in detecting previous extraction of a spice; that is, an exhausted spice will yield a small extract.

Procedure

Weigh 2 g of the spice sample into a 100-ml volumetric flask and fill to volume with ethanol. Stopper the flask, shake at 30-min intervals during an 8-hr period and then allow to stand for 16 hr without shaking. Filter the extract through a dry filter paper and transfer a 50-ml aliquot of the filtrate to a tared flat-bottomed, aluminum dish. Evaporate to dryness on a steam bath, then dry to a constant weight in a drying oven at 110°C. Cool and weigh the dish and contents. Calculate the result as a percentage of the original sample.

Cold-Water Extract

The percentage of cold-water extract, like that of alcohol extract, is useful in detecting previous extraction of a spice. The standards of identity for ginger specify the allowable percentage of cold-water extract. The water-soluble extract is determined in the same general way as the alcohol extract except that water, rather than alcohol, is used as the solvent.

Procedure

Place a 4-g sample in a 200-ml volumetric flask and fill to volume with water. Stopper the flask, shake at 30-min intervals during an 8-hr period, and then allow to stand for 16 hr without shaking. Filter the extract through a dry paper, then transfer a 50-ml aliquot of the filtrate to a tared flat-bottomed, aluminum dish. Evaporate to constant weight in a drying oven at 110°C. Cool and weigh the dish and contents. Express the results as a percentage of the original sample.

Crude Starch

Crude starch in spices is determined by first hydrolyzing the starch (either with acid or with the enzyme diastase followed by acid) and then measuring the resulting glucose by the Munson–Walker copper-reduction method, which was described in Chapter 4. With either method of hydrolysis, the spice sample is first extracted with ether and alcohol to remove soluble substances (e.g., simple sugars) that would react in the copper-reduction step.

The results obtained with the direct acid hydrolysis procedure are comparable to those obtained with the diastase hydrolysis procedure for spices that have an appreciable starch content and low fiber content (black and white pepper, ginger, nutmeg, and mace). However, in spices having little or no starch but appreciable amounts of fiber, tannin, and pentosans (e.g., cloves, cayenne pepper, allspice, and mustard), the direct acid values may be several times greater than the diastase hydrolysis values because these substances are also hydrolyzed by the dilute acid into copper-reducing substances.

In the following directions, each of the hydrolysis procedures is presented separately, then the details of the Munson–Walker determination are given.

Crude Starch by Direct Acid Hydrolysis

The spice sample is washed with ether and alcohol to remove soluble substances (e.g., simple sugars). The starch remaining in the sample is hydrolyzed with acid and the glucose determined by a copper reduction method. The glucose value is converted to starch value by the factor 0.925.

Reagents

1. HCl (sp gr 1.125)—Dilute 500 ml HCl to 1 liter with water.
2. 10% (w/v) NaOH solution.
3. Ethyl ether (anhydrous).
4. 10% (v/v) ethanol solution.

Procedure

Transfer a 4-g sample of well-mixed ground spice to a fine-porosity alundum crucible or to a funnel fitted with a paper of fine porosity and extract with five 10 ml portions of ethyl ether. Allow the ether to evaporate, then wash with 150 ml of 10% ethanol. (In the case of cassia, cassia buds, and cinnamon, omit the ethanol washing.)

Carefully wash the residue into a 500-ml flask with 200 ml of water, using a wash bottle, and gently rub the crucible or paper with a rubber policeman to aid in the quantitative transfer of the residue. Add 20 ml of HCl solution, attach a reflux condenser to the flask, and heat in a boiling water bath for 2.5 hr. Cool the solution and then nearly neutralize it with 10% NaOH (make alkaline to litmus with the NaOH, then add dilute HCl until just acid to litmus). Transfer this solution to a 500-ml volumetric flask, dilute to the mark, mix thoroughly, and filter the solution into a clean, dry flask.

Determine the glucose present in an aliquot of the filtrate by the Munson and Walker copper reduction procedure. The amount of glucose determined multiplied by 0.925 will give the amount of starch.

The final filtrate must be slightly acid to prevent tautomeric changes and decomposition of the sugar present. Yet, the filtrate should be only slightly acid so that no appreciable amount of the alkali in Fehling's solution will be neutralized during the copper-reduction step.

Starch by Diastase Hydrolysis

Products that contain starch along with other carbohydrate substances (e.g., pentosans) require a more complicated method of analysis. After removing the soluble carbohydrates present, the spice sample is subjected to hydrolysis by the enzyme diastase. Since diastase is specific for starch, it hydrolyzes starch to maltose. The maltose is separated by filtration from the other carbohydrate substances and is then converted to glucose via acid hydrolysis. The glucose is then determined by a copper reduction method.

Reagents

1. Malt diastase solution—Saturated solution of commercial malt diastase.
2. Ethyl ether (anhydrous).
3. 10% (v/v) Ethanol solution.

Procedure

Place a 4-g sample of well-mixed ground spice on a fine porosity alundum crucible or on a funnel fitted with a paper of fine porosity (Whatman No. 2) and extract with five 10-ml portions of ethyl ether. After the ether has evaporated, wash the residue with 500 ml of a 10% ethanol solution.

Quantitatively transfer the insoluble residue to a beaker using about 50 ml of water. Immerse the beaker in boiling water and stir constantly for about 15 min or until all the starch is gelatinized. Cool to 55°C, add 20 ml of the malt diastase solution, and incubate at this temperature for 1 hr in a water bath. Make a spot test for starch with iodine solution. If a blue color persists, repeat the malt diastase treatment.

Cool the solution, transfer to a 250-ml volumetric flask, dilute to the mark, mix thoroughly, then filter into a 500 ml flask. Add 20 ml of the HCl reagent, attach a reflux condenser to the flask, and digest for 2.5 hr in a boiling water bath. Cool, quantitatively transfer to a 500-ml volumetric flask, and nearly neutralize as described in the preceding procedure. Bring to the mark, mix thoroughly, and filter, collecting the filtrate in a clean dry flask.

Determine the glucose present in an aliquot of the filtrate by the Munson and Walker copper reduction procedure (Chapter 4). The amount of glucose determined (Table 12.6) multiplied by 0.925 will give the amount of starch present.

Munson–Walker Procedure

Pipette 25 ml of each of the Fehling's solution into a 400 ml beaker. Add to this 50 ml of the filtrate obtained either from direct acid hydrolysis or diastase hydrolysis. (If the expected starch content is more than 50%, use 35 ml of a filtrate plus 15 ml of water.) The total volume of the Fehling's solutions and sample must be 100 ml. Determine glucose present by the Munson–Walker method described in Chapter 4.

Crude Fiber

The term *crude fiber* refers to the dried residue remaining after successive treatments with boiling acid and alkali. The residue consists of cellulose, lignin, pentosans, and nitrogenous matter. The standards of identity for ginger, nutmeg, and mace specify the percentage of crude fiber allowable.

The crude fiber determination is of value in detecting the adulteration of spices with waste materials from the spices themselves (outer coatings, stems, etc.). When such materials are present, the crude fiber content is greater than in the unadulterated spice. Other adulterants such as nut shells, sawdust, and in some cases inferior varieties of the spices may also be detected by a higher crude fiber content.

Reagents and Materials

1. 0.255 *N* H_2SO_4—Dilute 1.25 g concentrated H_2SO_4 to 100 ml.
2. 0.312 *N* NaOH—Dissolve 1.25 g of carbonate-free NaOH in water and make up to 100 ml.
3. Erlenmeyer flasks (750 ml in capacity).
4. Filter cloths—Cut from butcher's linen or dress linen with approximately 45 threads to the inch, or use No. 40 filter cloth made by the National Filter Media Corp., New Haven, CT.
5. Prepared Gooch crucible.

Procedure

Carefully transfer the dried residue remaining in the extraction thimbles after the extraction of volatile and nonvolatile oils is carefully transferred to a 750-ml Erlenmeyer flask. Add 200 ml of boiling H_2SO_4 solution and one drop of diluted antifoam. Connect the flask with a reflux condenser and digest the sample at the boiling temperature for exactly 30 min. (The hot plate should be pre-adjusted so that the contents of flask will come to a boil within 1 min.) Rotate the flask at 5-min intervals to insure thorough wetting and mixing of the sample. The solid matter has a tendency to adhere to the sides of the flask, out of contact with the solution, and this must be avoided.

At the completion of the digestion period, remove the flask and filter the contents immediately through a linen filter supported in a fluted funnel, using a suction flask to speed filtration. Wash the residue with boiling water until free from acid. Bring a quantity of the NaOH solution to boiling and keep it at this temperature under a reflux condenser until used. Transfer as much as possible of the washed residue on the linen filter back to the Erlenmeyer flask by means of a spatula. Then spread out the filtering cloth in a wide-stemmed funnel inserted in the Erlenmeyer flask; wash the remainder of the residue into the flask by means of 200 ml of the boiling NaOH solution. A wash bottle marked to deliver 200 ml of the NaOH solution is helpful in effecting this transfer.

Immediately connect the flask to the reflux condenser and boil for 30 min with frequent rotation. At the end of boiling period, remove the flask and immediately filter the contents through a tared Gooch crucible. Wash the contents thoroughly with water and then with 15 ml of ethanol.

Dry the crucible and contents at 110°C until a constant weight is attained (about 1 hr). Cool in a desiccator and weigh. Then ignite the crucible and contents in a muffle furnace at a dull red heat (approximately 20 min). Cool in a desiccator and weigh.

Calculations

The loss in weight on incineration represents the weight of crude fiber in the sample.

$$\%\ \text{Crude fiber} = \frac{\text{Loss in weight (g)} \times 100}{\text{Wt or original sample (g)}}$$

Methods for Analyzing Specific Spices

In addition to the general analytical procedures described in the previous section, many methods applicable to specific spices have been developed. Several of these are described in this section.

Extractable Color in Paprika

Color is an important criterion of quality for paprika. It is bright red in color, used widely in the home for light-colored foods and is used in the food industry in the preparation of catsup, sauces, etc. Since paprika is used to impart color to a food, the spice assumes the role of a natural dye. Therefore, the food processor must be able to make an accurate estimation of color value to insure quality control of color in the finished product.

In this method (ASTA method 20.1), the color in paprika is extracted with acetone, and the absorbance measured at 460 nm. A standard color solution is used as a spectrophotometric check. The extractable color value for paprika is 120–130 ASTA units.

Reagents

1. 1.8 M H_2SO_4—Add 100 ml of concentrated sulfuric acid (sp gr 1.84) to approximately 500 ml distilled water, mix, then dilute to 1 liter.
2. Standard color solution—Dissolve 0.3005 g potassium dichromate and 34.96 g dried cobaltous ammonium sulfate in 1.8 M H_2SO_4 and dilute to 1 liter with 1.8 N H_2SO_4.
3. Acetone.

Procedure

Accurately weigh a sample of ground paprika into a 100-ml volumetric flask, dilute to volume with acetone, and stopper tightly. Shake the flask and allow to stand for 4 hours in the dark. Shake and allow to stand for 12 hours in the dark. Shake and allow to stand for 2 min for the particles to settle. Transfer a portion of the extract to an absorption cell (1-cm square matched cells with stopper) and determine the absorbance (A) at 460 nm using acetone as the blank. Determine A of the standard color solution at 460 nm using 1.8 M sulfuric acid as the blank.

Calculations

1. Instrument and cell length correction factor $= I_f$. $I_f = 0.0600/A_S$ where A_S = absorbance of the standard color solution at 460 nm. Determine A_S daily if spectrophotometer is left on.
2. Extractable color:

$$\text{ASTA color value} = \frac{A \text{ at 460 nm} \times 16.4 \times I_f}{\text{Sample wt (g)}}$$

The range of A should be 0.30 to 0.70. If the extract has an absorbance value >0.70, dilute with acetone to one-half the original concentration. If $A < 0.30$ repeat extraction using larger sample.

Nonvolatile Methylene Chloride Extract of Pepper

The United States Government Standards require that black pepper (*Piper nigrum*) contain not less than 6.75% of nonvolatile ether extract. The methylene chloride extraction method has been designated as the official method for determining whether or not a black pepper meets the standards.

Procedure

Weigh a 2-g sample of well-mixed ground pepper into a paper extraction thimble or an alundum crucible of porosity RA 360. Place the thimble in a continuous extraction apparatus and extract with methylene chloride for 20 hr. Transfer the extract to a tared aluminum dish and allow it to evaporate at room temperature, in a forced draft hood, until no methylene chloride is apparent. Then heat the extract to minimum weight at 110°C. The residue is the nonvolatile methylene chloride extract.

Calculate the nonvolatile methylene chloride extract as follows:

$$\% = \frac{\text{Wt of residue (g)}}{\text{Wt of sample (g)}} \times 100$$

Crude Piperine in Pepper

The principal pungent compound in pepper is piperine, an amide. The alkaloid is not volatile and exerts its effects on oral tissues. The Kjeldahl method is used to determine the nitrogen content (crude piperine) of the nonvolatile methylene chloride extract of pepper.

Apparatus and Reagents

1. Kjeldahl digestion and distillation apparatus.
2. Concentrated H_2SO_4 (sp gr 1.84), reagent grade.
3. Sodium sulfate (nitrogen free).
4. Mercuric oxide (nitrogen free).
5. Oxidation catalyst (0.7 g mercuric oxide, 0.3–0.5 g copper sulfate, 0.25 g selenium, or selenized granules).
6. Saturated boric acid solution.
7. Mixed indicator solution, 0.1 g methyl red and 0.5 g bromocresol green in 100 ml alcohol.
8. Metallic zinc.
9. NaOH solution—Dissolve 450 g of NaOH pellets in water and dilute to 1 liter (sp gr should be 1.36 or higher).
10. 0.1 *N* HCl (standardized).

Procedure

Dissolve the residue from the methylene chloride extraction of pepper in warm chloroform and transfer quantitatively into an 800-ml Kjeldahl flask. Remove the chloroform by evaporation on a steam bath. Add the oxidation catalyst, 15–18 g sodium sulfate, 25 ml of concentrated sulfuric acid, and a few boiling chips. Place the flask in an inclined position and heat carefully over low heat until the mixture no longer froths; then increase the heat so that the contents of the flask boil briskly. Digest the mixture until the solution is clear and continue the digestion for an additional 30 min to insure complete oxidation. Cool the digest, then add about 250 ml water and shake to dissolve the solid sulfates and cool below 25°C.

Transfer 25 ml of saturated boric acid solution into a 500-ml Erlenmeyer flask, add 3–5 drops of the mixed indicator solution, and place the flask under the condenser of the distilling unit; making sure that the tip of the condenser extends beneath the surface of the boric acid solution.

Hold the Kjeldahl flask containing the diluted and cooled digest in an inclined position and pour 80 ml of the NaOH solution into the flask so that it does not mix with the acid solution. Add a few pieces of metallic zinc to the flask and connect the flask to the condenser. Ignite the burner under the flask and quickly mix the contents of the flask by shaking, using a rotary motion. Distill at least 150 ml of liquid, remove the receiving flask so that the tip of the condenser tube is above the liquid in the flask, and wash off the condenser tube with distilled water. Then turn off the flame under the distilling flask and remove the receiving flask for titration.

Titrate the ammonia collected in the receiving flask with standard 0.1 *N* hydrochloric acid until the color changes from blue-gray to gray or pink. Perform a blank determination on the reagents.

Calculations

The sample titration minus the blank titration represents the volume of standard acid solution neutralized by the ammonia from the sample. The percentage nitrogen and percentage piperine are calculated as follows:

$$\%\ \text{Nitrogen} = \frac{(A-B) \times N \times 0.014}{\text{Wt of sample (g)}} \times 100$$

$$\%\ \text{Piperine} = \frac{(A-B) \times N \times 0.2853}{\text{Wt of sample (g)}} \times 100$$

where A = ml of acid for titration of sample; B = ml of acid for titration of blank; and N = normality of HCl.

Alcohol in Flavoring Extracts

The estimation of alcohol content in flavoring extracts is of value to the quality control-minded food processor. For example, vanilla extract is extracted from the vanilla bean, with or without the addition of sugar, dextrose, or glycerol. It contains in each 100 ml the soluble matters from not less than 10 g of vanilla bean. The finished extract should contain at least 35% alcohol by volume to keep the soluble matter in solution. Since the specific gravity of ethanol is considerably lower than that of water, the presence of ethanol in an aqueous solution reduces the specific gravity. In this method for determining the amount of ethanol in flavoring extracts, an appropriate quantity of the sample is codistilled with water. The specific gravity of the distillate, which contains only alcohol and water, is determined from a comparison of weights of equal volumes of distillate and water at t°C. The percentage of alcohol in the sample can be obtained from a standard alcoholometric table.

Procedure

Pipette 50 ml of the sample into 100 ml of recently boiled distilled water contained in a 500-ml round-bottomed flask. Add a few glass beads, then attach a condenser to the flask and distill with moderate heat into a 100-ml volumetric flask nearly to the mark; make to volume with water and mix. Cool the distillate to 10°–12°C.

Obtain the weight of a thoroughly clean and dried pycnometer. Completely fill the pycnometer with recently boiled distilled water that has been cooled to 10°–12°C. Force the stopper into the pycnometer and place the bottle in a water bath held at 15.6°C. After the pycnometer and contents have attained the temperature of the bath, quickly wipe off the drop of water at the top of pycnometer with a piece of filter paper. Remove the pycnometer from the bath and dry thoroughly with a cloth. Place the cap over the stopper as soon as possible to reduce the amount of evaporation, and weight to 0.1 mg. Dry pycnometer. Fill the pycnometer with the distillate to be tested, adjust to the correct temperature in a water bath, dry, and weigh.

Calculations

Specific gravity of distillate

$$= \frac{\text{Wt pycnometer and distillate at } t^0 - \text{Wt dry pycnometer}}{\text{Wt pycnometer and water at } t^0 - \text{Wt dry pycnometer}}$$

Obtain from Table 9.1 the percentage of alcohol by volume in the distillate at 15.56°C (60°F) corresponding to the measured specific gravity. If a 100-ml sample were used, the percentage of alcohol in the sample would be determined directly. However, since a 50-ml sample was used, it is necessary to multiply by 2 to obtain the percentage alcohol in the sample.

Table 9.1. Percentages by Volume of Ethyl Alcohol Corresponding to Relative Specific Gravity at 20° and 25°C Temperatures

% Vol of ethanol	Relative sp gr in air		% Vol of ethanol	Relative sp gr in air	
	At 20°/20°	At 25°/25°		At 20°/20°	At 25°/25°
0.00	1.0000	1.0000	44.00	0.9429	0.9403
1.00	0.9985	0.9985	45.00	0.9411	0.9385
2.00	0.9971	0.9970	46.00	0.9392	0.9366
3.00	0.9956	0.9956	47.00	0.9374	0.9348
4.00	0.9942	0.9941	48.00	0.9355	0.9329
5.00	0.9928	0.9927	49.00	0.9336	0.9309
6.00	0.9915	0.9914	50.00	0.9316	0.9289
7.00	0.9902	0.9901	51.00	0.9296	0.9269
8.00	0.9889	0.9888	52.00	0.9276	0.9248
9.00	0.9877	0.9875	53.00	0.9256	0.9228
10.00	0.9864	0.9863	54.00	0.9235	0.9207
11.00	0.9853	0.9850	55.00	0.9214	0.9186
12.00	0.9841	0.9838	56.00	0.9193	0.9164
13.00	0.9829	0.9826	57.00	0.9172	0.9142
14.00	0.9818	0.9814	58.00	0.9150	0.9120
15.00	0.9807	0.9802	59.00	0.9128	0.9098
16.00	0.9795	0.9790	60.00	0.9105	0.9076
17.00	0.9784	0.9779	61.00	0.9083	0.9053
18.00	0.9774	0.9767	62.00	0.9060	0.9030
19.00	0.9763	0.9756	63.00	0.9037	0.9007
20.00	0.9752	0.9744	64.00	0.9014	0.8983
21.00	0.9742	0.9733	65.00	0.8990	0.8959
22.00	0.9731	0.9722	66.00	0.8967	0.8936
23.00	0.9720	0.9708	67.00	0.8943	0.8912
24.00	0.9709	0.9698	68.00	0.8918	0.8887
25.00	0.9697	0.9686	69.00	0.8894	0.8863
26.00	0.9686	0.9673	70.00	0.8869	0.8838
27.00	0.9674	0.9661	71.00	0.8844	0.8812
28.00	0.9662	0.9648	72.00	0.8819	0.8787
29.00	0.9650	0.9635	73.00	0.8793	0.8761
30.00	0.9638	0.9622	74.00	0.8767	0.8735
31.00	0.9626	0.9609	75.00	0.8741	0.8709
32.00	0.9613	0.9595	76.00	0.8715	0.8682
33.00	0.9600	0.9581	77.00	0.8688	0.8655
34.00	0.9586	0.9567	78.00	0.8661	0.8628
35.00	0.9572	0.9552	79.00	0.8633	0.8600
36.00	0.9557	0.9537	80.00	0.8605	0.8573
37.00	0.9543	0.9521	81.00	0.8577	0.8544
38.00	0.9527	0.9505	82.00	0.8549	0.8516
39.00	0.9512	0.9489	83.00	0.8520	0.8487
40.00	0.9496	0.9473	84.00	0.8491	0.8457
41.00	0.9479	0.9456	85.00	0.8461	0.8428
42.00	0.9463	0.9439	86.00	0.8431	0.8398
43.00	0.9446	0.9421	87.00	0.8400	0.8367

Table 9.1. (*Continued*)

% Vol of ethanol	Relative sp gr in air		% Vol of ethanol	Relative sp gr in air	
	At 20°/20°	At 25°/25°		At 20°/20°	At 25°/25°
88.00	0.8369	0.8335	95.00	0.8126	0.8092
89.00	0.8337	0.8303	96.00	0.8087	0.8052
90.00	0.8305	0.8271	97.00	0.8045	0.8011
91.00	0.8271	0.8237	98.00	0.8002	0.7968
92.00	0.8237	0.8203	99.00	0.7955	0.7921
93.00	0.8201	0.8167	100.00	0.7905	0.7871
94.00	0.8164	0.8130			

Source: Based on National Bureau of Standards (1913).

Vanillin Content

Vanillin is the principal flavoring component in vanilla extract. The absorbance of alkaline vanillin solutions at 348 nm is the basis of the following method (AOAC 19.011 and 19.012) for determining it.

Preparation of Standard Curve

Dissolve 100 mg of pure vanillin in 5 ml of ethanol and dilute to 100 ml with water and mix. Transfer 5, 10, and 15 ml to 250-ml volumetric flasks, dilute to volume with water, and mix. Pipette 10 ml of each of these solutions into a 100-ml volumetric flask, add 2 ml of 0.1 *N* NaOH and approximately 80 ml water, mix, dilute to volume with water and mix again. Read absorbances of the alkaline standard solutions at 348 nm, using neutral standard solutions as reference blanks. Plot a standard curve.

Assay Procedure

If the sample to be assayed contains more than 300 mg vanillin per 100 ml, dilute with 35% alcohol to below this level. Pipette 10 ml of the sample into a 100-ml volumetric flask, dilute to the mark with water, and mix. Pipette 2 ml of this solution into each of two 100-ml volumetric flasks. Dilute one to volume with water and mix (neutral blank solution). To the other volumetric flask add 80 ml water and 2 ml 0.1 *N* NaOH, mix, dilute to volume, and again mix. Determine the absorbances at 348 nm, using the neutral solution as a reference blank. Determine the vanillin content from the standard curve.

Wichman Lead Number of Vanilla Extract

Imitation vanilla extracts are commercially available. These are prepared from vanilla, vanillin, and ethyl vanillate and are colored with caramel or coal-tar dye. Imitation extracts are cheaper but they lack the delicate flavor of the genuine extract in which minor components also play a part.

Vanilla extract contains organic and inorganic acids that form insoluble lead salts when treated with a neutral solution of lead acetate. In AOAC method 19.028, given here, these acids are precipitated with an excess of neutral lead acetate; the excess lead is precipitated with H_2SO_4 and determined gravimetrically. This is an empirical method, and the results are meaningful if the procedure is followed carefully. Lead numbers for vanilla extract range from 0.40 to 0.74 with an average of 0.54.

Reagents

1. Lead acetate solution—Dissolve 8 g lead acetate in 100 ml water; filter if not clear.
2. H_2SO_4 solution—Add concentrated H_2SO_4 to an equal volume of distilled water.

Procedure

Add 25 ml lead acetate solution, 175 ml boiled distilled water, and 50 ml of the singlefold extract to a 1-liter round bottom flask. Swirl to mix thoroughly and precipitate lead salt. Attach a condenser to the flask and distill, with moderate heat, until approximately 200 ml of distillate has been collected. (The distillate may be retained for the determination of alcohol by specific gravity.) Transfer the material remaining in the distillation flask quantitatively to a 100-ml volumetric flask and dilute to volume with CO_2-free distilled water. Mix, then filter. The filtrate contains the lead remaining after formation of the lead-salt complex. Conduct a blank determination substituting 5 drops of acetic acid for the sample and collecting only 150 ml of distillate.

Precipitate the soluble lead in the sample and blank filtrates in the following manner. Pipette 10 ml of each of the filtrates into separate 250-ml beakers, add 25 ml distilled water, 2 ml H_2SO_4 solution, and 100 ml of ethanol. Stir and allow to settle overnight. Filter with suction through a tared Gooch crucible (M porosity) and wash with ethanol. Ignite in a muffle furnace at 525°–550°C, cool in a desiccator, and weigh.

Calculate the lead number by subtracting the weight of lead sulfate in the sample from that of the lead sulfate in the blank determination and multiplying by 13.66.

Ash in Vanilla Extract

The procedure for determining ash in a flavoring extract is slightly different from the method for solid samples presented earlier. The directions given here are based on AOAC method 31.012.

Procedure

Measure 10 ml of extract into a 50- to 100-ml platinum dish and evaporate. Add a few drops of pure olive oil and heat slowly over a flame until the swelling stops. Transfer the dish to a muffle furnace at 525°C and heat until a white ash is obtained. Remove the dish from the muffle furnace, cool, and moisten the ash with a few drops of water. Dry on a steam bath, then on a hot plate, and finally reheat in the muffle furnace at 525°C to constant weight. Calculate the ash on the basis of grams per 100 ml of extract rather than on a percentage basis.

☐ SELECTED REFERENCES

ASTA. 1985. Official Analytical Methods. 3rd ed. Amer. Spice Trade Assoc., Englewood Cliffs, NJ.

AOAC. 1984. Official Methods of Analysis. 14th ed. S. Williams (Editor). Assoc. of Official Analytical Chemists, Arlington, VA.

FARRELL, K. T. 1985. Spices, Condiments, and Seasonings. AVI Publishing, Westport, CT.

HART, F. L. and FISHER, H. J. 1971. Modern Food Analysis. Springer-Verlag, New York.

NATIONAL BUREAU OF STANDARDS. OSBORNE, N. S. 1913. Density of ethyl alcohol and of its mixture with water. NBS Bulletin *9*, 424–425.

PARRY, J. W. 1969. Spices. Vols. I and II. Chemical Publishing Co., New York.

ROSENGARTEN, F. 1969. The Book of Spices. Livingston Publishing Co., Philadelphia.

UNITED STATED FEDERAL REGISTER. 1977. Code of Federal Regulations, Title 21. Washington, DC.

10

Coloring Agents and Color of Foods

☐ INTRODUCTION

Color is unconsciously incorporated into our impressions of foods because our eyes are constantly supplying the brain with color information that is instinctively associated with all other information about any given food. Radishes are "red," bananas are "yellow," and lettuce is "green." In fact, we associate a particular color with almost every food we come in contact with in our daily lives. Thus, color is a significant factor in consumer acceptance of foods, and consumers automatically associate certain color characteristics with fresh and wholesome quality. This evident characteristic of foods is included in the quality standards for fresh and processed fruits and vegetables, fats and oils, meats, dairy products, poultry, and eggs. In the processing, preservation, and storage of some food products, the "natural color" may be destroyed. This fact has spurred food processors to develop methods for retaining the best possible color in foods.

The grower of foodstuffs is in a position to exert an influence on the color of some products because hereditary varietal differences, maturity, and growing conditions (moisture, temperature, locality) have an effect on color. By selecting certain varieties of plants, the grower determines to some extent the color of his product. Also, by harvesting his crop at the proper stage of maturity, he can ensure the desired natural color of the raw product. Similarly, the producer of animal products can exert an influence on some products. For example, the type of feed fed to dairy animals and poultry will influence, to some degree, the yellowness of cream in cow's milk and yolk color of eggs.

The food processor has control of the natural color of a food only in so far as he can select his raw material. The extent to which this original natural color is retained during processing and subsequent storage is one important characteristic of a processing procedure. To minimize the loss of color requires close supervision of those stages in normal processing that can affect the color of a foodstuff. For example, products that are to be frozen or dehydrated are subjected to a blanching step in order to inactivate enzymes which could affect the color. Special care is taken to prevent overheating because excessive heat is conducive to color loss. Fruits to be dehydrated are frequently sulfured prior to dehydration to prevent formation of dark-colored products. The presence of trace metals (e.g., copper, iron) causes discoloration in some canned products.

Some natural food products (cakes, cookie fillings) are colorless or lightly colored. Other prepared food products, such as gelatin desserts, pudding mixes, drink powders, pectin jellies, candies, and carbonated beverages, are colorless as such. Since we associate a color with many of these latter products (e.g., an orange-flavored gelatin dessert, a synthetic mint jelly, or a grape-flavored soft drink), they often are artifically colored.

Before discussing the coloring agents that are used in foods and their analysis, we present an overview of the nature of light and color. This is a vast and complicated subject; only a rather brief summary can be given here.

☐ NATURE OF LIGHT AND COLOR

There is a very broad spectrum of electromagnetic radiant energy, of which visible radiant energy (light) is a narrow band (Fig. 10.1). At one end of the spectrum are cosmic rays and at the opposite end are electrical power waves. The most common way to specify visible light is in terms of wavelength, while in other parts of the spectrum, it is customary to express radiant energy in frequencies (number of wave cycles per second). The average wavelength of cosmic rays, is 1×10^{-5} nm (1 in. = 25.4 million nm); at the long end of the spectrum are electric power waves with an average wavelength of 3100 miles. The spectrum of electromagnetic radiant energy we call *light* ranges from

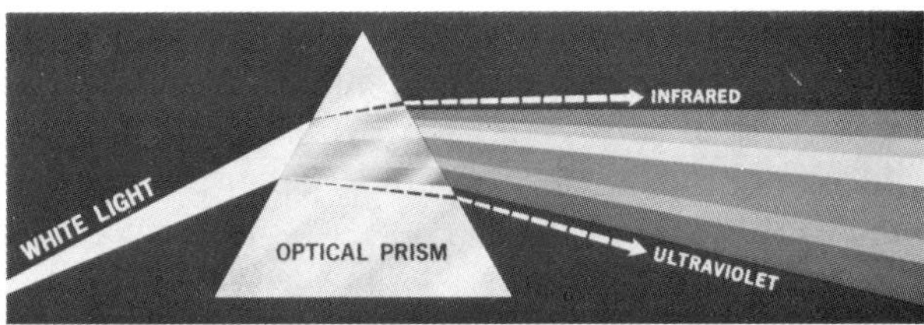

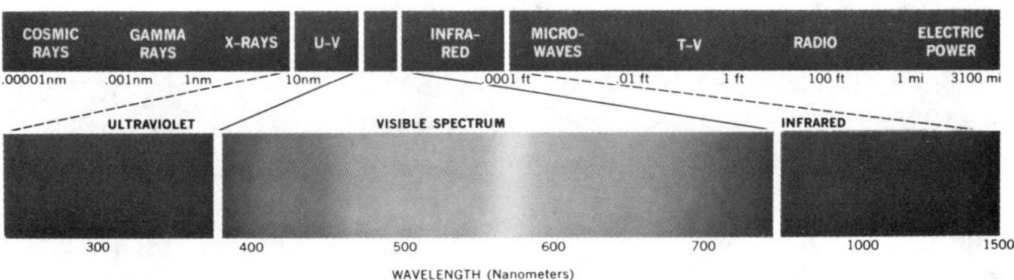

Figure 10.1. The spectrum of visible radiant energy covers just a small portion of the entire electromagnetic spectrum.

approximately 380 to 760 nm. Wavelengths shorter or longer than these do not stimulate the receptors in the eye (rods and cones) and, as a consequence, are not capable of initiating responses in the eye.

The sensation of color is due to an imbalance in the spectral distribution of visible radiant energy reaching the eye from light sources and objects. If the spectral distribution were equal, the light would be perceived as white. This unequal spectral distribution may be characteristic of the source of radiant energy, it may be the result of selective absorption by the object, or it may be related to the eye of the individual who observes the color. The "colors" composing white light can be easily demonstrated by passing white light through a prism; the emerging light is separated into different hues—(red, orange, yellow, green, blue, indigo, and violet). The number of imbalanced combinations of visible radiant energy is nearly limitless, which accounts for the plethora of terms we use to describe the color of objects.

A colored light source radiates more energy at certain wavelengths than others, while a white light source generally radiates similar amounts of energy at all visible wavelengths. Similarly, a colored object reflects or transmits some wavelengths more readily than other wavelengths. In either case, the color has both a qualitative and a quantitative characteristic. The qualitative characteristic relates to the wavelengths of visible radiant energy that are present; these are specified by the dominant wavelength and purity. The quantitative characteristic refers to the amount of visible radiant energy that is present at each wavelength, which is specified by luminance. These characteristics correspond in a general way to hue, saturation, and brightness—the three attributes by which color sensation is described.

The dominant wavelength is the wavelength of spectrally homogeneous radiant energy that appears to be most abundant. Purity is an expression of the proportion of the spectrally pure component in the mixture matching the chromaticity of the sample, or it is the percentage of color versus the percentage of white in any color. Luminance is the ratio of light leaving an object to that incident upon it. Chromaticity, which refers to the characteristics specified by dominant wavelength and purity, is discussed in more detail later.

To establish that color is the result of an imbalance of visible radiant energy, consider two objects illuminated with light from a perfectly balanced white light source. One object reflects light from the middle of the spectrum to a greater extent than that at the ends—it appears green. Another object reflects half of the energy at all wavelengths of the visible spectrim—it appears gray and produces no color sensation (all waves are still present, though only half as intense). Thus, the color perceived by an observer is due to a deficiency or an imbalance of radiant energy at individual wavelengths.

As noted already, the sensation of color involves three entities—the light source, the object, and the observer. Although the nature of the light source and the object determine the amount and spectral distribution of the radiant energy reaching the eye, the perception of color in terms of hue, saturation, and brightness resides in the observer. Thus, if a photocell rather than the human eye is the receptor of radiant energy, the measuring instrument does not record any sensation of color but only the amount of energy received at each particular wavelength. The observer, in effect, integrates the imbalanced spectral distribution of radiant energy that it receives into one sensation of color. No instrument can do this, which accounts for much of the difficulty in the measurement and specification of color.

Color in Light Sources

It is apparent from the previous discussion that colored light sources radiate much more energy at some wavelengths than others, while "white" light sources generally radiate energy equally at all visible wavelengths. Some light sources may be deficient in radiant energy at several wavelengths and still emit what is considered "white" light. This deficiency will affect the perception of colors and color differences.

Most people probably have noticed that the color of an object appears different depending on the light source illuminating the object. For example, daylight emphasizes "cool" colors (blue, green), whereas incandescent lights emphasize "warm" colors (red, yellow). As a rule, all incandescent and some fluorescent lamps produce light that tends to be strong in the red, yellow, or orange portion of the spectrum. There are, however, "cool" sources of light, such as clear mercury lamps and some fluorescent lamps, that produce light strong in the blue and green wavelengths. Some light sources have been manufactured with only one predominant color to achieve a specifically desired effect. The red lamps frequently used in taste panel study areas or in photographic dark rooms are examples.

Thus, different light sources emit colored light, i.e., light with an imbalanced spectral distribution. When light sources illuminate an object, the object can transmit or reflect only those wavelengths that are present in light. Because of this, the source of illumination affects the color of an object. In other words, the eye cannot perceive colors from an object unless the source emits light in the visible portion of the spectrum.

Color in Objects

It should now be apparent that for an object to appear colored it must somehow modify the light striking it to produce an unbalanced spectral distribution of radiant energy, which the eye perceives as color. There are two fundamental ways in which objects (and media) modify the radiant energy of a beam of white light—transmission and reflection. Because objects and media are selective with respect to the wavelengths of light they transmit or reflect, they are perceived as being colored. The spectral transmission diagram in Fig. 10.2a shows that a polished blue glass filter (e.g., a Kodak Wratten 49) will selectively absorb all wavelengths except blue from a light beam passing through it, thus the filter appears blue. To make CIE standard illuminants (B and C), light from a gas filled tungsten incandescent lamp at a color temperature of 2848°K is filtered through a cell containing an aqueous solution with specified levels of copper sulfate, mannite, pyridine, and sulfuric acid. The color and intensity of the light transmitted is dependent on the molecular composition of the substances through which the light passes.

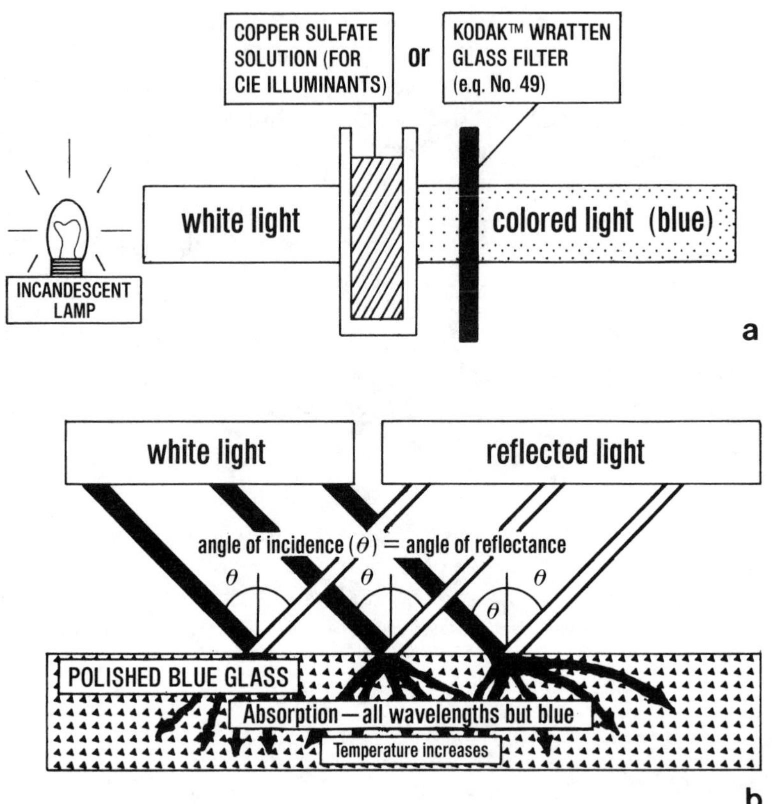

Figure 10.2. Color in objects. (a) Selective spectral transmission. (b) Spectral reflectance. *Note*: If the surface is smooth (a mirror), light will be reflected in one direction (spectral reflection); if surface is rough (e.g., blue paint) light will be uniformly reflected in all directions (diffuse).

Likewise, when a beam of light strikes the rough surface of an object, radiant energy is reflected in all directions, but only after it has been modified by the absorption characteristics of the surface of the object (Fig. 10.2b). Stated another way, the absorption of specific wavelengths of radiant energy by an object results in an imbalance of visible radiant energy reaching the eye; as a consequence, the object appears to be the color corresponding to the wavelengths that are selectively reflected.

One must conclude, therefore, that objects have color because of their ability to selectively transmit, reflect, or otherwise modify light.

Color Perception by an Observer

In the discussion so far, light has been defined primarily as an imbalance of radiant energy reaching the eye from light sources and physical objects. However, color also involves human interpretation of the neural impulses transmitted from the eye to the brain following stimulation of the receptors by the various wavelengths of light.

Perhaps color might be defined as "the perception resulting from the interaction of light source, object, eye, and brain." It is apparent that this definition of color requires the viewer to have normal color vision; the viewer cannot be color-blind, for the color-blind person cannot distinguish between various wavelengths of light. There are several types of color blindness. Usually, one has a weak red-green response. Total color blindness is rare, but to the totally color-blind person, objects are either black, white, or shades of gray. It is of interest to note that the wavelengths of light received by either a person with normal color vision or a person with color-blind vision are not changed by the receptors of the eye, rather it is the interpretation by the brain of what is seen that is different. Thus, color perception involves physics, physiology, and psychology.

☐ SPECIFICATION AND MEASUREMENT OF COLOR

Standards for many food products include a reference to color. Among these food products are butter, cheese, eggs, fruit and vegetables (fresh, canned, frozen, and dried), honey, maple syrup, cereal grains, and meats. The standards may be stated in terms of values for selected quality factors; for others, they may be stated by word description; and for some, standards are prepared in physical form. For example, the color of butter may be described as very light, light, medium, or high. Cheese may be uncolored, medium-colored, uniform, or high-colored. The yolks of eggs may vary from a deep reddish brown to a pale yellow. Citrus fruits are generally described in terms of percent color. Canned fruits and vegetables are generally graded on the basis of U.S. grades, which include color as a factor.

Thus, the specification of color is of great importance to the food industry for two reasons. First, people's reactions to color affect their individual preferences for food. Second, establishing and assuring compliance with standards of color for foods requires precise methods for measuring and specifying colors. Presumably, every colored object is susceptible to colorimetric specification, provided the color is stable long enough to be measured. Color measurements can be categorized into two groups: (1) those dealing with the determination of color via transmittance (e.g., how red is a red wine?); and (2) those dealing with opaque substances by reflectance (e.g., how red is a red apple?). Before we describe two common systems of color specification, the general relationships between colors are discussed.

Primary and Secondary Colors

The primary colors of light are red, green, and blue. These colors can be added to produce the secondary colors of light—magenta (red plus blue), cyan (green plus blue), and yellow (red plus green). A secondary color of light mixed in the right proportions with its opposite primary color will produce white light. For example, green and magenta, blue and yellow, and red and cyan are complementary colors of light. Thus, a mixture of these colors will yield white light. It is important to remember that light colors are additive; for example, one adds radiant energy in the visible spectrum (red and green) to produce the secondary color yellow, which when mixed with its opposite primary (blue) produces white light. These relationships are illustrated in Fig. 10.3, top.

In pigments, the opposite effect occurs. A primary pigment color is one that absorbs a primary color of light and reflects or transmits the other two (Fig. 10.3, bottom). Therefore, primary pigment colors (magenta, cyan, yellow) are subtractive in nature. Stated another way, the primary colors of light (green, red, blue) are the secondary colors in pigments. For example, a magenta filter absorbs green (transmits red and blue) and a cyan filter absorbs red (transmits blue and green). The two filters together transmit only blue, having subtracted the other two primary colors from the white light. If three pigment filters are superimposed, all light is absorbed, yielding black. It should be noted that the complementary pigment colors are the same as the complementary light colors—magenta and green, cyan and red, yellow and blue.

When black and white pigments are added to colored pigments, tints, shades, and tones are produced. The relationships are represented as a color triangle (Fig. 10.4). Tints are made by adding white to a pigment color (hue) whereas adding black to a pigment produces a shade. Gray is produced by mixing black and white pigments. A tone is produced when gray is added to a color pigment.

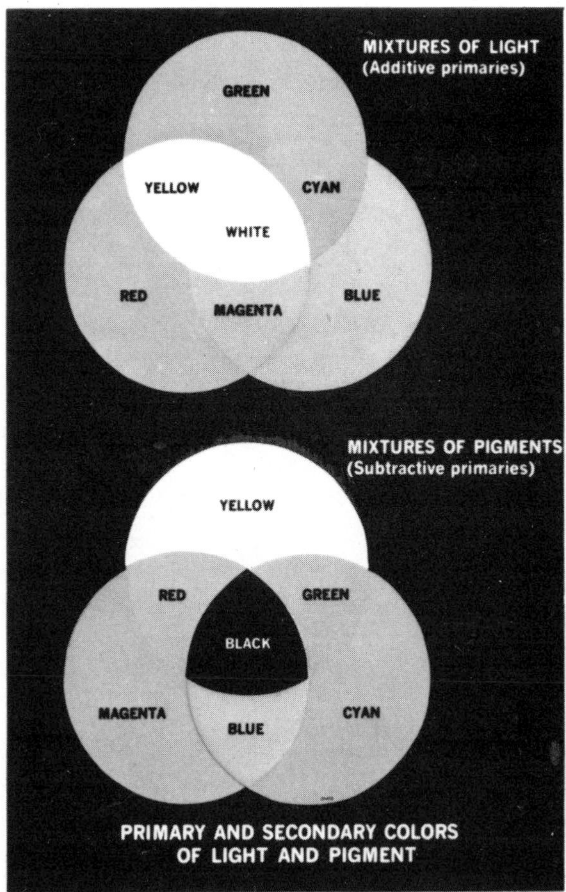

Figure 10.3. The primary light colors (*top*) are additive to give the secondary colors (e.g., red plus blue gives magenta), whereas the primary pigment colors (*bottom*) are subtractive (e.g., a magenta filter absorbs green and transmits red and blue).

Munsell Color System

In the Munsell system of notation, color is considered to have three dimensions (value, hue, and chroma), which can be represented spatially as a solid (Fig. 10.5). The central vertical axis (*value*) represents changes in gray from 0 (black) at the bottom to 10 (white) at the top. The colors of the spectrum are divided into 10 major *hues* as follows: red (R), yellow red (YR), yellow (Y), green yellow (GY), green (G), blue green (BG), blue (B), purple blue (PB), purple (P), and red purple (RP).

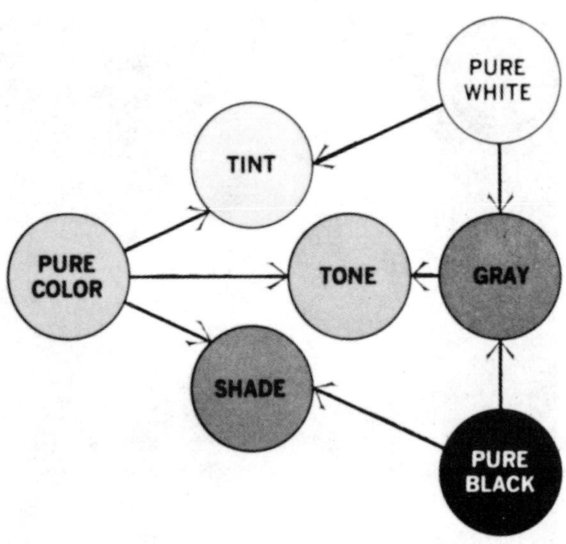

Figure 10.4. A color triangle illustrates the relationships among the tints, tones, and shades of a color.

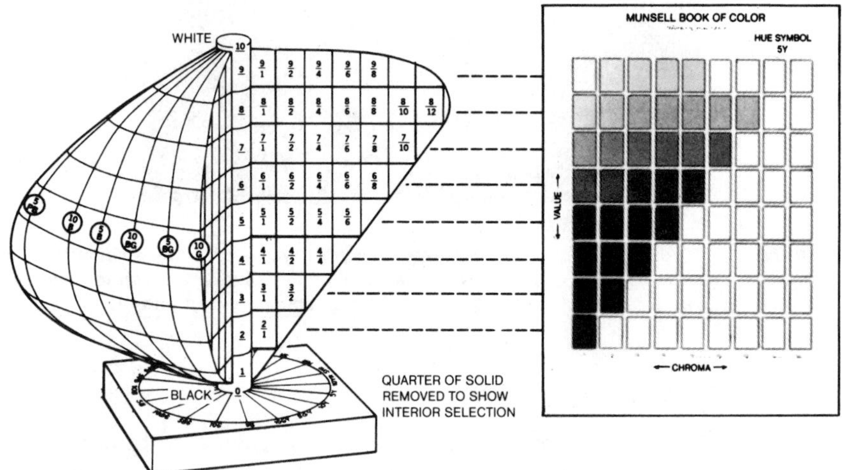

Figure 10.5. A Munsell color solid consists of scales for hue, value, and chroma. A sample color hue chart is shown on the right. A very grayed (low value) yellow would be identified by the notation 5Y 3/2.

Courtesy of Munsell Color, 2441 N. Calvert Street, Baltimore, MD 21218.

The hues are represented as vertical pie sections of the solid with the purest colors located around the equator. The purity of a color (*chroma*) is measured by the perpendicular distance from any point to the vertical axis. The purity of a color reflects its degree of difference from a gray of the same value. The values of purity range from 0 for gray (at the vertical axis) to 10 or more for the strong colors (at the periphery).

According to this system, each color can be specified by a three-part notation in the order Hue value/chroma (H V/C). When no number precedes the hue letter the number 5 is assumed; this R 2.6/13 may also be written 5R 2.6/13. A more orange color might be written as 10R 4/2. When value or chroma is reported separately, the position of the slant line indicates what is meant, e.g., 5/ means 5 value, and /5 means 5 chroma.

The specification of the color of a food product according to the Munsell system consists of two parts: (1) the percentage of the different specific colors that, when blended together, gives a composite color that exactly matches the sample and (2) the description of the color discs to match the sample. For example, the requirement for color in canned tomatoes according to the "FDA Definitions and Standards for Food" is

> Compare the color of the mixture, in full diffused daylight or its equivalent, with the blended color of combinations of the following concentric Munsell color discs of equal diameter, or the color equivalent of such discs:
>
> a. Red-Munsell 5R 2.6/13 (glossy finish);
> b. Yellow-Munsell 2.5YR 5/12 (glossy finish);
> c. Black-Munsell N 1/ (glossy finish);
> d. Gray-Munsell N 4 (matte finish).

According to this requirement, the color of canned tomatoes may be matched by varying the proportions of four particular colors: definite shades of red, yellow, neutral black, and neutral gray. Also, it means that the color produced will be equivalent to the color of the sample.

CIE Color System

The Commission Internationale de l'Ecclairage (CIE), an international body that makes recommendations on all matters concerning light and color, devised a method for the measurement and specification of color. This system defines exact conditions for the illumination (6 standard sources) and observation of samples (0° and 45°), mathematical forms in which results shall be expressed, and tables of figures that relate objective measurements to the visual response of a standard observer; so that results relate to what the sample looks like under the specified viewing conditions. There are two standard viewing conditions: a visual angle of 2° (1931 Standard Observer) and 10° (1964 Standard Observer). The 10° Standard Observer is used for more accurate color matching while the 2° Observer reflects the color matches made over a 2° visual field. These weighting values are listed in Judd and Wyszecki (1975). Coordinates made using either Standard Observer (2° or 10°) are analyzed in the same way using the appropriate 2° or 10° Chromaticity Diagrams. Since the 2° Observer is used most often in the food industry, these diagrams will be used here. The 10° Observer is used in the same way. Also, the CIE Standard Illuminant (usually C or D_{65}) must be given. Five of these (B, C, and the D series) simulate daylight under various conditions.

In the CIE color system, the relative percentage of each of the primary colors required to make a visual match with a sample are mathematically derived and then plotted on a chromaticity diagram (Fig. 10.6) as one chromaticity point. The dominant wavelength and purity can be determined from the chromaticity point.

To specify the chromaticity of a color, it is first necessary to measure, in a spectrophotometer, the light that is reflected or transmitted by the sample at intervals of 10–30 nm from 400 nm to 700 nm. These values then must be weighed by the values of the three theoretical primaries (red, blue, green); the resultant calculations delineate the percentage of the three primaries required to produce for the standard observer the color of the spectrum at a specific wavelength.

The sums of each of the red, green, and blue calculations are called *tristimulus values* of the specific color. These values are denoted by letters: X (for red—700 nm), Y (for green—546.1 nm), and Z (for blue—435.8 nm). Lowercase letters are used to denote *chromaticity coordinates*. Mathematically, the chromaticity coordinates are obtained as follows:

$$x = \frac{X}{X + Y + Z}$$

By substituting Y and Z in the numerator of similar equations, the chromaticity coordinates for y (green) and z (blue) may also be calculated. (The sum of x, y, and z equals unity (one) since the addends are fractions of the sum $X + Y + Z$). Thus, if two of the chromaticity coordinates are known, one can obtain the third merely by subtracting the sum of the two known coordinates from unity. A 10° Observer would have coordinates given as x_{10}, y_{10}, and z_{10}.

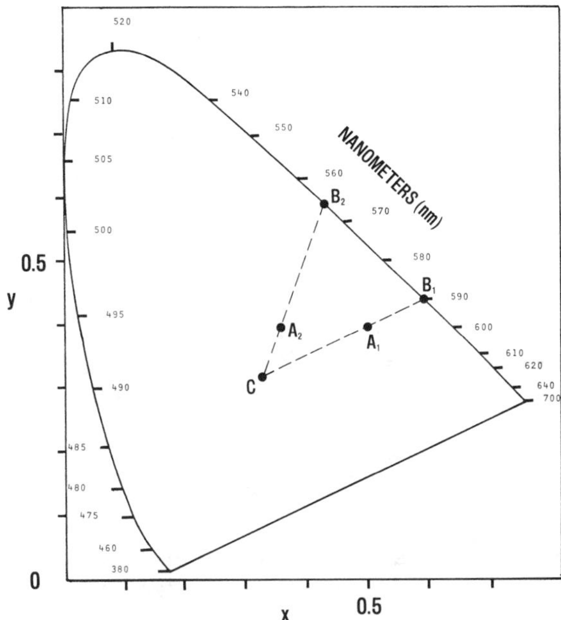

Figure 10.6. CIE chromaticity diagram (2° Observer) is used to determine the dominant wavelength and purity of a color. C = point of equal energy; A_1 and A_2 = chromaticity points of samples; B_1 and B_2 = corresponding dominant wavelengths. Purity is the ratio of the distances CA to CB.

When the x and y values for pure spectral colors are plotted against each other, a chromaticity diagram results. Each point on the curve in Fig. 10.6 represents the proportion of red (x) and green (y)—and hence also blue (z)—required to produce the color corresponding to that wavelength. In other words, when the chromaticity points of the pure spectral colors are connected in a smooth curve, they form a CIE chromaticity diagram.

When the x and y values derived from spectrophotometric measurements of a sample are plotted on a CIE diagram, the intersection point (i.e., the chromaticity point) rarely falls on the smooth curve that represents pure spectral colors because most sample colors are mixtures. As shown in Fig. 10.6, chromaticity points (A_1, A_2) usually fall within the boundaries of the pure spectral colors. After the chromaticity point of a color has been located, a line drawn from C, the point of equal energy through the color's chromaticity point (e.g., A_1) intersects the spectral energy locus at the dominant wavelength (B_1). The spectral energy locus represents 100% purity (i.e., pure spectral color) and the point of equal energy represents 0% purity (i.e., white). The ratio of the distance between the point of equal energy and the chromaticity point to the distance between the point of equal energy and the spectral energy locus is the purity of a color and is a measure of color intensity. For example, in Fig. 10.6, CA_1/CB_1 equals the purity of color A_1 whose dominant wavelength is B_1.

After the chromaticity coordinates (x, y) and the tristimulus value (Y) are measured, they can be converted to Munsell notation (H V/C) using charts published by the U.S. Agricultural Marketing Service. Standard curves relate Y to the Munsell value and once we know this, x and y can be plotted on the appropriate USDA chart to find the Munsell hue and chroma. Because of this, CIE measurements are usually given as (x, y, Y).

The procedure for determining the tristimulus values and chromaticity coordinates of a sample is explained in more detail in the section on quantitative analysis.

Hunter Color System

The Hunter (L, a, b) color system is a mathematical modification of the CIE system to make it easier to compare color differences with those made by the eye. That is, knowing the CIE coordinates $(x, y,$ and $z)$ one can convert them to Hunter coordinates (L, a, b) and through the CIE coordinates to Munsell notation (H V/C) and vice versa.

Here, three values are specified: lightness (L) which ranges between 0 (black) and 100 (white), a (red-green) where red is +, and b (yellow-blue) where yellow is + and blue is −. A different mathematical transformation made by the CIE in 1976 was done to expand the yellow and contract in the blue part of the diagram which gives the L^*, a^*, and b^* coordinates.

☐ COLORING AGENTS

It is more than likely that the first coloring agents added to foods were the so-called natural colors, e.g., annatto extract, saffron, paprika, and caramel. These color additives, in general, were not satisfactory in food manufacturing because of their heat and light instability as well as to their lack of uniformity. Consequently, as synthetic dyes and pigments became available, they virtually supplanted the naturally derived products.

In general, color additives may be divided into three categories: synthetic organic compounds (certified colors), naturally derived colorings (uncertified color additives), and mineral or synthetic inorganic colors (pigments and lakes). Of these, the synthetic organic compounds are the most extensively used; a few natural colors are routinely used in a limited number of products; and synthetic inorganic compounds are rarely used in food products.

Synthetic Organic Compounds (Coal Tar Dyes)

Of the great number of coal tar dyes known to science, only a few are permitted by law for use in foods. These dyes constitute the primary food colors, ones from which blends can be mixed to produce secondary colors (Table 10.1). The primary food colors can also be used to produce shades (intermediary colors).

All of the permitted dyes possess common names and in addition have numbers assigned to them. All are subject to FDA certification for purity. After passage of the Food, Drug, and Cosmetic Act of 1938, a new system for designation of these dyes was adopted. For each certified dye, the following are specified: (1) the use for which the color is permitted; (2) the predominant shade of color; and (3) a number to distinguish the color from others of the same shade. For example, FD&C Blue No. 1 is permitted in food, drugs, and cosmetics; blue is the predominant shade of color; and No. 1 distinguishes this dye from FD&C Blue No. 2.

The Color Additives Amendment (see Chapter 1) provisionally listed the food colors, both certified and uncertified, in use at the time, pending completion of feeding studies to establish their safety. Closing dates were established, but time extensions have been made when necessary to complete toxicological studies. It should be kept in mind that certified colors are synthetic materials, and certification concerns only the batch purity, and does not relate to the safety of a color. Thus, the difference between certified and uncertified colors relates to their purity, not their safety.

The most common color dyes are water soluble and are insoluble in most organic solvents (Table 10.2). In systems that must be anhydrous, propylene glycol and glycerin are used as solvents. These dyes have high tinctorial value; that is, only exceedingly small amounts of these color additives are needed to color foods. The certified food colors should be protected against intense light, oxidizing agents (ozone, chlorine, hypochlorite), reducing agents (sulfur dioxide, invert sugars, metallic ions, ascorbic acid), strong acids and alkalies, and excessive heat. Metals, particularly zinc, tin, iron, copper, and aluminum, are among the principle reducing agents.

The synthetic coloring agents approved for use in foods are listed in

Table 10.1. Examples of Color Formulations Based on FD&C Certified Food Colors

Shade	FD&C Dye		% of Blend
Orange	Yellow	#5	95
	Red	#40	5
Cherry	Red	#40	99
	Blue	#1	1
Strawberry	Red	#40	95
	Red	#3	5
Lemon	Yellow	#5	100
Lime	Yellow	#5	95
	Blue	#1	5
Grape	Red	#40	80
	Blue	#1	20
Raspberry	Red	#3	75
	Yellow	#6	20
	Blue	#1	5
Butterscotch	Yellow	#5	74
	Red	#40	24
	Blue	#1	2
Chocolate	Red	#40	52
	Yellow	#5	40
	Blue	#1	8
Caramel	Yellow	#5	64
	Red	#3	21
	Yellow	#6	9
	Blue	#1	6
Cinnamon	Yellow	#5	60
	Red	#40	35
	Blue	#1	5
Peach	Red	#40	60
	Yellow	#6	40
Cheddar cheese	Yellow	#5	55
	Yellow	#6	45
Tea, cola, or root beer	Yellow	#5	70
	Red	#40	25
	Blue	#2	5

Source: Warner-Jenkinson Co. (1984).

Table 10.3. The certified food colors permanently listed by the FDA include FD&C Blue No. 1, FD&C Red No. 3, FD&C Yellow No. 5, FD&C Green No. 3, Citrus Red No. 2, and FD&C No. 40. Provisional listing still applies on the lakes of these colors (with the exception of the lake of FD&C Red 40). The certified food colors provisionally listed include FD&C Blue No. 2, FD&C Yellow No. 6, and the lakes of all these colors. Citrus Red No. 2 is limited to coloring the skins of oranges

Table 10.2. Physical and Chemical Characteristics of Some Certified Food Colors

FD&C Name[a]	Hue	Stability to light	Oxidation[b]	Solubility[c] (g/100 ml) at 25°C Water	25% ETOH	Glycerin	Propylene glycol
Red No. 3	Bluish pink	Poor	Fair	9.0	8.0	20	20.0
Red No. 40	Yellowish red	Very good	Fair	22.0	9.5	3	1.5
Yellow No. 6	Reddish	Moderate	Fair	19.0	10.0	20	2.2
Yellow No. 5	Lemon yellow	Good	Fair	20.0	12.0	18	7.0
Green No. 3	Bluish green	Fair	Poor	20.0	20.0	20	20.0
Blue No. 1	Greenish blue	Fair	Poor	20.0	20.0	20	20.0
Blue No. 2	Deep blue	Very poor	Poor	1.6	0.5	1	0.1

Source: Warner-Jenkinson Co. (1984).
[a] See Table 10.3 for common names.
[b] Contacted color with oxidizers such as ozone or hypochlorites or reducers such as SO_2 and ascorbic acid.
[c] All of the listed dyes are insoluble in vegetable oil.

Table 10.3. Certified Food Colors[a]

FD&C name	Common name	Restriction
Permanently listed		
Red No. 3	Erythrosine	None
Red No. 4	Red No. 40	None
Red No. 40 Lake	Red No. 40 Lake	None
Yellow No. 5	Tartrazine	None
Green No. 3	Fast Green FCF	None
Blue No. 1	Brilliant Blue FCF	None
Blue No. 2	Indigotine	None
Citrus Red No. 2	Citrus Red	Orange skins (2 ppm)
Orange B	Orange B	Sausage casings (150 ppm w/w)
Provisionally listed		
Red No. 3 Lake	Red No. 3 Lake	None
Yellow No. 5 Lake	Yellow No. 5 Lake	None
Yellow No. 6 Lake	Sunset Yellow FCF	None
Green No. 3 Lake	Green No. 3	None
Blue No. 1 Lake	Blue No. 1 Lake	None
Blue No. 2 Lake	Blue No. 2 Lake	None

[a] FDA status as of January 1986

Table 10.4. Natural Organic Color Additives

Name	Source	Color principle	Color	Restriction on use
Alkanet	Root of *Alkanna tinctoria*	Alkannin	Red	Sausage casings, oleomargarines, & shortening
Annatto	Seed pods of *Bixa orellana*	Bixin (main color); orellin (minor color)	Both yellow	None
Beta-apo-8^1-carotenal	Fruts and vegetables	Carotene	Yellow-orange	Not to exceed 15 mg per pound or pint of food
Caramel	Heated sugar	High-molecular weight substances	Reddish-brown	None
Carotene	Plants and carrots	α, β, r, and k carotene	Yellow	None
Chlorophyll	Green leaves	Chlorophyll[a] (62% of color); chlorophyll[b] (23% of color); xanthophyll (10% of color)	Greenish-blue	Sausage casings, margarine, and shortening
Cochineal	Insect: Coccus cacti	Carminic acid	Red	None
Cocoa Red	Cocoa beans	Cocaonin	Red	None
Indigo	Plant genus Indigofera	Indigotin	Blue	None
Litmus	Lichens	Azolitmin	Red	None
Madder	Root of *Haematoxylon campechianum*	Haematoxylin	Red-brown	None
Paprika and its oleoresin	Fruit of *Capsicum annum* L.	Carotene and capsanthin	Red	Not for use on fresh meat or comminuted fresh meat products except chorizo sausage and Italian-brand sausage and other meats as specified by USDA
Quercitron	Inner bark of *Quercus nigra*	Quercitin	Yellow	None
Safflower	Flower of *Carthamus tinctoria*	Carthamin	Red	None
Saffron	Flower petals of *Crocus sativa*	Crocetin	Yellow	None
Turmeric	Rhizome of *Curcuma longa* L., *Lingberaceae*	Curcumin	Yellow	None

(maximum level of 2 ppm). The only application allowed for FD&C Red No. 4 is maraschino cherries at a level no greater than 150 ppm, while Orange B is approved only for sausage casings at a maximum concentration of 150 ppm.

The most common applications for certified food colors include soft drinks, dry mixes (cakes, gelatin desserts, pudding mixes, drink powders), baked goods (cookie fillings, bakery coatings), and confections (candy, jellies, candy cream centers, hard candy). It should be realized, however, that most foods in our diet (e.g., meats, vegetables, fruits, bread, and most fluid milk products) are not artificially colored.

Natural Organic Compounds

Although several natural organic compounds are approved for use as uncertified food colors (Table 10.4), only a few of these are used to any extent for coloring foods. The most widely used of the uncertified food colors are annatto extract and caramel.

Annatto extract is obtained from the seed of a tropical tree, *Biza orellano*. It is an oil-soluble color (bixin), but it can be converted to a water-soluble color (norbixin). It is a carotenoid derivative, and the shades produced by its use range from butter-yellow to peach. The extract finds usage in butter, margarine, cheese, popcorn oil, and salad dressings.

Caramel is produced by the controlled heat treatment of food-grade carbohydrates (sucrose, starch hydrolysates, molasses, and malt syrup). Caramel color contains a complex mixture of components, and its composition varies according to the method of manufacture. There are three types of caramel: (1) acid-proof caramel for used in carbonated beverages such as colas; (2) bakers and confectioners caramel for use in cookies, cakes, and rye bread; and (3) dry caramel for use in dry mixes.

The remaining natural organic colors are not used extensively and generally are used only for specific food products. For example, turmeric is used as a yellow color in some pickled products, and saffron, another yellow color, is used in the home in certain cooked dishes.

☐ QUALITATIVE ANALYSIS

There are times when an analyst may be called upon to make a qualitative study of the coloring agents present in food. This may involve determining whether the coloring agent is a certified food color, an uncertified food color (naturally present or added), or a lake. In some cases, actual identification of the coloring agent(s) present may be required.

All the certified food colors allowed in foods (except Citrus Red No. 2 and Orange B) are water soluble. They all have the property of dyeing wool in acid solutions. This kind of dye is relatively inexpensive and a small amount goes a long way. Consequently, there is little possibility that an analyst will find nonpermitted coal tar dyes in foods.

Since certified food colors are most commonly used coloring agents in foods, the analyst normally looks first for these dyes. If no certified food colors are identified, natural coloring agents are considered. Only those natural coloring agents that might contribute to the color of the foodstuff need to be considered; for example, one would consider as possibilities only yellow natural coloring matters if the foodstuff is yellow. If the color of the food cannot be identified as either a permitted certified food color (coal tar dye) or natural coloring substance, the possibility of a pigment or lake must be considered.

Differentiation between Coal Tar Dyes and Natural Organic Dyes

The certified food colors, coal tar dyes, normally dye animal fibers readily in dilute acid solution. The color can be removed easily and entirely by making the solution alkaline; if the stripped dye solution is again made acid, it will usually redye animal fibers.

In contrast, most organic dyes do not strip clean from an animal fiber, and as a consequence, the fiber is left with a brown to purple color. Too, natural organic dyes generally do not redye animal fibers after stripping. Because some vegetable colors, such as archil or turmeric, may impart a color to the fiber, strip, and redye as coal tar dyes do, this method is not absolutely conclusive in all cases.

Procedure

Comminute 50–100 g of the foodstuff in 250–500 ml of hot water; make the mixture slightly alkaline with ammonium hydroxide, then allow it to stand for 1–2 hr to strip the dye from the food material. Filter or decant through cheesecloth. Make the filtrate slightly acid with hydrochloric acid, then add a 6- to 8-in. length of animal fiber (white wool or silk) and bring the solution to a boil. Boil the acidic solution 20–30 min, then remove the dyed fiber from the solution and wash it thoroughly. Strip the color from the fiber by placing the dyed fiber in about 25 ml of a dilute ammonium hydroxide solution. Remove the stripped fiber, dilute the solution to about 250 ml, and again make it acidic with hydrochloric acid. Add a fresh piece of fiber to the solution and boil for 10–20 min. Remove the fiber, wash thoroughly with cold water, and then press dry. If the fiber is colored, the foodstuff probably contains a coal tar dye.

Identification of Individual FD&C Colors

The color changes that dyes undergo at different hydrogen ion concentrations (pH) can be used to identify the individual certified food colors. The method given here, however, is not suitable for mixtures of dyes.

Procedure

Following the procedure described in the previous method, prepare a foodstuff sample, strip the dye from it, and dye a piece of animal fiber with the extracted dye solution. Perform spot tests on the dyed animal fiber strip with each of the following reagents and observe the resultant color changes: concentrated hydrochloric acid, concentrated sulfuric acid, 10% sodium hydroxide solution, and 10% ammonium hydroxide solution. Identify the unknown dye by comparing its color reactions to the test reagents with those listed in Table 10.5. It is advisable, when identifying an unknown, to perform spot tests on fibers dyed to the same intensity with known dyes.

Table 10.5. Color Reactions Produced by Various Reagents on Fibers Dyed with FD&C Food Colors

	Color reaction with			
FD&C Name	Conc. HCl	Conc. H_2SO_4	10% NaOH	10% NH_4OH
Blue No. 1	Yellow	Yellow	No change	No change
Blue No. 2	Slightly darker	Darker	Greenish-yellow	Greenish-blue
Green No. 3	Orange	Green to brown	Blue	Blue
Red No. 3	Orange to yellow	Orange-yellow	No change	No change
Red No. 40	No change	Deeper red	Orange	Little change
Yellow No. 5	Slightly darker	Slightly darker	Little change	Little change
Yellow No. 6	Slightly redder	Slightly redder	Browner	No change
Orange 1	Violet	Violet	Red, dark	Red, dark

Source: AOAC (1950, pp. 658–659).

Comments

If the spot tests are inconclusive, the unknown sample may contain a mixture of dyes. If this is the case, then the individual dyes must be separated, by one of the procedures described later, before identification is possible.

Other techniques for identifying individual food colors are available. These include the reaction of coal tar dyes with reducing agents such as $TiCl_3$, $SnCl_2$, or Zn dust in acid solution and comparison of the absorption spectra of unknown dyes with those of standard dye solutions (Table 10.6).

Identification of Individual Uncertified Color Additives

The uncertified color additives (natural colors), as a group, do not dye wool fiber as readily as the certified food colors, and as noted previously, they generally do not redye fiber after stripping. Some of the natural colors contain complex mixtures of numerous components. The exact composition of a natural color may vary due to seasonal, geographical, and varietal variations. Dilute solutions of natural colors are sensitive to alkalies and some are sensitive to acids. Because of these characteristics, the natural coloring substances are more difficult to identify than are the certified food colors. Furthermore, there is no practical method for certifying the purity of natural colors, and as a consequence, the user must depend upon the integrity of the supplier for assurance of product quality.

The different natural colors react differently with a number of common reagents. By comparing the reactions of an unknown with those

Table 10.6. Wavelengths of Maximum Absorbance by FD&C Food Colors[a]

Dye	λ_{max} (nm)
FD&C Blue No. 1	630
FD&C Blue No. 2	610
FD&C Green No. 3	625
FD&C Red No. 3	527
FD&C Red No. 40	502
FD&C Yellow No. 5	428
FD&C Yellow No. 6	484
Citrus Red No. 2	515
Orange B	437

[a] Solvent is water for all dyes except Citrus Red No. 2 for which $CHCl_3$ is solvent.

of known color additives, the identity of an unknown can usually be deduced. In general, this procedure is not suitable for analysis of a mixture of natural colors.

Procedure

The first step in analyzing the natural colors is extraction of the coloring matter from the sample. The carotenoid pigments in leaves, fats, oils, carrots, tomatoes, paprika and egg yolk may be extracted by ether from neutral solutions. The coloring matters of alkanet, annatto, and turmeric are extracted from acid solutions by ether. Also extracted by ether from slightly acid solutions are the flavone coloring matters of Persian berries (rhamnetin) and Quercitron (quercitin). Slightly acidified amyl alcohol extracts the coloring matters in saffron and orchil. Amyl alcohol extracts relatively small amounts of the anthocyanins, which constitute the red colors of most fruits and of caramel.

Evaporate ether extracts to dryness on a steam bath; warm the residue with a small amount of ethanol and dilute with water. Dilute amyl alcohol extracts with ligroin (b.p. 90°–110°C) and then extract with water. Treat aliquots of the resulting solutions as described below and compare the reactions obtained with those in Table 10.7:

1. Add 1 or 2 drops of concentrated hydrochloric acid; then add an excess equal to three to four times the volume of the solution.
2. Add a drop of 10% sodium hydroxide solution.
3. Add a small crystal of sodium hyposulfite.
4. Add carefully, drop by drop, a freshly prepared 0.5% ferric chloride solution (colors are not obtained in some cases if an excess is used).
5. 10% Potassium or ammonium alum solution equal to one-fifth the volume of the aliquot.
6. Add, drop by drop, a 5% uranylacetate solution.
7. Evaporate an aliquot in a porcelain dish, allow the dish to cool thoroughly, and add to the dry residue 1 or 2 drops of concentrated sulfuric acid. Note the color when the acid wets the residue.

Identification of Pigments and Lakes

A pigment may be defined as a colored or white chemical compound that is capable of giving color and is insoluble in the solvent in which it is being applied. Pigments are most commonly used as facings on products. Wash the sample with water and allow the washings to settle. Examine and identify particles of coloring matter with the aid of a

Table 10.7. Reaction of Certain Natural Coloring Matters to Common Reagents

	Reactions with						
Coloring matter	Conc. HCl	10% NaOH	Sodium hyposulfite	0.5% FeCl	10% alum	5% Uranyl acetate	Conc. H_2SO_4
Annato	Remains orange	Violet	Little change	No marked change	No change	No change	Blue
Anthocyans of red fruit color	Deep red with excess of acid	Green, blue, slate, browner by oxidation	Anthocyanidins almost completely decolorized	No change	No change	No change	No change
Caramel	Little or no change	Little change, browner	Slightly paler	No change	No change	No change	No change
Carotene and xanthophyll	Little or no change, perhaps paler	Little or no change	Little change	No marked change	Little change	Little change	Blue reaction obtained with difficulty
Chlorophyll	More brownish	Brown phase reaction 34.020	More brownish	More brownish	More brownish	More brownish	More brownish
Flavone colors of Persian berries, etc	Intensely yellow with excess HCl	Bright yellow	Little change	Olive-green or black	More strongly yellow	Orange colorations	Yellow, orange
Saffron	Little or no change	Remains yellow	Little change	No marked change; browner perhaps	Little change	No change	Blue
Tumeric	Orange-red or carmine red with addition of excess of conc. HCl	Orange-brown	Little change	No marked change; perhaps browner	Little change	Somewhat browner	Red

Source: AOAC (1970).

microscope. Treat the insoluble material with reagents to determine its identity as follows:

1. Charcoal (or other forms of carbon) does not react with chemical reagents and will burn.
2. Ultramarine blue is stable to alkali treatment but decomposes and produces hydrogen sulfide on treatment with dilute hydrochloric acid solution.
3. Talcum can be tested for aluminum by fusing with cobaltous nitrate, a positive reaction giving a purple coloration.

A lake may be defined as a pigment prepared by precipitating a soluble dye (FD&C) onto an insoluble reactive or adsorptive substratum or diluent. Lakes may be identified by their color reactions with various reagents. Treatment with strong acids or alkalies will release the food color, which can then be identified by the procedure described already. Incineration of the lake will destroy the organic portion of the compound, leaving the metal component in the ash. The metal can then be identified by the standard techniques used in qualitative analysis.

Separation of Dyes in a Mixture

Among the methods available for separating the component food colors in a mixture are fractional dyeing, extraction with immiscible solvents, and chromatography. The first two are described here in a general manner, while detailed instructions are given for the chromatographic method.

Fractional Dyeing

The certified food colors dye animal fiber at different rates; consequently, by dyeing successive small pieces of fiber, the first pieces will be dyed principally by one dye and the latter pieces principally by another dye. By combining the relatively pure fractions, stripping the color from the fiber pieces, and redyeing, a reasonable separation can be effected.

Extraction with Immiscible Solvents

A mixture of dyes may be resolved by extraction of their aqueous solutions with organic solvents. Basic dyes are extracted from an alkaline solution by shaking with ethyl ether, while acid dyes are extracted from an acid solution by shaking with isoamyl alcohol. (See AOAC 34.007–34.008, 11th ed.)

Chromatographic Analysis

Chromatography is the preferred method for separating and identifying the component dyes in a mixture (see Chapter 3 for detailed discussion of various chromatographic techniques). This technique requires little equipment and is relatively easy and fast. By judicious selection of conditions, adsorbents, developing solvents, and eluting solvents, good separations of dye mixtures are possible.

The following procedure is adapted from Systematic Identification of Artificial Food Color Permitted in the United States, by R. L. Stanley and P. L. Kirk, from *Journal of Agricultural and Food Chemistry, 11,* 492–495. Copyright 1963 American Chemical Society. The component dyes are first isolated from a foodstuff by column chromatography on alumina columns; the dyes are then separated by ascending paper chromatography and finally identified by their R_f values in different developing solvents.

Reagents

1. Dilute HCl—Dissolve 1 volume of concentrated HCl with 9 volumes of water.
2. Dilute NH_4OH—Dilute 1 volume of concentrated NH_4OH with 9 volumes of water.
3. 1% NH_4OH—Dilute 41 ml of concentrated NH_4OH to 1 liter with water.
4. Pyridine–ethyl acetate–water developing solvent (1:2:2 by volume)—Shake this mixture in a separatory funnel and allow to separate. Discard the lower layer. (This developing solvent is best for multicomponent mixtures and mixtures of yellow and orange dyes.)
5. 3.9 *N* NH_4OH in isobutyl alcohol developing solvent—Mix equal volumes of concentrated NH_4OH and isobutyl alcohol in a separatory funnel, then allow to separate. Discard the lower layer, pipette 5 ml of the upper phase into 100 ml water, and titrate to a methyl red end-point with 1 *N* hydrochloric acid. On the basis of this titration, adjust the remainder of the upper phase to exactly 3.9 *N* with isobutyl alcohol. (This developing solvent is best for mixtures of blue and green dyes.)
6. Isoamyl alcohol–95% ethyl alcohol–concentrated NH_4OH–water (4:4:1:2) developing solvent—Mix indicated volumes. (This developing solvent is best for mixtures of red dyes.)

Apparatus

1. Chromatographic column with stopcock (20 × 300 mm)—Fill the column to a depth of 150 mm with alumina (150–250 mesh) that

has been activated by heating at 400°C for 1 hr. Wash packing with dilute HCl to remove fines and acidify the column.

2. Glass-stoppered graduated cylinders for paper chromatography (100 ml).

Preparation of Sample Dye Solution and Isolation by Column Chromatography

The sample should contain about 1 mg of each color present. Normally, a 50-g food sample will contain adequate color for subsequent analysis. Sample solutions from various types of foods are prepared as follows:

1. Dry Gelatin Desserts. Thoroughly mix 50 g of the sample with 100 ml of 95% ethanol and decant through a Buchner funnel with vacuum. Repeat the extraction until most of the color is removed. Then shake the sample with 100 ml of 1% aqueous NH_4OH and rapidly filter under vacuum. Combine the filtrates and adjust the pH with glacial acetic acid until the solution is strongly acidic.
2. Water-Soluble Foods. Dissolve solid samples in a minimum volume of 1% NH_4OH. Filter samples and acidify with glacial acetic acid.
3. Oil-Soluble Foods. Dissolve the sample in a minimum volume of ethyl ether and without filtering, extract with 1% NH_4OH. Acidify the aqueous extract with glacial acetic acid.
4. Solid Food Products Insoluble in Water and Fat Solvents. Cover samples with 80% ethanol containing 1% ammonia, grind in a food blender, and then filter under vacuum using Celite as a filter aid. Repeat until no additional color is extracted. Treat alimentary pastes and bakery goods in the same manner, but heat the ammoniacal alcohol to boiling before blending. Dilute the alcoholic extract with water to about 50% alcohol and acidify with glacial acetic acid.

Introduce the prepared sample solution into the column of activated alumina and apply pressure or vacuum. Wash the column with an equal volume of water, then elute the adsorbed dye with dilute NH_4OH. Two or three column volumes may be necessary to elute all the color.

Separation of Colors by Paper Chromatography

Three paper chromatograms are prepared for each sample and developed simultaneously in each of the three developing solvents. Spot a volume of the sample column eluate estimated to contain about 1 μg of dye to the base line of 1 × 12 in. strips of Whatman No. 1 paper. Place developing solvent in the bottom of a graduated cylinder, insert spotted strip, and secure each strip at the top with stopper so that the end of the paper protrudes from the cylinder.

After the solvent has ascended to within a few centimeters of the top, remove the paper chromatogram from the cylinder; note the position of the solvent front and the positions and colors of the sample component spots for each chromatogram. Dry the chromatograms and calculate the R_f value for each spot. Unknown dyes can be identified by comparing their R_f values with those of known dyes. Since R_f values are affected by many variables, it is best to run standard, known dyes with unknown samples.

Identification of Oil-Soluble Dyes

Citrus Red No. 2 and Orange B are the only oil-soluble dyes that may be used on food products in the United States. These may be used only on oranges and sausage casings, respectively. The following method is based on that of Przybylski *et al.* (1960).

Procedure

Wash the colorant from 10 oranges with 250 ml of chloroform. (Surface waxes, oils, and some natural pigments also wash off.) Combine the washings and make to a volume of 250 ml. Evaporate a 50-ml aliquot on a steam bath and dissolve the residue in 25 ml of petroleum ether.

Fill a 2.5 × 10 cm glass column with petroleum ether, then add sifted alumina into the column to a height of about 4 cm. Pass the petroleum ether aliquot of the sample into the column. Wash the column with 50 ml of petroleum ether, followed by 200 ml of carbon tetrachloride, and elute the adsorbed dye with 95% ethanol. Evaporate the eluate to dryness on a steam bath.

The oil-soluble colors can be separated by reverse-phase paper chromatography as follows. Soak a 7 × 22.5 in. strip of Whatman 3 MM paper in a 5% solution of light mineral oil in ethyl ether (w/v), drain, and air dry. Dissolve the eluted sample in a few drops of chloroform and spot as a narrow band, 2.5 in. from the bottom of the paper. Develop for 3 hr by descending chromatography with a developing solvent such as acetone–water (6:4), dioxane–water (6:4), or acetone–alcohol (7:3). Examine the chromatogram for coal tar colors.

Dry the chromatogram, extract the dye from the paper with ethanol, and obtain the absorption curve with a recording spectrophotometer. Compare the spectrum of the sample with spectra of the known dyes.

☐ QUANTITATIVE ANALYSIS

As discussed previously, any object (or food) will appear colored if there is an imbalance of visible radiant energy reaching the eye from light sources. The eye cannot perceive wavelengths of light but can see color. In contrast, an instrument, such as a spectrophotometer, cannot perceive color but can detect differences in the wavelength of radiation. Thus, it is possible to measure color in terms of a psychological response or by physical means (in terms of wavelength).

Psychological Measurements

The simplest way to measure or specify a color is by describing the visual appearance of the color. This procedure, which is frequently used in grading and classifying foodstuffs, is based on comparing the perceived color of a sample of food with a color standard the individual conveys in his mind. For example, the official U.S. grain standards for wheat recognize certain color designations (white, red, amber, dark, etc.). Butter is scored on a grading system in which 15 points of a total 100 points is allowed for color. The color may be described as very light, light medium, high, and abnormally high (a defect). Honey is graded into seven color classifications: water white, extra white, white, extra light amber, light amber, amber, and dark amber. Similarly, maple syrup is graded into three classes: light amber (Grade AA), medium amber (Grade A), and dark amber (Grade B).

An improvement over this method involves the comparison of the perceived color of a sample of food with that of a visible standard. These standards may be actual standard products, color charts, color solutions, or standard colored glasses. The use of Munsell color discs, as described earlier, to specify color standards for foods is an example of this approach. The use of a visible standard eliminates the need to refer to a hypothetical standard in one's mind and usually results in more consistent color specifications among different individuals. The color comparisons must be made by individuals who are not color blind and are sensitive to small differences in color.

Subjective methods such as these are used primarily for inspection purposes. Presently, spectrophotometric methods of colorimetric analyses are preferred because the color of a foodstuff can be standardized independently of physical color standards and is independent of variations in color perception between individuals.

Reflectance Spectrophotometry

Reflectance spectrophotometry involves the measurement of the amount of incident radiant energy of a specified wavelength that is reflected from an object. Reflectance may be expressed mathematically as follows:

$$R = R_0/R_w$$

where R_0 = intensity of light reflected from a sample and R_w = intensity of light reflected from a standard of known spectral reflectance at each wavelength. The standard is normally a white material, such as magnesium oxide, that reflects light equally at all wavelengths. Reflectance measurements for the purpose of color determinations are made in the spectral region from 380 to 750 nm.

Two types of reflectance spectrophotometers are available. Each type consists of a standard light source, a sample holder, and a photoreceptor. In one type of instrument, the light that reflects off the test sample is measured by a photoreceptor at a 45° angle from the sample. In the other type, the light that reflects off the test sample is collected in an integrating sphere such that nearly all of the reflected light is measured by the photoreceptor.

Reflectance readings over the visible spectrum must be converted into terms that bear resemblance to what the eye sees. Meaningful functions expressing how the eye sees color are the CIE tristimulus values *X, Y,* and *Z*, which were discussed already. Two methods are used to convert reflectance readings into CIE values: the weighed ordinant method and the selected ordinant method. Since computers are becoming more inexpensive, these methods are included with modern spectrophotometers to give these coordinates automatically. The procedure described in this section closely parallels the weighed ordinant method. The reflectance (R) of the test sample is "weighed" at preselected wavelengths throughout the visible spectrum and the energy reaching the detector at the corresponding wavelengths is converted to values representing how the eyes see color in the CIE system.

Apparatus

1. A spectrophotometer equipped with a color analyzer attachment (e.g., B & L Spectronic 20).
2. Magnesium carbonate block.
3. "Black body" fixture.

Procedure for Making Reflectance Measurements

Turn on instrument for 30 min to warm up before making any measurements. Measure the reflectance of the porcelain standard (cover of sphere) against the magnesium carbonate block at 30-nm increments from 415 to 685 nm. Plot the reflectance values of the porcelain surface against wavelength; this plot is used as an "adjusted" reference for sample measurements.

Measure the reflectance of the sample as follows:

1. Select desired wavelength.
2. Install "black body" fixture.
3. Open shutter, set 0% reflectance (100% T), then close shutter.
4. Remove "black body" fixture.
5. Close aperture cover tightly. Open shutter and set meter to the reflectance value of the porcelain standard for this wavelength, and then close shutter.
6. Install sample over opening. Open shutter, record the percent reflectance, and then close shutter.

Follow this procedure to take sample reflectance readings at 30-nm increments from 415 to 685 nm.

Computation of CIE Tristimulus Values and Chromaticity Coordinates

A sample form for computing CIE tristimulus values is shown in Fig. 10.7. In the left column opposite the standard wavelengths are written the values of the reflectance readings of the porcelain standard that correspond to a 100% reading for magnesium carbonate. For each sample reflectance reading, a short vertical line is drawn through top scale (%T) opposite the corresponding wavelength in each of the three sets (Total Y, Total X, and Total Z). This procedure is repeated for each wavelength at which reflectance readings are taken.

After marking all the reflectance readings on the top scales (%T), read the corresponding values on the bottom scales at each wavelength and record these values (to two decimal places) in the right-hand column. Add the numbers in the right-hand column in each set of values to obtain total X, Y, and Z. These are the tristimulus values for the sample.

Calculate the chromaticity coordinates x and y, from the following expressions:

$$x = \frac{X}{X + Y + Z}$$

$$y = \frac{Y}{X + Y + Z}$$

Determination of Dominant Wavelength and Purity

Since it is not easy to visualize a color or color difference from the tristimulus values or chromaticity coordinates, the general procedure is to determine the dominant wavelength and purity of the color. As discussed in the section on the CIE color system, this is done by plotting y versus x on a chromaticity diagram. In Fig. 10.8, for example, point

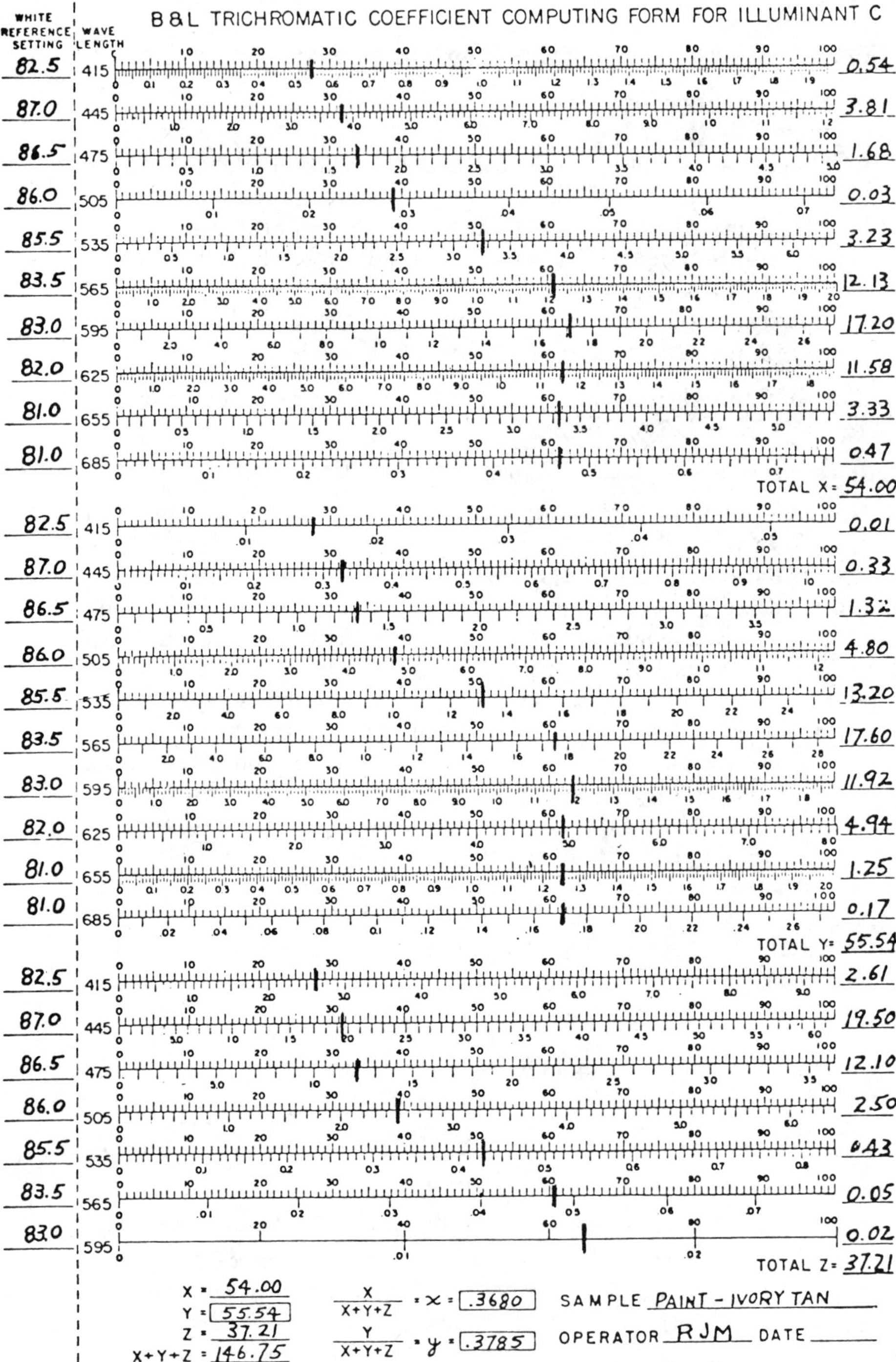

B&L TRICHROMATIC COEFFICIENT COMPUTING FORM FOR ILLUMINANT C

WHITE REFERENCE SETTING	WAVE LENGTH	
82.5	415	0.54
87.0	445	3.81
86.5	475	1.68
86.0	505	0.03
85.5	535	3.23
83.5	565	12.13
83.0	595	17.20
82.0	625	11.58
81.0	655	3.33
81.0	685	0.47
		TOTAL X = 54.00
82.5	415	0.01
87.0	445	0.33
86.5	475	1.32
86.0	505	4.80
85.5	535	13.20
83.5	565	17.60
83.0	595	11.92
82.0	625	4.94
81.0	655	1.25
81.0	685	0.17
		TOTAL Y = 55.54
82.5	415	2.61
87.0	445	19.50
86.5	475	12.10
86.0	505	2.50
85.5	535	0.43
83.5	565	0.05
83.0	595	0.02
		TOTAL Z = 37.21

X = 54.00
Y = 55.54
Z = 37.21
X+Y+Z = 146.75

$\frac{X}{X+Y+Z} = x = .3680$

$\frac{Y}{X+Y+Z} = y = .3785$

SAMPLE PAINT - IVORY TAN

OPERATOR RJM DATE ______

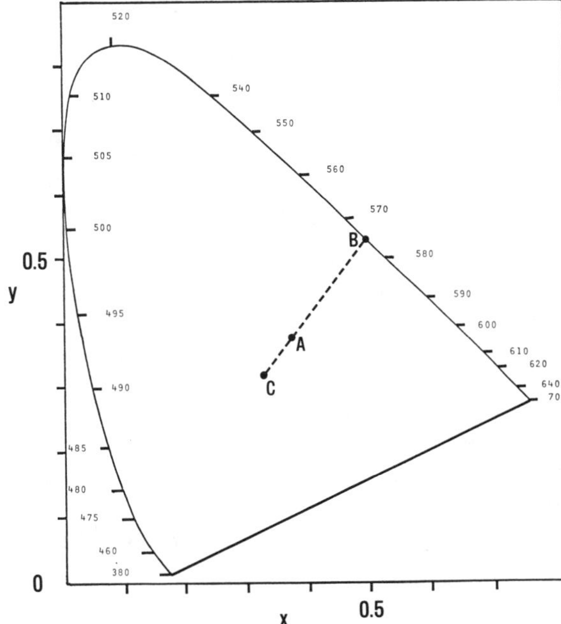

Figure 10.8. Chromaticity diagram (2° Observer) plot of x and y chromaticity coordinates for sample of ivory tan paint, based on data in Fig. 10.7.

C ($x = 0.3101, y = 0.3163$) is where the illuminant C (tungsten light) would fall, and point A ($x = 0.3680$, $y = 0.3785$) is the chromaticity point of the sample when lit by illuminant C. The line from point C through point A intercepts the curve representing the locus of the spectral colors at point B (about 575 nm). This is the dominant wavelength of the color and is analogous to the concept of hue. The ratio of the distance CA to the distance CB is called the purity of the color. The dominant wavelength and purity, together with brightness (Y value), give a better concept of a color or color difference than the tristimulus values.

Tristimulus Colorimetry

A simple tristimulus colorimeter consists of a light source, an object to be measured, a set of three filters that duplicate the responses of the three types of receptors in the human eye, and a photocell. White light from a standard source is shown on the surface of the test sample

Figure 10.7 Sample Bausch and Lomb form for computing CIE tristimulus values from sample reflectance readings.

to be measured. Light reflected at 45° is measured after it passes through a filter (*x*, *y*, or *z*).

Any tristimulus colorimeter that measures the Hunter color coordinates (Fig. 10.9) L, a_L, and b_L utilizing CIE illuminant C with a 45° illumination, 0° viewing, and a 2-in. (50-mm) diameter specimen area can be used. The following values are obtained with the tristimulus colorimeter: L = a vertical scale from 0 (black) to 100 (white); a_L = a horizontal scale representing $+a_L$ (redness) to $-a_L$ (greenness); and b_L = a horizontal scale representing $+b_L$ (yellowness) to $-b_L$ (blueness). These values can be converted into CIE tristimulus values and chromaticity coordinates from which the dominant wavelength and purity of a color can be determined.

Measurement Procedure

Samples of juices (e.g., grape and carrot juice) are used directly. For solids (e.g., tomato puree, spinach, potato chips, and ground beef), prepare a thick slurry of finely blended sample in water. Place an aliquot of the sample in specimen cell of the tristimulus colorimeter.

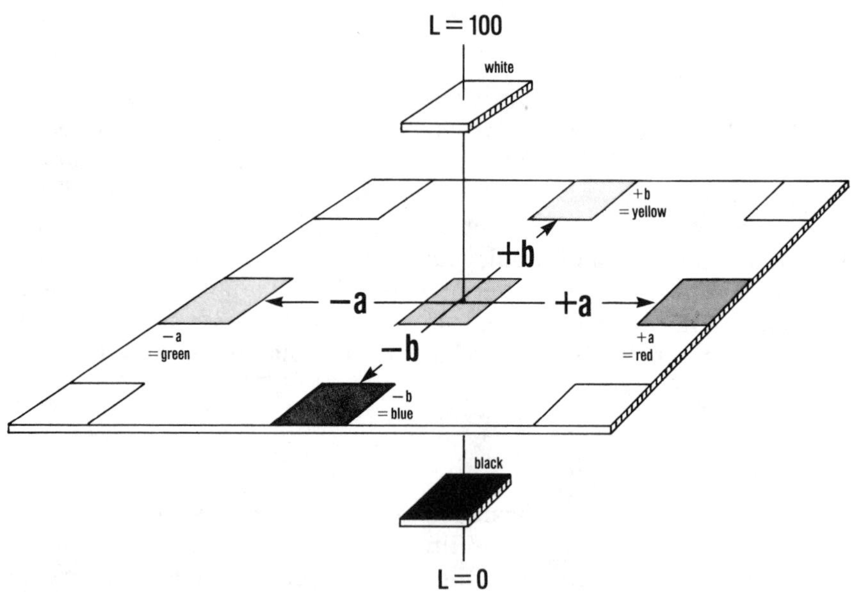

Figure. 10.9. Hunter color coordinates are basis of measurements with a tristimulus colorimeter. *Note*: The L,a,b coorcinates are mathematical transformations of the CIE coordinates (*x*, *y*, and *z*) and thus depend on the standard illuminant and standard observer.

Follow the manufacturer's instructions concerning operation of the tristimulus colorimeter to obtain measurements of L, a_L, and b_L for each sample.

Conversion to CIE System

Calculate the CIE tristimulus values (X, Y, Z) from the measured values for L, a_L, and b_L as follows:

$$Y = 0.01\ (L^2)$$

$$X = \frac{0.1(a_L L) + 0.175\ (L^2)}{17.85}$$

$$Z = \frac{0.07(L^2) - 0.1\ (b_L L)}{5.929}$$

Calculate the chromaticity coordinates (x,y,z) as described in the previous method. From a plot of y versus x on a chromaticity diagram, determine the dominant wavelength and purity of each sample (see Fig. 10.8).

Dominant wavelength and purity relate only to spectral colors produced by a single spectral stimulus. In the case of nonspectral colors (Fig. 10.10) such as purples and purplish reds, a complementary wave-

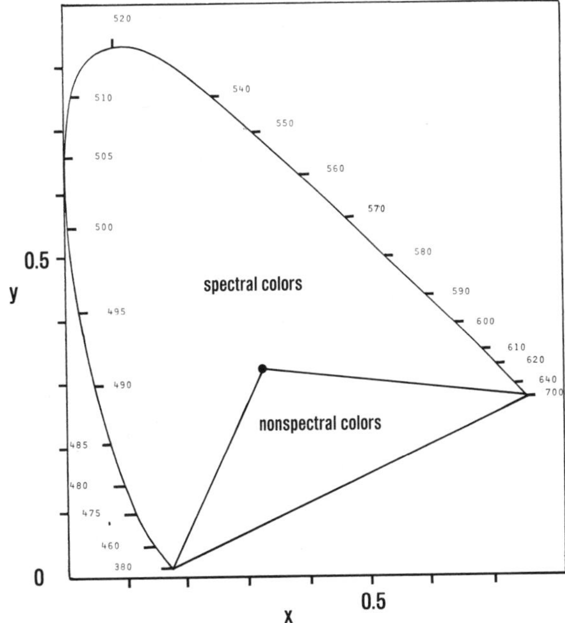

Figure 10.10. Division of chromaticity diagram into spectral and nonspectral areas.

length λ_c is specified rather than a dominant wavelength. The complementary wavelength (Fig. 10.11) is obtained by drawing a line from the sample chromaticity point D through the illuminant point C and beyond until it reaches the boundary of the color mixture diagram. This spectrum locus point E is designated the complementary wavelength, λ_c. Purity is calculated in a way similar to that used for spectral colors. The purity of a nonspectral color is the ratio of the distance between the illuminant and the sample (CD) to the distance between the illuminant and the nonspectral boundary (CF).

A microcomputer is used to calculate the chromaticity coordinates (x, y) and the tristimulus value (Y) from three components: (a) the object's measured reflectance/transmittance in the visible spectrum (380–770 nm), (b) the spectral irradiance of the standard illuminant at the same wavelengths, and (c) the color matching function for the x, y, and z values for either the 2° or 10° Observer. The second component is calculated using black-body radiation equations and depends on the source temperature and wavelength or is found in internally stored tables. Color matching functions have been tabulated by the CIE at 1 nm intervals for both Standard Observers and are also contained in memory. Standard colorimetry texts [e.g., Judd and Wyszecki (1975)] give the actual numerical values.

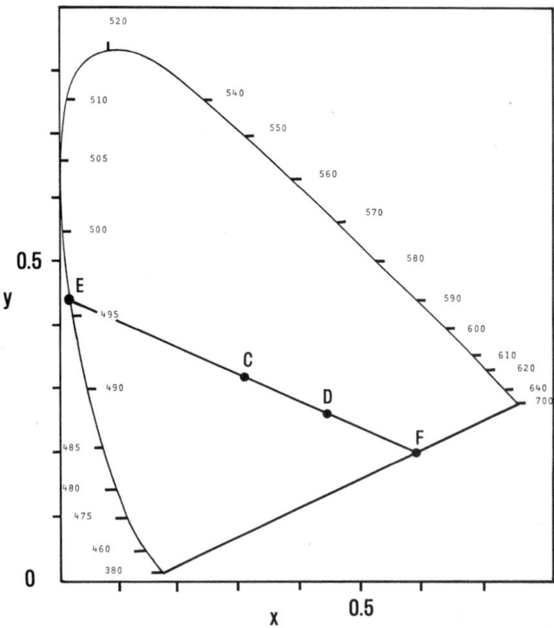

Figure 10.11. When the chromaticity point of a sample (D) falls in the nonspectral area, the complementary wavelength is specified. In this example, E is the complementary wavelength; the purity is the ratio of the distance CD to the distance CF.

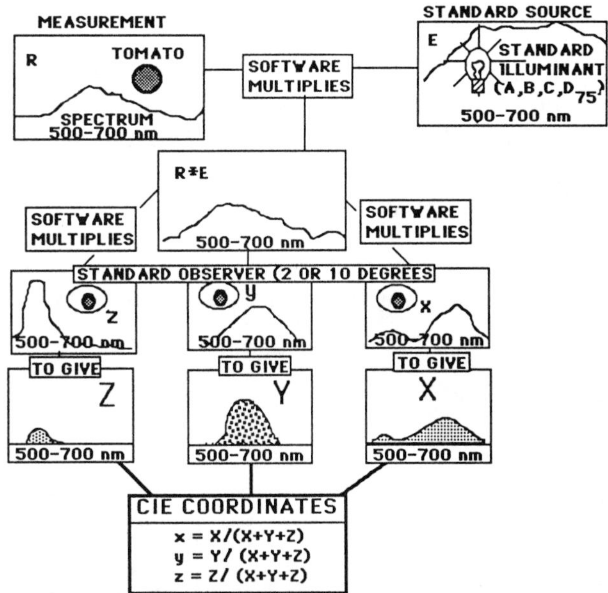

Figure 10.12. Diagrammatic sketch of how a computer calculates chromaticity coordinates and the converts them to specific color systems.

Table 10.8. Conversion of CIE Tristimulus Values (X, Y, Z) to Hunter L, a, b for Different CIE Standard Illuminants[a]

	CIE illuminant		
Constants	A "Light bulb"	B Noon sun	D_{65} Daylight
X_0	109.828	98.041	95.018
Y_0	100.00	100.00	100.00
Z_0	35.347	18.103	108.845
K_a	185	175	172
K_b	38	70	67

[a] K_a and K_b are chromaticity coefficients for illuminants. X_0, Y_0 and Z_0 are tristimulus values for perfect diffuser.

Hunter L, a, b: $L = 100\sqrt{Y/Y_0}$;

$$a = \frac{K_a\left[\dfrac{X}{X_0} - \dfrac{Y}{Y_0}\right]}{\sqrt{Y/Y_0}}; \qquad b = \frac{K_b\left[\dfrac{Y}{Y_0} - \dfrac{Z}{Z_0}\right]}{\sqrt{Y/Y_0}}$$

Diagrammatically, these values are calculated as shown in Fig. 10.12. The instrument digitally measures the reflectance (R) of the product in the visible region and these values are multiplied by the spectral irradiance (E) of the selected light source. The produce (ER) is multiplied by the color matching function for the desired standard observer (z, y, x) to give Z, Y, and X. The chromaticity coordinates (x, y, and z) are then calculated as shown before. Once these chromaticity coordinates are calculated there is little problem in mathematically converting them to other color systems (e.g., the Hunter L, a, b coordinates). Some color system interconversion equations are given in Table 10.8.

Hunter L, a, b Values for Tomato Juice
Reflectance 10 degrees CIE Standard Observer

Illuminant	Color Scale	Sample	Standard	Delta
	Hunter			
CIE Source C (Average Daylight)	L	25.53	98.84	-70.31
	a	10.33	-0.12	10.45
	b	6.66	0.11	6.55
	Hunter			
CIE Source A (Tungsten-Filament lamp)	L	30.54	98.84	-68.30
	a	14.27	-0.11	14.38
	b	4.86	0.05	4.81
	Hunter			
CIE Source B (Noon Sunlight)	L	29.06	98.84	-69.78
	a	11.91	-0.11	12.02
	b	6.18	0.09	6.09

Hunter L, a, b Plots
20 units / division

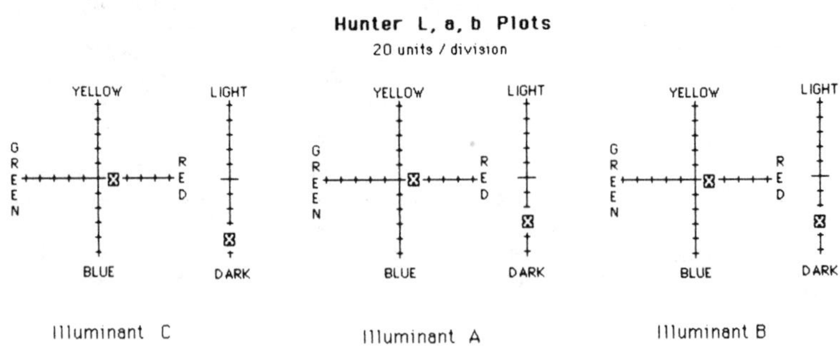

Figure 10.13. Hunter L, a, b value for tomato juice.

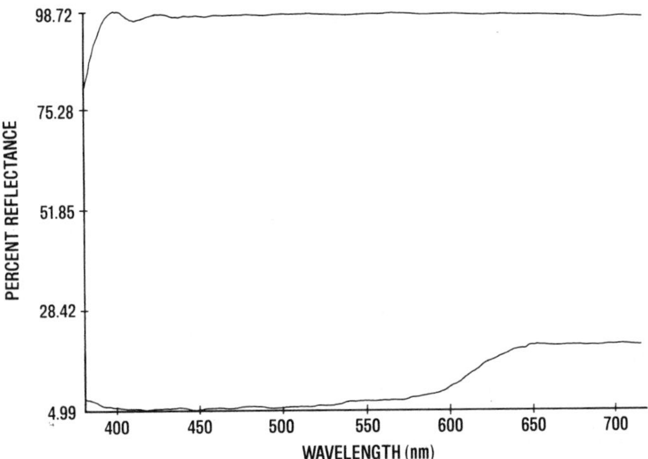

Figure 10.14. Spectral reflectance for tomato juice. Solid line indicates sample, dotted line indicates standard.

Apparatus

1. Digital color analysis system (e.g., Pacific Scientific Spectrogard Color System or Milton Roy Diano Match Mate Color Analysis System) and an appropriate minicomputer (Zenith Z-90 for Spectrogard or DEC 300 PC for Milton Roy). Procedures vary with the instrument manufacturer, software, and computer.
2. Tomato juice (any brand).

Procedure for Making Measurements

Turn on instrument and computer system. Load program diskette (which contains program, illuminance functions, and color weighting functions for 10-nm increments). Select mode (operate) and geometry (reflectance). After black calibration (set 0% reflectance), 100% reflectance is calibrated using a white tile. The setup menu displayed by the monitor (Observer, Color Scale, Illuminants) is then used to select the desired Observer (10%), Color Scale (Lab-Hunter), and Illuminants (C: average daylight; A: tungsten-filament and B: noon sunlight). Place sample in sample holder and collect spectral data (press SAMPLE). Place standard (red tile) in sample holder and collect spectral data (press STANDARD). Based on the setup menu entries, the Lh, Ah, and Bh values are given and plotted for the sample (tomato juice) and Standard (red tile) on the screen. This information can be printed (Fig. 10.13) along with the spectral data (Fig. 10.14) using the appropriate software commands. Other instruments have different procedures.

SELECTED REFERENCES

ANON. 1967. Grades of table maple syrup. Fed. Reg. *32*(100), 7579.

AOAC. 1950. Official Methods of Analysis. 7th ed. Assoc. of Official Analytical Chemists, Arlington, VA.

AOAC. 1970. Official Methods of Analysis. 11th ed. W. Horwitz (editor). Assoc. of Official Analytical Chemists, Arlington, VA.

AOAC. 1984. Official Methods of Analysis. 14th ed. S. Williams (editor). Assoc. of Official Analytical Chemists, Arlington, VA.

BAUSCH and LOMB. Spectronic 20—Spectrophotometer Operators Manual.

FRANCIS, F. J. and CLYDESDALE, F. M. 1975. Food Colorimetry: Theory and Applications. AVI Publishing Co., Westport, CT.

GENERAL ELECTRIC. 1978. Light and Color. Cleveland, OH.

JOSLYN, M. A. 1970. Methods in Food Analysis. Academic Press, New York.

JUDD, D. B. and WYSZECKI, G. 1975. Color in Business, Science and Industry. 3rd ed. John Wiley & Sons, New York.

MARMION, D. M. 1979. Handbook of U.S. Colorants for Foods, Drugs, and Cosmetics. John Wiley & Sons, New York.

MUNSELL COLOR COMPANY. 1941. Book of Color. Baltimore, MD.

NICKERSON, D. 1946. Color Measurement and Its Application to the Grading of Agricultural Products. U.S. Dept. of Agriculture, Misc. Publi. *580*.

PRZYBYLSKI, W., SMITH, R. B. and McKEOWN, G. G. 1960. Determination of coal tar colors on oranges. J. Assoc. Offic. Anal. Chem. *43*, 274–278.

STANLEY, R. L. and KIRK, P. L. 1963. Systematic identification of artificial food colors permitted in the United States. J. Agric. Food Chem. *11*, 492–495.

WARNER-JENKINSON CO. 1984. Certified Food Colors. St. Louis, MO 63178-4538.

11

Wheat and Wheat Products

☐ INTRODUCTION

Cereals are the dried seeds of the cultivated grasses, which belong to the family Gramineae. They include wheat, rye, barley, corn, oats, grain sorghum, and millet. Buckwheat, although not a true cereal, is usually included with them.

The cultivation and use of cereals antedates the recorded history of man. Excavations in early centers of civilization have indicated that one or another of the above cereals were known and used by the people of these cultures. The civilizations of Babylonia, Egypt, Greece, and Rome were founded on the production of wheat, barley, and the millets. The cultures of India, China, and Japan were dependent on the rice crop. The Inca, Maya, and Aztec cultures were based on the growing of corn (maize), which is apparently the only cereal indigenous to the Americas.

The cereals are, in general, the cheapest sources of food energy and can be grown almost anywhere in the world. They give high yields per acre, can be stored for relatively long periods of time, and can be transported cheaply.

In addition to being good sources of energy, the cereals are good sources of protein, the B vitamins, iron, and phosphorus. The protein of cereals is incomplete because it is low in lysine and tryptophan (see Proteins and Nutrition in Chapter 6). This deficiency is easily corrected by the addition of other foods to the diet such as meat, milk, eggs, or vegetable proteins. Cereals are poor sources of vitamins A and C, but these can be obtained from fruits and vegetables.

To a large extent climate determines the cereal species that can be grown in a region. Rice, the principal food of more than half of the world's population, is the basic cereal in the densely populated areas of Asia from India to Japan and the adjacent islands. Rye can be grown in colder climates and is used for breadmaking in the northern European countries. Corn, oats, and barley, which are grown in countries with a temperate climate, are used extensively as animal feeds and to a more limited extent as human food. Sorghum grain is grown in India and Africa. Millets are used as food primarily in eastern and southern Asia, parts of Africa, and parts of the Soviet Union. Wheat is grown in all countries lying in the temperate zone. In the United States, wheat is the most widely used cereal, hence it will be the only cereal discussed in detail in this volume.

□ CLASSIFICATION OF WHEATS

Wheat belongs to the grass family, Gramineae, and the genus *Triticum*. The known species and varieties of the genus *Triticum* are said to number over 30,000. They can be assembled into three groups (races), which are traced from separate original ancestors and which differ in their number of chromosomes, as shown in Table 11.1.

The classification shown in Table 11.1 suggests a possible explanation for differences in the flour from the three wheat groups. Einkorn is used only as an animal feed. Emmer wheats are used for macaroni, spaghetti, and other pasta products, and not for breadmaking. Wheat from the spelt group is used in baked goods and other cereal products. The principal wheats of commerce are *T. aestivum* and *T. compactum*, both in the spelt group, and *T. durum*, in the emmer group.

In addition to their classification into groups based on chromosome number, wheats are classified as hard or soft (referring to milling character) and strong or weak (referring to baking character).

The terms hard and soft wheat relate to the way the endosperm breaks during milling. In hard wheats, the endosperm tends to fracture along the lines of the cell boundaries, whereas in soft wheats the en-

Table 11.1. Classification of Wheat Species

Group	No. of chromosomes	Species names	Common name
Einkorn	7	*T. monococcum*	Einkorn
Emmer	14	*T. dicoccum*	Emmer
Emmer	14	*T. durum*	Macaroni wheat
Emmer	14	*T. turgidum*	Rivet, Cone
Emmer	14	*T. polonicum*	Polish wheat
Spelt	21	*T. aestivum*	Bread wheat
Spelt	21	*T. compactum*	Club
Spelt	21	*T. spelta*	Spelt
Spelt	21	*T. spharecoccum*	Indian dwarf
Spelt	21	*T. vulgare*	Bread wheat

Source: Tschermak (1914).

dosperm fractures in a random way. Hard wheats yield a coarse flour, consisting of regular-shaped particles, which is free-flowing and easily sifted. In contrast, the flour from soft wheats is very fine and consists of irregularly-shaped particles, which tend to adhere together; such flour sifts with some difficulty.

Strength is a characteristic of wheat associated with the ability of its flour to produce bread of large loaf volume, good crumb texture, and good keeping qualities. Strong wheats as a rule have a high protein content, whereas weak wheats have low protein contents. The flour from a weak wheat produces bread of small loaf volume, coarse crumb structure, and low protein content. Thus protein content is related to baking characteristics.

☐ GRAIN STANDARDS

The official grain standards of the United States divide wheat into seven commercial classes: hard red spring, durum, red durum, hard red winter, soft red winter, white, and mixed wheat. With the exception of mixed wheat, each of the classes is grown in certain areas or sections of the country depending largely upon climatic conditions. The amount of rainfall and temperature immediately preceding the ripening of a wheat crop affects the starch and protein content of the grain. In general, higher starch and lower protein contents occur when summers are cool and rainfall high, resulting in a relatively long growth period. Higher protein content occurs when summers are hot and dry.

Table 11.2. Classes and Subclasses of Wheat According to U.S. Grain Standards

Class	Subclass	Specifications
Hard Red Spring	Dark Northern Spring	75% or more of dark, hard, vitreous kernels
	Northern Spring	25–75% of dark, hard, vitreous kernels
	Red Spring	Less than 25% of dark hard, vitreous kernels
Durum	Hard Amber Durum	75% or more of hard, vitreous, amber kernels
	Amber Durum	60–75% of hard, vitreous, amber kernels
	Durum	Less than 60% of hard, vitreous, amber kernels
Red Durum	—	All varieties of Red Durum
Hard Red Winter	Dark Hard Winter	75% or more dark, hard vitreous kernels
	Hard Winter	40–75% of dark, hard, vitreous kernels
	Yellow Hard Winter	Less than 40% of dark, hard, vitreous kernels
Soft Red Winter	—	All varieties of soft red winter
White	Hard White	75% or more of hard kernels and not more than 10% of wheat of white club varieties
	Soft White	Less than 75% of hard kernels and not more than 10% of wheat of white club varieties
	White Club	Not more than 10% of other white wheats
	Western white	More than 10% of wheat of the white club varieties and not more than 10% of other white wheats
Mixed	—	All mixtures not provided for in other classes

Characteristics of Wheat Classes

The classes and subclasses of wheat provided for by federal standards are summarized in Table 11.2. The regions where the various classes are grown and their principal uses are described in the following sections.

Hard Red Spring Wheat

Hard red spring wheat is grown principally in North and South Dakota, Montana, Minnesota, and the Prairie Provinces of Canada. It may be grown in other areas to replace winter wheat that has failed due to winter-killing, drought, or other causes. Hard red spring wheat is divided into three subclasses according to its content of dark, hard, and vitreous kernels. These subclasses are dark northern spring wheat (75% or more), northern spring wheat (25–75%), and red spring wheat (less than 25%). Overall, HRS wheat accounts for about 20% of the total wheat acreage in the United States. This class of wheat is used for bread and rolls.

Durum and Red Durum Wheats

Durum and red durum wheats are grown in the north-central states, especially North and South Dakota. These classes of wheat account for approximately 5% of the wheat acreage of the United States. Durum wheat cannot be used to produce flour for bread making; rather it is milled to supply semolina for making macaroni, spaghetti, noodles, etc.

Durum wheat is divided into three subclasses: hard amber durum wheat, which has 75% or more of hard, vitreous kernels of amber color; amber durum wheat, which has 60–75% of hard, vitreous kernels of amber color; and durum wheat, which has less than 60% of hard, vitreous kernels of amber color. The red durum class includes all varieties of red durum wheat and has no subclasses.

Hard Red Winter Wheat (HRW)

Hard red winter wheat is grown principally in Texas, Oklahoma, Kansas, Colorado, Nebraska, Montana, South Dakota, and Minnesota. This class of wheat is grown wherever the winters are not too severe because it yields more per acre than spring wheat. This higher productivity occurs primarily because (1) winter wheat, which is sown in the fall, has a longer growing season than spring-sown wheat, and (2) winter wheat matures before the onset of hot weather, drought, and disease.

Hard red winter wheat is the most important class of wheat grown in the United States and accounts for about 50% of the total wheat acreage of the country. It is used for making bread and hard rolls. This class is divided into three subclasses: dark hard winter wheat with 75% or more of dark, hard, and vitreous kernels; hard winter wheat with 40–75% of dark, hard, and vitreous kernels; and yellow hard winter wheat with less than 40% of dark, hard, and vitreous kernels.

Soft Red Winter Wheat

Soft red winter wheat is grown primarily in the areas east and south of where hard red winter wheat is grown. The principle soft red winter wheat producing states include Missouri, Illinois, Indiana, Ohio, Pennsylvania, Virginia, West Virginia, Kentucky, North Carolina, and

South Carolina. A small acreage is also grown in the Pacific Northwest. This type of wheat accounts for approximately 15% of the total wheat acreage of the United States. The flour is used for cakes, cookies, crackers, and pastries.

White Wheat

The principal white wheat growing states include Michigan, New York, Washington, Oregon, Idaho, and California. The acreage devoted to white wheat is approximately 10% of the total wheat acreage. White wheat includes four subclasses: hard white must contain 75% or more of hard kernels and not more than 10% of the white club varieties; soft white contains less than 75% of hard kernels and not more than 10% of the white club varieties; white club consists of wheat of white club varieties and not more than 10% of other white wheat; and western white contains more than 10% of wheat of white club varieties and not more than 10% of other white wheat. The flour from white wheat is unsuitable for breadmaking, but is suitable for cakes, cookies, and pastries.

Mixed Wheat

The mixed wheat class includes all mixtures not provided for in the other classes.

Defects and Impurities

Wheat and other cereal grains are subject to government inspection and grading. In establishing the grain grades for wheat, certain defects and impurities in wheat have been identified as definitely affecting the yield and quality of the wheat. Consequently, it is important to consider some of the causes of these defects and their effect on the quality of wheat.

Damage Before Harvest

Before harvest, wheat may be infected with a fungal disease (rust) caused by a species of the genus *Puccinia*. Rust infection in wheat plants results in the production of kernels that are usually wrinkled, shrunken, and lightweight. Thus, the yield per acre of a rust-infected crop is lower than that of a healthy crop, and often the quality of its flour is reduced. This problem can best be controlled through use of rust-resistant strains. It should be noted, however, that rusts exist in many different forms and from time to time new rust forms occur to which hitherto resistant wheat strains may be susceptible.

Frosted wheat refers to wheat that has been damaged by heavy frost. This defect occurs only in the northern states and Canada, principally in areas growing spring wheat. The extent of the damage caused by heavy frost is dependent on the maturity of the kernels. If the grain is nearly ripened (stiff dough stage), the damage from freezing is slight. If the kernels are in the milk stage, the damage is greater and the milling quality of the grain is affected. Such kernels exhibit blistering of the bran and often show discoloration. Flour produced from heavy frosted grain produces bread of lower loaf volume, poorer texture, and darker color. The flour yield from such grain is reduced and the ash content is likely to be abnormally high.

Sprouted wheat, which may occur when wet weather follows wheat maturation, has a lower yield and produces flour with high α-amylase activity. Wheat that is low in α-amylase activity may have its diastatic activity increased by the addition of small amounts (2–3%) of sprouted wheat. However, excessive α-amylase activity seriously reduces the breadmaking quality of a flour.

Damage During Harvesting

The most obvious damage to wheat during harvesting is mechanical injury to the kernels due to incorrect setting of the threshing equipment. In addition, at the same time of combine harvesting, wheat should have a moisture content of less than 15%, otherwise mold develops. Artificial drying of grain may result in heat damage if the drying is too rapid and at too high a temperature. Heat damage is usually accompanied by discoloration of the kernels; however, the proteins of the grain may be heat-damaged without any discoloration of the kernels occurring.

Damage During Storage

Bin-burned wheat is wheat that has been damaged by spontaneous heating during storage. Such heating occurs when the normal respiration rate of the wheat kernels increases. This can be caused by an increased moisture content (especially above 15%) and higher temperature (above about 20°C). With increased respiration there is an increase in temperature, which, unless it is dissipated, will further increase the rate of respiration. The process is accumulative and temperatures may increase to a point where the kernels become brown and charred. Other factors, such as infestation by insects (which have a higher respiration rate than wheat kernels) and relatively large amounts of broken and shriveled kernels, also increase the respiration rate. Flour from bin-burned wheat produces bread with decreased loaf volume, poorer texture, darker color, and detrimental odor and flavor.

Impurities

The quantity and nature of the extraneous matter frequently encountered in wheat are an important criteria of quality. These impurities include the following:

- Plant matter—weed seeds and bulbs, other cereal grains, plant residues such as straw and chaff, and fungus (ergot, bunt)
- Animal matter—rodent excreta and hairs, insects, and mites
- Other foreign material—stones, dust, earth; string; metal objects (nails, bits of wire, etc.), and miscellaneous trash

These impurities together with damaged wheat grain are removed from wheat in the mill as screenings (*dockage* in the United States; *besatz* in Europe).

According to U.S. wheat standards, a distinction is made between dockage and foreign material. Dockage is defined as all matter that can be removed readily from wheat by prescribed mechanical means, while foreign material is defined as all matter other than wheat that remains after the removal of dockage and shrunken and broken kernels.

The International Association for Cereal Chemistry has adopted a system for evaluating the quality of wheat on the basis of besatz. Besatz in wheat includes all materials other than sound whole kernels of grain. Thus, in terms of U.S. standards, besatz includes dockage, foreign material, damaged kernels, shrunken kernels, and broken kernels.

Removal of impurities from wheat involves extra work and consequently affects the sale price of the wheat. In the United States it is an established marketing practice to deduct the weight of the dockage from the total weight of the product.

Grades of Wheat

After commercial wheat is placed into classes (e.g., hard red spring wheat) and subclasses (e.g., dark northern spring), it is further characterized by being assigned a grade designation, which is dependent upon the quality and condition of the wheat. Quality refers to such characteristics as plumpness, soundness, and cleanliness of the grain. The criteria of wheat quality include (a) test-weight per bushel, (b) content of damaged and shrunken kernels, (c) dockage, (d) foreign material, and (e) wheat of other classes. The criteria of wheat condition are (a) moisture content, (b) odor of the grain, and (c) freedom from smut, ergot, garlic, insects, staining, liming, or bleaching.

The requirements for the major wheat grades in the United States are presented in Table 11.3. Several special grades also are identified in the U.S. standards and are applied in addition to all other grade designations. *Ergoty wheat* contains more than 0.30% of ergot. *Light garlicky wheat* contains in a 1000-g portion two or more, but not more than six, green garlic bulblets, or an equivalent of dry or partly dry bulblets. *Garlicky wheat* contains in a 1000-g portion more than six green garlic bulblets. *Light smutty wheat* has an unmistakable odor of smut, or contains in a 250-g portion smut balls, portions of smut balls, or smut in excess of a quantity equal to 14 smut balls, but not in excess of a quantity equal to 14 large smut balls, nor in excess of a quantity

Table 11.3. Grade Requirements for Wheat Except Mixed Wheat

	U.S. Grades				
Requirements	No. 1	No. 2	No. 3	No. 4	No. 5
Minimum test wt. (lb/bu)					
Hard Red Spring or White Club Classes	58.0	57.0	55.0	53.0	50.0
All other classes and subclasses	60.0	58.0	56.0	54.0	51.0
Maximum limits defects (%)					
Heat-damaged kernels	0.1	0.2	0.5	1.0	3.0
Damaged kernels (total)	2.0	4.0	7.0	10.0	15.0
Foreign material	0.5	1.0	2.0	3.0	5.0
Shrunken and broken kernels	3.0	5.0	8.0	12.0	20.0
Defects, (total)	3.0	5.0	8.0	12.0	20.0
Wheat of other classes[a]					
Contrasting classes	1.0	2.0	3.0	10.0	10.0
Wheat of other classes (total)	3.0	5.0	10.0	10.0	10.0

Note: U.S. sample grade shall be wheat that does not meet the requirement for any of the grades from U.S. No. 1 to U.S. No. 5, inclusive; or that contains a quantity of smut so great that one or more of the grade requirements cannot be determined accurately; or that contains 8 or more stones; 2 or more pieces of glass; 3 or more Crotalaria seeds (*Crotalaria* sp.); 3 or more castor beans (*Ricinus communis*); 4 or more particles of an unknown foreign substance(s); or 2 or more rodent pellets or an equivalent quantity of other animal filth per 1000 g of wheat; or that has a musty, sour, or commercially objectionable foreign odor (except smut or garlic odor); or that is otherwise of low quality.

[a] Unclassed wheat of any grade may contain not more than 10% of wheat of other classes.

equal to 30 smut balls of average size. *Smutty wheat* contains in a 250-g portion smut balls, portions of smut balls, or spores of smut in excess of a quantity equal to 30 smut balls of average size. *Weevily wheat* is infested with live weevils or other insects injurious to stored grain. *Treated wheat* is wheat that has been scoured, limed, washed, sulfured, or treated in such a manner that the true quality is not reflected by either the U.S. numerical grade or the U.S. Sample grade designation alone.

☐ ANALYSIS OF WHEAT

The initial evaluation of the quality of wheat is often based only on the physical criteria discussed in the previous section. Today's food technology, however, requires more precise chemical criteria of wheat quality. Among the common analyses are determination of moisture content, crude protein, ash, and crude fat. The validity of any of these analyses depends on proper sampling, which is discussed before the analytical procedures are presented.

Sampling

A double-tube compartment probe should be used to sample bulk wheat (and other whole grains) in a railcar, truck, wagon, or any container in which the grain is about the same depth as it is in a carload. Samples are removed with the probe from five or more places, well distributed throughout the container. For example, samples taken with five probings from a railroad car should be procured as follows: probe 1 in center of car; probe 2 from 3 to 5 ft back from door post toward center of the car and approximately 2 ft out from one side of the car; probe 3 from 3 to 5 ft from the same end of the car and approximately 2 ft from the opposite side of the car as in probe 2; probes 4 and 5 are obtained in the same manner as probes 2 and 3 except from the opposite ends and sides of the car.

For sampling grain being loaded aboard a boat, barge, or vessel, obtain samples from the loading spout by use of a "Pelican" at regular and frequent intervals so as to procure an average and representative sample.

For sampling sacked grain, obtain samples from many individual sacks, selected at random, so as to obtain an average and representative sample of the entire lot.

Reduce the amount of grain sample to one of laboratory size, preferably with the aid of a Boerner sampler, or riffle, or any other device that will give representative portions.

Preparation of Sample for Analysis

Grind 30–40 g of a representative wheat sample on a Burr mill or Wiley Mill (18- or 20-mesh screen) to a particle size so that at least 50% will pass through a 36 GG (grits gauze) sieve. Mix the ground sample thoroughly and place in an airtight container of such size that the sample completely fills the container. Store in a cool place.

If the sample of wheat is damp (contains more than 13% moisture), it is preferable to air-dry it before grinding because moisture may be lost in the sample grinding process. It is necessary, however, to determine the percentage of moisture lost in air-drying. (AACC 62-70)

Moisture

A knowledge of the moisture content of wheat is useful in a number of ways. Wheat must have a relatively low moisture content to be stored without damage. Wheat with a high moisture content may "heat" on storage and bin-burn or become "sour" or moldy. On the other hand, if wheat is too dry, the kernels break easily during handling; such broken kernels are removed in cleaning operations and as a consequence are of little milling value. In addition, very dry kernels are at times difficult to temper to the moisture level required for milling.

The amount of dry matter in wheat is inversely related to the moisture content of the wheat. As a consequence, in purchasing wheat for flour manufacture it is economically advantageous to purchase wheat with a lower moisture content than the resulting milled products may contain when marketed. For example, a difference of 1% moisture in 2000 bu of wheat is equivalent to about 1200 lb of dry wheat; at 14% moisture, this amounts to 23.1 bu, or enough wheat to make approximately 5 barrels of flour.

The method for determining moisture content depends on the moisture content of the grain. For samples containing less than 13% moisture, a one-stage method is suitable, but for samples with more than 13% moisture a two-stage method is required.

One-Stage Method

Weigh accurately 2- to 3-g portions of the prepared wheat sample into two or more dried, tared metal dishes about 55 mm in diameter.

Place uncovered dishes in an oven provided with ventilation ports and set to maintain a temperature of 130° ± 3°C. The thermometer should be placed so that the tip of the bulb is level with the top of the moisture dishes, but not directly over any dish. Dry for exactly 1 hr after the oven regains a temperature of 130°C. Cover the dishes while they are still in the oven, transfer them to a desiccator, and weigh after the dishes have reached room temperature (45–60 min).

Two-Stage Method

Fill several tared moisture dishes with an unground wheat sample, cover, and weigh. Remove the covers and place the dishes in warm, well-ventilated places (preferably on top of a heated oven) so that after 14–16 hr the wheat will have reached an air-dry condition. Replace the covers, cool, and again weigh the dishes. Calculate the percentage of moisture lost in air-drying. Grind the air-dried sample to the appropriate particle size and proceed as directed in the one-stage procedure. Calculate the percentage moisture in the original sample from the following equation:

$$\text{T.M.} = A + \frac{(100 - A)B}{100}$$

where T.M. = % total moisture; A = % moisture lost in air-drying; and B = % moisture in air-dried sample as determined by oven-drying at 130°C.

Crude Protein

The protein content of wheat varies from about 8% to about 22%, with an average of about 14%. The variation in composition is due in part to the variety and class of wheat, as well as to environmental factors (e.g., rainfall during the period of grain development and soil nitrogen). Flour for the production of yeast-leavened bread requires wheat with a protein content of at least 12%, whereas flour for other uses is made from wheats of lower protein content. Consequently, the crude protein content is an important determinant of the uses for a particular wheat. In some cases, wheat of high-protein content may be blended with low-protein wheat to obtain a wheat with the proper protein content for a particular use.

The crude protein content of wheat is determined on a 1-g sample by the Kjeldahl nitrogen procedure described in Chapter 2. The percentage of crude protein is obtained by multiplying the percentage of nitrogen by the factor 5.7. (For other grains, the customary factor of 6.25 is used.)

Ash

Accurately weigh 2–5 g of a ground wheat sample into a previously ignited and weighed porcelain or platinum crucible. Carefully ignite the sample over a burner and heat until the sample is thoroughly charred. Transfer the crucible and contents to a muffle furnace held at a dull red heat (550°–600°) and continue the ashing process until the ash is white or gray in appearance. Cool in a desiccator and weigh. Reheat in the muffle furnace for 1-hr intervals until a constant weight is obtained. Calculate the percentage of ash on the original sample and on a 15% moisture basis.

Crude Fat

The "crude fat" method will determine as fat all materials extracted by ethyl ether. In addition to the true fat (triglycerides), the extract may include phospholipids, sterols, fat-soluble pigments, free fatty acids, etc. For this reason the term "crude fat" or "ether extract" can be used interchangeably.

The material to be analyzed must be thoroughly dried in an oven and the solvent must be anhydrous so that water-soluble materials are not extracted and determined as fat. Two types of apparatus are in general use: continuous extraction and intermittent extraction. Intermittent extractors; for example, Soxhlet extractors, are more efficient than continuous extractors.

Procedure

For wheat and flour dry a 2-g sample preferably in a vacuum oven at 100°C, to remove moisture (or use residue from moisture determination). Transfer the dried sample to an extraction thimble and plug the top of the thimble with fat-free cotton. (For a fine product such as flour, an equal weight of clean, dry sand may be mixed with the sample in the thimble.) Drop the thimble and its contents into a Soxhlet extractor. Pour approximately 75 ml of anhydrous ether through the sample in the extractor and into a weighed flask below the extractor. Extract at a rate of 2–3 drops per second for 16 hr to remove all the ether-extractable materials. At the conclusion of the extraction period, detach from the apparatus the previously tared Soxhlet flask containing the extracted material in ether. Remove ether by cautious evaporation, dry the flash and its contents in a drying oven at 100°C for 45 min, cool in a dessicator, and weigh the flask. The weight of ether extract is the increase in weight of the flask and its contents over the original weight of the flask. Calculate the percentage of crude fat in the original sample and on a 15% moisture basis.

Crude Fiber

Crude fiber, along with the ash content, is associated with the amount of bran in a wheat sample. Consequently, these two chemical values can be used as indicators of wheat quality because small kernels or shriveled kernels usually have more bran, on a percentage basis, than large kernels.

Using the procedure described in Chapter 2 (Crude Fiber Determination in Grains), determine crude fiber on a sample previously extracted in the crude fat determination. Calculate percentage of crude fiber in the original sample and on a 15% moisture basis.

☐ PRODUCTION OF WHEAT FLOUR

Whole wheat grain can be used in various ways but it is usually ground in preparation for use in foods. Discussion of grinding (or milling) requires some knowledge of the grain structure.

The wheat grain is in general composed of three main parts: the embryo (germ), endosperm, and bran. The embryo, or germ, from which the root and leaf of the plant are formed, contains a relatively high percentage of oil. Due to the oil content, germ fragments tend to flatten out when passed between rollers in the milling process. The endosperm consists mainly of potential food for the developing embryo plant. It forms 81–84% of the grain and consists of an aleurone layer and starch cells surrounded by protein. The aleurone layer is removed in the milling process and goes along with the bran. The starch endosperm is friable and tends to pulverize when passed between rolls in the milling process. The bran, which forms a protective covering for the grain, is composed of two pericarp layers, one testa layer, and a perisperm layer. The branny structure is fibrous and tough so that when tempered with moisture it does not readily break up in the milling process. The bran forms roughly 14–16% of the grain.

Milling is a process for removing the bran and germ of the wheat kernel from the starchy endosperm, then pulverizing the pure endosperm to yield flour. Historically, the purpose of milling has been to produce flour of greater refinement. On the average, only 72–75% of the potential flour in the endosperm is obtained by the flour-milling process. The term *flour extraction*, used as an index of the efficiency of the milling process, can be defined as the proportion of flour by weight obtained from a known quantity of grain. The flour extraction is also used to define grades of flour. Thus, if the flour extraction is 75% or less, a white flour is implied; as this proportion approaches 100%; it indicates an increasing amount of non-endosperm material. A flour extraction of 100% implies a whole-wheat flour. The various steps in the production of white flour by roller-milling are described in the following sections.

Cleaning

Wheat as delivered to the mill generally contains extraneous matter such as straw, chaff, pebbles, bits of metal, soil, and seeds of various kinds which must be removed from the wheat before it goes into the storage bins at the mill. Coarse and fine materials are removed by passing the wheat through sieves. The wheat is then dried to a moisture content of less than 14.5%.

As grain is needed for milling, the wheat is withdrawn from the concrete storage bins and brought into the cleaning house at the mill. Here it is passed through sieves of varying sizes to remove the coarse and fine impurities. Wheat chaff is removed by air currents, magnets remove bits of metal. Another cleaning operation removes the cereal grains other than wheat, and the weed seeds. Next, the wheat is scoured by passing it through an emery-lined cylinder to remove by friction, the surface dirt on the wheat kernel.

The main impurities in hard wheat from the Great Plains are weed seeds. These are removed by a graduated series of sieves with or without airblast and/or special seed separators. Smut is more prevalent in wheat grown in the Pacific northwest. Smut is removed from wheat by washing the grain or scouring it with lime, followed by thorough washing. Garlic is a common impurity in wheat from the eastern states. Removal of garlic bulblets is accomplished by aspiration.

The final cleaning step in the cleaning operation is a water wash. This step dissolves the dirt and permits grit and bits of metal to sink. Following washing, the excess water on the surface of the wheat kernel is removed by centrifuging.

Tempering

The objective of wheat tempering is primarily to improve the physical state of the grain for milling. Tempering involves adding water to wheat to raise the moisture content (to 15–19% for hard wheat and 14.5–17% for soft wheat) by allowing the wheat to stand in water for 18–72 hr. During this interval, the outer layers of the bran quickly absorb water, and the bran becomes tough and rubbery. The net result is a bran that is easily separated from the endosperm during milling. The conditions of tempering (amount of water, length of time in the tempering bin, and temperature) vary depending upon the original moisture content of the wheat, the plumpness of the kernel, and the hardness of the grain.

Breaking the Wheat Kernel

In the first step in the actual milling process, tempered wheat is crushed between fluted rollers called break rolls. These rolls are mounted in pairs and they rotate toward each other at different rates of speed. The rapidly moving roll rotates about 2.5 times faster than the slower one. The narrow gap, or *nip*, between the rolls can be adjusted. The speed differential between the two rolls produces a shearing rather than grinding effect. The *first break grind* shears the grain into chunks of different sizes and textures, which are separated by sieving or bolting into fine particles (flour), intermediate-size particles (either semolina, middlings dunst, or bran snips), and coarse particles (break, chop, or stock). The break stocks are fed to a second break; this process may continue through five or six breaks. The number of flutes on the rolls becomes progressively greater at each successive break (12, 14, 18, 26, etc., to the linear inch), and the roll gap becomes progressively narrower (0.020–0.003 in.). The coarse fraction (chop) from the last break grind yields no additional endosperm and is principally bran.

Sifting or Bolting

Whenever the product has passed through a set of break rolls, it must be sifted to segregate it into fractions of different particle size. This sieving process is carried out in a plansifter, a machine consisting of flat sieves piled in tiers one above the other. The action of the sifter is rotary and revolves in a plane horizontal to the floor. The product to be sifted is fed in at the top and drops from sieve to sieve while oversized particles travel across the sieve to a collecting trough and are removed. A plansifter may incorporate four or five separate compartments with sieves of different mesh size, thereby delivering fractions of different particle size. These fractions include (1) coarse fragments that are reground and eventually give bran, (2) fine particles that constitute a low-grade flour, (3) coarse granular fragments (coarse middlings) that are later reduced to fine middlings by the separation, by sifting, of germ and shorts; and (4) fine granular fragments (fine middlings) that are pulverized to high-grade flour.

Sieves employed in the sifting (or bolting) process may be constructed of woven wire, silk, or nylon. Wire is generally used in separating coarse branny particles, whereas the other sieves are made of silk or nylon. Coarse bolting cloths have 30, 45, 50, or 70 mesh/linear in. These are used in sifting out fine branny material from the middlings and also for grading middlings into different sizes. The cloths through which flour is bolted have 129–196 mesh/in.; the finest cloths have openings of about 0.06 mm.

Purification

During the purification step minute-sized bran particls are removed from the segregated products obtained by bolting. A purifier is essentially an oscillating sieve, inclined downwards with the sieves becoming progressively coarser in mesh from head to tail. The purifier takes the middlings from the sifter and further segregates the particles according to size. Of greater importance however, are the air currents forced in the direction from floor to ceiling. The middlings are fed onto the head end of the purifier and are moved down by oscillating motion towards the tail, or coarse, sieve section. Particles of endosperm, because of their density, fall through the sieves against the direction of the air flow; the light branny material is lifted by the air current and collected in trays suspended above the sieves. The latter go to millfeed. Particles intermediate in density (consisting of bran and bran plus endosperm) are not lifted, nor do they fall through the sieves; rather they remain floating on the sieves and are removed as overs and returned to the break mills. The "clean" middlings are reduced to flour.

Reduction System

The reduction system consists of two parts: the roll mills and the sifting machines. The purified middlings are pulverized to the desired fineness by passage through a series of smooth rolls. These rolls are graduated so that the product of successive reductions becomes finer and finer. After each reduction, the resulting product is sent to sifters to separate the flour from the middlings and chop. The oversized material is sent back to the reduction rolls for further processing. Flour produced from the first reduction of the purified middlings is more highly refined than that produced by subsequent grinding of the residues. Subsequent grindings yield lower-grade flour because the quantity of bran fragment in the flour becomes greater. The remnant of the reduction process is used for animal feed.

Air Classification

Some flour mills add an air classifier to their mill stream in order to further process flour into fractions of widely different protein content. In the normal roller-milling process, the minimum size separation on a sieve is about 60 μm, whereas air classification separates flour particles differing in size by 10–40 μm.

In this process, finished flour from the roller mill is further reduced in size by special grinders (Turbo grinders). Following this, the product is separated by air currents into fractions according to size, shape, and specific gravity (flow-dynamic properties). Each of these fractions has specific chemical, physical, and baking properties. Thus, it is possible to separate a flour into fractions with (1) very high protein content; (2) normal protein content for bread flour; (3) lower-protein cookie and cracker flour; and (4) low-protein cake flour. In essence, then, air classification is a milling process designed to manufacture flours for specific uses.

Grading

The flour streams from the various grinding machines in the break and reduction systems differ in chemical composition because they contain different proportions of endosperm, germ, and bran. Thus, the quality of each flour is distinctive. Flour grade is based on the degree of refinement; the more refined grades contain lower amounts of bran and germ than less refined grades. The many flour streams produced in a mill may be blended together to yield a straight-run flow, or different grades of flour may be produced by selecting and blending specific flour streams, in definite proportions, to form a flour with particular baking characteristics. Grade designations used to describe various flours include straight, patent, standard patent, short or fancy patent, long patent, clear, low grade, red dog, and others. Straight flour is a combination of all the flour streams. Patent flours are made from the more refined streams and account for 70–95% of the total flour produced. Clear flour is a lower grade than patent flour; the flour from the last reduction, referred to as red dog, contains relatively large amounts of bran and germ and is used as animal feed.

Flour quality (or grade) may be determined by chemical tests (e.g., ash, protein, moisture determinations) or by degree of extraction. The complex nature of flour and the manner in which the flour will be used makes it difficult to prescribe one method over the other.

Chemical Tests

Of the chemical tests, the ash determination is most commonly used to characterize grade. Patent flours have lower ash contents than either straight or clear flours. However, the ash content of a flour not only varies with refinement but also with the kind of wheat (hard winter or soft winter), differences in grades of the same kind of wheat, and variations in milling conditions (grinding, tempering, fluctuations in temperature and humidity). The protein content of a flour has a large influence on its baking performance. A high protein content is desirable for bread flour, whereas a relatively low protein content is desirable for cake and cookie flours. Moisture content can be used as an indicator of storage ability (maximum 15%).

Degree of Extraction

Normally, during milling about 70% of the wheat kernel is manifested as flour of some sort. The remaining wheat is recovered as animal feed and consists of bran and shorts. The miller produces flours of varying qualities by selecting (compounding) the proper flour streams. In compounding a patent flour, the miller takes the flour streams of highest refinement (first endosperm fractions) and adds to it progressively less refined streams until he obtains a patent flour with a predetermined degree of extraction. For example, a 40% patent flour is a flour compounded from the most refined streams until it represents 40% of the total flour produced; the other 60% would go into lower-grade flours. Such a 40% patent would be classified as an extra short or fancy patent flour (extra short or fancy patents correspond to 40–60% extractions; above 60% extraction, flours are classified as long patents). As the percentage of extraction increases, say, to 80% (short patent flour), the refinement has become much less. The patent flours constitute 70–95% of the total flour produced. The remaining flour is called a clear flour. If all flour streams are combined, the resulting product is called a straight-grade flour.

At first glances, it might appear that expressing grade in terms of extraction would be definite. This would be true if all millers operated their mills alike and if there were no differences in their milling equipment. Realistically this is not attainable; therefore, a 40% patent flour produced in one mill will not necessarily be identical with a 40% patent produced in some other mill.

Bleaching

Bleaching, although not actually a part of milling, is normally performed by the miller before packaging. Flour color is determined by the degree of granulation and the amount of dirt, foreign matter, bran particles, and carotenoid pigments present. Of these, only carotenoid pigments are affected by bleaching. Bleaching of these pigments, via oxidation, occurs rapidly when the flour is exposed to the atmosphere and more slowly when flour is stored in bulk. The bleaching process involves the agitation of flour with such gases as chlorine, nitrosyl chloride, or nitrogen peroxide. Solid bleaching agents such as benzoyl peroxide also may be added to flour.

☐ STANDARDS FOR WHEAT FLOURS AND RELATED PRODUCTS

As discussed in Chapter 1, the Food, Drug, and Cosmetic Act of 1938 authorized the establishment of food definitions and standards of identity. In their final form, these definitions and standards of identity become part of the Code of Federal Regulations. Those relating to food, which are administered by the FDA, are recorded in Title 21 of the Code (21 CFR). A summary of the U.S. standards of identity for wheat flours and related products is presented in Table 11.4.

Flour, white flour, wheat flour, and plain flour (21 CFR 137.105) is prepared by grinding and bolting cleaned wheat other than durum wheat and red durum wheat. To compensate for any natural deficiency of enzymes, malted wheat, malted wheat flour, malted barley flour, or any combination of two or more of these may be added, but the quantity of malted barley flour so used cannot be more than 0.25%. One of the cloths through which the flour is bolted has openings not larger than those of woven cloth designated "149 micron (No. 100)" in Table 1 *Annual Book of ASTM Standards, Part 30* published in 1972 by the American Society for Testing and Materials. The flour is freed from bran coat, or bran coat and germ, to such extent that the percentage of ash (calculated to a moisture-free basis) is not more than the sum of 1/20 of the percentage of protein (calculated to a moisture-free basis) and 0.35. Its moisture content is not more than 15%. Unless such addition conceals damage or inferiority of the flour or makes it appear

Table 11.4. U.S. Standards of Identity for Wheat Flours and Related Products

Product	Source	Granulation	Moisture (maximum)	Ash (moisture-free basis)	Optional ingredients
Flour, white flour, wheat flour, plain flour (21 CFR 137.105)	Wheat other than durum wheat and red durum wheat	Not less than 98% passes through No. 70 sieve	15%	Not more than $\frac{1}{20}$ the percent protein (moisture-free basis) plus 0.35	Bleaching agents (see text); acetone peroxide; azodicarbonamide (45 ppm max); defatted wheat germ (5% max); malted wheat, malted wheat flour, and malted barley flour, alone or in combination (0.25% max)
Bromated flour (21 CFR 137.155)	Same as white flour (137.105) except for addition of potassium bromate not exceeding 50 ppm				
Enriched bromated flour (21 CFR 137.160)	Same as bromated flour (137.155) plus enriched flour (137.165)				
Enriched flour (21 CFR 137.165)	Same as white flour (137.105) except for addition of 2.9 mg thiamin, 1.8 mg riboflavin, 24 mg niacin, and 40 mg iron per pound; may contain calcium (960 mg/lb) and monocalcium phosphate (0.25–0.75% by weight)				
Phosphated flour (21 CFR 137.175)	Same as wheat flour (137.105) except for addition of monocalcium phosphate (0.25–0.75% by weight)				
Self-rising flour (21 CFR 137.180)	Same as white flour (137.105) except for addition of sodium bicarbonate and one or more of the acid-reacting substances monocalcium phosphate, sodium acid phosphate, and sodium aluminum phosphate at 4.5 parts max/100 parts flour (not less than 0.5% CO_2 evolved)				
Enriched self-rising flour (21 CFR 137.185)	Same as self-rising flour (137.180) plus enriched flour (137.165)				

(*continued*)

Table 11.4. *(continued)*

Product	Source	Granulation	Moisture (maximum)	Ash (moisture-free basis)	Optional ingredients
Whole wheat flour, graham flour, entire wheat flour (21 CFR 137.200)	Wheat other than durum wheat and red durum wheat (proportion of natural constituents unaltered)	Not less than 90% passes through No. 8 sieve and not less than 50% through No. 20 sieve	15%	Same as in wheat	Malted wheat, malted wheat flour, malted barley flour, or any combination of two or more of these (0.75% max); azocarbonamide (max 45 ppm); chlorine or chlorine dioxide
Cracked wheat (21 CFR 137.190)	Same as for whole wheat flour	Not less than 90% through No. 8 sieve and not more than 20% through No. 20 sieve	15%	Same as in wheat	
Crushed wheat (21 CFR 137.195)	Same as for whole wheat flour	40% or more passes through No. 8 sieve and less than 50% through No. 20 sieve	15%	Same as in wheat	

Durum flour (21 CFR 137.220)	Durum wheat	Not less than 98% passes through No. 70 sieve	15%	$\leq$1.5%
Whole durum flour (21 CFR 137.225)	Durum wheat	Not less than 90% passes through No. 8 sieve and not less than 50% through No. 20 sieve	15%	$\leq$15%
Farina (21 CFR 137.300)	Wheat other than durum wheat and red durum wheat	Passes through No. 20 sieve, but not more than 3% passes through No. 100 sieve	15%	0.6%
Enriched farina (21 CFR 137.305)	Same as farina (137.300) except for addition of not less than 2.0–2.5 mg thiamin, 1.2–1.5 mg riboflavin, 16.0–20.0 mg niacin, and 13 mg iron per pound			
Semolina (21 CFR 137.320)	Durum wheat	Passes through No. 20 sieve, but not more than 3% passes through No. 100 sieve	15%	0.92%

better or of greater value than it is, one or any combination of two or more of the following optional bleaching ingredients may be added in a quantity not more than sufficient for bleaching or, in case such ingredient has an artificial aging effect, in a quantity not more than sufficient for bleaching and such artificial aging effect: (1) oxides of nitrogen; (2) chlorine; (3) nitrosyl chloride; (4) chlorine dioxide; and (5) one part by weight of benzoyl peroxide mixed with not more than six parts by weight of one of any mixture of two or more of the following: potassium alum, calcium sulfate, magnesium carbonate, sodium aluminum sulfate, dicalcium phosphate, tricalcium phosphate, starch, calcium carbonate.

Enriched flour (21 CFR 137.165) conforms to the definition and standard of identity for white flour except that it contains in each pound 2.9 mg of thiamin, 1.8 mg of riboflavin, 24 mg of niacin, and 40 mg of iron. It may contain added calcium in such quantity that the total calcium content is 960 mg/lb. Enriched flour may be acidified with monocalcium phosphate within the limits prescribed for phosphated flour and may contain not more than 5% by weight of wheat germ or partly defatted wheat germ.

Bromated flour (21 CFR 137.155) conforms to the definition and standard of identity for white flour except that potassium bromate is added in a quantity not exceeding 50 ppm of the finished bromate flour. Potassium bromate is added only to flours whose baking qualities are improved by such addition.

Durum flour (21 CFR 137.220) is prepared by grinding and bolting cleaned durum wheat. When tested for granulation, as described previously for white flour, not less than 98% of durum flour passes through a No. 70 sieve. It is freed from bran coat, or bran coat and germ to such an extent that the percentage of ash (calculated to a moisture-free basis) is not more than 1.5%. Its moisture content is not more than 15%.

Self-rising flour (21 CFR 137.180) is a mixture of flour, sodium bicarbonate, and one or more of the acid-reacting substances monocalcium phosphate, sodium acid pyrophosphate, and sodium aluminum phosphate (4.5 parts maximum per 100 parts flour). It is seasoned with salt.

Phosphated flour (21 CFR 137.175) conforms to the definition and standards for white flour except that monocalcium phosphate is added in a quantity not less than 0.25% and not more than 0.75% by weight of the finished phosphated flour.

Whole wheat flour, graham flour, and entire wheat flour (21 CFR 137.200) is prepared by so grinding cleaned wheat, other than durum wheat and red durum wheat, that not less than 90% passes through a No. 8 sieve and not less than 50% passes through a No. 20 sieve. The proportions of the natural constituents of such wheat, other than moisture, remain unaltered. To compensate for any natural deficiency of enzymes, malted wheat, malted wheat flour, malted barley flour, or any combination of two or more of these may be used; but the quantity of malted barley flour must not exceed 0.75%. Bleaching is permitted unless it conceals damage or inferiority.

Crushed wheat (21 CFR 137.195) is prepared by crushing wheat other than durum wheat so that 40% or more passes through a No. 8 sieve and less than 50% passes through a No. 20 sieve. It must not contain more than 15% moisture, and the proportions of natural constituents other than moisture must not be altered.

Cracked wheat (21 CFR 137.190) is prepared by cracking or cutting into angular fragments cleaned wheat other than durum so that not less than 90% passes through a No. 8 sieve, and not more than 20% passes through a No. 20 sieve. The proportions of the natural constituents other than moisture may not be altered. The maximum moisture content is 15%.

Farina (21 CFR 137.300) is prepared by grinding and bolting cleaved wheat other than durum to such fineness that it passes through a No. 20 sieve, but not more than 3% passes through a No. 100 sieve. It is freed from bran and germ so that the percentage ash calculated on a moisture-free basis is not more than 0.6 %. Its moisture content is limited to a maximum of 15%.

Semolina (21 CFR 137.320) is prepared by grinding and bolting cleaned durum wheat so that it passes through a No. 20 sieve, but not more than 3% passes through a No. 100 sieve. It is freed from bran and germ so that the percentage ash calculated on a moisture-free basis is not more than 0.92%. Its moisture content is limited to a maximum of 15%.

☐ COMPONENTS OF FLOUR

The properties of a flour depend not only on its physical characteristics (in particular, granulation) but also on its chemical composition. The moisture and ash (mineral matter) contents are specified in the standards of identify for flour products (Table 11.4). Too high a moisture content in flour is conducive to molding and souring. The ash content is used as an indicator of flour grade (refinement) because the endosperm yields relatively little ash, whereas the bran, aleurone, and germ yield much more. The other important components of flour are proteins, lipids, carbohydrates, acids, and enzymes. The relationship of these components to the quality and potential uses of a flour are described in the following sections.

Proteins

Protein quality and quantity are primary criteria in assessing the potential end use of a flour. Protein quality is related specifically to the physicochemical properties of the gluten portion of the flour protein; protein quantity relates to the crude protein content in the flour.

Wheat contains small amounts of albumin and globulin, but the two most important proteins in wheat are glutenin and gliadin, which are present in wheat endosperm in approximately the same concentrations. Glutenin is a glutelin, and gliadin is a prolamine or alcohol-soluble protein. When these protein fractions are mechanically agitated in the presence of water, they form gluten, which is capable of retaining gases; this property makes a leavened product possible. No other cereal contains as large a proportion of these two protein fractions as does wheat. Thus, when other cereal flours are used for making leavened products, they are mixed with enough wheat flour so that the resulting product will rise and have a respectable texture.

The nature of the gluten required in different types of baked products varies greatly. For breadmaking, strong, tenacious glutens in a dough are essential. Strong glutens resist extension and must exhibit tolerance or resistance to breakdown during mechanical mixing in commercial dough mixers. In general, flour from hard wheats contains a stronger gluten and has a higher gluten concentration than flour from other types of wheat. Generally, hard red spring wheat flour contains a stronger gluten and has a higher gluten content than hard red winter wheat flour. Sometimes the gluten in flour from hard red spring wheat flour is too strong and an oxidizing agent has to be added to "mellow" the gluten. Soft wheat flours contain weak glutens that are easily extended and can stretch long distances. The latter glutens are better for making cookies, pastries, and cakes.

Gluten strength is measured in several ways. Washing the gluten from flour with an automatic device and examination of the resulting gluten dough ball gives some indication of gluten quality. A gluten that is elastic but fairly tough is described as *strong*, while one that is sticky and not very elastic is called *weak*. Gluten strength may be measured with a Farinograph, a machine in which flour is mixed into a dough under standard conditions. The resistance to this mixing procedure over time is measured and provides an indicator of gluten strength. Generally speaking a desirable flour will rise to a peak fairly rapidly and will breakdown slowly. Another method for determining gluten strength is the MacMichael test, which involves measuring the viscosity of a flour and weak lactic acid suspension in a special viscometer. The resistance of the swelled gluten to shear is a measure of flour strength; that is, the stronger a gluten is, the greater its resistance to shear.

Flours from soft wheat (i.e., weak-gluten flour) are used for making cakes (layer or ordinary white), cookies, crackers, and pie crusts. For products such as pound and fruit cakes which contain large quantities of sugar and shortening, a stronger-gluten flour often is used. As noted already, breadmaking requires strong-gluten flours.

Lipids

The lipid content of flour is approximately 1%, whereas the wheat kernel contains approximately 2% lipids. The lipid content of flour varies with the blend of flour and the variety of wheat used. Lipid content is especially valuable as a criterion of grade in self-rising flours for which the ash percent has no meaning because chemical leavening agents are added. The lipids present in flour have little significance relative to its use. They include neutral glycerides as well as phospholipids and sterols.

Carbohydrates

The principal component of flour is starch, averaging about 70–75% of the flour. Flour in bread is an important source of calories in the diets of a majority of the world's inhabitants. When starch is acted upon by diastase during the breadmaking process, it is hydrolyzed to maltose, which in turn furnishes yeast food and contributes to crust color and taste in bread. Dextrins and cellulose also are present in small amounts in flour, as well as lignans and pentosans (especially in the bran). The water-soluble simple sugars present in flour include raffinose, sucrose, maltose, and dextrose.

Acids

The titratable acidity in flour is low and varies with the type and grade of flour. If flour is stored under adverse conditions with respect to moisture and temperature, its titratable acidity will increase as a consequence of microbiological spoilage. Consequently, titratable acidity can be used as an indicator of flour soundness. The hydrogen ion concentration is higher in higher-grade flours than in lower-grade flours.

Enzymes

Of several enzymes present in flour, the most important are the amylases, which convert starch to utilizable sugars. Flour normally contains a very limited amount of α-amylase but ample amounts of β-amylase. These two enzymes differ, in their ability to produce soluble sugars. α-Amylase readily attacks both damaged and undamaged starch granules, whereas β-amylase attacks only damaged starch granules and then only at a low rate. Therefore, it is desirable to have sufficient α-amylase present in yeast-leavened products to ensure that adequate amounts of sugars are formed for gas production.

In wheat, starch is encapsulated in granules in the cells of the kernel; during milling these cells are ruptured, resulting in release of the starch granules. Grinding also causes structural damage to some of the starch granules so that starch is released from the granule case. The freed starch serves as the substrate for the amylolytic enzymes to produce sugars, which are necessary for the metabolism of yeast during fermentation. The level at which these damaged granules occur is important in determining the baking quality of a flour, i.e., some starch damage is desirable (about 40%), but too much damage reduces baking quality. Malting or germination of grain increases the enzyme activity of grain. To insure proper amylase activity, millers are permitted to add germinated wheat and/or malted barley flour to wheat flour as mentioned in the section on flour standards.

A lipoxidase also occurs in flour. Its activity may be responsible to some extent for destroying flour pigments during natural aging and perhaps affects the flavor of dough.

☐ ANALYSIS OF FLOUR

In addition to proximate analyses for moisture, crude fat, ash, crude protein, and crude fiber, flour is often analyzed for gluten, starch, pigments, diastatic activity, and pentosans. As with all analytical determinations, proper and representative sampling is necessary to achieve valid results.

Sampling of Flour in Sacks

The American Association of Cereal Chemists' "Cereal Laboratory Methods" describes the following procedure (AACC 64-60) for obtaining flour samples.

Procedure

Remove samples from a number of sacks equivalent to the square root of the total sacks in the lot, but never less than 10, i.e., sample 10 sacks out of 100 or less, 15 out of 225, etc. In selecting the sacks to be sampled from a lot, it is necessary to consider their exposure. For example, if 10 sacks are to be sampled, 4 are selected from the most exposed of the lot, 3 from the next less exposed, 2 from the next, and 1 from the least exposed part of the lot.

Samples are obtained from the sacks selected for sampling by a definite procedure. Insert a cylindrical, pointed, polished metal trier (0.5 in. in diameter, with a slit at least one-third of the circumference) from one corner of the top diagonally to the center of the sack and remove the core of material. Obtain a second core of material in the same manner from the other top corner to the center of the sack. Combine the two cores of material in a clean, dry, airtight container that has stood open for a few minutes near the flour to be sampled and seal immediately. A separate container is used for each sack sampled.

Before the sample is opened for analysis, alternately invert and roll each container several times to secure a homogeneous mixture. Keep the sample tightly sealed at all times, and when opening the container for analysis avoid extreme temperatures and humidities in the room. If it is necessary to store a sample for a period of time before analysis, the container should be of such size that the sample completely fills it.

Moisture

Dry flour is very hygroscopic; therefore, every precaution must be taken during handling to prevent absorption of moisture, which will result in erroneous moisture values.

The vacuum oven method described in Chapter 2 is suitable for determining the moisture content of flour. Use approximately 2 g of a well-mixed sample for this determination.

Crude Fat

Two methods for determining crude fat are given here. In the first method, the sample is first subjected to acid hydrolysis to release bound lipids and then the total lipids are extracted with ethyl ether and petroleum ether: in the second, the dry sample is continuously extracted with anhydrous ether. This method will extract only ether-soluble lipids. In both methods, the extraction solvents are then evaporated and the extracted fat dried to constant weight.

Acid Hydrolysis Method

This method (AACC 30-10) is suitable for determining crude fat in flour, bread, and baked goods not containing fruit.

Place a 2-g sample of flour or similar product in a 50-ml beaker. Moisten with 2–3 ml of ethanol and stir to prevent lumping on addition of acid. Add 25 ml HCl (25 parts HCl + 11 parts H_2O), mix, and set beaker in a water bath held at 70°–80°C for 30–40 min, stirring frequently. Add 10 ml of ethanol and cool.

Transfer the mixture to a Rohrig or Mojonnier fat-extraction apparatus; rinse the beaker with 25 ml ethyl ether (three portions) and add rinsings to extraction apparatus. Stopper with a cork, Neoprene, or other synthetic stopper not affected by the solvent and shake vigorously for 1 min. Add 25 ml of redistilled petroleum ether and shake vigorously for 1 min. Let stand until upper liquid is practically clear. Decant as much as possible of the ether–fat solution through a funnel, containing a wad of cotton, into a weighed 125-ml wide-mouthed Erlenmeyer flask. (Before weighing flask, oven-dry it at 100°C, then cool to room temperature.)

Re-extract the aqueous layer in the extraction flask twice, each time using 15 ml each of ethyl ether and petroleum ether. Allow the layers to separate and then decant the ether solutions into the same tared flask as before. Wash off the stopper and rim of the extraction flask, funnel and the funnel stem with a few milliliters of a mixture of equal volumes of the two ethers.

Evaporate the ethers slowly on a steam bath, then dry the flask and contents in an oven at 100°C to constant weight. Remove the flask from the oven, let stand in air to constant weight (ca. 30 min.), and weigh.

Run a blank determination on reagents, and subtract blank value from sample value. Express results as percentage fat by acid hydrolysis.

Ether Extraction Method

Transfer the dried flour sample from the moisture determination to an extraction thimble and proceed as described in the method given on p. 505 in Crude Fat under Analysis of Wheat.

Crude Fiber

Follow the procedure described in Chapter 2 for determing crude fiber, using as a sample the ether-extracted residue from the crude fat determination. Calculate the percentage of crude fiber present in the sample on the original weight basis.

Ash

Accurately weigh 3–5 g of a well-mixed flour sample into a tared ashing dish (preferably platinum or silica) that has been previously ignited. Transfer the dish to an electric muffle furnace equipped with an indicating pyrometer and a thermostatic heat control. Initially, the muffle furnace should not be over 425°C; gradually increase the temperature to 550°C for soft wheat flours or 575°–590°C for hard wheat flours. Incinerate until a white or gray ash results, then transfer the dish to a desiccator, cool to room temperature, and weigh immediately. Calculate as percent ash.

Crude Protein

Crude protein in flour, as in wheat, is determined by the Kjeldahl nitrogen method outlined in Chapter 2. The percentage of crude protein equals the percentage of nitrogen times 5.7. For grains other than wheat, and flours made from them, the conversion factor is 6.25.

Crude Gluten

The term gluten refers to a complex of proteins in flour that is formed by the combination of glutenin and gliadin. As discussed already, strong-gluten flour is necessary for breadmaking, and wheat is the only cereal whose flour yields significant quantities of gluten. Consequently, flour millers have used the gluten content as an index of the protein quality of flour.

Procedure

Weigh 25.00 g flour into a cup or porcelain dish and add sufficient water (ca. 15 ml) to form a firm dough ball. With a spatula, work the mixture into a dough ball, taking care that none of the dough adheres to the container or spatula. Cover the dough ball with water and allow to stand at room temperature for 1 hr. Then knead gently in a stream of tap water over bolting cloth (approximately 60-grit gauze) until the starch and all soluble matter are removed. This washing procedure requires approximately 12 min. To determine if the gluten is starch free, allow 1 or 2 drops of the wash water, obtained by squeezing, to drop into a beaker containing clear water; if starch is present, the water will become cloudy.

Cover the washed gluten with water and allow to stand for 1 hr. Then remove the gluten from the water, press as dry as possible between one's hands, roll into a ball, place in a previously tared flat-bottomed dish, and weigh (moist gluten).

Transfer the gluten to a drying oven, dry to constant weight at 100°C (approximately 24 hr), cool, and weigh (dry gluten). Calculate the percentage of moist gluten and the percentage of dry gluten present in the flour.

Starch

The starch content of flour may be determined by a polarization method (AOAC 14.031).

Reagents

1. Calcium chloride solution—Dissolve two parts of $CaCl_2 \cdot 6H_2O$ in one part of water and adjust to a density of 1.30 at 20°C. Add phenolphthalein indicator and then 0.1 *N* NaOH until solution is faintly pink.
2. 65% Ethanol (by weight; $d_4^{20} = 0.88$).
3. 0.8% Acetic acid—2 ml glacial acetic acid diluted to 250 ml.

Procedure

Weigh exactly 2.000 g of the flour into a 50-ml round-bottom centrifuge tube having a lip. Wash with ether to remove lipid material, then add 10 ml of the 65% ethanol solution and stir thoroughly with a glass rod. Centrifuge the suspension and pour off the liquid. Repeat washing the residue until 60 ml of the ethanol solution have been used, stirring after each addition of the wash liquid with the same glass rod.

Stir the residue with 10 ml distilled water and pour the suspension into a 250-ml Erlenmeyer flask. Transfer any material remaining in the tube to the flask by washing with a total of 60 ml of the calcium chloride solution containing 2 ml of 0.8% acetic acid. Support the flask on a wire gauze over a burner, transfer the glass rod to the flask, and, with frequent stirring, quickly bring the mixture to its boiling point. Boil briskly from 15 to 17 min, being careful to prevent foaming and burning. During the boiling, rub down particles on the sides of the flask with the glass rod.

At the completion of boiling, quickly cool the solution in cold water and pour it into a 100-ml volumetric flask. Transfer the remaining material in the Erlenmeyer flask to the volumetric flask by rinsing with the calcium chloride solution contained in a wash bottle having a medium jet. Dilute to volume with the rinse solution (the addition of a drop of ethanol, if necessary, will destroy froth).

Thoroughly mix the contents of the volumetric flask and pour about 10 ml of the contents on a fluted filter (Whatman 42 or 44), making sure that the paper is wetted. Permit the filter to run dry and discard the filtrate. Continue the filtration through the filter, collecting 40–50 ml of filtrate in a dry flask.

Fill a 200-mm saccharimeter tube with the clear filtrate and take 10 readings on the saccharimeter. Refill the saccharimeter tube with a fresh sample of filtrate and take a second set of 10 readings. Average the readings to obtain the average degree (°S). The percentage of starch equals °S × 4.2586. This formula holds only if this procedure is followed exactly, i.e., a sample of exactly 2 g, dilution to 100 ml, and readings in a 200-mm tube.

Hydrogen-Ion Activity

Weigh a 10.0-g sample of the flour into a clean, dry Erlenmeyer flask and add 100 ml of recently boiled distilled water cooled to 25°C. Shake the flask and contents thoroughly until the flour particles are evenly suspended and no lumps remain in the mixture. Let it stand for 30 min with frequent shaking, then an additional 10 min without shaking to allow the solid material to settle out. Decant the supernatant liquid and determine the pH with a suitable pH meter at 25°C.

Pigments

The carotenoid pigments in flour can be determined spectrophotometrically (AACC 14-50) by comparing the absorbance of a sample at 440 nm with that of standard β-carotene solutions.

Weigh an 8.0-g sample of flour into a 125-ml glass-stoppered Erlenmeyer flask and add 40 ml of water-saturated *n*-butyl alcohol. Stopper tightly, shake flask and contents thoroughly, and let stand for 15–30 min in the dark. Reshake well, and filter through a folded 12.5-cm No. 1 Whatman filter paper into a test tube or flask. Refilter if extract is not entirely clear.

Fill a 1-cm spectrophotometer absorption cell with the extract and fill a duplicate cell with water-saturated *n*-butyl alcohol. Measure the absorbance at 440 nm. Prepare several solutions of β-carotene of known concentrations in water-saturated *n*-butyl alcohol, measure their absorbance at 440 nm, and plot a standard curve of absorbance versus concentration. Determine the β-carotene concentration in the flour extract from the standard curve.

Diastatic Activity

It is desirable to have sufficient amylase activity in yeast-leavened products for the formation of fermentable sugars. Millers are permitted to add germinated wheat (and, or malted barley to wheat flour to insure proper amylase activity). Diastatic activity refers to the combined activity of α- and β-amylase in converting starch to fermentable sugar. In the procedure given here (AACC 22-15), the fermentable sugar (maltose) resulting from the enzymatic breakdown of endogenous starch, when a flour sample is incubated under standard conditions, is taken as a measure of diastatic activity. Maltose is determined by oxidizing it with excess potassium ferricyanide and then titrating the excess ferricyanide with standard thiosulfate solution.

Reagents

1. Buffer solution (pH 4.6–4.8)—Dissolve 3 ml glacial acetic acid and 4.1 g anhydrous sodium acetate in water; dilute to a final volume of 1 liter.
2. 3.68 *N* H_2SO_4—Dilute 10 ml concentrated H_2SO_2 (sp gr 1.84) to 100 ml; adjust the concentration if necessary.
3. 12% Sodium tungstate solution—Dissolve 12.0 g $Na_2WO_4 \cdot 2H_2O$ in water and dilute to 100 ml with water.
4. 0.1 *N* Alkaline ferricyanide reagent—Dissolve 33 g pure dry $K_3Fe(CN)_6$ and 44 g anhydrous Na_2CO_3 in water; diluting to 1 liter with water. Standardize by adding 25 ml of the acetic acid–salt solution and 1 ml of soluble starch–KI solution to 10 ml of the alkaline ferricyanide solution and titrate with standard 0.1 *N* thiosulfate solution. Exactly 10 ml of solution should be required to abolish the blue color.
5. Acetic acid–salt solution—Dissolve 70 g KCl and 40 g $ZnSO_4 \cdot 7H_2O$ in 750 ml water; then slowly add 200 ml glacial acetic acid and dilute to 1 liter with water.
6. Soluble starch—KI solution—Suspend 2 g soluble starch in small amount of cold water and pour slowly into boiling water with constant stirring. Cool thoroughly, add 50 g KI, and dilute to 100 ml with water. To this solution add 1 drop of saturated NaOH solution.

Procedure

Put 5 g of flour and a teaspoon of ignited quartz into a 125 ml Erlenmeyer flask and mix by rotating the flask. Add 46 ml of the buffer solution at 30°C and rotate the flask until all of the flour is in suspension. (The flask and all the ingredients should be individually brought to 30°C before being mixed together.) Place the flask in a water bath, maintained at 30°C, for exactly 1 hr; shake the flask by rotation every 15 min.

Remove the flask at the end of 1 hr, add 2 ml of 3.58 N H_2SO_4, and mix thoroughly. Immediately add 2 ml of the sodium tungstate solution, mix thoroughly, allow to stand 2 min, and then filter through Whatman No. 4 or equivalent filter paper, discarding the first 8 or 10 drops of filtrate.

Thoroughly mix the filtrate, then determine maltose in the following manner: Into a test tube of approximately 50-ml capacity (2 × 20 cm), pipette 5 ml of the filtrate and 10 ml of the alkaline ferricyanide reagent, mix, and immerse the test tube in a vigorously boiling water bath. The surface of the liquid in the test tube should be 3–4 cm below the surface of the boiling water. (Delay between filtering of the extract and treatment in the boiling water bath should not exceed 15–20 min. Further delay may cause an error from sucrose hydrolysis in acid solution.) After exactly 20 min, remove the tube from the boiling water bath, cool it under running water, and pour contents into a 125-ml Erlenmeyer flask. Rinse the test tube with 25 ml of the acetic acid–salt solution and add rinsings to the Erlenmeyer flask. Mix the contents by rotating the flask, add 1 ml of the soluble starch–KI solution, and titrate with standard 0.1 N thiosulfate solution to the complete disappearance of the blue color. (A 10-ml microburette is recommended for the titration.)

Calculations

Calculate the milliliters of ferricyanide reduced by subtracting the milliliters of thiosulfate required from the thiosulfate equivalent of the ferricyanide reagent. Refer to Table 11.5 to determine the milligrams of maltose produced by 10 g of flour in 1 hr at 30°C from the milliliters of thiosulfate required in the titration.

Comments

The above directions are applicable to all ordinary flours for which the amount of maltose produced from 10 g of flour in 1 hr does not exceed 600 mg. If the solution in the test tube is colorless after treatment in the boiling water bath and gives no blue color after the addition of the starch–KI solution, there is an excess of reducing sugar present.

Table 11.5. Thiosulfate–Maltose Conversion Chart

0.1 *N* Thiosulfate	Maltose per 10 g flour	0.1 *N* Thiosulfate	Maltose per 10 g flour	0.1 *N* Thiosulfate	Maltose per 10 g flour
ml	mg	ml	mg	ml	mg
0.10	618	3.40	373	6.70	166
0.20	608	3.50	367	6.80	161
0.30	598	3.60	360	6.90	156
0.40	588	3.70	353	7.00	151
0.50	578	3.80	347	7.10	145
0.60	568	3.90	341	7.20	140
0.70	558	4.00	334	7.30	135
0.80	550	4.10	328	7.40	130
0.90	542	4.20	322	7.50	126
1.00	534	4.30	315	7.60	121
1.10	527	4.40	308	7.70	116
1.20	519	4.50	302	7.80	111
1.30	512	4.60	295	7.90	106
1.40	505	4.70	288	8.00	101
1.50	499	4.80	282	8.10	96
1.60	492	4.90	276	8.20	90
1.70	485	5.00	270	8.30	85
1.80	478	5.10	264	8.40	80
1.90	472	5.20	257	8.50	76
2.00	465	5.30	251	8.60	71
2.10	458	5.40	244	8.70	65
2.20	451	5.50	237	8.80	60
2.30	445	5.60	231	8.90	56
2.40	438	5.70	225	9.00	51
2.50	431	5.80	218	9.10	46
2.60	425	5.90	213	9.20	41
2.70	418	6.00	207	9.30	36
2.80	412	6.10	201	9.40	31
2.90	406	6.20	195	9.50	25
3.00	398	6.30	188	9.60	20
3.10	392	6.40	182	9.70	15
3.20	385	6.50	176	9.80	10
3.30	379	6.60	171	9.90	5

Source: Sandstedt (1937).

In this case, repeat the determination using a smaller aliquot of the filtrate (e.g., 1, 2, or 3 ml instead of 5 ml). However, in such cases, dilute to 5 ml before determining maltose and multiply the milligrams of maltose found by the appropriate dilution factor.

It is unnecessary when assaying normal sound flour to make a blank determination, which indicates the maltose or reducing sugar originally present in the flour. The quantity of reducing sugars present as such in flour from sound wheat is so small and so constant that it may be neglected for all practical purposes. However, if a blank determination is desired, the procedure is as follows:

Combine the 5 ml of 95% ethanol (by volume); 50 ml of acid buffer solution (dissolve 3 ml glacial acetic acid, 4.1 g anhydrous sodium acetate, and 4.5 ml sulfuric acid with sp gr 1.84, and dilute to 1 liter with water); and 2 ml of the sodium tungstate solution. To 5 ml of this mixture (used in place of the 5 ml of flour filtrate), add 10 ml of the ferricyanide solution, and proceed as in the preceding determination of maltose. It should require 10 ml of the thiosulfate solution to discharge the blue starch–iodine color. If the titration (thiosulfate equivalent) falls within 10 ($\pm$ 0.05) ml, the reagents need not be discarded, but the appropriate correction should be made in the maltose calculations.

Pentosans

Wheat flour contains 2–3% pentosans but they have a pronounced effect on the water absorption of flour. The effect of pentosans on dough-development time is contradictory but it is generally accepted that soluble pentosans have a helpful effect on bread quality, the insoluble part is harmful; both fractions have an unfavorable effect on cookie dough. The procedure given here (AACC 52-10) involves acid hydrolysis of the pentosans, and then using iodine indirectly for the analysis of bromide–bromate solution, which is a stronger oxidizing agent than iodine. In this reaction the iodide ion is added in considerable excess to insure complete reduction of the bromide–bromate solution.

Reagents

1. Bromide–bromate solution—Dissolve 3.0 g potassium bromate and 50.0 g potassium bromide in water and dilute to 1 liter with water.
2. 12% HCl (1 part HCl + 2 parts H_2O) by volume.
3. 0.1 *N* Sodium thiosulfate.
4. 10% KI solution.

Procedure

Weigh 0.500 g of flour into a 500-ml distilling flask containing a few glass beads or boiling chips; then add 125 ml of the HCl solution. Place 360 ml of the HCl solution in a 500-ml separatory funnel and attach it to the distilling flask so that acid can be introduced into the flask during distillation. Heat the flask on a glycerol bath, distilling it at the rate of 30 ml/10 min at 150°C. After 30 ml have distilled over, slowly add 30 ml of acid from the separatory flask, allowing 1–2 min for the addition. Again distill at the same rate, adding another 30 ml of acid each time 30 ml of distillate have been collected, until 360 ml have been collected. Stopper the receiving flask and place immediately in an ice bath and cool to 0°C.

Add 50 ml of the bromide–bromate solution, using a volumetric pipette, to the sample in the ice bath, stopper immediately and let it react for exactly 4 min. Add 10 ml of the KI solution, restopper, and shake gently. Remove flask from the ice bath and titrate the released iodine with 0.1 *N* sodium thiosulfate using starch solution as an indicator. Determine a blank on the bromide–bromate solution, acidifying with 5.0 ml concentrated HCl before adding the KI solution.

$$\text{Pentosan (g)} = \text{ml of } 0.1\ N \text{ thiosulfate soln} \times 0.0082$$

$$\text{Pentosan (\%)} = \frac{(\text{ml blank} - \text{ml titration})(N \text{ of } Na_2S_2O_3) \times 0.082 \times 100}{\text{sample weight}}$$

☐ END PRODUCTS OF FLOUR

Bread

Most of the flour milled in the United States is utilized for the manufacture of bread. The process involves a number of steps including formulation and mixing, fermentation of the dough, molding and panning the dough into loaves, and baking of the dough. Some variations in the first steps may occur depending upon whether a straight-dough or sponge-dough procedure is used, but the remaining steps (fermentation by means of yeast and aeration of the dough) are essentially the same for all doughs.

Standards of identity have been promulgated by the FDA (21 CFR 136.10) for bread, white bread, enriched white bread, milk bread, raisin bread, and whole wheat bread. The corresponding rolls and buns are included in these definitions. The word *bread* when used in the name of a food means the unit weighs one-half pound or more after cooling, while *rolls* and *buns* are defined as units weighing less than one-half pound after cooling.

The definitions and standards for bakery products are too voluminous to be reproduced here. The ingredients must include flour, water, yeast, and salt, and may include any of a number of optional ingredients listed in the standards. All bread, rolls, and buns must contain not less than 62% total solids. All ingredients from which a bakery product is fabricated must be safe and suitable.

Ingredients

Bread in the United States is typically made with flour, water, sugar, yeast, shortening, and salt. Other ingredients, which may be added, include yeast foods, gluten, raisins, oxidizing agents, and enriching agents to improve the nutritional value of the bread. When these ingredients are mixed in correct amounts to make a dough, the protein in the flour combines with the added water to form gluten, which is responsible for gas retention, and carbon dioxide is formed by the action of the enzymes in the yeast upon the sugars. The end product is a bread having good loaf volume, an attractive shape and color, proper texture, palatability, and general consumer acceptance.

Flour constitutes the basic ingredient of bread, and percentages of all other ingredients are calculated on the basis of flour (100 parts flour) in the formula. A good breadmaking flour has the following properties:

1. Protein adequate in quantity and quality to provide the gluten that forms the skeletal framework responsible for gas retention.
2. Sufficient damaged starch to supply sugar during fermentation (gassing properties) and amylase activity.
3. Satisfactory moisture content (not higher than 14%)
4. Satisfactory color

Water is necessary to develop and hydrate the gluten during the mixing process. Water is also absorbed by the starch and as a consequence serves as the medium in which the enzymes and their substances are brought together. The amount of water that is required to make a dough of standard consistency is directly related to the amount of protein and damaged starch present in the flour. For example, flours from strong wheat (higher protein content) or hard wheat (higher damaged starch content) require more water than flours from weak wheat or soft wheat to make a dough having the same consistency. When the dough is placed in the oven for baking, the heat causes some of the water to pass off as vapor; this along with expansion of the fermentation gases causes a rapid rise in loaf volume (oven spring).

Sugar is required in bread doughs for yeast growth and development. The small amounts of sugar naturally present in flour are soon consumed by the yeast. Continued yeast growth and development then depends on the sugar produced by amylolytic action from starch. Good crust color is due to a nonenzymatic browning reaction (Maillard reaction) in which protein reacts with reducing sugars. Thus, sufficient sugar (6–8%) must be added in the bread formula to provide for the fermentation process and also to assure that enough sugar remains at the time of baking to give the brown color of the crust and an acceptable taste to the finished product.

Yeast produces carbon dioxide and alcohol via the fermentation process. The carbon dioxide is entrapped by the gluten, which enables the bread to rise, and most of the alcohol is volatilized at time of baking. The amount of yeast (baker's) used is related inversely to the time of fermentation and to the temperature of the dough.

Salt enhances the flavor of bread. It also toughens the gluten and gives a less viscous dough. In addition, salt controls yeast fermentation; as a consequence, its addition is sometimes delayed until the dough has been partly fermented.

Fat as a shortening agent contributes to the general acceptability of bread. Fat is also necessary for good loaf volume and texture because the interaction of fat with flour components has an effect on gas retention. An excessive amount of fat will decrease loaf volume.

Commercial Breadmaking

Mixing or Forming the Dough. In the straight-dough system, the formula ingredients are mixed in a dough mixer, using water at a temperature that will bring the mix to approximately 27°C. Optional ingredients that may be added include milk solids, yeast foods, vitamins and minerals, malt, and rope preventive. Mixing develops the gluten and thoroughly blends the ingredients throughout the dough. This process is continued until a smooth dough, having an optimum elasticity, is obtained. When sufficiently mixed, the dough is transferred from the mixer into a dough trough and set aside while fermentation proceeds.

In the sponge-dough procedure, which is commonly employed in the United States, only a portion of the flour (50–75%) is mixed at first with all the yeast, sugar, and sufficient water to make a dough. The first dough, or sponge, is allowed to ferment for 3.5–5.0 hr. The sponge is returned to the mixer and punched down. The remainder of the flour, water, and all of the other ingredients are added and the mixing continued until a smooth homogeneous dough results. It is then transferred from the dough mixer into a dough trough.

Fermentation. The dough trough containing the dough is placed in a fermentation room maintained at 27°C with a relative humidity of 75–

80%. The dough is allowed to ferment and rise to a point at which, when the fingers are inserted approximately 4 in. into the dough and quickly withdrawn, the dough will fall. The dough is folded over punched down or knocked back, which accomplishes several things: (1) carbon dioxide is expelled and also evenly distributed throughout the dough; (2) the rise in temperature resulting from fermentation is equalized throughout the dough; and (3) the gluten is developed by stretching and working. In the straight-dough process, the dough may be punched two or three times during the fermentation time, whereas in the sponge-dough process, the dough may be punched just once or not at all.

Dividing, Molding, and Panning. The first step in dough make-up is one of dividing the dough into loaf-size pieces via a dough divider. These pieces of dough are sent through a rounder to form a skin around the dough piece and thus minimize gas diffusion. The rounded pieces of dough, because the dough has been subjected to mechanical abuse during the dividing and rounding steps, is given a rest period (intermediate proof) of 10–15 min to recover its pliability and extensibility. The dough pieces are then sent to a molder where they are molded into a loaf and placed into a pan.

Pan Proofing. The panned loaves are sent to a proof cabinet for the last fermentation. Proofing requires 30–60 min, a relative humidity of 80–85%, and a box temperature of 35°–38°C. The loaves more than double in size during pan proofing.

Baking. The proofed loaves are sent to the oven where they are baked at temperatures of 190°–232°C. One-pound loaves usually take 30 min at 218°C.

Cakes

The texture of cakes is softer and more crumbly than that of bread. These characteristics are influenced greatly by the type of flour used. For cakemaking, the flour should be milled from a soft, low-protein wheat with low α-amylase activity.

Ingredients in the cake formulas include flour, sugar, shortening, chemical leavening agents, eggs, and liquid. The flour contributes to the structure of a cake, but gluten is not developed as in bread. The stability of a cake depends largely on the presence of uniformly swollen starch granules; hence the starch granules should be undamaged during milling and undamaged by amylolytic enzymes. The sugar in cakes contributes to flavor, slows gluten development, and increases the coagulation temperature of the protein, thereby allowing an increased time for stretching cell walls. Thus, sugar contributes to flavor, cake volume, and cake tenderness. Shortening also contributes to tenderness. Cakes are leavened by chemical leavening agents and eggs rather than by yeast fermentation.

In general, cake ingredients are mixed with beaters or paddles. The sugar and shortening are first creamed to a fluffy foam. Eggs are then added. The flour, together with the leavening agent and salt, is added and mixing is continued just sufficiently to produce a uniformly blended batter. The batter is then transferred to a depositor, which deposits the required amount of batter in each pan. The pans are run into ovens and baked at a lower temperature than for bread. The usual baking temperature is 150°C, but this may vary with the type of cake.

Crackers

Cracker production involves both yeast fermentation and chemical leavening. A moderately strong, soft wheat flour is required for cracker production. This kind of flour produces a dough having more extensibility and less spring than a flour used for bread dough. Spring, or recovery, is not required because the dough piece should retain its size and shape after being stamped out.

Most crackers are made by a sponge-dough method. A relatively stiff dough is made, both in the sponge and in the dough stages. The sponge contains flour, water, yeast, and some sugar to supplement the fermentable carbohydrates. Small amounts of proteolytic enzymes may also be added. The sponge may be set for periods up to 18 hr. This fermentation destroys most of the gluten and builds up a fairly high acidity through the production of organic acids. The fermented dough is then mixed with additional flour, soda, and other ingredients (e.g., cheese, seeds, spices) to the proper consistency.

The dough is allowed to stand for a short time and is then passed through a roll. The flattened dough ribbon is folded several times and again passed through the roll. This rolling and folding produces a laminated structure in the product. The dough then passes through gauge rollers to produce the desired thickness, then it passes under a stamping machine that cuts the dough into proper shape. Docker pins puncture the dough to allow carbon dioxide to escape through the openings. The stamped and cut dough pieces may pass under a salting machine that applies a light sprinking of salt on the top surface. The pieces are baked on a mesh band for a short time (3–7 min) at a relatively high temperature (285°–312°C).

□ DETERMINATION OF BAKING QUALITY

The baking quality of a particular wheat flour is assessed by judging the resulting bread in terms of physical criteria (e.g., volume, form, moisture absorbed) and organoleptic criteria (e.g., color, texture, taste, aroma). In addition, the response of the dough to varying mixing and fermentation times is evaluated. In order to compare the baking quality of different flours, all factors involved in making bread may be standardized, so that only differences in the baking quality of the flour will be reflected in the quality of the resulting bread.

In this section, we describe the standard conditions for a straight-dough baking test (AACC 10-10) and explain how the resulting bread is scored. Although many of the judgments involved in scoring bread are subjective, with practice an analyst can become accomplished in making consistent assessments of the baking quality of flour.

Apparatus

1. Mixer—Conventional experimental dough mixer designed to mix a quantity of dough containing between 100 and 500 g of flour. Calibrate the mixer by one of the methods described by Landis and Freilich (1935) for each weight of dough to be used. Swanson type mixers are recommended.
2. Thermometers—For dough, AACC thermometers graduated from 15° to 40°C (Sherwood 1930); for ovens, thermometers graduated from 100° to 260°C.
3. Fermentation bowls—Aluminum or granite ware "oatmeal bowls" with top diameter of 14.5 cm, bottom diameter of 5 cm, and depth of 6.5 cm.
4. Fermentation cabinet capable of maintaining a temperature of 30° to 0.5°C and a relative humidity of at least 75°. Cabinets similar to those described by Bailey (1930A) or Larmour *et al.* (1931) are recommended.
5. Punching and molding machines similar to those described by Heald (1939) or Merritt *et al.* (1940) are recommended.
6. Baking pans constructed of 2 XX Tin-Either tall- or low-form tins with the following inside dimensions may be used. Low-form tins: top-length, 11.5 cm; width, 7.0 cm; bottom-length, 9.5 cm;

width, 5.5 cm; depth, 5.0 cm. Tall-form tins: top-length, 10.5 cm; width, 6.0 cm; bottom-length, 9.3 cm; width, 5.3 cm; depth, ends, 6.8 cm; sides, 8.5 cm.

7. Baking oven capable of maintaining a temperature of 230° ± 5°C. A rotating plate or reel should be used to carry the pans. Types similar to the one described by Finney and Barmore (1939) are recommended.
8. Miscellaneous—Rough and fine balances, scoops, spatulas, pipettes, burettes, flasks, beakers, etc.

Reagents

1. Yeast suspension—Suspend 12 g of fresh refrigerated compressed yeast in water and make up to 100 ml. This suspension should be freshly prepared for each series of tests.
2. Salt–sugar solution—Dissolve 4 g NaCl and 20 g sucrose and make up to 100 ml.

Baking Procedure

Mixing. Place in the mixing bowl a quantity of flour equivalent to 100 g on a 14% moisture basis (86 g dry matter); add 25 ml yeast suspension, and 25 ml of the salt–sugar solution. Now add sufficient additional water to bring the dough to the desired consistency after mixing; note volume of water added. Multiples of this charge may be used according to the capacity of the mixer.

Start the mixer and mix for 2 min if a Swanson type is used, or until the standard consistency is obtained. The temperature of the solutions, flour, room, and mixing bowl should be so adjusted that the final dough temperature is 30°C, but heating of the yeast suspension should be avoided.

To determine the mixing response, mix separate portions of dough for varying lengths of time, keeping the final dough temperature as close to 30°C as possible; proceed with remainder of baking procedure as described; and note the changing character of the resulting loaves. Report the optimum mixing time and mixing tolerance under the specified conditions.

Fermentation. Remove dough from mixing bowl; if a multiple charge was used, scale to appropriate weight. Two methods of scaling are employed by baking technologists, namely, to constant dough weight (usually 150–175 g, depending upon the type of flour under test and the pan form used) or to constant flour weight (usually 100 g on a 14% moisture basis).

Round up the dough by folding 20 times in the hands; put dough in fermentation bowl and place bowl in fermentation cabinet. Ferment for a total of 180 min. Give the dough the first punch after 105 min and the second after an additional 50 min; mold after an additional 25 min.

To determine the fermentation response, vary the fermentation time on separate portions of dough above and below the recommended times; proceed with baking procedure as described; and note the character of the resulting loaves. Report optimum mixing time and mixing tolerance under the specified conditions.

Punching (Machine Method). Remove the dough from fermentation bowl, lightly seal over the wet side of the dough by drawing the dry edges together with the thumb and fingers, elongate dough ball slightly, and pass it once through the sheeting rolls set with a clearance of $\frac{9}{32}$ in. Roll up the dough sheet lightly without rounding or sealing edges. Replace in fermentation bowl and return to cabinet. This roll setting has been found satisfactory for most flours but may have to be modified slightly to obtain optimum results with extreme types of flours.

Punching (Hand Method). Remove dough from fermentation bowl and place, wet side down, on a piece of canvas belting provided with a wooden track giving $\frac{5}{16}$-in. clearance. With a rolling pin, roll once each way from the center of the dough. Invert dough, overlap opposite ends, replace it seam down in the fermentation bowl, and return to the cabinet.

Molding (Machine Method). Sheet the dough for molding as described under Punching, passing the dough twice through the rolls, first with a setting of $\frac{9}{32}$ in. and second with a setting of $\frac{3}{16}$ in. Lightly roll up dough and place in molder with the outside end of the dough sheet as far away from the revolving drum as possible. Run dough through molder and place in baking pan, seam down. Pans may be very lightly greased when absolutely necessary to prevent the loaves from sticking.

Molding (Sheet Roll Method). Sheet the dough as just described under Molding (Machine Method). Lightly roll up dough, seal the seam, roll lightly under the palm of the hand, and place seam down in the pan. The length of the dough should not exceed that of the pan before the final light rolling. Pans may be very lightly greased when absolutely necessary to prevent the loaves from sticking.

Molding (Hand Method). Sheet dough as described under Punching (Hand Method). Overlap the opposite ends of the dough sheet, invert, and turn parallel with the long axis of the wooden track. Roll once each way from the center, turn dough over, and starting at the more remote end, roll it up by hand toward the operator. Seal the seam, roll the dough lightly under the palm of the hand and place, seam down, in the baking pan. The length of the dough should not exceed that of the pan prior to the final light rolling. Pans may be lightly greased when absolutely necessary to prevent loaves from sticking.

Proofing. Proof the dough at 30°C and at least 75% relative humidity. Two methods of proofing are employed by baking technologists, namely, to constant time (usually 35 min) or to constant height (9.5 cm has been suggested). The method used depends upon the purpose of the test and the interpretation to be placed upon the results.

Baking. Bake for 25 min at a temperature of 230° (±5°)C, as determined by an oven thermometer placed at the level of the top of the baking pan and 5 cm from it on the side next to the axis of rotation of the shelf. Precise control of temperature is essential. An open pan of water should be placed in the oven. To provide more uniform oven conditions bake a series of "dummy" loaves just before and following the experimental series.

Measurement and Scoring

Calculate absorption on a 14% moisture basis from the following relation:

$$\%\ \text{Absorption} = W - (100 - F)$$

where W = total ml water added and F = weight of flour taken. In calculating total water added, the 25 ml of yeast suspension is equivalent to 22.5 ml of water, and the 25 ml of the salt–sugar solution is equivalent to 21.5 ml of water.

One hour after removal from the oven, weigh the loaf and measure length × width × height to determine the volume in cubic inches. A standard one-pound loaf of bread is judged to have a volume of 125 to 155 cubic inches. Place the loaves in a fairly airtight cabinet until they are scored for external and internal characteristics or one can use a volume measuring apparatus similar to that described by Malloch and Cook (1930) or by Binnington and Geddes (1938).

In scoring bread, the various characteristics are judged relative to those of a hypothetical standard loaf. Thus, this scoring procedure is based primarily on personal, subjective judgments; that is, raters compare the loaves under evaluation with the ideal loaf they visualize. Clearly, no absolute values are possible for characteristics such as color of crust, texture, aroma, and grain. Instead, arbitrary maximum values are assigned to the loaf characteristics that are typically evaluated (Table 11.6). Loaves under evaluation are then scored in terms of these maximum point values. For example, if the crust color of an experimental loaf is judged to be only about half as good as the ideal, it would be scored 4 on this characteristic. The following descriptions of the characteristics of an "ideal" loaf of bread provide guidance in scoring.

Color of Crust. The color of crust should be an appetizing golden brown. The sides, ends, and bottom of the loaf should all be well baked, and although not necessarily as deeply colored as the top of the loaf, they should not be pale.

Evenness of Color. The coloring should be uniform, free from spots and streaks. A dark or burned bottom, or a highly colored top with pale sides, may be mentioned as poor qualities. The intensity of the color is scored separately (color of crust) and should not be confused with evenness of color, which relates to uniformity alone.

Symmetry of Form. The ideal loaf should be symmetrical without low ends or overlapping. The most common faults that detract from symmetry form include flat tops caused by weak gluten, bulging sides or ends, and improper loaf dimensions. Well-defined, even slightly rounded tops are the most desirable.

Table 11.6 Score Card for Bread

Characteristic		Point value
External appearance		
Color of crust		8
Evenness of color		3
Symmetry of form		3
Break and shred		3
Volume		10
Character of crust		3
	TOTAL	30
Internal appearance		
Grain		10
Color of crumb		10
Texture		15
Aroma		15
Taste		20
	TOTAL	70
	TOTAL BREAD SCORE	100

Break and Shred. These terms refer to the rupture in the crust that occurs along the top, sides, and ends of the loaf during oven spring. The loaf should have a uniform break, with a well-defined shred on one or both sides. A split-top loaf is an exceptional case in that the shred occurs only at the center of the top and at the ends of the loaf. The shred in a split-top loaf should be prominent. A ragged break or shell crust detracts from the appearance of the loaf.

Volume. The space (volume) occupied by a loaf is an indication of flour strength. Volume may be either too large, too small, or satisfactory. Large volume is undesirable if it results in an open grain and a weak texture. Small volume may be due to either improper fermentation or to poor gluten quality. On the average, a round-top loaf weighing 1 lb should have a volume of not less than 2 liters.

Character of Crust. A thin, tender crust is desirable. This attribute is obtained by the use of an appropriate amount of shortening and sugar combined with correct fermentation. Oven conditions also influence the character of crust. For example, an underheated oven yields a loaf having a tough crust with a dull color; an overheated oven yields a loaf having a leathery crust and highly glazed appearance; and an oven with improper heat distribution yields a loaf with an uneven thickness in the crust. The top and bottom crust should not be too thick.

Grain. The grain of the crumb is the structure formed by the strands of gluten, including the area they surround. Uniformity of size and small cells with thin walls are most desirable. Coarseness, thick cell walls, uneven cell size, and large holes are indicative of poor grain.

Color of Crumb. A soft white, free from dark streaks or spots, is the most desirable crumb color. Extreme whiteness and yellowness are both to be avoided. Color depends primarily upon the natural color of the endosperm of wheat and of the bran particles present in the flour. To a lesser extent, crumb color can also be affected by the grain of the crumb, the color being lighter with increasing fineness or grain. Crumb color should be judged only in evenly distributed natural light, or with a "daylight" lamp if artificial light is used. Evaluation of crumb color should be made on portions cut at the same part of the loaf as for the grain evaluation. Too, the evaluation should be made on freshly cut surfaces because the crumb tends to darken on exposure of its cut surface.

Texture. Texture is determined by pressing the finger tips lightly on the cut surface and moving them gently over the surface. The sensation produced can be characterized as velvety, silky, soft, elastic, lumpy, doughy, rough, or harsh. The ideal texture is free from lumps and harshness and gives a smooth silky surface. Doughiness or crumbliness are undesirable features.

Odor. Odor is determined by the sense of smell. The air in the cell structure carries the odor. Thus, to determine odor a half loaf is held close to the nose and the air squeezed out of it during the act of smelling. The odor of bread should be appetizing. A rich, sweet, fresh, malty odor is the desirable one. Undesirable odors may be described as sour, cheesy, yeasty, musty, moldy, ropy, rancid, etc.

Taste. The sensation of taste results from the combined stimulation of both the organs of taste and smell; that is, it is more than just the simple sensation of salty, sour, sweet, or bitter (the four basic tastes). The desirable taste for bread is very similar to the desirable odor—sweet, pleasant, nutty, wheatlike, appetizing. Part of the taste sensation also includes the chewing quality of the bread. A small piece of bread should break down easily in the mouth on mastication without forming doughy lumps that are difficult to chew and swallow.

Total Bread Score. The sum of the point values for the individual characteristics is an index that reflects the degree to which the test loaf differs from the standard reference loaf.

SELECTED REFERENCES

AACC. 1962. Cereal Laboratory Methods. 7th ed. Am. Assoc. of Cereal Chemists, St. Paul, Mn.

AOAC. 1984. Official Methods of Analysis. 14th ed. S. Williams (editor). Assoc. of Official Analytical Chemists, Arlington, VA.

BAILEY, C. H. 1930A. A new type of fermentation cabinet. Cereal Chem. *7*, 341.

BAILEY, C. H. 1930B. Calibration of loaf volume measuring devices with metal models. Cereal Chem. *7*, 346.

BINNINGTON, D. S. and GEDDES, W. F. 1938. An improved wide-range volume measuring apparatus for small loaves. Cereal Chem. *15*, 235.

FINNEY, K. F. and BARMORE, M. A. 1939. Maintaining a uniform temperature in an experimental baking oven. Cereal Chem. *16*, 289.

FOOD and DRUGS, Code of Federal Regulations. Title 21, Parts 100 to 199. 1977 ed.

HART, F. L. and FISHER, H. J. 1971. Modern Food Analysis. Springer-Verlag, New York.

HEALD, W. L. 1939. Hand punching and hand molding. Cereal Chem. *16*, 24.

KENT, N. L. 1975. Technology of Cereals with Special Reference to Wheat. 2nd ed. Pergamon Press. New York.

KOTSCHEVAR, L. H. 1975. Quality Food Production. Cahners Publishing Co., Boston.

LANDIS, Q. and FREILICH, J. 1935. Studies on test dough mixer calibration. Cereal Chem. *12*, 665.

LARMOUR, R. K. 1931. A fermentation and proofing cabinet giving low temperature variability. Cereal Chem. *8*, 233.

MALLOCH, J. G. and COOK, W. H. 1930. A volume-measuring apparatus for small loaves. Cereal Chem. *7*, 307.

MATZ, S. A. (Editor) 1959. Chemistry and Technology of Cereals as Food and Feed. AVI Publishing Co., Westport, CT.

MERRITT, P. P., MARKLEY, M. C., and ROTHHOLZ, E. 1940. Studies of the usefulness of a motor-driven sheeter in test baking. Cereal Chem. *17*, 384.

POMERANZ, T. (editor) 1971. Wheat Chemistry and Technology. 2nd ed. Am. Assoc. of Cereal Chemists, St. Paul, MN.

PYLER, E. J. 1973. Baking Science and Technology. Vols. 1 and 2. Siebel Publishing Co., Chicago, IL.

SANDSTEDT, R. M. 1937. The adaptation of the ferricyanide maltose method to high diastatic flours. Cereal Chem. *14*, 603.

SHERWOOD, R. C. 1930. Report of sub-committee on dough thermometers. Cereal Chem. *7*, 362.

TSCHERMAK, E. VON. 1914. Die Verwertung der Bastardierung fur phylogenetische Fragen in der Getreidegruppe. Z. für Pflanzenzuchtung *2*, 292.

USDA. 1978. The Official U.S. Standards for Grain. Government Printing Office, Washington, DC.

12

Milk and Milk Products

☐ INTRODUCTION

Milk is the normal secretion of the mammary glands for the feeding of the young of mammals. In the United States, the primary milk in commerce is cow's milk. The responsibility for insuring the quality of this food product rests primarily with the individual states rather than with the federal government. Consequently, legal definitions of milk are not uniform within the United States. The U.S. Public Health Service (USPHS) has encouraged the adoption of adequate and uniform state and local control legislation. The following definition from its Grade A Pasteurized Milk Ordinance (1965) has been adopted by a majority of the states: "Milk is hereby defined to be the lacteal secretion, practically free from colostrum, obtained by the complete milking of one or more healthy cows, which contains not less than 8¼% milk solids-not-fat and not less than 3¼% milk fat."

The composition of milk is rather complex. Of the major components of milk, the one that most readily distinguishes milk from other foods is lactose, which occurs naturally only in milk. Similarly, casein occurs only in milk. The proteins of milk are of great importance and contribute essential amino acids necessary for normal health and growth. Milk fat is a palatable mixture of glycerides of fatty acids. The components of milk are in complex equilibrium, and much is yet to be learned regarding the forms and combinations in which they exist.

Lactose and some of the mineral salts are in true solution; the milk proteins as well as some of the calcium phosphate exist in colloidal solution; the fat exists in macroscopic dispersion in the milk plasma. Besides these major constituents, milk contains other constituents such as vitamins, enzymes, pigments, and lactic acid.

☐ COMPOSITION OF MILK

The average gross composition of cow's milk is as follows: water, 87%; fat, 3.7%; lactose, 4.9%; proteins, 3.5%; and ash, 0.7%. However, the composition of milk from individual cows may vary considerably from these average values. Breed differences, the time of year, the time of day, individual differences, the age of the cow, the period of lactation, the portions of any one milking, feeding, etc., are some of the factors that may contribute to variation in the composition of milk.

Variations in milk composition due to breed are of the greatest magnitude. Jersey, Guernsey, and Ayrshire cows give milk richer in fat than Holstein cows. Usually milk obtained in the fall and early winter is richer than that obtained in the spring and summer. The morning milk is usually richer in fat than evening milk, at times by almost 2%. The first portions (or *fore* milk) of milk drawn in the milking process are lower in fat than the last portions (or *strippings*). Milk obtained the first few days after calving is known as colostrum and differs materially from normal milk. During the final stages of lactation, when the daily production of milk is decreasing, there is some increase in the percentages of fat and casein. The differences in milk composition occur mainly in the relative amounts of fat, protein, and water. The percentages of ash and lactose remain fairly constant in normal milk regardless of the amounts of other components.

Fat

Fat and associated lipid material are present in milk as an emulsion of small individual globules in an aqueous phase. These fat globules are predominantly in the liquid state at 37° and range in number from 2.5 to 5 $\times$ 10^9/ml of milk. The fat globules vary in size from about 0.1 to 10 μm in diameter, an average about 3 μm in diameter (1 μm is approximately 1/25,000 in.).

Surrounding each milk fat globule is a membrane composed of cholesterol, phospholipids, proteins, and other surface-active molecules. The fat globule membrane appears to have two laters: an inner layer that contains a phospholipid–protein complex, and an outer layer that contains enzymes originating in the mammary gland. The membrane serves to prevent individual fat globules from combining with others to form a larger globule. Mechanical agitation can lead to a disruption of the membrane and thus cause the individual fat globules to coalesce to form larger globules and eventually to form butter. Disruption of the globule membrane may also contribute to the development of rancid flavor in milk.

Milk fat is made up of approximately 12.5% glycerol and 85.5% fatty acids, by weight. The predominant glycerides are triglycerides (98–99%); diglycerides account for 0.05%; and monoglycerides for 0.04% of the total. These glycerides contain more than 16 different fatty acids combined with glycerol in a number of different ways. The principal fatty acids contain an even number of carbon atoms (Table 12.1). Milk fat is unique in that it is the only natural food fat that contains butyric acid in its glyceride structure. This fatty acid, as well as the C_6, C_8, and C_{10} saturated fatty acids, have a strong characteristic odor. Their release from triglycerides during lipolysis is the cause of rancid odor and flavor in milk. The polyunsaturated fatty acids may give rise to an off-flavor identified as oxidized flavor. Of the fatty acids in milk fat, 60–75% are saturated, 25–35% are unsaturated, and approximately 4% are polyunsaturated. Oleic and palmitic acids make up over 50% of the fatty acids found in milk fat.

Milk fat contains a number of minor components. Among these are

Table 12.1. Fatty Acid Content of Milk Fats

Fatty acid	No. carbon atoms	Weight percentage (g/100 g milk fat)					
		Cow	Goat	Buffalo	Ewe	Mare	Human
Butyric	4	3.1	2.6	5.8	3.0	0.4	0.4
Caproic	6	1.0	2.3	0.6	2.0	0.9	0.1
Caprylic	8	1.2	2.7	0.9	2.0	2.6	0.3
Capric	10	2.6	8.4	1.0	5.0	5.5	1.7
Lauric	12	2.2	4.5	1.6	4.1	5.6	5.8
Myristic	14	10.5	11.1	9.0	10.0	7.0	8.6
Palmitic	16	26.3	28.9	35.2	24.0	16.1	22.6
Stearic	18	13.2	7.8	15.3	13.0	2.9	7.7
Arachidic	20	1.2	0.4	0.1	1.0	0.3	1.0
Oleic	18:1	32.2	27.0	20.5	26.0	19.0	36.4
Linoleic	18:2	1.6	2.6	1.5	5.0	8.0	8.3
Linolenic	18:3	—	—	—	—	16.0	0.4

Source: National Research Council (1953) for cow, goat, buffalo, mare, and human data; Hilditch and Williams (1964) for ewe data.

free fatty acids, cholesterol, phospholipids, and the fat soluble vitamins.

Proteins

Casein, lactalbumin, and lactoglobulin are the principal proteins of milk (Table 12.2). Casein occurs in milk as a colloidal calcium caseinate–calcium phosphate complex and forms about 3% of the whole milk. Lactalbumin constitutes about 0.10–0.45% and lactoglobulin 0.20–0.40% of milk. In addition to these proteins there is also a small amount (approximately 0.2%) of heat-stable milk serum proteins.

Milk and milk products furnish approximately 25% of the total proteins in the diets of people in the United States and Canada. These proteins are high quality becauses they contain all the essential amino acids and in adequate amounts. The amino acid composition of milk proteins is shown in Table 12.3.

Casein

Casein may be defined as the fraction of skim milk that is precipitated by acidifying to pH 4.6–4.7, leaving the milk serum proteins in solution. Casein is composed of three electrophoretic components, which are designated as α, β, and γ in order of decreasing mobility. Associated with the α-casein complex is a portion known as κ-casein, which is specifically hydrolyzed from casein via the action of rennin to give para-κ-caseinate. The para-κ-caseinate is insoluble in the pres-

Table 12.2. Principal Proteins in Milk

Protein component	Concentration (gm/100 ml)	Approximate % of total protein
Caseins	2.0–3.5	80
α-Casein		45
β-Casein		23
κ-Casein		8
γ-Casein		4
Whey proteins	0.5–0.7	20
α-Lactalbumin		4
β-Lactoglobulin		9
Protease–peptone		4
Blood protein		
Serum albumin		1
Immunoglobulin		2

Source: Adapted from Rose (1970).

Table 12.3. Amino Acid Composition of the Major Milk Proteins

Amino acid	Weight percentage (g/100 g protein)			
	Total milk proteins	Casein	α-Lactalbumin	β-Lactoglobulin
Essential				
Arginine	3.5	3.8	3.4	2.9
Histidine	2.7	2.3	1.6	1.6
Isoleucine	6.5	6.1	5.1	6.8
Leucine	9.9	10.8	14.1	15.5
Lysine	8.0	6.8	7.3	11.3
Methionine	2.4	2.9	2.4	3.2
Phenylalanine	5.1	5.5	4.1	3.7
Threonine	4.7	4.4	5.0	5.3
Valine	6.7	6.6	5.0	5.8
Nonessential				
Alanine	3.6	2.3	2.6	6.8
Aspartic acid	7.5	5.8	9.6	11.0
Cystine	0.9	0.3	3.1	4.1
Glutamic acid	21.7	21.7	15.2	19.8
Glycine	2.1	0.4	0.0	1.5
Proline	9.2	9.8	4.0	4.7
Serine	5.2	5.4	4.0	4.3
Tyrosine	4.9	6.0	4.0	3.7

Source: National Research Council Publication (1953).

ence of calcium ions and as a consequence it precipitates as a clot or curd. This phenomenon is the basis of cheese manufacture.

It is noteworthy that milk of ruminants (buffalo, cow, ewe, and goat), which contains at least 2.5% casein, gives a cheese-type curd with rennet, whereas milk of nonruminants (human, mare), which contains less than 2.5%, does not give a cheese-type curd with rennet.

Whey Proteins

The soluble proteins retained in the supernatant (whey), following the removal of casein from skim milk, include lactalbumin and lactoglobulin. Unlike casein, the whey proteins are unaffected by dilute acids or rennin. Both of these proteins are coagulated by heat, but the degree of such coagulation is determined by the pH of the solution, temperature, time of holding, and salt concentration.

The components of the albumin and globulin fractions of milk are shown in Table 12.2, and the amino acid composition of the individual proteins is given in Table 12.3.

Each of the whey proteins has a different sensitivity towards heat denaturation. The immunoglobulins are the most sensitive, β-lactoglobulin and serum albumin are less sensitive, and α-lactalbumin is the least sensitive to heat denaturation.

Other Proteins

After the removal of casein, albumin, and globulin from skim milk, the filtrate still contains a fraction that may be precipitated by saturation with ammonium sulfate. This fraction, referred to as the proteose–peptone fraction, consists of material of smaller molecular size than true proteins.

Nonprotein Nitrogen

In addition to the constituents summarized in Table 12.2, a variety of nonprotein nitrogen-containing constituents are present in milk. These substances include amino acids, certain vitamins, urea, uric acid, ammonia, creatine, and creatinine. Little is known about this fraction, but it probably is the end product of nitrogen metabolism in the cow's body and in the synthesis of milk in the udder. Normally this fraction amounts to about 5% of the total nitrogen in milk, but physiological disturbances such as mastitis result in a higher level of nonprotein nitrogen.

Lactose (Milk Sugar)

The characteristic carbohydrate of milk is lactose, a disaccharide composed of one molecule of D-glucose and one of D-galactose, which is synthesized in the mammary gland.

The lactose content of cow's milk is fairly constant, ranging from 4 to 5%. In contrast, milk from cows with mastitis have lactose contents as low as 2.7%. The more lactose a milk contains, the sweeter its taste. Cow's milk, with a mean lactose content of 4.6%, tastes faintly sweet. It is interesting to note that although sweet (nonacid) whey has about the same concentration of lactose as the milk from which it is made, it is much sweeter than milk, due undoubtedly to the removal of casein.

Aqueous solutions of lactose reduce Fehling's solution. After hydrolysis, the reducing power of lactose is doubled due to the two active carbonyl groups. Lactose is soluble in hot acetic acid and insoluble in alcohol and ether. The solubility of lactose in water is much lower than that of other common sugars (at 100°C, it is $\frac{1}{3}$ as soluble; at 0°C it is only $\frac{1}{14}$ as soluble). In equal concentration, lactose is only about $\frac{1}{3}$ as sweet as sucrose.

Lactose exists in two stereoisomeric forms, α and β. These two forms have quite different physical properties, but no matter which form is dissolved in water the solution soon contains both forms in an equilibrium with one another. When this equilibrium is reached, approximately 60% of the lactose is in the β form.

Mutarotation manifests itself in the solubility behavior of lactose. For example, if an excess of α-lactose monohydrate is stirred into 100 g of ice water (0°) about 5.0 g of lactose will dissolve rapidly (Table 12.4). This value is called the *initial solubility*, and it represents the maximum amount of α-lactose that will dissolve before any of it is converted to the β form. If stirring is continued, an additional amount of a α-lactose dissolves because it replaces that portion of α-lactose that has changed over to the β form. Eventually the solution will contain 11.9 g of lactose: 5.0 g in the form of α-lactose and 6.9 g in the form of β-lactose. The value 11.9 is called the *final solubility*, and it represents the true solubility of α-lactose and its increasing solubility due to mutarotation.

At equilibrium the solution just described is saturated with respect to α-lactose, but additional β-lactose could be dissolved in it due to the greater solubility of the β form. However, additional β-lactose will upset the equilibrium because with time the β-lactose undergoes mutarotation to form α-lactose. Since the solution is already saturated

Table 12.4. Solubilities of Lactose at Various Temperatures

Temperature (°C)	Solubility (g/100 g of water)			
	Initial solubility		Final solubility	Supersolubility
	α Form	β Form		
0	5.0	45.1	11.9	25
15	7.1		16.9	38
25	8.6		21.6	50
39	12.6		31.5	74
49	17.8		42.4	
64	26.2		65.8	
74	34.4		86.2	
89	55.7		139.2	
100		94.7	157.6	

Source: Whittier (1944).

with respect to α-lactose, the α-lactose formed by mutarotation of the β-lactose will crystallize to reestablish equilibrium. If one could prevent α-lactose from crystallizing, one theoretically could prepare a solution containing enough α-lactose to be in equilibrium with 45.1 g of β-lactose. This would be the final solubility of β-lactose. Needless to say such a solution would be highly unstable. The α-lactose would soon crystallize, and the β-lactose would continue to undergo mutarotation until only enough β was left to maintain equilibrium with the 5.0 g of a α-lactose that would stay in solution.

Thus, whether one starts with α- or β-lactose, at equilibrium the amount of lactose (α + β forms) will correspond to the final solubilities shown in Table 12.4. Since the β form is more soluble, any excess crystallizes out of solution as α-lactose. This explains why lactose crystals in dairy products are in the α form. These hard crystals produce a gritty sensation on the tongue, which is referred to as sandiness.

The lactose of commerce is α-lactose (monohydrate). It is prepared by crystallization from solution at room temperature. If the temperature of crystallization is increased, the solubilities of both the α and β forms increase; however, the solubility of α-lactose increases more rapidly with temperature than does that of β-lactose (Table 12.4). At approximately 93°C the solubilities of both forms are equal. Above 93°C the solubility of β-lactose is less than that of α-lactose; as a consequence one can prepare the β form (anhydride) by evaporating a solution of lactose to dryness above 93.5°C. The anhydride is obtained by dehydrating the α monohydrate at a temperature above 65°C. The α and β anhydrides change to the monohydrates in the presence of small amounts of water at temperatures below 93.5°C.

Lactose in milk is readily fermented by a number of bacteria to yield lactic acid. The changes that accompany the conversion of lactose into lactic acid is associated with the souring of milk. The odor of sour milk is not due to lactic acid but to other volatile products formed during fermentation.

Lactose, in common with other disaccharides, must be hydrolyzed by specific enzymes in the digestive tract into monosaccharides before absorption and utilization by the body. Lactase, the enzyme responsible for the hydrolysis of lactose to D-glucose and D-galactose, occurs in the intestinal mucosa of mammals. When lactose intake exceeds the amount that lactase can hydrolyze, symptoms of lactose intolerance (flatulence, cramps, diarrhea) may ensue. In recent years numerous clinical studies have indicated that the ability to utilize lactose may be a problem for a major portion of the world's population because of insufficient intestinal lactase activity.

Ash (Mineral Matter)

When the water of a milk sample is removed by evaporation and the residue heated to 485°C for 16 hr in a muffle furnace, most of the residue will be burned. That which remains is designated the *ash* or mineral matter of the milk. The average ash content of milk is 0.72%. It is important to remember that several of the acid radicals (citrates, carbonates, etc.) are lost during the ashing process, leaving the metallic elements as oxides. Furthermore, some of the sulfur and phosphorus present in proteins, phospholipids, etc., is oxidized and included in the ash. Likewise, the calcium and magnesium linked to casein is liberated to become a part of the ash. Potassium and sodium occur as chlorides and to a lesser degree as phosphates and citrates. The net result is that the ash from milk is always alkaline because the basic elements are in excess of the acidic radicals.

Major Constituents

The major constituents of the ash of cow's milk are presented in Table 12.5. The major portion of the ash is composed of the chlorides and oxides of potassium, calcium, and phosphorus. It is of interest to note that these three elements are in greater concentration in milk than in blood; thus, the mammary gland exerts a selective action to concentrate these elements. Too, a fact frequently overlooked is the greater weight percentage of the potassium over that of calcium. This kind of information can be used to detect injured udders because in the injured udder the composition of the milk becomes more like that of the blood. The changes in chloride content and in acidity are easily measured.

Table 12.5. Composition of the Ash of Cow's Milk

	Weight percentage (g/100 g ash)	
Element	Range	Average
Calcium (CaO)	20.01–27.32	22.37
Phosphorus (P_2O_5)	21.57–29.33	25.67
Potassium (K_2O)	23.63–30.33	26.30
Sodium (Na_2O)	5.82–11.92	9.03
Magnesium (MgO)	2.25–3.12	2.63
Iron (Fe_2O_3)	0.05–0.40	0.20
Chlorine (Cl^-)	13.57–16.38	14.44
Sulfur (SO_3)	Trace–3.84	2.69

Source: Webb and Johnson (1975).

Milk from cows suffering from mastitis contains a higher sodium and chloride content than normal milk. A chloride content in milk in excess of 0.14% suggests the possibility of mastitic milk. All cases of high chloride content in milk are not due to disease. Milk secreted toward the end of lactation and colostrum secreted at the beginning of lactation contain more sodium and chloride than normal milk.

Small variations in the concentrations of Ca and Mg relative to the concentrations of citrate and phosphate are known to affect the ability of milk to resist coagulation when heated. The ratio between the different components (Ca + Mg: citrate + phosphate) seems to be more important than the actual amounts present. The term *salt balance* is used to refer to the ratio of these components. In general, the most common manifestation of a disturbed salt balance is a deficiency of citrate and phosphate ions (particularly when cows are off pasture), which would suggest that in such circumstances addition of sodium citrate or phosphate would correct the balance through its sequestering effect on calcium ions and render the milk heat stable.

Milk contains more calcium than most foods because of the distribution of calcium in milk. About two-thirds of the total calcium in milk is in colloidal form as calcium caseinate, citrate, and phosphate; the remaining third is in true solution. Only because of this phenomenon, is it possible for milk to have an osmotic pressure equal to that of the blood. When milk is heated or pasteurized, the equilibrium between the colloidal and soluble phases is altered in the direction of the colloidal phase. In cheesemaking, this change in the ratio between soluble and colloidal calcium is an important factor in the time taken by rennet to coagulate the milk.

Trace Elements

In addition to the ash components listed in Table 12.5, some of the elements that have been reported in milk are zinc, copper, flourine, manganese, iodine, aluminum, barium, boron, cobalt, lithium, strontium, titanium, silicon, tin, and molybdenum. Some of these elements are constituents of enzymes; for example, iron occurs in peroxidase and molybdenum in xanthine oxidase. It is interesting to note that many of these trace elements are attached to proteins of the fat-globule membrane.

Vitamins

The vitamins found in milk include vitamins A, D, E, and K, ascorbic acid, thiamin, riboflavin, niacin, pantothenic acid, pyridoxine, folic acid, and cyanocobalamin. Ruminants derive all the water-soluble vitamins via synthesis by rumen microbial flora, but are dependent on an exogenous supply of the fat-soluble vitamins. The quantities of water-soluble vitamins and of vitamin K in milk are little-affected by feed, season, breed, or stage of lactation, whereas the quantities of vitamins A, D, and E are influenced by feed. All the vitamins appear to be well absorbed from milk products and make important contributions to human diets.

Fat-Soluble Vitamins

Vitamin A. The amount of vitamin A in milk varies with the carotene content of the feed. Normally, vitamin A potency is highest when the cow is on succulent pastures in the spring and lowest when the cow is hay-fed during the winter season. The vitamin A potency of milk can be increased to a level that approaches summer milk by feeding rations high in carotene content.

Because vitamin A is heat stable, there is no loss of vitamin A when milk is pasteurized, evaporated, or dried. It is, however, quite susceptible to oxidation especially in the presence of unsaturated fats under oxidizing conditions.

The ability to convert carotene into vitamin A before it is secreted into the milk varies with different breeds. The Guernsey cow (or Jersey) secretes a large proportion of vitamin A as carotene, while other breeds such as Holstein-Friesian and Ayrshire produce a milk low in carotene but equally high in vitamin A on a fat basis. Consequently the latter milks have less color than either Jersey or Guernsey milk.

Vitamin D. The amount of vitamin D secreted in milk is directly related to its supply in the feed and to the direct exposure of the animal to sunlight. Vitamin D in milk is highest during the summer months and lowest in winter. The vitamin D level in milk may be raised by simply adding to milk a dispersible preparation of the vitamin. This generally is accomplished by adding a concentrate to the milk, prior to pasteurization, to increase the potency to at least 400 IU of vitamin D per quart of milk. Pasteurization of milk and sterilization of evaporated milk do not destroy vitamin D.

Vitamin E. The tocopherols comprise a family of substituted quinones and occur in most plants. These compounds function as antioxidants both in the animal body and in milk. Alpha-tocopherol is present in milk, but milk is a poor source of the vitamin. The concentration of the vitamin in milk is also related to that in the diet of the animal—

the potency being highest during summer grazing and lowest during winter feeding.

Vitamin K. Vitamin K occurs in two forms, both related to 2-methyl-1,4-naphtoquinone. Vitamin K_1 occurs in the green leafy tissues of plants; vitamin K_2 is synthesized by bacteria in the intestinal tract. The concentration of vitamin K in milk, unlike that of the other fat-soluble vitamins, is not dependent upon the supply in the feed. Milk is a relatively poor source of vitamin K.

Water-Soluble Vitamins

As noted already, all the water-soluble vitamins are synthesized by bacteria in the rumen. As a consequence, ruminants are largely independent of an exogenous supply, and the content of water-soluble vitamins in milk is largely unaffected by feed.

Thiamin. The thiamin in milk occurs both free and phosphorylated. The thiamin content of cow's milk has been reported to be 27–90 μg/100 ml; for pooled milk a representative value would be 42–45 μg/100 ml (0.44–0.48 mg/qt).

Milk is a fair source of thiamin. Under acid conditions, thiamin is relatively heat stable, but under the more neutral conditions in milk (approximately pH 6.6), it undergoes decomposition. Pasteurization results in a loss of 10–20% of the thiamin originally present in milk. Short time–high temperature pasteurization of milk results in thiamin losses of only about 3%.

Riboflavin. Milk is an excellent source of riboflavin. As with the other water-soluble vitamins, the riboflavin content of milk is relatively independent of diet. However, some effect due to feed seems to occur, since summer milk contains 13–30% more riboflavin than winter milk. There is also a tendency for some breeds of cows to produce milk that contains more riboflavin than do other breeds. For example, Jersey and Guernsey cows produce milk with 20–50% more riboflavin than the Ayrshire and Holstein breeds. A representative value for the riboflavin content of milk seems to be about 1.9 mg/qt.

Riboflavin is relatively heat stable; thus, milk can be pasteurized or sterilized with little loss in riboflavin content. However, if milk is exposed to light, especially direct sunlight, riboflavin is destroyed. Consequently, milk must be protected from bright light to protect the riboflavin. Fluorescent light is less harmful than sunlight.

Niacin. Milk is a relatively poor source of niacin, but the niacin that is present is readily available, while in other foods the niacin is frequently not available. There is no loss in the niacin content in milk upon exposure to light, pasteurization, evaporation, or drying.

Ascorbic Acid. The ascorbic acid content of cow's milk is about 20 mg/liter and is fairly uniform throughout the year. Ascorbic acid is a very labile vitamin. Heat, exposure to sunlight, and the presence of trace elements (Fe^{3+}, Cu^{2+}) have an adverse effect on the vitamin. Milk held under refrigeration (5°C) will show a progressive loss of ascorbic acid.

Other Vitamins. Milk contains several other water-soluble vitamins, including pantothenic acid, pyridoxine, folic acid, and vitamin B_{12}. The levels of all of these in milk are independent of feed and breed of cow. Pasteurization has little or no effect on the content of these vitamins in milk.

Enzymes

A number of enzymes occur in milk. Many of them find their way into milk through passage from blood or from the breakdown of mammary cells. Too, bacteria growing in milk contribute enzymes to those originally present. This is significant in that excessive amounts of these enzymes may be used as tests to indicate milk quality and proper pasteurization.

Lipases

Lipases are enzymes that catalyze the hydrolysis of lipids. Under certain conditions the lipases present in milk cause hydrolysis of triglycerides to fatty acids and glycerol. The free fatty acids impart a characteristic odor and flavor to milk or milk products. This is desirable in some instances such as in cheese (e.g., Roquefort) while in other instances (e.g., milk and cream) it is undesirable.

Certain milking operations render raw milk susceptible to lipolysis. Agitation and foaming of milk when it is pumped through pipelines (risers) where air may pass through the milk causing agitation and foaming. Variation in temperature is conducive to lipolysis. Milk should not be alternately cooled to 5°C, warmed, then cooled as may happen in farm milk tanks.

Homogenization renders milk more susceptible to lipolysis because it causes a reduction in the size of the fat globules. Lipases are inactivated by heat. Lipase activity in milk generally increases with the pH of milk and the period of lactation.

Phosphatases

Several phosphatases, which hydrolyze esters of phosphoric acid, are present in milk. The principal one, alkaline phosphatase, is inactivated at the temperatures used in pasteurizing milk; as a consequence, the presence of alkaline phosphatase in heated milk is useful as a test for proper pasteurization (or of contamination with raw milk).

Other Enzymes

Other enzymes found in milk include lactoperoxidase, catalase, amylase, protease, and xanthine oxidase. These enzymes are of limited importance because their substrates are either not present or are present in relatively small amounts in milk.

☐ GRADES OF MILK

Milk generally is designated by grades. These refer to the care with which the milk is produced and handled, as well as to the quality of the milk from the standpoint of bacterial count, purity, cleanliness, etc.

In a general way, there are two division of market milk—raw and pasteurized. Raw milk may be defined as milk that is delivered in its natural state to the creamery for processing and without any treatment other than cooling. Pasteurized milk is milk that has been heated to temperatures adequate to destroy all disease-producing bacteria which might occur in the raw milk. The grades of milk are grade A raw milk for pasteurization, grade A pasteurized milk, and certified milk.

Grade A Raw Milk

Raw milk produced by dairy farms that meet the required chemical, bacteriological, and temperature standards is designated grade A milk. The milk must be cooled to 10°C, or less, and maintained there until processed. The bacterial content of grade A raw milk from an individual producer must not exceed 100,000/ml before being combined with milk from other producers. Prior to pasteurization, the bacterial content of commingled milk must not exceed 300,000/ml. The antibiotic content must be less than 0.05 units/ml as determined by the *Bacillus subtilis* method or its equivalent.

Grade A Pasteurized Milk

Grade A raw milk that has been pasteurized, cooled to 10°C or less and placed in a container in a dairy plant meeting state and local sanitary regulations is designated grade A pasteurized milk. Such milk must pass a phosphatase test (less than 1 μg/ml by Scharer Rapid Test or equivalent); the bacterial limit must not exceed 20,000/ml or a coliform count must not exceed 10/ml.

Certified Milk

Certified milk is a copyrighted name given to raw and pasteurized milk produced under very rigid sanitary conditions and containing a minimum bacterial count. This milk must conform to the requirements established for the product. The standards for certified raw milk require that it have a bacterial limit of 10,000/ml or a coliform limit of 10/ml. Certified pasteurized milk has a bacterial limit of 500/ml or a coliform limit of 5/ml.

☐ CLASSES OF MILK

An economic classification also may be applied to milk. It is a classification in which milk production and sales competition is considered. The enabling legislation for this classification, which is intended to stabilize milk markets and prices, is the Federal Milk Orders Law. (Federal Milk Orders apply only to specified areas). This law requires that the minimum price paid to the milk producer be based on the "classified use" of the milk. Milk used for fluid purposes is classified as class 1; it is the highest priced milk because of the cost differential in meeting the standards for grade A milk. Milk and milk products included in class 1 are whole milk, skim milk, low-fat milk, chocolate milk, and fortified skim milk. Milk that is used for any other purposes is placed in class 2. This group includes manufactured products such as butter, cheese, dry milk, evaporated milk, and ice cream.

☐ SOME COMMERCIAL MILK PRODUCTS

As noted earlier, the legal definitions and standards of identity for milk and milk products are not strictly uniform throughout the United States because, in general, these standards are promulgated by the individual state governments rather than the federal government. However, despite some variations in the details, general descriptions of the various types of milk and milk products can be stated. These are summarized in this section.

Raw milk is defined as the lacteal secretion, practically free from colostrum, obtained by the complete milking of one or more healthy cows, that contains not less than $8\frac{1}{4}$% milk solids-not-fat and not less than $3\frac{1}{4}$% milk fat.

Pasteurized milk is milk that has been subjected to a temperature not lower than 63°C for not less than 30 min, or to a temperature not lower than 72°C for not less than 15 sec, and then promptly cooled to 10°C or lower. The process must be accomplished in equipment that is properly operated and approved by the health authorities.

Homogenized milk is milk in which the fat globules have been subdivided into smaller globules by a mechanical process. According to the USPHS, homogenized milk must not show visible evidence of cream separation after 48 hr of quiescent storage at 7°C; and the fat percentage of 100 ml of the milk in a quart bottle must not differ by more than 10% from the fat percentage of the remaining milk after mixing.

Homogenization is accomplished either by pumping milk under high pressure through a small opening of definite size between a valve and its seat or by passing milk through closely approximated vibrating plates. The pressure generally is 2000–4000 lb/in.^{2}. With either method, there is a reduction in size of the fat globules and the ability of newly formed flobules to coalesce is practically destroyed owing to the presence of an adsorbed film of protein, which forms around them. Homogenization prevents the formation of larger individual fat globules, but it does not prevent the smaller globules formed during the process of homogenization from clumping or adhering together. This phenomenon is reflected in an increase in the viscosity of the product following homogenization.

Homogenization is not limited to milk itself but it is also used in the manufacture of ice cream, evaporated milk, cream cheese, cultured buttermilk, cultured sour cream, and fortified milk.

Vitamin D milk is milk in which the vitamin D potency has been increased by an approved method to at least 400 USP units/qt.

Soft-curd milk is milk with a curd tension of 30 g or less. The curd tension of milk normally ranges from 50 to 90 g. Human milk gives a fine flocculent curd when precipitated in the stomach, whereas cow's milk gives a curd of greater toughness (higher curd tension). A number of authorities consider a soft-curd milk to be more digestible and preferable for infant feeding.

Soft-curd milk may be produced in a number of ways. One of the simplest procedures is to heat or boil milk beyond pasteurization temperatures. Other methods for lowering curd tension include the removal of calcium from milk by base-exchange; partial digestion of the milk proteins by the action of proteolytic enzymes; addition of salts of hexametaphosphate, pyrophosphate, and citrate; subjecting milk to high-frequency vibration; and dilution with water.

Reconstituted or recombined milk is milk that has been made by recombining milk constituents with potable water so that it conforms to the standards for milk fat and solids-not-fat of normal milk.

Skim milk is milk from which sufficient milk fat has been removed to reduce its milk fat content to less than 0.50%. The removal of milk fat is usually accomplished by a centrifugal separator.

Low-fat milk is milk from which sufficient milk fat has been removed to reduce its milk fat content to not less than 0.50% and not more than 2.0%.

Cream (sweet cream) is the sweet, fatty liquid separated from milk, with or without the addition of skim milk, that contains not less than 18% milk fat. Cream is separated from the remainder of milk either by rising to the surface on standing or by use of a centrifugal separator.

Light cream, coffee cream, or table cream is cream that contains not less than 18%, but less than 30% milk fat.

Whipping cream is cream that contains not less than 30% milk fat.

Light whipping cream is cream that contains not less than 30% but less than 36% milk fat.

Heavy cream or heavy whipping cream is cream that contains not less than 36% milk fat.

Sour cream or cultured sour cream is a fluid or semifluid cream resulting from the souring, by lactic acid-producing bacteria or a similar culture, of pasteurized cream; it contains not less than 18% milk fat and not less than 0.50% acidity expressed as lactic acid.

Butter is that product obtained by churning cream (30–33% milk fat) until granules of butter are formed, and the granules are then forced into a compact mass.

Buttermilk originally referred to the fluid product resulting from the manufacture of butter from milk or cream. However, most of the buttermilk now used for food is cultured buttermilk. This is made by inoculating pasteurized skim milk or pasteurized low-fat skim milk with a lactic acid-producing culture and incubating at a temperature around 21°C until the desired acidity is reached (12–15 hr). The curdled milk is cooled to 4°C, with agitation, to produce a smooth product. It is most stable at acidities between 0.7 and 0.9% with a storage temperature of 0.6°C.

Evaporated milk is whole milk from which water has been removed by evaporation. The milk fat and total milk solids content of evaporated milk is not less than 7.5 and 25.5%, respectively. Evaporated milk contains added Vitamin D in such quantity that each fluid ounce contains 25 IU Vitamin A may be added such that each fluid ounce contains not less than 125 IU. Safe and suitable emulsifiers and stabilizers, with or without dioctyl sodium sulfosuccinate, may be used as a solubilizing agent. The product is sealed in a container and so processed by heat, either before or after sealing, so as to prevent spoilage.

Sweetened condensed milk is made by evaporating a mixture of milk and suitable nutritive sweetener (sucrose or a combination of sucrose and dextrose) so that the finished food contains not less than 8.5% milk fat and not less than 28% total milk solids. The amount of added sweetener must be sufficient to prevent spoilage. The food is pasteurized and may be homogenized.

Dried milk products are made from milk, skim milk, cream, whey, buttermilk, and other liquid milk products. These products may be dried by either spray drying or drum drying. In drum drying, a thin film of the liquid adheres to the revolving drums of the dryer and the film is dried before a complete revolution of the drum is made. The time for one complete revolution is 6–30 sec, depending upon the degree to which the product was concentrated by evaporation. The dried product is removed from the drum by means of a knife and is then ground to a powder.

Ice cream is a frozen dairy product made from milk products in combination with a stabilizer, sweetening agent, and flavoring; it sometimes includes coloring matter, fruit, and nuts. The definitions of ice cream frequently specify that freezing must be accomplished with agitation of the ingredients. They also usually specify the percentage of milk fat that must be present in the product; the percentage of milk solids-not-fat, fruits, nuts, cocoa or chocolate, or confections to be included; and the percentage of solids that must be contained in a given volume of product.

The milk products used in making ice cream are cream, milk, butter, butter oil, concentrated milk, evaporated milk, sweetened condensed milk, dried milk, skim milk, condensed skim milk, sweet cream buttermilk, condensed sweet cream buttermilk, dried sweet cream buttermilk, concentrated cheese whey, and others. Stabilizers include gelatin, eggs, sodium alginate, gum acacia, and others. Sweetening agents include sugar, dextrose, corn syrup, maple syrup, maple sugar, honey, brown sugar, molasses, and others.

Cheese may be defined as the food product made by modifying or ripening the separated curd obtained by coagulating the casein of milk with rennet, lactic acid, or other suitable enzymes or acids. The FDA has issued definitions and standards of identity for about 60 kinds of cheeses. Since there are so many different types of cheese, it would be impracticable to attempt to discuss them in this book.

☐ ANALYSIS OF FLUID MILK PRODUCTS

Sampling and Preparation of Sample

The required sample size depends on the type of analysis to be performed. For most analyses, a sample of 200–250 ml is sufficient; a sample of 50–60 ml is sufficient for fat determinations.

In order to secure a representative sample for analysis, it is necessary to collect milk from one or more containers as prepared for sale. Thoroughly mix the milk by pouring it from its container into another vessel and back again several times. If this is not feasible, the milk should be stirred thoroughly. Shaking the milk is to be avoided, as this tends to cause separation of the milk fat. If cream has formed, detach all of it from the sides of the container and stir until the liquid is evenly emulsified.

Place samples in nonabsorbent, airtight containers, and keep them cold, but above freezing temperature, until analyzed. Tablets containing a $HgCl_2$, $K_2Cr_2O_7$ or another suitable preservative at a concentration of at least 0.5 g active ingredient/250 ml milk, but having a total weight not exceeding 1 g, may be used provided the preservative does not interfere in the tests to be made; 0.1 ml of 36% formaldehyde solution per 30 ml of milk may also be added as a preservative. If a phosphatase test is to be made, chloroform is the only permissible preservative.

Bring the sample to about 20°C, then thoroughly mix the sample by pouring the milk from its container into a clean receptacle and back again several times. Promptly weigh or measure the test portion. If lumps of cream do not disperse, warm the sample in a water bath to about 38°C and continue mixing until the sample is homogeneous, using a rubber policeman if necessary to remove any cream adhering to the walls of the container or stopper. Sampling and the preparation of samples are described in AOAC methods of 16.019 and 16.020.

Determination of Total Solids

Instead of determining the percentage of moisture, it is customary in the case of fluid whole milk to determine the percentage of dry matter (total solids). By subtracting the percentage of total solids from 100, the percentage of water present in the sample can be obtained.

Weighing Method

A weighed quantity of milk is transferred into a tared flat dish (porcelain or aluminum) containing a layer of sand. The dish and contents are dried to constant weight, then cooled in a dessicator. The increase in weight corresponds to the total solids in the sample. The purpose of the sand is to increase the drying surface of the container. The dried solids should be practically white. Any appreciable browning indicates that some of the lactose has been caramelized and that the results will be somewhat in error.

Procedure

Add 10–15 g of white sand to a flat aluminum or porcelain dish not over 5 cm in diameter. Heat the dish and sand at 100°C in a drying oven until a constant weight is obtained. Weigh 5 g of prepared milk sample into the tared dish. Dry on a steam bath for 15 min and then in an air oven at 98°–100°C for 3 hr. Cool in a desiccator and weigh as rapidly as possible. Report the percentage of residue a total solids.

Lactometer Method

The constituents of normal milk generally are present in a fairly constant ratio. Because of this, the total solids can be calculated with a fair degree of accuracy if the specific gravity and fat percentage of a milk sample are known (AOAC method 16.033).

Procedure

Bring a suitable milk sample to 16°C, then determine the specific gravity of the milk with a Quévenne lactometer. Determine the percentage of fat in the sample by one of the methods described later. Calculate the total solids from the following formula:

$$\text{Total solids (TS):} = L + 1.2f + 0.14$$

where L = lactometer reading and f = percentage of fat in the milk. For example, with a lactometer reading of 32.0 (sp gr 1.032) and a fat content of 4.0%, the percentage of total solids would be 12.94 [32/4 + 1.2(4) + 0.14 = 12.94%].

Determination of Ash

The organic matter in a weighed sample of milk is oxidized by combustion in a muffle furnace. The remaining residue, the ash, is weighed and the percentage of ash calculated.

Milk contains appreciable quantities of sodium and potassium chlorides. As a consequence, ashing should be done at a temperature of 500°C in order to prevent volatilization of the chlorides. A muffle furnace permits a much more accurate control of the ashing procedure and is preferred to a gas flame.

Procedure

Heat a crucible of suitable size at a temperature of 500°C for 1 hr, cool in a dessicator, and weigh. Pipette 10 ml of a well-mixed milk sample into the crucible and weigh. Evaporate to dryness on a steam bath. Transfer the crucible and dried sample to a muffle furnace held at 500°C and heat until the ash is carbon free (white ash). Cool in a dessicator, weigh, and calculate the percentage of ash.

Remarks

The milk is evaporated on the steam bath to eliminate the bulk of the water present.

Determination of Crude Fat

The fat content of milk is determined either volumetrically as in the Babcock method, or gravimetrically as in the Mojonnier method. The Babcock method is a rapid method for general dairy control testing whereas the Mojonnier method is used for determining the fat content of dairy products including whole milk, skim milk, buttermilk, cream, powdered milks, and ice cream. The latter method is the more accurate of the two. Most workers have observed that the Babcock method ranges from 0.05–0.10% higher than the Mojonnier method.

Babcock Method

The Babcock method is based on the principle that when milk is treated with concentrated sulfuric acid, the proteins of the milk are first denatured and then hydrolyzed, permitting the liberated fat to coalesce. After centrifuging to effect a more complete separation of the fat from the aqueous phase, water is added, forcing the fat column up into the calibrated neck of the special flask so that the volume of the fat may be read, and from this the percentage of fat is determined. The details of the Babcock method were presented in Chapter 2.

Mojonnier Method

The Mojonnier method originated from the Roese-Gottlieb method (AOAC 16.064) and is now widely used for the estimation of fat in dairy products. In this method the fat is released from other milk constituents by treatment with ammonium hydroxide and ethanol, then dissolved in a mixture of ether and petroleum ether. The ether layer, containing the fat, is decanted into a tared aluminum dish, the ethers evaporated, and the dish again weighed. The difference in weight is the amount of fat extracted.

Procedure

Weigh approximately 10 g of well-mixed milk (at 20°C) into a Mojonnier extraction flask. Add 1.25 ml of concentrated ammonium hydroxide (2 ml if sample is sour) and mix thoroughly in the small bulb of the flask. Add 10 ml of 95% ethanol, stopper the flask and, holding it horizontally with a finger on the stopper, shake thoroughly. Add 25 ml of ethyl ether and shake vigorously for 1 min; add 25 ml of petroleum ether and again shake for 1 min. Centrifuge flask or let stand until upper layer is clear.

Remove the stopper and carefully decant off as much as possible of the ether layer into a tared aluminum dish. By holding the flask in a horizontal position with the small bulb down, an excellent separation of the layers may be made.

Repeat extraction of the liquid remaining by adding successively 5 ml ethanol, 15 ml ethyl ether, and 15 ml of petroleum ether, shaking thoroughly after each addition as before. Let stand until clear and, if the dividing line is not at the upper level of the constriction between the small and large bulb, add water to bring it to that point. Decant off the ether layer into the tared aluminum dish containing the first ether extraction. Evaporate off the ether layers on a steam bath and dry in an oven at 135°C for 5 min. Cool in a desiccator and weigh. Calculate the percentage of fat in the sample from the weight of fat obtained.

Determination of Total Acidity

The acidity of freshly drawn milk, called *apparent acidity*, is due to the presence of casein, phosphates, albumin, carbon dioxide, and citrates. Any increase in acidity over the apparent acidity is due to the fermentation of lactose to lactic acid by the bacteria present in the milk. This acidity is called *real acidity*. When sufficient lactic acid is produced, the proteins coagulate and the milk is termed sour.

Procedure

Weigh 20 g (about 20 ml) of a well-mixed milk sample into a porcelain casserole (or beaker placed on a white background). Dilute with an equal volume of carbon-dioxide free water (recently boiled and cooled). Add a few drops of phenolphthalein indicator and titrate the milk solution with standard 0.1 *N* NaOH until the first pink color appears and persists for 1 min. Report the acidity as percentage lactic acid by weight (1 ml of 0.1 *N* NaOH is equivalent to 0.090 g lactic acid).

Determination of Total Protein

The Kjeldahl method is normally used for determining the total protein content of milk. Casein and lactalbumin, the major proteins of milk, contain approximately 15.7% nitrogen. Consequently, the factor for converting nitrogen to protein is 6.38 (100/15.7) rather than the customary 6.25.

Procedure

Transfer an accurately weighed sample of approximately 5 g of well-mixed milk into an 800-ml Kjeldahl flask. Add the various Kjeldahl reagents, digest, distill, and titrate the distillate as described in Chapter 2. Conduct a blank determination along with the samples. Calculate the percentage of protein present in the milk as follows:

$$\%\text{Protein} = \text{Sample titration (ml)} - \text{Blank titration (ml)} \times \text{Normality of solution} \times 0.14 \text{ (meq of N)} \times 6.38$$

Determination of Lactose (Milk Sugar)

The optical activity of lactose solutions is the basis of the polarimetric determination of lactose (see Polarimetric Method in Chapter 4).

Polarimetric Method

A portion of milk equal in weight to twice the normal weight of lactose (normal weight, 32.9 g/100 ml solution) is treated with a clarifying agent, diluted to definite volume, and the percentage lactose determined on the filtrate by means of a polarimeter.

Reagents

1. Mercuric nitrate solution—Dissolve a quantity of metallic mercury in twice its weight of nitric acid and dilute this volume with five volumes of water.
2. Mercuric iodide solution—Dissolve 33.2 g potassium iodide and 13.5 mercuric chloride in 200 ml acetic acid and 640 ml of water.
3. Phosphotungstic acid solution—Dissolve 5 g phosphotungstic acid in 95 ml water.

Procedure

Transfer 65.8-g samples of well-mixed milk into 100-ml and 200-ml volumetric flasks. Add to each flask 20 ml of the mercuric nitrate solution *or* 30 ml of the mercuric iodide solution. To the 200-ml flask, add 15 ml of the phosphotungstic acid solution and dilute to the mark with water. Dilute the contents of the 100-ml volumetric flask to the mark with the phosphotungstic acid solution. Shake both flasks frequently for a period of 15 min; then filter the contents of each flask through a dry filter, discarding the first 10 ml of filtrate. Collect the filtrates, then place each in a 400-mm polarimeter tube and determine their readings in a polarimeter.

Calculations

In theory, the polarimeter readings for the 100-ml solution should be twice as great as those for the 200-ml solution because it is twice as concentrated. However, the readings obtained for each solution are in error due to the volume of the precipitate formed by the clarifying agent. Since both solutions contain the same quantity of milk, they both should give the same volume of precipitate. However, the effect of precipitate error will be twice as great in the 100-ml solution as in the 200-ml solution. Because of this relationship, the precipitate error may be calculated from the polarimeter readings for the two solutions, and then applied as a correction to the readings.

For example, assume the polarimeter readings obtained for the milk sample were 20.20 for the 100-ml solution and 9.85 for the 200-ml solution. If no error occurred due to the formation of a precipitate, the reading for the 200-ml solution should be exactly one-half that of the 100-ml solution. However, when one multiplies the reading for the 200-ml solution, a value of 19.70 is obtained, which is less than the reading of 20.20 for the 100-ml solution. This difference ($20.20 - 19.70 = 0.5$) represents one-half the error of the precipitate.

Thus, the precipitate error and corrected polarimeter reading can be obtained from the following expressions:

$$2S_2 = S_1 - \tfrac{1}{2}E$$

$$E = 2S_1 - 4S_2$$

$$S_c = S_1 - E = S_1 - (2S_1 - 4S_2) = 4S_2 - S_1$$

where S_2 = reading of 200-ml solutions; S_1 = reading of 100-ml solution; E = precipitate error; and S_c = corrected reading for 100-ml solution.

Since the sample weight is twice the normal weight and is viewed in a 400-mm polarimeter tube, rather than the normal 200-mm polarimeter tube, the corrected reading (S_c) must be divided by four to obtain the percentage of lactose in the sample.

Munson–Walker Method

The reducing property of lactose is the basis for several methods of determining lactose in milk; these methods basically differ only in the procedure by which the cuprous oxide (Cu_2O) formed in the reduction step is measured. In all of these reduction methods, a protein-precipitating agent is added to a weighed portion of milk. The precipitated protein and absorbed fat is filtered off, leaving a filtrate containing lactose and salts in solution. An aliquot of the filtrate is heated with an alkaline copper sulfate solution (Fehling's solution) for a definite time period to bring about the reduction of Cu^{2+} to Cu^{+} ions. The amount of Cu_2O precipitate formed is proportional to the amount of lactose present under the conditions of the experiment. In the Munson–Walker method, the precipitated Cu_2O is determined gravimetrically.

Reagents

1. Fehling's A solution—Dissolve 34.64 g of copper sulfate crystals ($CuSO_4 \cdot 5H_2O$) in water and dilute to 500 ml.
2. Fehling's B solution—Dissolve 173 g of Rochelle salts (sodium potassium tartrate) and 50 g of sodium hydroxide in water and dilute to 500 ml. Let stand for a few days, filter, and store in a clean dry bottle with a rubber stopper.

Precipitation of Proteins

Transfer 25 g of a well-mixed milk sample, or an accurately measured volume corresponding to this weight, into a 500-ml volumetric flask. Add 400 ml of water, 10 ml of Fehling's A solution, and 44 ml of 0.1 N NaOH. Make up to volume, shake thoroughly, allow the precipitate to settle, and filter through a dry filter. The solution should have an acid reaction and contain copper in solution. Exact measurements of the copper sulfate and NaOH solution are necessary to obtain a solution that will filter rapidly and give a clear filtrate. Lactose is determined in an aliquot of this filtrate from which proteins and fat have been removed.

Copper Reduction

Transfer 25 ml of each of the Fehling's solutions to a 400-ml beaker. Add to this 50 ml of the sugar solution to be tested; if a smaller volume of sample is used, add sufficient water to make a final volume of 100 ml in the beaker. Cover the beaker with a watch glass and heat on an asbestos gauze over a Bunsen burner that has been previously regulated so that the solution will come to a boil in exactly 4 min. Boil for 2 min and filter immediately through a tared porcelain filtering crucible (e.g., Selas Porcelain, Porosity M). Thoroughly wash the precipitated Cu_2O with 300 ml hot water using a suction flask. Empty the hot water out of the suction flask and continue washing the crucible, first with 10 ml 95% ethanol and then with 10 ml ethyl ether, using suction.

Determination of Reduced Copper

Dry the crucible and precipitate for 45 min at 100°C, cool for 20 min, and weigh. Reheat for a 30-min period, cool, and reweigh. Determine the weight of lactose corresponding to the weight of Cu_2O obtained from the Munson and Walker table (Table 12.6; AOAC 43.012).

Iodide–Thiosulfate Method

The Cu_2O formed in the reduction step of the Munson–Walker method just described may be determined by titration with standard thiosulfate. This method depends on the oxidation of Cu_2O with nitric acid to give a soluble cupric salt. When treated with an excess of potassium iodide in an acid medium, the cupric ions are reduced and a corresponding amount of iodide ion is oxidized to molecular iodine. The iodine liberated is then titrated with standardized 0.1 *N* sodium thiosulfate; from this titration value, the corresponding amount of cuprous oxide is calculated. The following equations represent the reactions involved:

$$3Cu_2O + 2HNO_3 \rightarrow 6CuO + 2NO + H_2O$$

$$CuO + 2HNO_3 \rightarrow Cu(NO_3)_2 + H_2O$$

$$2Cu(NO_3)_2 + 4KI \rightarrow Cu_2I_2 + 4KNO_3$$

$$I_2 + 2Na_2S_2O_3 \rightarrow 2NaI + Na_2S_4O_6$$

Table 12.6. Munson and Walker Table for Calculating Glucose, Invert Sugar, Lactose, and Maltose (in milligrams)[a]

Cuprous oxide (Cu_2O)	Glucose	Invert sugar	Lactose + H_2O	Maltose + H_2O
10	4.0	4.5	6.3	6.2
12	4.9	5.4	7.5	7.9
14	5.7	6.3	8.8	9.5
16	6.6	7.2	10.0	11.2
18	7.5	8.1	11.3	12.9
20	8.3	8.9	12.5	14.6
22	9.2	9.8	13.8	16.2
24	10.0	10.7	15.0	17.9
26	10.9	11.6	16.3	19.6
28	11.8	12.5	17.6	21.2
30	12.6	13.4	18.8	22.9
32	13.5	14.3	20.1	24.6
34	14.3	15.2	21.4	26.2
36	15.2	16.1	22.8	27.9
38	16.1	16.9	24.2	29.6
40	16.9	17.8	25.5	31.3
42	17.8	18.7	26.9	32.9
44	18.7	19.6	28.3	34.6
46	19.6	20.5	29.6	36.3
48	20.4	21.4	31.0	37.9
50	21.3	22.3	32.3	39.6
52	22.2	23.2	33.7	41.3
54	23.0	24.1	35.1	42.9
56	23.9	25.0	36.4	44.6
58	24.8	25.9	37.8	46.3
60	25.6	26.8	39.2	48.0
62	26.5	27.7	40.5	29.6
64	27.4	28.6	41.9	51.3
66	28.3	29.5	43.8	53.0
68	29.2	30.4	44.7	54.6
70	30.0	31.3	46.0	56.3
72	30.9	32.3	47.4	58.0
74	31.8	33.2	48.8	59.6
76	32.7	34.1	50.1	61.3
78	33.6	35.0	51.5	63.0
80	34.4	35.9	52.9	64.6
82	35.3	36.8	54.2	66.3
84	36.2	37.7	55.6	68.0
86	37.1	38.6	57.0	69.7
88	38.0	39.5	58.4	71.3
90	38.9	40.4	59.7	73.0
92	39.8	41.4	61.1	74.7
94	40.6	42.3	62.5	76.3
96	41.5	43.2	63.8	78.0
98	42.4	44.1	65.2	79.7

Table 12.6. (*continued*)

Cuprous oxide (Cu_2O)	Glucose	Invert sugar	Lactose + H_2O	Maltose + H_2O
100	43.3	45.0	66.6	81.3
102	44.2	46.0	68.0	83.0
104	45.1	46.9	69.3	84.7
106	46.0	47.8	70.7	86.3
108	46.9	48.7	72.1	88.0
110	47.8	49.6	73.5	89.7
112	48.7	50.6	74.8	91.3
114	49.6	51.5	76.2	93.0
116	50.5	52.4	77.6	94.7
118	51.4	53.3	79.0	96.4
120	52.3	54.3	80.3	98.0
122	53.2	55.2	81.7	99.7
124	54.1	56.1	83.1	101.4
126	55.0	57.0	84.5	103.0
128	55.9	58.0	85.8	104.7
130	56.8	58.9	87.2	106.4
132	57.7	59.8	88.6	108.0
134	58.6	60.8	90.0	109.7
136	59.5	61.7	91.3	111.4
138	60.4	62.6	92.7	113.0
140	61.3	63.6	94.1	114.7
142	62.2	64.5	95.5	116.4
144	63.1	65.4	96.8	118.0
146	64.0	66.4	98.2	119.7
148	65.0	67.3	99.6	121.4
150	65.9	68.3	101.0	123.0
152	66.8	69.2	102.3	124.7
154	67.7	70.1	103.7	126.4
156	68.6	71.1	105.1	128.0
158	69.5	72.0	106.5	129.7
160	70.4	73.0	107.9	131.4
162	71.4	73.9	109.2	133.0
164	72.3	74.9	110.6	134.7
166	73.2	75.8	112.0	136.4
168	74.1	76.8	113.4	138.0
170	75.1	77.7	114.8	139.7
172	76.0	78.7	116.1	141.4
174	76.9	79.6	117.5	143.0
176	77.8	80.6	118.9	144.7
178	78.8	81.5	120.3	146.4
180	79.7	82.5	121.6	148.0
182	80.6	83.4	123.1	149.7
184	81.5	84.4	124.3	151.4
186	82.5	85.3	125.8	153.0
188	83.4	86.3	127.2	154.7

Table 12.6. (*continued*)

Cuprous oxide (Cu_2O)	Glucose	Invert sugar	Lactose + H_2O	Maltose + H_2O
190	84.3	87.2	128.5	156.4
192	85.3	88.2	129.9	158.0
194	86.2	89.2	131.3	159.7
196	87.1	90.1	132.7	161.4
198	88.1	91.1	134.1	163.0
200	89.0	92.0	135.4	164.7
202	89.9	93.0	136.8	166.4
204	90.9	94.0	138.2	168.0
206	91.8	94.9	139.6	169.7
208	92.8	95.9	141.0	171.4
210	93.7	96.9	142.3	173.0
212	94.6	97.8	143.7	174.7
214	95.6	98.8	145.1	176.4
216	96.5	99.8	146.5	178.0
218	97.5	100.8	147.9	179.7
220	98.4	101.7	149.3	181.4
222	99.4	102.7	150.7	183.0
224	100.3	103.7	152.0	184.7
226	101.3	104.6	153.4	186.4
228	102.2	105.6	154.8	188.0
230	103.2	106.6	156.2	189.7
232	104.1	107.6	157.6	191.3
234	105.1	108.6	159.0	193.0
236	106.0	109.5	160.3	194.7
238	107.0	110.5	161.7	196.3
240	108.0	111.5	163.1	198.0
242	108.9	112.5	164.5	199.7
244	109.9	113.5	165.9	201.3
246	110.8	114.5	167.3	203.0
248	111.8	115.4	168.7	204.7
250	112.8	116.4	170.1	206.3
252	113.7	117.4	171.5	208.0
254	114.7	118.4	172.8	209.7
256	115.7	119.4	174.2	211.3
258	116.6	120.4	175.6	213.0
260	117.6	121.4	177.0	214.7
262	118.6	122.4	178.4	216.3
264	119.5	123.4	179.8	218.0
266	120.5	124.4	181.2	219.7
268	121.5	125.4	182.6	221.3
270	122.5	126.4	184.0	223.0
272	123.4	127.4	185.3	224.6
274	124.4	128.4	186.7	226.3
276	125.4	129.4	188.1	228.0
278	126.4	130.4	189.5	229.6

Table 12.6. (*continued*)

Cuprous oxide (Cu_2O)	Glucose	Invert sugar	Lactose + H_2O	Maltose + H_2O
280	127.3	131.4	190.9	231.3
282	128.3	132.4	192.3	233.0
284	129.3	133.4	193.7	234.6
286	130.3	134.4	195.1	236.3
288	131.3	135.4	196.5	238.0
290	132.3	136.4	197.8	239.6
292	133.2	137.4	199.2	241.3
294	134.2	138.4	200.6	242.9
296	135.2	139.4	202.0	244.6
298	136.2	140.5	203.4	246.3
300	137.2	141.5	204.8	247.9
302	138.2	142.5	206.2	249.6
304	139.2	143.5	207.6	251.3
306	140.2	144.5	209.0	252.9
308	141.2	145.5	210.4	254.6
310	142.2	146.6	211.8	256.3
312	143.2	147.6	213.2	257.9
314	144.2	148.6	214.6	259.6
316	145.2	149.6	216.0	261.2
318	146.2	150.7	217.3	262.9
320	147.2	151.7	218.7	264.6
322	148.2	152.7	220.1	266.2
324	149.2	153.7	221.5	267.9
326	150.2	154.8	222.9	269.6
328	151.2	155.8	224.3	271.2
330	152.2	156.8	225.7	272.9
332	153.2	157.9	227.1	274.6
334	154.2	158.9	228.5	276.2
336	155.2	159.9	229.9	277.9
338	156.3	161.0	231.3	279.5
340	157.3	162.0	232.7	281.2
342	158.3	163.1	234.1	282.9
344	159.3	164.1	235.5	284.5
346	160.3	165.1	236.9	286.2
348	161.4	166.2	238.3	287.9
350	162.4	167.2	239.7	289.5
352	163.4	168.3	241.1	291.2
354	164.4	169.3	242.5	292.8
356	165.4	170.4	243.9	294.5
358	166.5	171.4	245.3	296.2
360	167.5	172.5	246.7	297.8
362	168.5	173.5	248.1	299.5
364	169.6	174.6	249.5	301.2
366	170.6	175.6	250.9	302.8
368	171.6	176.7	252.3	304.5

Table 12.6. *(continued)*

Cuprous oxide (Cu_2O)	Glucose	Invert sugar	Lactose + H_2O	Maltose + H_2O
370	172.7	177.7	253.7	306.1
372	173.7	178.8	255.1	307.8
374	174.7	179.8	256.5	309.5
376	175.8	180.9	257.9	311.1
378	176.8	182.0	259.3	312.8
380	177.9	183.0	260.7	314.5
382	178.9	184.1	262.1	316.1
384	180.0	185.2	263.5	317.8
386	181.0	186.2	264.9	329.4
388	182.0	187.3	266.5	321.1
390	183.1	188.4	267.7	322.8
392	184.1	189.4	269.1	324.4
394	185.2	190.5	270.5	326.1
396	186.2	191.6	271.9	327.7
398	187.3	192.7	273.3	329.4
400	188.4	193.7	274.7	331.1
402	189.4	194.8	276.1	332.7
404	190.5	195.9	277.5	334.4
406	191.5	197.0	278.9	336.0
408	192.6	198.1	280.3	337.7
410	193.7	199.1	281.7	339.4
412	194.7	200.2	283.2	341.0
414	195.8	201.3	284.6	342.7
416	196.8	202.4	286.0	344.4
418	197.9	203.5	287.4	346.0
420	199.0	204.6	288.8	347.7
422	200.1	205.7	290.2	349.3
424	201.1	206.7	291.6	351.0
426	202.2	207.8	293.0	352.7
428	203.3	208.9	294.4	354.3
430	204.4	210.0	295.8	356.0
432	205.5	211.1	297.2	357.6
434	206.5	212.2	298.6	359.3
436	207.6	213.3	300.0	361.0
438	208.7	214.4	301.4	362.6
440	209.8	215.5	302.8	364.3
442	210.9	216.6	304.2	365.9
444	212.0	217.8	305.6	367.6
446	213.1	218.9	307.0	369.3
448	214.1	220.0	308.4	370.9
450	215.2	221.1	309.9	372.6
452	216.3	222.2	311.3	374.2
454	217.4	223.3	312.7	375.9
456	218.5	224.4	314.1	377.6
458	219.6	225.5	315.5	379.2

Table 12.6. (continued)

Cuprous oxide (Cu_2O)	Glucose	Invert sugar	Lactose + H_2O	Maltose + H_2O
460	220.7	226.7	316.9	380.9
462	221.8	227.8	318.3	382.5
464	222.9	228.9	319.7	384.2
466	224.0	230.0	321.1	385.9
468	225.1	231.2	322.5	387.5
470	226.2	232.3	323.9	389.2
472	227.4	233.4	325.3	390.8
474	228.3	234.5	326.8	392.5
476	229.6	235.7	328.2	394.2
478	230.7	236.8	329.6	395.8
480	231.8	237.9	331.0	397.5
482	232.9	239.1	332.4	399.1
484	234.1	240.2	333.8	400.8
486	235.2	241.4	335.2	402.4
488	236.3	242.5	336.6	404.1
490	237.4	243.6	338.0	405.8

Source: Bureau of Standards, 1918.
[a] Applicable when Cu_2O is weighed directly.

Reagents

1. Nitric acid solution (1:1 by volume)—To a measured volume of water add an equal volume of concentrated nitric acid.
2. Sodium acetate solution—Dissolve 575 g sodium acetate trihydrate in water and dilute to 1 liter.
3. KI solution—Dissolve 42 g KI in water and dilute to 100 ml.

Standardization of Sodium Thiosulfate Solution

Weigh 37–40 g sodium thiosulfate ($Na_2S_2O_3 \cdot 5H_2O$) into a 1-liter volumetric flask and make up to the mark with water. (This should give a 0.15–0.16 *N* solution, which is recommended for determination of reduced copper.) Accurately weigh 0.2–0.4 g pure electrolytic copper and transfer to a 300-ml Erlenmeyer flask graduated roughly at 20-ml intervals. Dissolve the copper in 5 ml of the nitric acid solution, dilute to 20 or 30 ml, boil to expel the red fumes, add a slight excess of saturated bromine water, and boil until the bromine is completely removed. Cool, and add 10 ml of the sodium acetate solution to buffer the nitric acid and liberate some acetic acid. Add 10 ml of the KI solution, then slowly and with constant agitation titrate with the sodium thiosulfate solution until the brown color becomes faint. Add starch indicator and continue titration until the blue starch–iodine color just disappears. Add approximately 2 g of ammonium thiocyanate and stir until completely dissolved. Continue titration to the disappearance of the blue color.

From the weight of copper and the volume of thiosulfate solution used, calculate the normality of the thiosulfate solution, and also its equivalence in milligrams of copper, as follows:

$$\text{ml} \times N\ Na_2S_2O_3 \times \text{meq of Cu} \times 1000 = \text{mg Cu}$$

$$\text{ml} \times N\ Na_2S_2O_3 \times 0.06357 \times 1000 = \text{mg Cu}$$

or

$$\text{Normality of } Na_2S_2O_3 = \frac{\text{g Cu} \times 1000}{\text{ml } Na_2S_2O_3 \times 63.57}$$

Titration of Reduced Copper

Treat a milk sample as described under the Precipitation of Proteins and Copper Reduction steps in the Munson–Walker method to obtain the Cu_2O corresponding to the lactose in the sample.

Set a short-stem funnel on the opening of a 300-ml Erlenmeyer flask that is roughly marked at 20-ml intervals. Set the filtering crucible containing the thoroughly washed Cu_2O in the funnel and cover the crucible with a watch glass. Pipette 5 ml of the nitric acid solution under the watch glass to dissolve the Cu_2O. Carefully wash the watch glass, crucible, and funnel with small quantities of water, directing the washings into the Erlenmeyer flask. Boil to expel the red fumes, add a slight excess of saturated bromine water, and boil until the bromine is completely removed. Proceed as in the previous Standardization step, starting with the addition of 10 ml sodium acetate solution. From the volume of sodium thiosulfate required to titrate the reduced copper, determine the weight of lactose in the sample from the Munson and Walker table (Table 12.6).

Oxidation and Titration with Potassium Permanganate

The Cu_2O formed by reduction with the lactose in milk can be determined by titration with a standard potassium permanganate solution. In this method, the Cu_2O is treated with a solution of ferric sulfate and acidified with 4 N sulfuric acid. The cuprous ion is oxidized in the reaction and a corresponding quantity of ferric ion is reduced to the ferrous state. The ferrous ion is then titrated with a standard permanganate solution. The reactions involved are

$$Cu_2O + Fe_2(SO_4)_3 + H_2SO_4 \rightarrow 2CuSO_4 + 2FeSO_4 + H_2O$$

$$10FeSO_4 + 2KMnO_4 + 8H_2SO_4 \rightarrow 5Fe_2(SO_4)_3 + K_2SO_4 + 2MnSO_4 + 8H_2O$$

Reagents

1. 0.15 to 0.16 *N* Potassium permanganate—Dissolve 4.98 g potassium permanganate in 1 liter of water and boil for several minutes. Allow the solution to stand for several days, then heat to boiling and filter through a prepared asbestos filter using suction. Store in a bottle with a glass stopper. Standardize as directed in the section that follows.
2. 4 *N* Sulfuric acid—To 500 ml of water, add 115 ml concentrated sulfuric acid and dilute to 1 liter with water.
3. Ferric sulfate solution—Dissolve either 135 g $FeNH_4(SO_4)_2 \cdot 12H_2O$ or 55 g anhydrous ferric sulfate in water and dilute to 1 liter. (The anhydrous ferric solution dissolves quite slowly.) To 50 ml of this solution add 20 ml of 4 *N* sulfuric acid and titrate with the permanganate solution to the first perceptible color change. Use this titration figure as a zero-point correction in subsequent analytical titrations for reduced copper.

Standardization of Permanganate Solution (Jackson and McDonald 1941)

The potassium permanganate solution is standardized against sodium oxalate ($Na_2C_2O_4$), which is a primary standard. Accurately weigh 0.35 g sodium oxalate into a 400-ml beaker, add 200 ml of water, and heat to 80°–90°C. Remove the beaker from the flame, add 10 ml dilute sulfuric acid (1:1 by volume), and titrate at once with the potassium permanganate, stirring vigorously throughout the titration. The end point of the titration is taken as the first pink coloration that persists for 30 sec. Calculate the normality of the potassium permanganate solution as follows:

$$N = \frac{\text{g } Na_2C_2O_4}{\text{ml} \times 0.0670}$$

Titration of Reduced Copper

Treat a milk sample as described under the Precipitation of Proteins and Copper Reduction steps in the Munson–Walker method to obtain the Cu_2O corresponding to the lactose in the sample.

Wash the precipitated cuprous oxide thoroughly, then transfer the crucible containing the precipitate to a beaker. Add 50 ml of the ferric sulfate solution and disintegrate the asbestos pad, stirring thoroughly until all the precipitate has dissolved. Examine for complete solution by holding the beaker above eye level. (If desired the crucible may now be left in the beaker until completion of the titration. If not, remove the crucible from the solution with the aid of a stirring rod and wash the crucible, collecting the washings in the beaker.) Add 20 ml of the 4 *N* sulfuric acid solution and titrate with the standard potassium permanganate solution to the same end point as in the titration for the zero-point correction on the ferric sulfate solution. Deduct the zero-point correction from the sample titration value. Calculate the milligrams of copper equivalent to the corrected titration volume; obtain weight of reducing sugar that is equivalent to the weight of copper from Hammond's table (Table 12.7).

Folin–Wu Colorimetric Method

In the Folin–Wu method for determining lactose, a milk sample is first freed from the suspended components, then heated with an alkaline copper solution in a Folin–Wu sugar tube to prevent reoxidation. The reduced Cu_2O is oxidized by phosphomolybdic acid and the blue color is measured in a colorimeter. The concentration of lactose is determined by comparison with a sugar solution of known concentration treated in the same manner as the sample.

Reagents

1. 10% Sodium tungstate solution—Dissolve 10 g sodium tungstate in 90 g water.
2. Alkaline copper solution—Dissolve 40 g anhydrous sodium carbonate in 400 ml water and transfer to a 1-liter volumetric flask. Dissolve in this solution, first 7.5 grams tartaric acid, and then 4.5 g crystalline copper sulfate. Dilute to volume and mix thoroughly. If a sediment forms on standing, decant and use the clear supernatant.
3. Phosphomolybdic acid solution—Place 35 g molybdic acid, 5 g sodium tungstate, 200 ml 10% NaOH, and 200 ml water in a beaker. Boil vigorously for 20–40 min to remove most of the ammonia present in the molybdic acid. Cool and transfer to a 500-ml volumetric flask, adding sufficient water to bring the volume to about 350 ml. Add 125 ml phosphoric acid (85%), dilute to volume with water, and mix.
4. 0.66 *N* Sulfuric acid solution.

Table 12.7. Hammond Table for Calculating Glucose, Fructose, Lactose, and Invert Sugar (in milligrams)[a]

Copper (Cu)	Glucose	Fructose	Lactose + H_2O	Invert sugar
10	4.6	5.1	7.7	5.2
12	5.6	6.1	9.3	6.2
14	6.5	7.2	10.8	7.2
16	7.5	8.3	12.3	8.2
18	8.5	9.3	13.8	9.2
20	9.4	10.4	15.4	10.2
22	10.4	11.5	16.9	11.2
24	11.4	12.5	18.4	12.3
26	12.3	13.6	19.9	13.3
28	13.3	14.7	21.5	14.3
30	14.3	15.8	23.0	15.3
32	15.3	16.8	24.5	16.3
34	16.2	17.9	26.1	17.3
36	17.2	19.0	27.6	18.3
38	18.2	20.1	29.1	19.4
40	19.2	21.1	30.6	20.4
42	20.1	22.2	32.2	21.4
44	21.1	23.3	33.7	22.4
46	22.1	24.4	35.2	23.5
48	23.1	25.4	36.8	24.5
50	24.1	26.5	38.3	25.5
52	25.1	27.6	39.8	26.5
54	26.1	28.7	41.4	27.6
56	27.0	29.8	42.9	28.6
58	28.0	30.9	44.4	29.6
60	29.0	31.9	46.0	30.6
62	30.0	33.0	47.5	31.7
64	31.0	34.1	49.0	32.7
66	32.0	35.2	50.6	33.7
68	33.0	36.3	52.1	34.8
70	34.0	37.4	53.6	35.8
72	35.0	38.5	55.2	36.8
74	36.0	39.6	56.7	37.9
76	37.0	40.7	58.2	38.9
78	38.0	41.7	59.8	40.0
80	39.0	42.8	61.3	41.0
82	40.0	43.9	62.8	42.0
84	41.0	45.0	64.4	43.1
86	42.0	46.1	65.9	44.1
88	43.0	47.2	67.4	45.2
90	44.0	48.3	69.0	46.2
92	45.0	49.4	70.5	47.3
94	46.0	50.5	72.1	48.3
96	47.0	51.6	73.6	49.4
98	48.0	52.7	75.1	50.4

Table 12.7. *(continued)*

Copper (Cu)	Glucose	Fructose	Lactose + H_2O	Invert sugar
100	49.0	53.8	76.7	51.5
102	50.0	54.9	78.2	52.5
104	51.1	56.0	79.7	53.6
106	52.1	57.1	81.3	54.6
108	53.1	58.2	82.8	55.7
110	54.1	59.3	84.4	56.7
112	55.1	60.4	85.9	57.8
114	56.1	61.6	87.4	58.9
116	57.2	62.7	89.0	59.9
118	58.2	63.8	90.5	61.0
120	59.2	64.9	92.1	62.0
122	60.2	66.0	93.6	63.1
124	61.3	67.1	95.2	64.2
126	62.3	68.2	96.7	65.2
128	63.0	69.3	98.2	66.3
130	64.3	70.4	99.8	67.4
132	65.4	71.6	101.3	68.4
134	66.4	72.7	102.9	69.5
136	67.4	73.8	104.4	70.6
138	68.5	74.9	106.0	71.6
140	69.5	76.0	107.5	72.7
142	70.5	77.1	109.0	73.8
144	71.6	78.3	110.6	74.9
146	72.6	79.4	112.1	75.9
148	73.7	80.5	113.7	77.0
150	74.7	81.6	115.2	78.1
152	75.7	82.8	116.8	79.2
154	76.8	83.9	118.3	80.3
156	77.8	85.0	119.9	81.3
158	78.9	86.1	121.4	82.4
160	79.9	87.3	122.9	83.5
162	81.0	88.4	124.5	84.6
164	82.0	89.5	126.0	85.7
166	83.1	90.6	127.6	86.8
168	84.1	91.8	129.1	87.8
170	85.2	92.9	130.7	88.9
172	86.2	94.0	132.2	90.0
174	87.3	95.2	133.8	91.1
176	88.3	96.3	135.3	92.2
178	89.4	97.4	136.9	93.3
180	90.4	98.6	138.4	94.4
182	91.5	99.7	140.0	95.5
184	92.6	100.9	141.5	96.6
186	93.6	102.0	143.1	97.7
188	94.7	103.1	144.6	98.8
190	95.7	104.3	146.2	99.9

Table 12.7. (*continued*)

Copper (Cu)	Glucose	Fructose	Lactose + H_2O	Invert sugar
192	96.8	105.4	147.7	101.0
194	97.9	106.6	149.3	102.1
196	98.9	107.7	150.8	103.2
198	100.0	108.8	152.4	104.3
200	101.1	110.0	153.9	105.4
202	102.2	111.1	155.5	106.5
204	103.2	112.3	157.0	107.6
206	104.3	113.4	158.6	108.7
208	105.4	114.6	160.2	109.8
210	106.5	115.7	161.7	110.9
212	107.5	116.9	163.3	112.1
214	108.6	118.0	164.8	113.2
216	109.7	119.2	166.4	114.3
218	110.8	120.3	167.9	115.4
220	111.9	121.5	169.5	116.5
222	112.9	122.6	171.0	117.6
224	114.0	123.8	172.6	118.8
226	115.1	125.0	174.2	119.9
228	116.2	126.1	175.7	121.0
230	117.3	127.3	177.3	122.1
232	118.4	128.4	178.8	123.3
234	119.5	129.6	180.4	124.4
236	120.6	130.8	181.9	125.5
238	121.7	131.9	183.5	126.6
240	122.7	133.1	185.1	127.8
242	123.8	134.2	186.6	128.9
244	124.9	135.4	188.2	130.0
246	126.0	136.6	189.7	131.2
248	127.1	137.7	191.3	132.3
250	128.2	138.9	192.9	133.4
252	129.3	140.1	194.4	134.6
254	130.4	141.3	196.0	135.7
256	131.6	142.4	197.5	136.8
258	132.7	143.6	199.1	138.0
260	133.8	144.8	200.7	139.1
262	134.9	145.9	202.2	140.3
264	136.0	147.1	203.8	141.4
266	137.1	148.3	205.3	142.6
268	138.2	149.5	206.9	143.7
270	139.3	150.6	208.5	144.8
272	140.4	151.8	210.0	146.0
274	141.6	153.0	211.6	147.1
276	142.7	154.2	213.2	148.3
278	143.8	155.4	214.7	149.4
280	144.9	156.5	216.3	150.6
282	146.0	157.7	217.9	151.8

Table 12.7. *(continued)*

Copper (Cu)	Glucose	Fructose	Lactose + H_2O	Invert sugar
284	147.2	158.9	219.4	152.9
286	148.3	160.1	221.0	154.1
288	149.4	161.3	222.6	155.2
290	150.5	162.5	224.1	156.4
292	151.7	163.7	225.7	157.5
294	152.8	164.9	227.3	158.7
296	153.9	166.0	228.8	159.9
298	155.1	167.2	230.4	161.0
300	156.2	168.4	232.0	162.2
302	157.3	169.6	233.5	163.4
304	158.5	170.8	235.1	164.5
306	159.6	172.0	236.7	165.7
308	160.7	173.2	238.2	166.9
310	161.9	174.4	239.8	168.0
312	163.0	175.6	241.4	169.2
314	164.2	176.8	243.0	170.4
316	165.3	178.0	244.5	171.6
318	166.5	179.2	246.1	172.8
320	167.6	180.4	247.7	173.9
322	168.8	181.6	249.2	175.1
324	169.9	182.8	250.8	176.3
326	171.1	184.0	252.4	177.5
328	172.2	185.2	253.9	178.7
330	173.4	186.4	255.5	179.8
332	174.5	187.6	257.1	181.0
334	175.7	188.8	258.7	182.2
336	176.8	190.1	260.2	183.4
338	178.0	191.3	261.8	184.6
340	179.2	192.5	263.4	185.8
342	180.3	193.7	265.0	187.0
344	181.5	194.9	266.6	188.2
346	182.7	196.1	268.1	189.4
348	183.8	197.3	269.7	190.6
350	185.0	198.5	271.3	191.8
352	186.2	199.8	272.9	193.0
354	187.3	201.0	274.4	194.2
356	188.5	202.2	276.0	195.4
358	189.7	203.4	277.6	196.6
360	190.9	204.7	279.2	197.8
362	192.0	205.9	280.8	199.0
364	193.2	207.1	282.4	200.2
366	194.4	208.3	284.0	201.4
368	195.6	209.6	285.6	202.6
370	196.8	210.8	287.1	203.8
372	198.0	212.0	288.7	205.0
374	199.1	213.3	290.3	206.3

Table 12.7. (*continued*)

Copper (Cu)	Glucose	Fructose	Lactose + H_2O	Invert sugar
376	200.3	214.5	291.9	207.5
378	201.5	215.7	293.5	208.7
380	202.7	217.0	295.0	209.9
382	203.9	218.2	296.6	211.1
384	205.1	219.5	298.2	212.4
386	206.3	220.7	299.8	213.6
388	207.5	221.9	301.4	214.8
390	208.7	223.2	303.0	216.0
392	209.9	224.4	304.6	217.3
394	211.1	225.7	306.2	218.5
396	212.3	226.9	307.8	219.8
398	213.5	228.2	309.4	221.0
400	214.7	229.4	311.0	222.2
402	215.9	230.7	312.6	223.5
404	217.1	232.0	314.2	224.7
406	218.4	233.2	315.9	226.0
408	219.6	234.5	317.5	227.2
410	220.8	235.8	319.1	228.5
412	222.0	237.1	320.7	229.7
414	223.3	238.4	322.4	231.0
416	224.5	239.7	324.0	232.3
418	225.7	241.0	325.7	233.6
420	227.0	242.2	327.4	234.8
422	228.2	243.6	329.1	236.1
424	229.5	244.9	330.8	237.5
426	230.7	246.3	332.6	238.8
428	232.0	247.8	334.4	240.2
430	233.3	249.2	336.3	241.5
432	234.7	250.8	338.3	243.0
434	236.1	252.7	340.7	244.7

Source: Natl Bureau Standards (1940, 1948).
[a] Applicable when Cu is determined by analysis.

5. Standard lactose solution—Dissolve exactly 0.3000 g of pure lactose monohydrate in water and dilute to a 1 liter.

Procedure

Pipette accurately 1.00 ml of milk into a 100-ml flask, add 2.0 ml of the sodium tungstate solution, and then, drop by drop, add 2.0 ml of the sulfuric acid. Mix thoroughly and allow to stand for 5 min. Dilute to volume with water, mix thoroughly, and filter, discarding the first 5–10 ml of filtrate.

Pipette accurately 1.0 ml of the filtrate and 1.0 ml water into a Folin–Wu sugar tube. Pipette, into a similar tube, exactly 2.0 ml of the standard lactose solution. Add 2.0 ml of the alkaline copper solution to each tube, and place the tubes in a beaker of boiling water for 6 min. Transfer to a cold water bath and cool with shaking for 2–3 min. Add 4.0 ml of the phosphomolybdic acid solution. Let stand for 1 min; then dilute to the 25 ml mark with a 1:4 dilution of the phosphomolybdic acid solution and mix thoroughly. Let stand for 10–15 min, transfer to a colorimeter tube, and read within 15 min in a colorimeter set at 420 nm.

Run parallel determinations on water as a blank and on the standard solution. The intensity of the colored solutions is directly proportional to the amount of lactose present. Calculate the milligrams of lactose in 1.0 ml of the milk sample as follows:

$$c = c_s \times A/A_s$$

where c = mg/ml of milk sample; c_s = mg/ml of standard lactose solution; and A and A_s = absorbance of milk sample and standard glucose solutions, respectively. If the specific gravity of the milk is known, calculate the percentage lactose in the milk.

Detection of Added Water

The mineral matter and lactose of milk are the least variable components, while fat and protein are the most variable components of milk. Consequently, the addition of water to milk will dilute the dissolved minerals and lactose in the aqueous portion (i.e., serum) of the milk. The clear serum is obtained by precipitating the proteins with either calcium chloride, acetic acid, or copper sulfate and filtering off the precipitated proteins; the fat is carried down with the precipitated proteins. Detection of added water can be made by measuring the refraction of the clear serum or determining the solids content of the clear serum. These two methods have been largely replaced by the cryoscope method, which is capable of detecting a smaller percentage of added water.

Copper Sulfate Serum Refraction Method

Minimum values for the refraction, specific gravity, and total solids content of the copper serum of normal milk have been established experimentally. The addition of 10% or more of water can be detected by a deviation in the refraction, specific gravity, or total solids of a milk sample from the minimum values. This method (AOAC 16.098) should be applied only to fresh milk.

Procedure

Dissolve 72.5 g copper sulfate ($CuSO_4 \cdot 5H_2O$) in water and dilute to 1 liter. Adjust the concentration of this solution so that it gives a reading of 36 on the scale of the immersion refractometer at 20°C, or has a specific gravity of 1.0443_4^{20}. Add 4 volumes of milk to 1 volume of the copper sulfate solution, shake thoroughly, and filter. Discard the first filtrate that comes through. Collect the clear filtrate and determine either the refraction at 20°C with an immersion refractometer, the specific gravity (20°/4°), or the total solids content.

Minimum values for the copper serum of fresh milk are immersion refractometer reading at 20°C of 36; a specific gravity of 1.0245_4^{20}, and a total solids content of 5.28%.

Freezing Point Method

The freezing point of milk is a function of its osmotic pressure, which is determined by the components in true solution—lactose and soluble salts. Fat has no effect on the freezing point, and the proteins exert little effect. Under normal circumstances, the freezing point of milk is relatively constant because it remains nearly isotonic with blood. However, the addition of water to milk reduces the concentration of dissolved substances and proportionately raises the freezing point. Thus, accurate measurements of the freezing point of milk can be used to detect the presence of added water. There are two limitations on the freezing point method. It is impossible to use this method to detect the addition of separated milk or the removal of fat. Further, the freezing point of milk is affected by the development of acidity. More mineral matter is solubilized with increasing acidity, which results in an increased osmotic pressure and correspondingly lower freezing point. Therefore, the increase in the freezing point caused by watering may escape detection if the milk is acid (sour).

Apparatus

Freezing point determinations can be made with a Thermistor instrument (Advanced Instruments, Inc., Needham Hts., MA) or a Fiske cryoscope (Fiske Associates, Inc., Uxbridge, MA). The apparatus consists of a cooling bath, an air agitator, a thermistor probe, a seeding rod, a wheatstone bridge, and a galvanometer measuring circuit. The observed freezing point values are read from a temperature dial, calibrated in degrees Hortvet, that balances the bridge.

Procedure

Operate the instrument in accordance with the instructions issued by the manufacturer. Use the same procedure for both standards and sample to obtain valid freezing point values. The sample tubes and pipettes must be clean and dry. Prechill the tubes containing 2 ml of sample by placing them in the sample cooling bath, sample well, or operating head; lower the operating head to position the sample in cryoscope cooling bath. Locate thermistor probe, both horizontally and vertically, at the midpoint of the sample. Supercool to the same temperature each time. After seeding, a long steady temperature plateau must be obtained. Read the freezing point value where the galvanometer spot or meter needle ceases to move to the right. Follow the same technique with standard solutions; the spot or needle will become steady and move to the left sooner with the standards than with milk samples.

The presence of added water is indicated if the freezing point is above −0.525°C. This applies to either raw or pasteurized milk that has not been subjected to vacuum pasteurization; vacuum pasteurization raises the freezing point approximately 0.005°C. The percentage of added water can be calculated as $W = 100(T - T_1)/T$, in which T = average freezing point of normal milk (−0.525) and T_1 = freezing point of a given sample.

Comments

If the titratable acidity is over 0.18%, results by this method may underestimate the amount of added water. A tolerance of 3% may be allowed on results for added water.

Residual Phosphatase Test for Proper Pasteurization

The test most commonly used for the detection of improperly pasteurized milk is the phosphatase test (AOAC 16.112). It depends on the fact that alkaline phosphatase, an enzyme present in raw milk, is progressively inactivated at the temperatures employed in pasteurization. Essentially the determination is based on the hydrolysis of disodium phenyl phosphate by the enzyme. The phenolate is then quantitated by any one of several colorimetric reactions. Furthermore, the determination can be made with great accuracy, so that the addition of 0.1 ml of unpasteurized milk to 99.9 ml of pasteurized milk can be detected. With modifications, the phosphatase test may be applied to the different types of dairy products.

Reagents

1. Barium buffer—Dissolve 25.0 g of CP barium hydroxide ($Ba(OH)_2 \cdot 8H_2O$; fresh, not deteriorated) in distilled water and dilute to 500 ml. In another flask dissolve 11.0 g of CP boric acid (H_3BO_3) and dilute to 500 ml. Warm each solution to 50°C, mix the two solutions together, stir, cool to approximately 20°C, filter, and keep filtrate (pH 10.6) in a tightly stoppered container. (For use with milk dilute 500 ml of this buffer with 500 ml distilled water.)
2. Substrate solution—Dilute 500 ml of barium buffer with 500 ml distilled water. Dissolve 0.10 g phenol-free crystalline disodium phenyl phosphate in 100 ml of the diluted barium buffer.
3. Zinc–copper precipitant—Dissolve 3.0 g zinc sulfate ($ZnSO_4 \cdot 7H_2O$) and 0.6 g copper sulfate ($CuSO_4 \cdot 5H_2O$) in water and dilute to 100 ml with water.
4. Color development buffer—Dissolve 6.0 sodium metaborate ($NaBO_2$) and 20 g sodium chloride in water and dilute to 1 liter with water.
5. Color dilution buffer—Dilute 100 ml of the color development buffer to 1 liter with water.
6. BQC solution (Gibbs reagent)—Dissolve 0.040 g BQC (2,6-dibromoquinonechloroimide) in 10 ml absolute ethyl or methyl alcohol and transfer to a dark-colored dropping bottle. If kept in the ice tray of a refrigerator, this solution will remain stable for at least a month. Do not use it when it begins to turn brown.
7. *n*-Butyl alcohol (bp 116°–118°C)—Adjust the pH by mixing a liter of the alcohol with 50 ml of the color development buffer.
8. 0.05% Copper sulfate—Dissolve 0.05 g copper sulfate in water and dilute to 100 ml with water.
9. Phenol standard solutions—Transfer 1.0 g of phenol, accurately weighed, into a 1-liter volumetric flask, dissolve in water, and dilute to the mark with water. Thus, 1 ml of this stock solution contains 1 mg of phenol. (The solution is stable for several months if kept in a refrigerator.) Dilute 10 ml of this stock solution to a volume of 1 liter with water to give 10 μg phenol/ml. Finally, dilute aliquots (5, 10, 30, and 50 ml) of this diluted stock solution to a volume of 100 ml with water, giving solutions containing 0.5, 1.0, 3.0 and 5.0 ug phenol/ml, respectively. Store these standard solutions in a refrigerator.

Standard Curve

Measure appropriate quantities of the phenol standard solutions into a series of tubes (preferably graduated at 5.0 and 10.0 ml) to provide a range of standards containing 0 (control or blank), 0.5, 1.0, 3.0, 5.0, 10.0, and up to 30 or 40 μg of phenol. To improve the stability of the standards and increase the brightness of blue solutions, add 1.0 ml of the copper sulfate solution to each tube. Then add 5.0 ml of the color dilution buffer and dilute to 10 ml with water. Add 2 drops of the BQC solution, mix, and allow to stand for 30 min at room temperature to develop the blue color. Extract each tube with butyl alcohol as described under Assay Procedure. Determine absorbance of extract at 650 nm. Plot absorbance versus micrograms of phenol equivalents.

Assay Procedure

Pipette a 1.0-ml sample of milk into each of 3 test tubes and reserve one of these tubes for a blank or control. Heat the blank sample for about a minute to at least 85°–90°C in beaker of boiling water (cover the beaker so that the entire tube is heated to approximately 85°C), remove the tube, and cool to room temperature.

Add 10 ml of the substrate solution to each tube, stopper, and mix. Incubate the tubes in a water bath at 37°–38°C for 1 hr, occasionally shaking the tubes. Place the tubes in a beaker of boiling water for nearly a minute, heating to approximately 85°C, and then cool to room temperature.

Pipette 1 ml of the zinc–copper precipitant into each tube and mix thoroughly. Filter through a 5-cm funnel, using a 9-cm Whatman No. 42 or No. 2 filter paper. Collect 5 ml of the filtrate in a tube graduated at 5.0 and 10.0 ml. Add 5 ml of the color development buffer; then add 2 drops of the BQC solution, mix, and allow to stand for 30 min at room temperature to develop the color.

Add 5 ml butyl alcohol to each tube; invert tube slowly several times. Centrifuge to break emulsion and to remove water suspended in the alcohol layer. After centrifuging, remove the butyl alcohol by pipette, filter into a photometer tube, and determine absorbance in a photometer at 650 nm.

Subtract the absorbance reading of the blank from those of the samples, then average the corrected sample absorbance values and multiply by 2.4 (to correct for dilution during assay). Determine the micrograms of phenol equivalents per milliter of sample from the standard curve. A value greater than 4 μg/ml of sample indicates underpasteurization. With modifications, the phosphatase test may be applied to the different types of dairy products (AOAC 16.112).

Determination of Minerals

Of the mineral elements commonly present in milk (Table 12.5), calcium is of greatest interest because milk is an important source of calcium in human diets. Since iron is added to mineral-fortified milk, analysis for this mineral is sometimes required to determine that the level of added iron meets the standard. Both calcium and iron can be determined by atomic absorption spectrophotometry (see Chapter 3 for a full discussion of this technique). Specific chemical assays have also been developed for calcium and iron. The other major cations in milk (sodium and potassium), as well as many trace elements, also can be determined by atomic absorption spectrophotometry. Because this technique is very sensitive, it is particularly useful in detecting small amounts of contaminants (e.g., lead, mercury) that are toxic to humans.

Metals Analysis by Atomic Absorption

As noted, atomic absorption spectrophotometry can be used to determine the amounts of many metals (Ca, Fe, Mg, Mn, K, Na, Zn, Pb, and Hg) in food products. The usual procedure is to remove organic matter, which may interfere with the metal analysis, by either dry ashing or wet digestion and then to dissolve the residue in dilute acid. The solution is then aspirated directly into the flame of an atomic absorption spectrophotometer, and the absorption of the metal to be analyzed is determined at a specific wavelength.

In this section, general directions are given for metal analysis by atomic absorption. See Chapter 3 for information about selecting the best operating conditions for an atomic absorption spectrophotometer.

Standard Solutions

To prepare a stock standard solution (1000 mg/liter), weigh out an appropriate quantity of a salt of the metal to be determined, dissolve in 25 ml of 3 *N* hydrochloric acid, and dilute to 250 ml with water. Prepare a series of standard solutions (0.05–5 μg metal/ml) by diluting the stock standard solution with water (if wet digestion is used) or 0.3 *N* hydrochloric acid (if dry ashing is used). Add other salts where necessary.

Standard solutions can be prepared from certified atomic absorption standards supplied by several chemical companies. These standards, which contain an exact unit weight of an ionic species, are diluted to a definite volume to give an appropriate concentration.

Wet Ashing

To 50 ml of whole milk in a 250-ml beaker add 20 ml of 10% acetic acid, cover with speed-vap covers, and dry overnight to dryness at 80°C. Add 20 ml of concentrated nitric acid to each sample and boil slowly to near dryness. Thereafter, add 15 ml concentrated perchloric acid, 3 ml concentrated nitric acid, and 3 ml demineralized water and boil until near dryness; repeat this procedure until samples remain clear and colorless. Boil to dryness.

Dry Ashing

To 50 ml of whole milk in a 100-ml Coors (Vycor) high-form ceramic crucible, add 20 ml of 10% acetic acid, cover with speedy-vap covers, and dry overnight to dryness in a forced draft oven at 80°C. Place the samples in a muffle furnace and ash at 400°C for 12–15 hr.

After ashing, add 2 ml of concentrated hydrochloric acid and 2 ml demineralized water to each sample, heat on a steam bath, and then transfer contents to a 50-ml volumetric flask. If calcium is to be determined, add 2.5 ml of 10% (w/v) lanthanum chloride solution per 50 ml of sample solution. If iron is to be determined, add an internal standard of 5 ppm to the volumetric flask. Then add demineralized water to the volume of the flask. Prepare a blank by taking the same amount of reagents through the various steps.

Atomic Absorption Measurements

Standardize the instrument according to the manufacturer's instructions. Prepare a standard curve for each mineral to be determined. Aspirate the samples directly into the flame. Dilute with deionized water if necessary to bring the concentration of the metal of interest into a range suitable for atomic absorption. While running the samples, periodically check to see that the standardization values remain constant. Read the micrograms of metal in the unknown sample directly from the standard curve and subtract the value of the reagent blank.

Chemical Analysis for Calcium

Calcium is precipitated at ph 4.0 as the oxalate, which is dissolved in an acid and the liberated oxalic acid is then titrated with standard potassium permanganate. The quantity can then be calculated from the equation representing the reaction:

$$2KMnO_4 + 5CaC_2O_4 + 8H_2SO_4 \rightleftharpoons 2MnSo_4 + K_2SO_4 + 5CaSO_4 + 10CO_2\uparrow\ 8H_2O$$

Reagents

1. Dilute ammonia—Mix one volume ammonia solution with two volumes of water.
2. Dilute sulfuric acid—Slowly add 20 ml of concentrated sulfuric acid to 180 ml of water with stirring, then cool the solution.

3. Dilute acetic acid—Mix one volume of glacial acetic acid with two volumes of water.
4. 5 *N* Hydrochloric acid—Slowly add 4.25 ml of hydrochloric acid to 500 ml of water with stirring, cool, then dilute to 1 liter.
5. Saturated ammonium oxalate solution.
6. Bromocresol green (0.1% solution in 95% alcohol).
7. 0.05 *N* Potassium permanganate solution—Dissolve 1.58 g of potassium permanganate in 1 liter of water.

Procedure

To 25 ml of milk, add 10 ml of 10% acetic acid, cover with speed-vap covers, and dry overnight at 80°C in a forced-air oven. Place sample in a muffle furnace and ash at 400°C for 12–15 hr. Cool the ash, add 10 ml of 5 *N* hydrochloric acid, heat the solution just to boiling, dilute with 10 ml water, and filter into a 250-ml beaker. Repeat the acid extraction, dilution, and filtration of the ash. Add 1 drop of bromocresol green indicator solution to the filtrate and make it alkaline with the dilute ammonia. Acidify the solution with the dilute acetic acid, then add 0.5 ml of glacial acetic acid. Heat the solution to boiling and slowly add 10 ml of the saturated ammonium oxalate solution. Then add dilute ammonia to the hot solution until it becomes yellow-green ($pH \cong 3.8$) and allow to stand for at least 4 hr. Filter the solution through a Whatman No. 44 filter paper, checking the first few milliliters of filtrate for unprecipitated calcium by adding ammonium oxalate solution. Wash the filter paper with 75–100 ml of a very dilute ammonia solution (1 volume ammonia solution + 50 volumes water). Repeat washing. Pierce the paper with a pointed glass rod and wash the precipitate into an Erlenmeyer flask with a fine spray of cold water, then with 10 ml of warm (60°C) dilute sulfuric acid and finally with hot water to give a final volume of approximately 150 ml.

Titrate the solution with 0.05 *N* potassium permanganate at 75°C until a pink color persists for 30 sec. Run at least two blank determinations. Average the results obtained for the blanks, and subtract this value from the volume of potassium permanganate used in each titration.

Calculations

Calculate the quantity of calcium by means of the following equation:

$$\text{mg calcium} = \text{ml } KMnO_4 \text{ (corrected for blanks)} \times \text{normality of } KMnO_4 \times \text{meq Ca}$$

(1 ml of 0.05 *N* $KMnO_4$ = 1 mg calcium).

Notes

The calcium concentration of normal milk averages 125 mg/100 ml. The oxalate precipitation step eliminates phosphate interference.

Chemical Analysis for Iron

The basis for this determination is the formation of a colored complex between ferrous ions and 2,2-dipyridyl. The sample is first treated with acid and heated to precipitate the proteins. The ph of the supernatant is adjusted and the 2,2-dipyridyl reagent is added in the presence of a reducing agent. The absorbance of the resulting colored complex at 522 nm is then read and the corresponding concentration of iron is determined from a standard curve prepared from solutions of known concentrations of iron.

Regulations generally require that vitamin–mineral fortified milk contain 10 mg of iron/qt. Unfortified raw and market whole milks contain 0.1 mg/100 ml.

Reagents

1. 2,2-Dipyridyl solution—Dissolve 0.2 g of the reagent in 5 ml of glacial acetic acid and dilute to 100 ml with distilled water.
2. Ammonium hydroxide solution—Dilute 420 ml concentrated ammonium hydroxide to 1 liter with water.
3. Buffer solution—Mix 27.2 ml of glacial acetic acid and 33.4 g of sodium acetate in water and dilute to 250 ml.
4. 6 *N* Hydrochloric acid—Dilute 495.0 ml of concentrated hydrochloric acid with water to 1 liter.
5. 0.1% *p*-Nitrophenol indicator solution (aqueous).
6. 25% Trichloroacetic acid (aqueous).
7. Standard iron solution—Dissolve 0.351 g of $Fe(NH_4)_2(SO_4)_2$ $6H_2O$ in water, add 2 drops of 6 *N* HCl, and dilute the mixture with water to 1 liter. Dilute 10 ml of this solution to 1 liter to give a concentration of 0.51 mg Fe/ml.

Procedure

To 50 ml of whole milk in a 15-ml centrifuge tube, add 5 drops of mercaptoacetic acid, 2 ml of the trichloroacetic acid solution, and 1 ml concentrated hydrochloric acid. Also prepare a blank with all the reagents plus 5 ml of water instead of the milk. Stir well and place in a water bath (90°–95°C) for 5 min. Decant the supernatant liquid into a 25-ml volumetric flask. Wash the precipitate with a mixture of 2 ml distilled water, 1 ml trichloroacetic acid solution, and 1 ml concentrated hydrochloric acid, then repeat the heating, centrifuging, and decantation steps.

To the combined supernatants, add 1 drop of the *p*-nitrophenol indicator and slowly add the ammonium hydroxide solution until the solution becomes yellow; make acid with 1 or 2 drops of 6 *N* hydrochloric acid or until the yellow color disappears. Add 1 ml of the buffer solution, dilute to 25 ml, and mix. Pipette a 5- or 10-ml aliquot (estimated to contain 1–2 μg Fe) into a flask, and dilute if necessary to 10 ml with water. Add 2 drops mercaptoacetic acid (to reduce any ferric iron present) and 1 ml of the 2,2-dipyridyl solution, mix, and read the absorbance in a spectrophotometer at 522 nm. Correct for the absorbance of the reagent blank, and determine the amount of iron by reference to a standard curve. If a 5-ml aliquot was taken, this value is in terms of μg/ml Fe in the milk. For a 10-ml aliquot, divide by 2 to obtain μg Fe/ml.

Preparation of Standard Curve

Dilute aliquots (1, 2, 3, 4, and 5 ml) of the standard iron solution to 5 ml with water, and treat as outlined under Procedure (except the centrifugation may be omitted), using 5-ml aliquots for the color development step. Correct for the absorbance of the reagent blank, and plot the absorbances of the various solutions against micrograms of Fe.

☐ ANALYSIS OF DRIED MILK PRODUCTS

The terms applied to the various types of dried milk are designed to disclose the fat content of the dried product, the nature of the milk from which the product was made, and the type of drying process used in the manufacture. Whole milk, skim milk, cream, buttermilk, and whey are dried commercially. Dried milk, made from whole milk, must contain not less than 26% milk fat, and the moisture content must not exceed 5%. A greater moisture content than this is conducive to spoilage by microorganisms, whereas a very low moisture content is conducive to lipid oxidation.

Dried milk is used in a wide variety of food products. This has resulted in the development of methods for the analysis of dried milk itself and also of numerous products for their dried milk content. In general, the methods of analysis for dried milk are the same as for liquid whole milk, allowing for the difference in solids content.

Determination of Moisture

Although moisture in dried milks may be determined by drying to constant weight at 100°C for 5 hr at 100 mm Hg, the toluene distillation method is more generally used since the high sugar content of dried milk (36.8%) causes some difficulty when drying at the above conditions. Consult Chapter 2 for discussion of the moisture determination by distillation methods.

Procedure

Connect a 300-ml flask of Pyrex or other resistant glass by means of a Bidwell-Sterling moisture receiver to a 500-mm Liebig condenser. Calibrate the receiver, 5 ml capacity, by distilling known amounts of water into a graduated column, and estimating the column of water to 0.01 ml. Clean the apparatus with chromic acid cleaning mixture, rinse thoroughly with water, then alcohol, and dry in an oven to prevent undue amount of water from adhering to inner surfaces during determination.

Place sufficient dry sand in a 300-ml Erlenmeyer flask to cover the bottom of the flask and add 75–100 ml of toluene. Introduce 25.0 g of dried milk into the flask and set up the apparatus as described in the preceding paragraph. Pour toluene through the top of the condenser until the collecting tube is filled and distill; when the distillation is complete, remove the heat source, disconnect the collecting tube, and centrifuge it. After allowing the collection tube to attain room temperature, read the volume of water distilled. Assuming the specific gravity of water is 1.00, calculate the percentage of water present in the sample.

Determination of Total Protein

Weigh accurately and rapidly a 1-g sample of the dry milk onto a 11-mm filter paper, wrap it tightly, and transfer to a Kjeldahl digestion flask. Proceed as directed for the protein determination of fresh milk. Multiply the percentage of nitrogen by the factor 6.38 to convert to percentage of protein.

Determination of Ash

Weigh accurately and rapidly, 1 g dried milk into a porcelain crucible. Carefully char over a burner at low heat, then transfer to a muffle furnace and ignite to constant weight at 550°C. Calculate the percentage of ash.

Determination of Crude Fat

Crude fat in dried milk is determined by the Mojonnier method, which was described in the previous section on analysis of liquid milk. However, dried milk must first by solubilized before it can be extracted.

Procedure

Weigh accurately and rapidly, a 1-g sample of dried milk into a Mojonnier fat-extraction flask. Add 10 ml of warm water and shake vigorously until homogenous. Add 1.0–1.25 ml of concentrated ammonium hydroxide and heat in a water bath at 60°–70°C for 15 min with occasional shaking, and then cool. Proceed as in the Mojonnier method, starting with the addition of 10 ml of alcohol.

SELECTED REFERENCES

AAMMC. 1982. Methods and Standards for the Production of Certified Milk. American Association of Medical Milk Commissions, Milwaukee, WI.

AOAC. 1984. Official Methods of Analysis. 14th ed. S. Williams (editor). Assoc. of Official Analytical Chemists, Arlington, VA.

BUEGAMER, W. R., MICHAUD, L. and ELVEHJEM, C. A. 1945. A simplified method for the determination of iron in milk. J. Biol. Chem. *158*, 573–576.

COX, H. E. and PEARSON, D. 1962. Chemical Analysis of Foods. Chemical Publishing Co., New York.

HAND, D. B. and SHARP, P. F. 1939. The ribolflavin content of cow's milk. J. Dairy Sci. *22*, 779–783.

HART, L. F. and FISHER, H. J. 1971. Modern Food Analysis. Springer-Verlag, New York.

HILDITCH, T. P. and WILLIAMS, P. N. 1964. The Chemical Constitution of Natural Fats, 4th ed. John Wiley & Sons, New York.

JACKSON, R. F. and MCDONALD, E. J. 1941. Errors of Munson and Walker's reducing sugar tables and the precision of their method. J. Assoc. Off. Agric. Chem. *24*, 767–788.

NATIONAL ARCHIVES OF THE UNITED STATES. Code of Federal Regulations: Title 21—Food and Drugs. Washington, DC.

NATL. BUREAU STANDARDS. J. Res. Natl. Bur. Stand. 1940. Hammond table. *24*, 589–596.

NATL. BUREAU STANDARDS. J. Res. Natl. Bur. Stand. 1948. Hammond table *41*, 217–220.

NATIONAL RESEARCH COUNCIL. 1953. The Composition of Milks. Publication No. *253*.

PATTON, S. and JENSEN, R. G. 1976. Biomedical Aspects of Lactation. Pergamon Press, Oxford.

PEARSON, D. 1973. Laboratory Techniques in Food Analysis. John Wiley & Sons, New York.

ROSE, D. (Chairman). 1970. Nomenclature of the proteins of cow's milk, 3rd rev. J. Dairy Sci. *53*, 1–17.

U.S. BUREAU OF STANDARDS. 1918. Munson and Walker table. Circ. 44, 2nd ed., p. 129.

USDA. 1968. The Federal Milk Marketing Order Program. Marketing Order Program. Marketing Bulletin No. *27*. U.S. Dept. of Agriculture, Washington, DC.

USDA. 1968. Government's Role in Pricing Fluid Milk in the United States. Agr. Econ. Rept. *152*. U.S. Dept. Agriculture, Washington, DC.

U.S. DEPARTMENT OF HEALTH, EDUCATION, AND WELFARE. 1967. Grade "A" Pasteurized Milk Ordinance—1965 Recommendations of the United States Public Health Service. Public Health Service Publication No. *229*.

WEBB, B. H. and JOHNSON, A. H. 1975. Fundamentals of Dairy Chemistry. AVI Publishing Co., Westport, CT.

WHITTIER, E. O. 1944. Lactose and its utilization: A review. J. Dairy Sci. *27*, 505–537.

13

Meat, Poultry, and Fish

☐ MEAT

In this discussion, meat is defined as the flesh of cattle, swine, sheep or goats that is consumed for food. It consists of muscle fibers held together by connective tissue and interspersed with nerves and blood vessels, which normally accompany the tissue. A single muscle contains a number of fiber bundles, held together by a sheath of connective tissue, the epimysium. Groups of fibers are associated into fiber bundles which, in turn, are surrounded by a sheath of connective tissue, the perimysium. The basic unit of muscle is the fiber, and it is bounded by the endomysium and sarcolemma.

A muscle fiber comprises a number of long, thin, cylindrical rods known as *myofibrils*, the essential contractile units of muscle, which are separated from one another by a highly specialized network of tubules, the *sarcoplasmic reticulum*. Myofibrils are bathed in an aqueous fluid (sarcoplasm), which is about 75–80% water and contains mitochondria, enzymes, glycogen, adenosine triphosphate, creatine, phosphate and myoglobin. Each myofibril consists of two sets of filaments: a thick set of filaments containing the protein *myosin* and a thin set of filaments containing the protein *actin*. These two sets of filaments are arranged within the myofibril. The thin filaments (I bands) extend longitudinally from the Z lines, while the thick filaments (A band) span the gaps between the tips of the opposing actin units. Contraction and relaxation of striated muscle appears to be related to the interaction between actin, myosin, and ATP. The energy released by the calcium-activated enzymic dephosphorylation of adenosine triphosphate (ATP) brings about muscle contraction through the sliding action of the actin filaments over the myosin filaments, forming contractile actomyosin. On relaxation, the reverse process occurs.

Muscle fibers are not directly attached to the bones which they move; rather movement is transmitted via the connective tissue (endomysium, perimysium, and epimysium) to the tendons, which in turn are attached to the skeleton. Thus, contractile movement would be limited to the myofibrillar level were it not for the force-transmissive capabilities of the connective tissue. Collagen is the major component of the connective tissue, and it plays an important role in the textural properties of meat.

The conversion of muscle to meat involves a number of biochemical and biophysical changes, which appear to occur in stages. In the first stage (*prerigor stage*) immediately after death of animal, the muscle tissue is soft and pliable. As postmortem continues, and glycogen is converted to lactic acid, resulting in a concomitant decrease in pH, which reaches an ultimate value of pH 5.5 (after almost 24 hr). There also is a decrease in creatine phosphate, and as a consequence, a decrease in ATP because as the reserve creatine phosphate is depleted, the capacity to resynthesize ATP is lessened. The second postmortem stage is known as *rigor mortis*. During this stage, the myofibrillar proteins—actin and myosin—gradually associate to form actomyosin. This phenomenon occurs as a result of a decreasing level of ATP. It is the actomyosin that is responsible for the inextensible properties of muscle manifested as rigor mortis; i.e., the muscle ceases to be elastic and tends to stiffen. During the third stage, referred to as the *postrigor stage*, meat gradually tenderizes, becoming more acceptable as the aging process continues.

Composition of Meat

The major components of meat are water (56–72%), protein (15–22%), fat (5–34%), and soluble nonprotein substances (ca. 3.5%). This latter component includes carbohydrates, inorganic salts, soluble nitrogenous substances, trace metals, vitamins, etc. The actual composition of a piece of meat depends on the species, breed, age, sex, and diet of the animal from which it is derived and on its anatomical location. Average proximate analyses for different cuts of meat are presented in Table 13.1. The composition of meat is also influenced by how it is trimmed, curing and processing, and the cooking method.

Table 13.1. Proximate Analysis and Energy Values of Edible Portions of Meat

Type	Cut of meat (good grade)	Water (%)	Protein (%)	Fat (%)	Ash (%)	Energy (cal)
Beef	Chuck	56	18	25	0.8	303
	Flank	72	22	5	1.0	139
	Loin (short)	50	15	34	0.7	370
	Rib (blade)	57	18	25	0.8	300
	Round	67	20	12	0.9	197
	Rump	59	18	21	0.8	271
	Hamburger	60	18	21	0.7	268
Veal	Chuck	70	19	10	1.0	173
	Loin	69	19	11	1.0	181
	Rib	66	19	14	1.0	207
Pork	Ham	57	16	27	0.7	308
	Loin	57	17	25	0.9	298
	Spareribs	52	15	33	0.7	361
Lamb	Leg	66	18	15	1.4	209
	Loin	59	17	23	1.3	276
	Rib	56	16	27	1.2	312
	Shoulder	61	16	22	1.2	265

Source: USDA (1963).

Proteins

Since protein is the main component of meat (except for water), meat is generally considered to be a protein food. The myofibrillar proteins (myosin and actin) are the most abundant proteins of muscle. They are globulins and are soluble in concentrated salt solutions. Also present in muscle tissue, in lesser amounts, are the proteins of connective tissue (soluble in concentrated salt solutions) and the sarcoplasmic proteins (soluble in water or dilute salt solutions).

Myosin constitutes about 7.5% of the total muscle protein. This protein, in form, is slender and greatly elongated, the ratio of diameter to length being about 1:100. The molecular weight of myosin is approximately 500,000. It contains large amounts of glutamic acid, aspartic acid, and dibasic amino acids.

Actin constitutes approximately 2.5% of the total muscle protein. It can exist either as a monomer, G-actin, or as a polymer, F-actin. G-actin is globular in form and has a molecular weight of approximately 70,000. F-actin is formed by the end-to-end polymerization of G-actin units to form a chain. The formation of F-actin from G-actin occurs in the presence of salts and small amounts of ATP. The union of F-actin with myosin forms the contractile actomyosin of active muscle and the inextensible actomyosin of muscle during rigor mortis.

Collagen, another protein in meat, is an important component of connective tissue, tendons, and bones. This protein contains the highest amount of hydroxyproline of any common protein. Collagen also contains appreciable amounts of dibasic and diacidic amino acids and little or no tryptophan and cystine. When heated in water at 60°–70°C, the collagen fibers shorten in length. When the temperature is raised above 80°C, collagen is converted into water-soluble gelatin.

Reticulum is so closely associated with collagen that it is difficult to isolate from collagen. It is insoluble in concentrated salt solutions and does not yield gelatin on boiling.

Elastin occurs in ligaments. It differs from collagen in having only a small amount of hydroxyproline (1.6%), few polar amino acids, and much higher levels of glycine and alanine. Elastin contains a chromophoric group, which gives the protein its characteristic yellow color. It is not hydrolyzed on boiling with water.

Sarcoplasmic proteins include a protein fraction that can be extracted from muscle with cold water. This fraction includes a number of proteins, including myogen (5.6%), myoglobin (0.4%), and hemoglobin (0.04%).

Myogen is a very complex mixture of proteins, many of which are the enzymes essential for glycolysis. The majority of these proteins have been characterized, but the individual proteins will not be considered here.

Myoglobin is the principle pigment in the muscle cells responsible for the color of fresh and cooked meat. Myoglobin and hemoglobin function similarly to form complexes with molecular oxygen. There is, however, a difference in their metabolic function: myoglobin serves as a storage site for oxygen within the cells, while hemoglobin serves as the oxygen carrier of the blood.

Myoglobin, which is composed of one heme moiety and one protein moiety (globin), has a molecular weight of 17,000. In contrast, hemoglobin is composed of four heme moieties and one protein moiety, with a molecular weight of 68,000. The heme moiety is composed of an iron atom and a porphyrin. Myoglobin has a greater affinity for oxygen than hemoglobin which facilitates the transfer of oxygen from the blood capillaries to the cells.

Metmyoglobin and oxymyoglobin are formed when myoglobin is exposed to oxygen. The relative amounts of these two forms depend upon the partial pressure of oxygen, with metmyoglobin formation occurring at low oxygen pressures. The formation of oxymyoglobin occurs readily when myoglobin is exposed to oxygen. Oxymyoglobin has a very bright red color, whereas myoglobin has a more purplish red color. The bright red color of the surface of meat is due to the shift of myoglobin to oxymyoglobin in the presence of a plentiful supply of oxygen. In contrast, the interior of the meat typically has a dark purple color, because the myoglobin is in the reduced state. When the meat is cut and the resulting surface exposed to air, the color becomes bright red as oxymyoglobin is formed. Finally, when the tissue supply of oxidizable substrates is consumed, the reducing power of the muscle meat is lost and the iron of the heme pigment is oxidized to the brown metmyoglobin.

Intramuscular Fats

The fat of adipose tissue, a specialized type of connective tissue, generally consists almost entirely of mixed triglycerides. In contrast, intramuscular fat contains triglycerides, phospholipids (phosphoglycerides, plasmologens, and sphingomyelin), glycolipids, and nonsaponifiable components such as cholesterol and the vitamins A, D, E, and K. The amounts and kind of fatty acids present varies with the species of animal and to some extent on the feed of the animal. The fatty acid composition of some animal fats is given in Table 5.2. Oleic, palmitic, and stearic acids are the most common fatty acids found in the fat of meat animals. Also, in the majority of fats that have been examined by partial degradation with esterase, saturated fatty acids were found preferentially in the terminal positions of the glycerol molecule and unsaturated fatty acids were found preferentially in the C-2 position of the glycerol molecule.

Some Commercial Meat Products

Meat is legally defined as the properly dressed flesh derived from cattle, swine, sheep, or goats sufficiently mature and in good health at the time of slaughter, including the part of the muscle of any cattle, sheep, swine, or goat which is skeletal or which is found in the tongue, in the diaphragm, in the heart, or in the esophagus, with or without the accompanying and overlying fat, and the portions of bone, skin, sinew, nerve, and blood vessels which normally accompany the muscle tissue and which are not separated from it in the process of dressing. Meat does not include the muscle found in the lips, snout, or ears. *Fresh meat* is meat that has undergone no substantial change in character since the time of slaughter. *Meat by-products* are any part, other than meat, derived from cattle, sheep, swine, or goats and capable of use as human food. *Meat food product* is any substance capable of use as human food that is made wholly or in part from any meat or other portion of the carcass of any cattle, sheep, swine, or goats except those exempted from definition as a meat food product, those that contain meat or other portions of the carcass only in relatively small proportion, or those historically considered by consumers as products of the meat food industry. However, the meat or other carcass portions in products exempted from definition must not be adulterated, and such articles must not be represented as meat food products. *Prepared meat* is the product obtained by subjecting meat to comminuting, drying, curing, smoking, cooking, seasoning, flavoring, or any combination of such processes. The definitions and standards of identity for common prepared meat products are summarized in the following paragraphs.

Cured meat is the product obtained by subjecting meat to a process of salting, by the use of common salt, with or without the use of one or more of the following: salts of nitrites and nitrates, sugar, dextrose, syrup, honey, or spice. The ingredients used in a dry cure can also be dissolved in water to form a brine cure. The meat can either be immersed in the curing brine or the brine can be injected into the meat with needles (e.g., artery-pumped, stitch-pumped).

Dry salt meat is prepared meat that has been cured by the application of a dry cure consisting of common salt, with or without the use of one or more of the following: sugar, dextrose, a syrup, honey, or sodium nitrate and sodium nitrite.

Corned beef is made by soaking meat (usually brisket, plate, or round cut) in a solution of common salt, with or without one or more of the following substances: sugar, dextrose, invert sugar, honey, corn syrup, solids, sodium nitrate, sodium nitrite, potassium nitrate, potassium nitrite, flavorings, or spices. The salt solution may be injected into the meat, but does not have to be.

Smoked meat is prepared by subjecting meat (fresh, dried, cured) to the action of smoke produced by the slow combustion of hardwoods or sawdust derived from hardwoods.

Canned meat is fresh meat or prepared meat, packed in hermetically sealed containers, with or without subsequent heating for the purpose of sterilization.

Sausage meat is prepared with fresh meat or frozen meat or both and does not include by-products. It is sometimes comminuted and can be sold in bulk or packed in casings.

Sausages are defined as coarse or finely comminuted meat food products prepared from one or more kinds of meat, containing various amounts of water, that are usually seasoned with condiments and frequently are cured.

Some of the products included in this group are veal loaf, peppered loaf, summer sausage, bologna, liver sausage, souse, pickle and pimento loaf, and braunschweiger.

Sausages may contain binders and extenders such as cereal, vegetable starch, soy flour, soybean concentrate, isolated soy protein, nonfat dry milk, or dried milk. The finished product shall not contain more than 3.5% of these additives. Water may be added, to facilitate chopping or mixing, in an amount not to exceed 3.5% except that frankfurters, vienna sausage, bologna, knockwurst and others may contain up to 10% water. Such products may contain raw or cooked poultry meat not in excess of 15% of the total ingredients. A list of ingredients, in order of their content by weight, must be shown on the package. Sausages can be processed in a variety of ways: cooked, cured, and smoked.

Grades of Meat

Meat grades are an indication of meat quality. A meat processor who wishes to sell graded meat must request and pay for grading, which is performed by U.S. Department of Agriculture personnel. The various grades established by the USDA are shown in Table 13.2 in order of descending quality. There are eight quality grades for beef, six for veal, four for pork, and five for lamb and yearling mutton. The top three grades—Prime, Choice, and Good—are the same for beef, veal, lamb, and mutton. The remaining grades for beef are Standard, Commercial, Utility, Cutter, and Canner; for veal, Standard, Utility, and Cull; for lamb and mutton, Utility and Cull. The pork grades are designated by

Table 13.2. USDA Quality Grades of Meat

Beef	Veal	Pork	Lamb	Mutton
Prime	Prime	No. 1	Prime	Prime
Choice	Choice	No. 2	Choice	Choice
Good	Good	No. 3	Good	Good
Standard	Standard	No. 4	Utility	Utility
Commercial	Utility		Cull	Cull
Utility	Cull			
Cutter				
Canner				

numbers. The lower grades—Utility, Cutter, Canner, and Cull (No. 4 for pork)—are generally used in meat products such as frankfurters.

☐ POULTRY

Most poultry (chickens, turkeys, ducks, and geese) is purchased ready to cook; that is, the feathers and viscera have been removed and the carcass washed. Poultry is sold whole or cut into pieces, chilled or frozen.

Chickens are by far the most common type of poultry in the United States. Chickens are classed by age, which determines the use for which they are most suitable. Broilers or fryers are usually 9–12 weeks old with a ready-to-cook weight of 1.5–2.5 lb; roasters are between 3 and 5 months of age and their ready-to-cook weight ranges from 2.5 to 4.5 lb; stewing chickens are mature hens or male chickens usually over 10 months of age with ready-to-cook weights of 2.5–5 lb.

Turkeys of excellent quality are now available. Turkey roasters are usually 5–6.5 months of age. Their weight varies according to the breed and sex, ranging from 5 to 24 lb.

The grading of poultry by the U.S. Department of Agriculture is an optional service. There are three grades: Grade A, attractive, well finished, full fleshed and meaty; Grade B, good table quality but slightly lacking in fleshing, meatiness, and finish; and Grade C, less meat in proportion to bone, poor fat covering, and crooked or misshapen bones.

Poultry by-products include the skin, fat, gizzard, heart, liver, or any combination thereof, of any poultry. However, the term should not be used in the labeling of other poultry and/or meat products in which poultry skin, fat, gizzard, hearts, or liver are used as ingredients.

Poultry and poultry products may be used in traditional meat products, but they cannot be used as a replacement in whole or part for the required meat content. The labels of products that contain poultry or poultry products in addition to the specified meat content must list poultry as an ingredient. If a meat product, for which a minimum meat content is specified, does not meet the standard because poultry or poultry products have replaced some of the required meat, the product must bear a descriptive label (e.g., beef and chicken stew).

Composition of Poultry

The proximate analysis and calorie contents of several of the food birds is shown in Table 13.3. Young poultry (fryers) are lower in fat

Table 13.3. Proximate Analysis and Energy Values of Edible Portions of Poultry

Type (class)	Water (%)	Protein (%)	Fat (%)	Ash (%)	Energy (cal)
Chicken (fryer)	76	19	5	0.8	124
Chicken (roaster)	63	18	18	0.9	239
Chicken (hens & cocks)	57	17	25	0.9	298
Duck	54	16	29	1.0	326
Goose	51	16	32	0.9	354
Turkey (all classes)	64	20	15	1.0	218

Source: USDA (1963).

and, therefore, in calories than most meats (cf. Table 13.1). Older birds (roasters) contain more fat than fryers, but much of it is located under the skin where it can easily be removed if desired. Ducks and geese are higher in fat than either chickens or turkeys. Turkey meat is similar to chicken meat in composition.

☐ ANALYSIS OF POULTRY AND MEAT PRODUCTS

Various analyses are useful in determining the quality of meat and meat products and to follow significant changes that take place during their processing and storage. Too, chemical analyses are required for determining conformity with governmental regulations.

The type of analysis required depends upon the nature of the problem under investigation. For example, proximate analyses to determine the major components of a product are common. This kind of analysis provides basic information for ascertaining conformity to certain governmental regulations. However, such determinations are of little value in showing the stability of a product.

Proper interpretation of the results obtained from a chemical analysis is as important as the method *per se*. Some of the factors to be considered are the nature of the product to be analyzed, the normal variability of the product under investigation, and the accuracy of the method used.

Preliminary Treatment and Preparation of Sample

It is extremely difficult to obtain a representative sample from many meat products, and as a consequence, there are no general rules that one can use for sampling all products. Sampling procedures may be found in AOAC 24.001 and Laboratory Methods of the Meat Industry (AMI 1954).

The analysis of a meat sample may involve the whole sample, including skin, bones, and fat in addition to lean meat; or it may involve only the edible portions of the sample; or only the lean meat. If the entire sample is to be analyzed, the bones are first removed, pulverized, and added back to the rest of the sample, which is then ground in a meat grinder to a homogeneous state. For an analysis of the edible portion only, the bone and other edible parts are removed as completely as possible from the sample. The sample is then passed rapidly through a meat grinder, three or more times, mixing thoroughly after each grinding, until a homogeneous sample is obtained. For an analysis of the lean meat, all bone, skin, and visible fat is removed, after which the meat is passed through a meat grinder. Proper grinding and mixing of the sample is mandatory; for most analyses, it is impossible to grind a sample too fine or mix it too thoroughly.

Samples must be prepared rapidly to avoid moisture loss and chemical changes that may occur. It is desirable to maintain a temperature below 7°C, if possible, during the grinding and mixing of a sample in order to minimize detrimental changes to the sample. Following sample preparation, the sample should be placed in a vaportight container (sample must completely fill the container) and then stored at a refrigerator temperature of 4.5°C or less until analyses can be made.

If prepared samples must be transported an appreciable distance before analyses are made, they should first be frozen and then shipped frozen with dry ice. Likewise, if long periods of storage are necessary before analysis, the sample should be frozen in order to prevent spoilage. Before an aliquot of the sample is removed for analysis, the entire sample is allowed to reach room temperature; then the container is opened, and the contents remixed to a homogeneous composition.

Moisture Determination

Weigh accurately a sample containing 3–5 g of prepared meat sample into an aluminum moisture dish approximately 55 mm in diameter and 25 mm deep, and equipped with a cover. When all samples have been weighed, place the dishes in the oven with lids slightly ajar. Dry to a constant weight (approximately 5 hr) at 95°–100°C, under a pressure not to exceed 100 mm Hg. At the end of the heating period, remove dishes from oven, close the lids, and cool in a desiccator. Weigh the dishes as soon as they reach room temperature, and calculate the percentage moisture from the loss in weight.

Crude Fat Determination

Weigh 3–4 g of prepared meat sample into a thimble containing a small amount of sand, then mix the contents with a glass rod. Place the thimble, with rod, in a beaker and dry for 6 hr at 100°–102°C, or for 1.5 hr at 125°C. Remove the sample from the oven, cool in a desiccator, and reweigh. If desired, the percentage of moisture in the sample may be calculated from the loss in weight.

Place the thimble in a Soxhlet extraction apparatus containing approximately 75 ml of anhydrous ether and extract for 16 hr (2–3 drops/sec). At the end of the extraction period, remove the thimble and dry for 30 min at 100°C. Remove, cool in a desiccator, and reweigh. The loss in weight due to extraction is the fat content. Calculate the percentage of fat (AOAC 24.005). The ethyl ether must be anhydrous and the sample to be extracted must be free of moisture; otherwise, water-soluble components will be extracted and reported as fat. Samples should be dried at temperatures below 125°C in order to avoid changes that may interfere with the extraction of the fat.

Crude Protein Determination

Weigh a piece of qualitative filter paper (9 cm) and record the weight as a tare. Accurately weigh approximately 2 g of the prepared sample onto the filter paper.

Close the paper around the sample and drop it into a Kjeldahl flask. Carry out the Kjeldahl determination of crude protein as described in Chapter 2. Multiply the percentage of nitrogen determined by the factor 6.25 to convert it to protein.

Ash Determination

Accurately weigh portions of the prepared sample (approximately 5 g lean meat or 6 g fat meat) into previously ignited and weighed porcelain crucibles. Heat the crucibles in an oven at 100°C until all moisture is volatilized. Add a few drops of olive oil, then heat slowly over a burner until the swelling ceases. Place the crucible and contents in a muffle furnace held at a temperature of 525°C, and allow to remain until all evidence of carbon disappears. Remove the crucible containing the ash, allow to cool, and then moisten the residue with distilled water. Place the crucible and its contents on a steam bath and evaporate to dryness. Return the crucible and dried residue to the muffle furnace and re-ash at 525°C to constant weight. Calculate the percentage ash in the sample.

Determination of Added Water

Federal regulations permit the addition of water to processed meat products. The permitted levels of added moisture are as follows: fresh sausage, 3%; cooked sausage, 10%; "water-added" cured and smoked and/or cooked pork products, 10%; chopped ham, 3%; and luncheon meat, 3%. Water is added to facilitate grinding, chopping, and mixing or to make the product palatable.

For several years the standard procedure for determining if water had been added to meat products assumed that meat normally contained four times as much water as protein. Therefore, multiply the percentage of protein by four (normal protein multiplier) and subtract the resulting product from the percentage of water; any remainder is "added water." The protein (P) and water percentages (W) are determined by the usual procedures, as described already.

More recently this formula has been questioned and it was concluded that specific protein multipliers must be used to estimate the amount of added water. The formula recommended is as follows:

$$\text{Added water} = W - (P \times \text{Specific protein multiplier})$$

Examples of specific protein multipliers are smoked ham, 3.79; canned ham, 3.83; and pork shoulder picnic, 3.93. It is suggested that a protein multiplier of 3.8 will give a good estimate of added water. It should be emphasized that total protein must be corrected if protein had been added in the formulation.

Determination of Curing Ingredients

The process for curing meat products involves the use of salt, nitrate, and/or nitrite. An analysis of these ingredients often is necessary to determine whether or not a product meets certain governmental regulatory requirements. Cereal flours, milk powder, and curing adjuncts are also utilized in the production of prepared meat products. Usually these adjuncts are present in small amounts and an analysis for these substances is sometimes required both in quality control and/or meeting government specifications.

Sodium Chloride

In this procedure, a weighed sample is treated with an amount of silver nitrate to react with all the chloride present and also leave an excess of the standard silver nitrate. The organic matter is oxidized by heating with nitric acid and potassium permanganate. The precipitated silver chloride is prevented from further reaction by the addition of ether. The excess silver nitrate is then titrated with potassium thiocyanate, using ferric alum as an indicator.

Reagents

1. 0.5 *N* Silver nitrate—Weigh out 85.0 g of silver nitrate (equiv. wt. 169.87), dissolve and dilute to 1 liter volume with distilled water.
2. 0.1 *N* Potassium thiocyanate—Weigh out 9.718 g of potassium thiocyanate (equiv. wt. 97.18), dissolve, and dilute to 1 liter volume with distilled water.
3. Ferric alum indicator (saturated solution of $FeNH_4(SO_4)_2 \cdot 12H_2O$.

Procedure

Weigh 2.5–3 g of prepared sample into a 300-ml Erlenmeyer flask. Add a measured volume of standard 0.5 *N* silver nitrate solution that is in excess of the volume required to react with the chloride present in the sample (5 ml or more will be required, depending on the chloride content of the sample). Add 15 ml of concentrated nitric acid and boil until the sample is digested (usually about 10 min). Add small portions of a saturated solution of potassium permanganate, boiling under a hood after each addition until the permanganate color disappears and the solution becomes colorless or nearly so. Add 25 ml distilled water, boil for 5 min, cool, and dilute to a volume of approximately 150 ml with distilled water.

Add 25 ml of ether, and shake. Then add 5 ml of ferric alum indicator solution and titrate the excess silver nitrate with 0.1 *N* potassium thiocyanate solution to a permanent light brown end point. Titrate in a similar fashion the amount of 0.5 *N* silver nitrate used to precipitate the sample. The difference between these titrations is equivalent to the amount of salt in the sample.

$$\%\ \text{NaCl} = (\text{Blank titration} - \text{Sample titration}) \times N \text{ of KSCN} \times \frac{58}{1000} \times \frac{100}{\text{Sample wt}}$$

Nitrite

A maximum concentration of 200 ppm of sodium nitrite is permitted in finished sausage products.

The determination of nitrites in cured meats is relatively simple. A water extract of the sample is prepared, then reacted with a color-developing reagent, and the resulting color measured spectrophotometrically. The nitrite concentration is determined by reference to a standard curve previously prepared with solutions containing known amounts of nitrite (AOAC 24.044).

Reagents

1. NED reagent—Dissolve 0.2 g of *N*-(1-naphthyl) ethylenediamine in 150 ml of a 15% solution of acetic acid. Filter, if necessary, and store in a glass-stoppered brown glass bottle.
2. Sulfanilamide reagent—Dissolve 0.5 g sulfanilamide in 150 ml of a 15% solution of acetic acid. Filter, if necessary, and store in a glass-stoppered brown glass bottle.
3. Standard sodium nitrite solutions—Prepare stock solution (1000 ppm) by dissolving 1.000 g sodium nitrite in 1 liter. Dilute 100 ml of stock solution to 1 liter with distilled water. Dilute 10 ml of the intermediate solution to 1 liter with distilled water to give working solution (1 ppm sodium nitrite).

Preparation of Standard Curve

Take suitable aliquots of the nitrite working solution (10, 20, 30, and 40 ml), place in a series of 50-ml volumetric flasks, add 2.5 ml of the sulfanilamide reagent, and mix. Allow to stand for 5 min, then add 2.5 ml of the NED reagent, dilute to volume, mix, and let color develop for 15 min. Measure the absorbance in a 1-cm cuvette at 540 nm against a blank of 45 ml water, 2.5 ml of sulfanilamide reagent, and 25 ml of NED reagent. Plot the absorbance values against the concentrations of sodium nitrite.

Assay Procedure

Weigh 5 g of a finely comminuted and thoroughly mixed meat sample into a 50-ml beaker. Add approximately 40 ml water that has been heated to 80°C. Mix thoroughly to break up all clumps, and quantitatively transfer to a 500-ml volumetric flask. Add all washings to the volumetric flask and adjust the volume (with 80°C water) to approximately 300 ml. Transfer the flask to a steam bath, and let stand for 2 hr, with occasional shaking. Cool to room temperature, dilute to volume with water, and remix. Filter through a No. 42 Whatman paper, transfer an aliquot (containing 5–50 ppm sodium nitrite) to a 50-ml volumetric flask, mix, and allow to stand for 5 min. Add 2.5 ml NED reagent, then dilute to volume, and mix. After 15 min, measure the absorbances at 540 nm. Determine the concentration of sodium nitrite from the standard curve and then calculate the nitrite content of the sample, taking dilutions into account.

Nonfat Dry Milk Determination

Federal regulations permit the addition of not more than 10.5% of binders and extenders to meat food products. Included in the list of binders are nonfat dry milk and dried milk. Additions of up to 2% corn syrup solids and/or dextrose are allowed. In this method, the amount of added nonfat dry milk or dried milk is determined by analyzing the product for its lactose content. This procedure (AOAC 24.072) involves dissolving the sugars, precipitating the proteins, and then obtaining a filtrate to ferment all sugars except lactose. Lactose is then determined by a titration procedure. Nonfat milk solids can be determined from the lactose value because the lactose content of milk is relatively constant. This procedure can be used in the presence of maltose.

Reagents

1. Yeast medium—Dissolve each of the following in a small amount of water: 1 g anhydrous magnesium sulfate, 2.0 g ammonium chloride, 1.0 g anhydrous dibasic potassium phosphate, 0.5 g potassium chloride, 0.02 g ferrous sulfate, 0.7 g peptone, and 20.0 g technical maltose. Add these solutions, in the order given to an Erlenmeyer flask containing approximately 500 ml of water. Dilute to 1 liter, warm filter, bring filtrate to a boil, and let cool to room temperature.
2. Acclimated yeast suspension—Suspend two cakes of bakers yeast in about 50 ml of water. Collect the cells by centrifugation and wash twice with 50 ml portions of water. Let cool to room temperature. Add the washed yeast to the medium and incubate for 24 hr at 30°C, stirring frequently during the first few hours. Separate yeast by decanting and centrifugation, wash twice with water, and add to a fresh liter of the medium. Incubate for an additional 24 hr, then separate cells from medium, wash thoroughly with water, and dilute to 100 ml. Store in refrigerator at 4°C and shake well before using.
3. Benedict's solution—Dissolve 16 g copper sulfate in 125–150

ml water. Dissolve 150 g sodium citrate, 130 g anhydrous sodium carbonate, and 10 g sodium bicarbonate in approximately 650 ml hot water. Combine the two solutions, cool, dilute to 1 liter, and filter.

4. Lactose standard solution—Dissolve 1.5789 g lactose in water and dilute to 1 liter to give a solution of 1.5 mg lactose per ml.
5. Iodine standard solution—Mix 5.08 g iodine with 10.2 g potassium iodide, dissolve in a small volume of water, filter, and dilute to 1 liter.
6. Sodium thiosulfate standard solution—Dissolve 9.92 g sodium thiosulfate in water and dilute to 1 liter.
7. Dilute acetic acid solution—Dilute 240 ml acetic acid to 1 liter with water.
8. Dilute phosphoric acid solution—Dilute 240 ml phosphoric acid to 1 liter with water.
9. Citric acid–phosphate buffer (pH 4.8)—Mix 10.14 ml of 0.1 *M* citric acid (19.21 g/liter) and 9.86 ml of 0.2 *M* dibasic sodium phosphate (28.4 g anhydrous/liter), and adjust to pH 4.8, using a pH meter. Store in refrigerator and discard if solution becomes turbid.
10. Starch solution—Solubilize 2.5 g soluble starch and about 10 mg mercuric chloride in 500 ml boiling water.
11. Dilute hydrochloric acid—One volume concentrated HCl + four volumes water.
12. Phosphotungstic acid—20% w/v.

Procedure

Place a 10-g sample in a 100-ml volumetric sugar flask, add 50 ml distilled water, and stir or shake to break up any lumps. Warm on a steam bath for 30 min. Cool to room temperature, add 20 ml dilute HCl and dilute to volume using bottom of the fat layer as the meniscus. Add 5 ml of 20% phosphotungstic acid solution, mix well, and let stand for a few minutes. Filter through a moistened filter paper. Pipette 40 ml of the filtrate into a 50-ml volumetric flask and neutralize to pH 4.8–5.0 (acid side of chlorophenol red indicator). Dilute to volume and mix.

Transfer about 40 ml of this solution to a 100 ml centrifuge tube to which 5 ml of yeast suspension has been added and from which water has been separated. Mix yeast and sample well and incubate for 3 hr at 30°C, stirring frequently. Centrifuge and determine lactose.

Pipette 10 ml of the clear supernatant into a 300 ml Erlenmeyer flask, add 20 ml of Benedict solution, bring to a boil in 3–5 min, and boil for exactly 3 min. Remove from heat, cool rapidly, add 100 ml water, and then 10 ml of the dilute acetic acid solution slowly while swirling. Add an excess (15–20 ml) of the standard iodine solution, and agitate to dissolve the cuprous oxide. Let flask stand for exactly 5 min, add 20 ml of the dilute phosphoric acid solution, and titrate the excess iodine with standard sodium thiosulfate solution, using starch solution as an indicator.

Determine iodine:sodium thiosulfate ratio by using 10 ml of distilled water, adding 20 ml of Benedict's solution, and carrying through the determination described in the preceding paragraph.

$$I_2:Na_2S_2O_3 \text{ ratio} = \frac{\text{Volume } I_2 \text{ (ml)}}{\text{ml } Na_2S_2O_3} = A$$

Determine lactose:iodine ratio by using 10 ml of the standard lactose solution, add 20 ml of Benedict's solution, and carrying through the determination as above.

$$\text{Lactose:Iodine} = \frac{15 \text{ mg lactose}}{\text{ml } I_2 - (\text{ml } Na_2S_2O_3)(A)} = B$$

Calculations

$$\% \text{ Lactose} = 100 \, [\text{ml } I_2 \text{ added to flask} - (A)(\text{ml } Na_2S_2O_3 \text{ for back titration})][B]/C$$

where A and B are ratios defined in the preceding equations and C = mg of sample in aliquot (consider the volume of the original sample solution as 200 ml to correct for volume occupied by sample).

$$\% \text{ Nonfat dry milk or dry milk} = (\text{Lactose} \times 2) - \text{Correction}$$

where Correction is 0.4% in absence of corn syrup or corn syrup solids, and 0.8% in the presence of corn syrup or corn syrup solids.

☐ SEAFOODS

The two major groups of fish are the finfish and the shellfish. Finfish are marketed in various forms, including whole or round fish, drawn fish (internal organs removed), dressed fish (scales, head, tail, and fins removed), steaks (cross-sectional slices from large dressed fish), and fillets (sides of a fish cut lengthwise away from backbone). Shellfish can be subdivided into a group having a soft body protected by a shell (oysters, clams, scallops), and another group that has a segmented crustlike shell (lobsters, shrimp, crabs).

Table 13.4. Proximate Analysis and Energy Values of Fish

Type	Water (%)	Protein (%)	Fat (%)	Ash (%)	Energy (cal)
Bass, black	79	19	1.2	1.2	93
Catfish, freshwater	78	18	3.0	1.3	103
Cod	81	18	0.3	1.2	78
Flounder	81	17	0.8	1.2	79
Haddock	81	18	0.1	1.4	79
Halibut	77	21	1.2	1.4	100
Mackerel	67	19	12.2	1.6	191
Perch	76	19	4.0	1.2	118
Salmon	64	23	13.4	1.4	217
Shad	70	19	10.0	1.3	170
Trout, brook	78	19	2.0	1.2	101
Tuna, bluefin	71	25	4.1	1.3	145

Source: USDA (1963).

Most of the fish consumed in the United States are either fresh, frozen, or canned. A relatively small percentage of the harvest is consumed in the cured form, but in many countries cured fish are consumed in large quantities.

The U.S. Department of the Interior, through the Bureau of Commercial Fisheries, makes available an inspection service for all types of the processed fishery products—fresh, frozen, canned, and cured. It is offered upon a fee-for-service basis.

Grade standards have been established by the Bureau of Commercial Fisheries, U.S. Department of the Interior, for 15 processed fishery products. These grades are Grade A, B, C, and substandard of which Grade A is the most widely sold. Grade A products are of best quality, are uniform in size, practically free from blemishes and defects, and possess a good flavor. Grade B products are of good quality, but they are not as uniform in size or as free from blemishes or defects as Grade A products. Grade B is the general commercial grade. Grade C products are of good quality; they are just as wholesome and maybe as nutritious as higher-grade products (see Table 13.4).

☐ ANALYSIS OF SEAFOODS

Preliminary Treatment and Preparation of Sample

To prevent loss of moisture during preparation and subsequent handling, use as large a sample as practical. Keep ground material in a container with an airtight cover. Begin all analytical determinations as soon as practical. If any delay occurs, chill the sample to inhibit decomposition.

Fresh Fish

Clean, scale, and eviscerate fish in the usual way. In the case of small fish (6 in. long or less), use 5–10 whole fish. For large fish, cut from each of at least three fish, three cross-sectional slices, 1 in. thick: one slice first aft of pectoral fins, one slice halfway between first slice and vent, and one slice just aft of the vent. (Skin and bones may be removed if desired.) For fat and fat-soluble components include the skin since many fish store large amounts of fat directly beneath the skin.

Pass the sample rapidly through a meat chopper, three times. Remove unground material from the chopper after each grinding and mix thoroughly with the ground material. The meat chopper should have holes as small as practical ($\frac{1}{16}$ and $\frac{1}{8}$ in. in diam.) and should not leak around the handle end. As an alternative procedure for small fish, a high-speed blender may be used. Blend several minutes, stopping the blender frequently to scrape down the sides of the cup.

Canned Fish, Shellfish, and Other Marine Products

Blend the entire contents of the can in a blender or pass through a meat chopper three times.

For products packed in oil, sauce, broth, or water, drain 2 min on a No. 8 sieve before blending or grinding. The oil and brine may be analyzed separately or may be reincorporated with the solids.

For fish packed in salt or brine, drain the brine and rinse off adhering salt crystals with saturated salt solution. Drain again after 2 min, and then proceed as for fresh fish. Dried smoked or dried salt fish are prepared in the same manner as fresh fish.

Shellfish Other than Oysters, Clams, and Scallops

If shellfish are received in the shell, wash the shells with water to remove all loose salt and dirt; drain well. Separate the edible portions in the usual way and prepare samples for analysis as described for canned fish, shellfish, and other marine products.

Shell Oysters, Shell Clams, and Scallops

Wash shellfish with water, then drain well. Shuck enough into a clean dry container to yield about 1 pint of drained meats. Transfer the meats to a skimmer (a flat-bottomed metal pan or tray with 2 in. sides, if not less than 300 in^2 in area, with perforations 0.25 in. in diameter arranged 1.25 in. apart in a square pattern), pick out pieces of shell, drain 2 min on the skimmer, and proceed as described for canned fish, shellfish, and other marine products.

Total Solids Determination

Quickly, yet accurately, weigh 5.0–7.0 g of homogenized sample into a covered, weighed flat-bottom aluminum moisture dish (about 9 cm in diameter). Spread samples evenly over the bottom of the dish. Then *either* (1) evaporate just to dryness on a steam bath and dry 3 hr in an oven at 100°C; or (2) insert directly into a preheated forced-draft oven at full draft, and dry 1.5 hr at 100°C. Cover the dish, cool in a desiccator, and weigh promptly.

Ash

Accurately weigh portions of the prepared sample (*ca.* 5 g) into tared porcelain crucibles and proceed as described for the ash determination of poultry and meat products, using a temperature of about 550°C. The difference in weight before and after ashing equals total solids.

Crude Fat Determination

Accurately weigh 8 g of a well-mixed sample into a 50-ml beaker and add 2 ml of concentrated hydrochloric acid. Using a stirring rod, break up coagulated lumps until the mixture is homogeneous. Add an additional 6 ml of concentrated hydrochloric acid and mix, cover with a watch glass and heat on a steam bath for 90 min, stirring occasionally with the stirring rod. Cool the solution and transfer to a Mojonnier fat-extraction tube. Rinse the beaker and rod with 7 ml of alcohol, transfer the alcohol to the Mojonnier tube and mix. Rinse the beaker with 25 ml of ethyl ether, added in three portions; add rinsings to the Mojonnier tube. Stopper the tube and shake vigorously for 1 min. Add 25 ml of petroleum ether (b.p. 30°–60°C) to the extraction tube and repeat the shaking. Let stand until upper liquid is virtually clear or centrifuge 20 min at about 600 rpm.

Decant as much as possible of the ether-fat solution through a filter, consisting of a pledget of cotton (packed just firmly enough in the stem of the funnel to permit the ether to pass freely) into a weighed 125 ml beaker or flask. (This container must previously have been dried and tared.)

Re-extract the liquid remaining in the extraction tube, each time using only 15 ml of each ether. Shake well on the addition of each ether. Decant the clear ether solutions through the filter into the flask as before. Wash the tip of the extraction tube, funnel, and tip of stem with a few ml of equal-volume mixture of the two ethers. Evaporate the combined extracts slowly on a steam bath, then dry the fat in a 100° oven to constant weight (about 30 min) and weigh. (Owing to the size of the flask and the nature of the material, cool sample in air rather than in a desiccator.) Correct this weight by a blank determination on the reagents. Report as percentage fat by acid hydrolysis.

Sodium Chloride Determination

The method for determining sodium chloride in seafoods is similar to that described earlier for poultry and meat products. The sample is treated first with an excess of silver nitrate, followed by the concentrated nitric acid, then wet-ashed, and then the excess silver nitrate back-titrated with potassium thiocyanate.

Reagents

1. 0.1 *N* Silver nitrate—Weigh out 17.0 g of silver nitrate (equivalent weight, 169.87), dissolve in distilled water, and dilute to 1 liter. Standardize against 0.1 *N* sodium chloride (5.845 g/l) according to this procedure (Volhard).
2. 0.1 *N* Potassium thiocyanate—Weigh out 9.718 g of potassium thiocyanate (equivalent weight, 97.18), dissolve in distilled water, and dilute to 1 liter. Check strength of this solution as follows: pipette 25 ml of 0.1 *N* silver nitrate solution into an Erlenmeyer flask, add 80 ml of distilled water, 15 ml of aqueous nitric acid (v/v, 1:1), and 2 ml of the ferric alum indicator. Titrate with potassium thiocyanate to the end point. The ratio of the volumes should be 1:1.
3. Ferric alum indicator (saturated solution of $FeNH_4(SO_4)_2 \cdot 12 H_2O$).

Procedure

Weigh 10 g of the sample into a 250-ml Erlenmeyer flask. Add 25.0 ml of 0.1 *N* silver nitrate solution, swirl sample and solution to insure mixing, and then add 20 ml of concentrated nitric acid. Add boiling chips and boil gently on a hot plate until sample dissolves. Add potassium permanganate (5%) in small amounts; continue boiling until color disappears. Add 25 ml of water, boil for 5 min, cool, and dilute to approximately 150 ml with water. Add 5 ml ether, 2 ml ferric alum indicator, and shake to coagulate the precipitated silver chloride. Titrate the excess silver nitrate with the standard potassium cyanate solution to the permanent end point (light brown).

Calculations

$$\%\text{NaCl} = \frac{(\text{ml } 0.1\ N\ \text{AgNO}_3 - \text{ml KSCN})(0.1\ N)(5.85)}{\text{Sample weight}}$$

or

$$\%\text{NaCl} = \frac{(\text{ml Blank} - \text{ml Sample})(N\ \text{KSCN})(5.85)}{\text{Sample weight}}$$

Determination of Trimethylamine Nitrogen

The spoilage of fish is due to enzymatic and bacterial action, which results in the production of various volatile compounds such as trimethylamine (TMA), dimethylamine (DMA), ammonia, and volatile acids.

In this colorimetric procedure (Dyer 1945) a sample containing trimethylamine is made alkaline with potassium carbonate in the presence of formaldehyde, then extracted with toluene and reacted with a toluene solution of picric acid to form trimethylamine picrate. The yellow color is measured at 410 nm. The test is generally used as a means for differentiating between acceptable and nonacceptable saltwater fish.

Reagents

1. Trichloroacetic acid (7.5% aqueous solution)—Can cause severe burns to skin and respiratory tract. Use rubber gloves, eye protection, and effective fume removal device to remove vapors generated.
2. Toluene—Shake with 1 *N* sulfuric acid, distill, and dry over anhydrous sodium sulfate.
3. Picric acid solutions—Dissolve 2 g dry picric acid in 100 ml of dry toluene; then dilute 1 ml of this stock solution to 100 ml with dry toluene to give working solution.
4. Potassium carbonate solution—Dissolve 100 g potassium carbonate in 100 ml water.
5. Formaldehyde solution—Shake 1 liter formalin (40%) with 100 g magnesium carbonate, and filter. Dilute 10 ml to 100 ml with water.

6. Trimethylamine (TMA) standard solution—(1) To 0.682 grams of trimethylamine hydrochloride, add 1 ml concentrated hydrochloric acid and dilute to 100 ml with water to give stock solution. Determine the basic N content of 5-ml aliquots by adding 6 ml of 10% NaOH and distilling into 10 ml of 4% boric acid in a micro-Kjeldahl distillation apparatus; titrate the distillate with 0.1 *N* sulfuric acid solution, using methyl red–methylene blue indicator. Add 1 ml of this stock TMA solution to 1 ml concentrated hydrochloric acid and dilute to 100 ml with water to give working solution (0.01 mg TMA-nitrogen/ml).

Procedure

Weigh 100 g minced or chopped, well-mixed sample. Add 200 ml 7.5% trichloroacetic acid and blend. Centrifuge blended solution at 2000–3000 rpm until supernatant is practically clear.

Pipette an aliquot (preferably containing 0.01–0.03 mg TMA-N) into a 20 × 150 mm Pyrex glass-stoppered test tube and dilute to 4.0 ml with water. For standards, use 1.0, 2.0, and 3.0 ml of the TMA working standard solution, diluting to 4.0 ml with water; for blank, use 4.0 ml water. Add 1 ml of the formaldehyde solution, 10 ml toluene from automatic pipette, and 3 ml potassium carbonate solution. Stopper tube and shake vigorously by hand approximately 40 times. Pipette 7–9 ml of the toluene layer into a small test tube containing approximately 0.1 g anhydrous sodium sulfate. Avoid removing droplets of aqueous layer. Stopper the tube and shake well to dry toluene.

Pipette 5 ml of the toluene extract into a dry colorimeter tube. Add 5 ml picric acid solution and mix by swirling gently. Determine the absorbance at 410 nm of the sample and TMA standards against a water blank subjected to the entire procedure. The color should be stable. If the original sample aliquot contains more than 0.03 mg TMA-N, dilute the supernatant obtained in the centrifugation step with trichloroacetic acid and repeat the determination.

Calculations

Calculate the TMA-N content as follows:

$$\text{mg TMA-N/100 g sample} = \frac{A}{A_1} \times C_1 \times V_1 \times 300$$

where A and A_1 are absorbance of sample and standard, respectively; C_1 is mg TMA-N/ml standard solution; and V_1 is volume of standard solution. This calculation is based on a 1-ml sample aliquot. If a dilution is made or a larger aliquot is used, an appropriate correction must be made. In the calculations, use the A_1 of the standard solution nearest to the A of the sample.

Comments

Do not use stopcock grease; a mixture of sugar and glycerol ground together may be used if necessary. Do not wash the tubes with soap or other detergent. Rinse them with water and occasionally clean them with nitric acid.

Determination of Total Volatile Bases

Spoilage of fish results in the formation of volatile bases; in particular, trimethylamine (TMA) and ammonia. TMA, which originates from the reduction of trimethylamine oxide (TMAO), comprises about 95% of the total while ammonia arises from protein breakdown. Freshwater fish contains no TMAO so that the volatile bases formed during fish spoilage consist primarily of ammonia. Consequently, total volatile bases (TVBs) can be used as an index of fish spoilage for both freshwater and saltwater fish while the TMA method can be used as an index of fish spoilage in saltwater fish.

The following procedure is based on the method proposed by Lücke and Geidel (1935). The volatile bases are isolated from the sample by distillation, the distillate is trapped in a boric acid solution, and the volatiles measured by titration with a standard acid.

Procedure

Set up a Kjeldahl distillation apparatus with a 500- to 700-ml receiving flask. Weigh 10 g of minced or chopped, well-mixed fish. Add 100 ml water, blend, and quantitatively transfer into the distilling flask with 200 ml water. Add 2 g magnesium oxide and an antifoaming agent (e.g., octyl alcohol). Add 25 ml 2% boric acid to the receiving flask and a few drops of methyl red indicator. Connect the apparatus making sure the receiver tube is below the surface of the boric acid solution. Heat the distilling flask so that the liquid boils in exactly 10 min and distill for exactly 25 min at the same rate of heating. Wash down the condenser with distilled water, and titrate the distillate with 0.1 N sulfuric acid. Calculate the TVB as mg N per 100 g of sample as follows:

$$\text{TVB} = \frac{(\text{ml Acid} - \text{ml Blank}) \times \text{Normality} \times 14 \times 100}{\text{g Sample}}$$

☐ SELECTED REFERENCES

AMERICAN MEAT INSTITUTE FOUNDATION. 1960. The Science of Meat Products. W.H. Freeman and Co. San Francisco.

AMI. 1954. Laboratory Methods of the Meat Industry. Amer. Meat Institute, Chicago.

AOAC. 1984. Official Methods of Analysis. 14th ed. S. Williams (editor). Assoc. of Official Analytical Chemists. Arlington, VA.

DE HOLL, J. C. Encyclopedia of Labeling Meat and Meat Products. Library of Congress Card No. 74-80935. Meat Plant Magazine, St. Louis.

DYER, W. J. 1945. Amines in Fish Muscle, 1. Colorimetric Determination of Trimethylamine as the Picrate Salt. J. Fish. Res. Bd. Can. *6*, 351–358.

FOODS AND DRUGS. 1982. Code of Federal Regulations, Title 21, Parts 100–199.

HART, L. F. and FISHER, H. J. 1971. Modern Food Analysis. Springer-Verlag, New York.

LAWRIE, R. A. 1974. Meat Science. 2nd ed. Pergamon Press, New York.

LÜCKE, F. and GEIDEL, W. 1935. Bestimmung des flüchtigen basischen Stickstoff in Fischen als masstab für ihren Frischezustand. Z. Untersuch. Lebensm. *70*, 441–458.

ROMANS, J. R., JONES, K. W., COSTELLO, W. J., CARLSON, C. W. and ZIEGLER, P. T. 1977. The Meat We Eat. 13th ed. The Interstate Printers and Publishers, Inc. Danville, IL.

USDA. 1963. Composition of Foods, Agriculture Handbook No. 8, Agriculture Research Service, U.S. Dept. of Agriculture. Washington, DC.

USDA. 1980. Food Safety and Quality Service, Meat and Poultry Inspection Program. U.S. Dept. of Agriculture. Washington, DC.

USDA. 1984. Chemistry Laboratory Guidebook, Food Safety and Inspection Service, U.S. Dept. of Agriculture. Washington, DC.

14

Food Deterioration, Preservation, and Contamination

INTRODUCTION

Since prehistorical times, humans have sought to keep excess food obtained in times of plenty for use during periods when obtaining food was difficult. However, because of its chemical composition, food is subject to various deteriorative processes. Thus, of necessity, man learned early to preserve food, i.e., treat it so as to prevent or retard its natural deterioration. During the past few hundred years, the development of increasingly effective preservation methods, coupled with the advent of efficient means of transportation, has permitted the growth of towns and cities. The food supply for today's towns and cities is no longer dependent on production from the immediate surrounding area but rather is drawn from all regions of a country, and to some extent even from foreign countries. Without effective food preservation methods and adequate transportation modern urban life would be impossible.

During the past 20–30 years, the problem of food contamination has received considerable attention. Although some contaminants may occur naturally in foods, there is no question that increasing amounts of potentially hazardous chemicals are entering the environment every year and that some of these find their way into the food supply, and ultimately into living organisms including man. In this regard, those in the food industry are most concerned about materials used in the production of food (e.g., insecticides, fumigants, growth regulators, antibiotics) residues of which may be present in food products when they are eaten by consumers. Another troubling possibility is contamination of food by heavy metals (especially lead and mercury) and various industrial chemicals many of which are toxic at relatively low concentrations. Such materials if present in the air, soil, or water eventually may enter food crops, fish and shellfish used for food, and livestock.

☐ FOOD DETERIORATION

Foods may spoil through the action of several agents: (1) yeasts, molds, and fungi; (2) bacteria; (3) enzymes; (4) oxidation; (5) absorption of odors; and (6) insect and animal contamination. The character of the food material will determine largely which type of spoilage occurs. An acid condition is favorable to yeasts and molds but unfavorable to bacteria. Thus, fruit products and acid vegetables are susceptible primarily to yeast fermentation or mold growth, and nonacid vegetables and meats are susceptible to the action of bacteria. All food products are susceptible to enzymatic activity. Oxidative deterioration occurs in foods that contain organic compounds with unsaturated linkages (e.g., fats, pigments, flavoring materials, and essential oils). Fatty foods also readily absorb odors from their surroundings. Most food products are subject to insect and animal contamination. A proper evaluation of these factors involved in food spoilage is essential to the proper processing, transportation, storage, handling, and merchandising of food products.

Spoilage of Sugar Products

Soluble carbohydrates are very susceptible to attack by microorganisms because they are so readily available as food. In general, microorganisms commonly ferment sugars by the following mechanisms:

1. *Complete oxidation*, in which sugar is completely oxidized to carbon dioxide and water. Most molds, a number of bacteria, and a few yeasts will do this.
2. *Partial oxidation*, a more common type of oxidation, in which intermediate compounds are formed that can be further oxidized to carbon dioxide and water by other organisms and in some instances by the same organism.

3. *Alcoholic fermentation*, brought about mainly through the action of yeasts on sugars, although bacteria and molds produce some alcohol.
4. *Lactic fermentation* brought about by a number of organisms widely distributed in nature. These hardy organisms have the ability to withstand adverse conditions that inhibit and eliminate many other organisms.
5. *Gassy fermentation*, occurring as commonly in foods as lactic fermentation and produced by a great variety of bacteria and yeasts. In this fermentation, sugars are broken down to acids, alcohols, and carbon dioxide or hydrogen gases.
6. *Butyric fermentation*, produced principally by anaerobic organisms. It is undesirable in foods, since it produces off-flavors. In addition to butyric acid, other acids, alcohols, carbon dioxide, and hydrogen gases are produced.

In general, then, carbohydrate foods deteriorate through fermentative changes involving the production of acids (particularly lactic acid), alcohols, and gases (carbon dioxide and hydrogen). If a carbohydrate food is in a closed container, the acidity may build up to a degree that it suppresses any proteolytic activity and the only noticeable change in the food is its high acid content. The acidity will eventually increase to the point where it interferes with the acid-tolerant bacteria. If air has free access to the food, however, the acids produced through fermentation may be removed. Molds can grow in highly acid media and, in the presence of air, can destroy or remove acids. Thus in the presence of carbohydrates, bacteria cause only acid fermentations with inhibition of proteolytic decomposition. Molds, on the other hand, show considerable proteolytic activity even though sugars and acids are present.

Spoilage of Cereal Grain Product

The cereal grain itself is subject to several types of spoilage. For example, if grain with a high moisture content is placed in storage bins, the grain will "heat," causing the kernels to acquire off-odors and colors, and the quality of the protein will be impaired. Molding may also occur if the product is too moist. Insects and rodents will attack whole grain, as well as flour and cereal products, at every opportunity.

Bread may spoil in several ways. Through drying out, the bread becomes stale. In the summer months, bread may deteriorate through the action of molds and bacteria. Bread as it comes from the oven is sterile on its surface; however, in preparing sliced bread, which is now the common practice, new surfaces are produced that can be infected by mold. The rope organism, *Bacillus mesentericus*, is a spore-former, and the spores survive the baking temperature in the interior of the loaf. Bread containing these spores will, under favorable conditions of moisture and temperature, develop a yellow-brown, sticky crumb with an offensive odor. The name *rope* is derived from the characteristic tiny, silklike strands produced when the infected crumb is pulled apart. Fortunately, rope bacteria are harmless when eaten, so that infected bread may be eaten if the bacteria have not developed to the point where they affect the palatability of the bread.

Relatively dry cereal products (crackers, cookies, breakfast cereals, etc.) may undergo oxidative deterioration in storage.

Spoilage of Fats and Fat Products

Fats deteriorate through atmospheric oxidation, enzymatic activity, mold growth, and flavor reversion. Fats and fat products may also deteriorate through the absorption of odors from their environment (e.g., the acquisition of a paint flavor if they are in a room that is being painted).

Spoilage of Milk Products

Liquid milk and milk products may deteriorate through lactic acid fermentation or souring. Fat deterioration may cause the development of various off-flavors (cardboard, tallowy, oxidized, rancid, etc.). Cooked flavors may result in products that have been subjected to high heat treatments. Dried dairy products may undergo fat deterioration as well as the browning reaction.

Butter may develop hydrolytic or oxidative rancidity the same as any other fat product. Molds may also grow on the surfaces of butter. Fishy butter is produced by the decomposition of lecithin to give trimethylamine. This type of spoilage is catalyzed by the presence of salt, acidity, and metals in butter, as well as the overworking of butter.

Spoilage of Fresh Fruits and Vegetables

Insect infestation may occur in the growing produce (maggots in blueberries, corn-ear worm, etc.) and so be present in the food when harvested. Mold may also occur in some produce when weather conditions are favorable. The mold count is incorporated in the standards for tomato products, since it gives an indication of the quality of tomatoes used in the product.

Enzymatic changes occur in fruits and vegetables during storage. Sweet corn, when freshly picked, contains appreciable sugar in the kernels; in a matter of a few hours the sugar has disappeared. Potatoes held in storage will have a higher sugar content at low temperatures than at higher temperatures. Consequently, potatoes to be used for chip manufacture are placed at a higher temperature for a period of time previous to chipping to reduce the sugar content to a minimum. A high sugar content in potatoes produces a dark-colored chip. Fruits and vegetables that are to be preserved by freezing or dehydration are subjected to blanching, which immediately inactivates the enzymes present.

In the manufacture of tomato products (tomato juice, ketchup, tomato puree, etc.), it is virtually impossible to prevent some contamination of the product with mold. The mold grows in cracks in the tomato and also in the internal portions of the tomato, so that it is not eliminated in the cleaning and sorting operations. Inasmuch as this fact is recognized by industry and the government, certain tolerances have been established for the amount of mold permitted in these products. Climatic conditions influence the prevalence of mold. Wet, humid weather when the tomatoes are ripening is conducive to mold growth and so aggravates processing problems.

Spoilage of Meat, Poultry, and Eggs

Primarily protein foods are subject to putrefaction rather than fermentation. Bacterial putrefaction in meat and poultry, which is induced by moisture and heat, begins at the surface and penetrates into the deeper parts. As the putrefaction develops, offensive smelling products are produced. Molds will grow on the surfaces of meats if the temperature is conducive. Normally, molds can be wiped off without leaving any bad effects except perhaps a musty odor.

Eggs in storage deteriorate in quality due to several factors:

1. *Loss of water*—Because egg shells are porous, eggs lose water in storage to an extent depending on the temperature and other storage conditions. This loss may amount to 3.5% in 6 months. As a result of this drying of egg contents, the air cell increases in size.
2. *Thinning of white and yolk*—A loss of carbon dioxide by the egg, leads to thinning of the white and yolk. Fresh egg whites have

a pH of approximately 7.6 and yolks about 6.0. As soon as the egg is laid, it begins to lose carbon dioxide, which, in turn, causes an increase in the pH. The egg white becomes more watery and water passes from the white to the yolk on storage. All these changes may be retarded by preventing the escape of carbon dioxide.

3. *Off-flavors*—Cold-storage flavors develop in eggs after the seventh month in storage. Eggs with a high pH develop more undesirable flavors than those with a low pH. Packing materials around the eggs vary in the amount of flavor they transmit to eggs.

Infestation by Insects

One of the big problems in food production, storage, processing, and marketing is the control of insect pests. It has been estimated that 2000–5000 different kinds of insects cause economic farm losses in the United States amounting to more than $3 billion annually. The farmer, however, is not the only one to suffer. Those who process and market foods also incur tremendous economic losses through insect depredation.

No material of vegetable or animal origin is completely free from insect manifestation; thus food manufacturers may obtain foodstuffs contaminated with or despoiled by insects as well as face contamination by insects in the processing plant. Food merchandisers may face losses resulting from growth and development of insects from insect eggs in products or further contamination by insects before products reach consumers. Actually, the greatest damage to food by insects is not necessarily the amount they consume but that which is made unfit or undesirable by the presence of the insects.

People do not want to eat food upon which insects have been living. The presence of insects in a particular food product may prejudice people against that and other products manufactured and distributed by the same company, thereby causing loss of good will. This damage to good will may represent a serious economic loss when one considers the money expended by a company in developing and advertising a product. Thus, it is to the advantage of all segments of the food industry to exercise close control over insect pests.

The control of insect pests is not a simple task. Insects possess characteristics (e.g., ready adaptation to environment, rapid multiplication, short period of infancy, flight in many species, ability to conceal themselves, and varied feeding habits) that make their complete control virtually impossible. Changes in food preparation and marketing procedures have made insect control more difficult in recent years. For example, the widespread transportation of foods has contributed to the spread of insects. Likewise, the rather recent policy of the government and private individuals of holding certain foodstuffs for long periods for economic purposes has created an additional insect hazard.

FOOD PRESERVATION METHODS

The preservation of foods is intended to reduce or inhibit microbial growth, enzymatic activity, and atmospheric oxidation. These factors responsible for food spoilage may be controlled by food processing procedures or by chemical additives (food additives). Only a relatively few chemical additives are used in food preservation and their use must comply with FDA regulations under the Food Additives Amendment to the Food, Drug, and Cosmetic Act (Chapter 1).

Removal of Water

One of the main methods for preserving food is the removal of water, usually by drying, dehydration, or evaporation. Although the terms are sometimes used interchangeably, *drying* normally refers to sun drying or air drying, *evaporation* refers to drying by the application of artificial heat but under natural draft, and *dehydration* implies drying by the controlled application of artificial heat with mechanical circulation of air or under a vacuum.

The objective of all three drying procedures is the same, i.e., the removal of water without injuring the food material with respect to its use. To accomplish this is not simple. Volatile components, responsible for food flavor, may be volatilized and lost. Easily oxidizable materials may undergo a drastic oxidation, catalyzed by the heat requird for drying, with the production of obnoxious off-flavors or substances, which will undergo deterioration in storage. Food pigments, in general, may be destroyed completely or converted into other products having undesirable colors during drying. In addition, such common food components as sugars may undergo caramelization, producing an off-colored product. Of great importance is the effect of the drying method on the proteins present in the food. Protein denaturation caused by heat will influence rehydration of a dried product and its subsequent palatability or commercial use. For example, the protein of dried wheat may be altered to such an extent that flour produced from the wheat will no longer produce acceptable bread. The vitamin content of a food is, in general, reduced by drying. This is especially true for those vitamins (e.g., vitamin A and ascorbic acid) that are easily destroyed by heat in the presence of air, through oxidation. The sulfuring of fruit previous to drying aids in reducing the amount of ascorbic acid destroyed during drying. In general, the proper drying of a food requires thorough knowledge by the operator of the nature of the material to be dried.

Sun drying is limited to fruits and has been practiced in California and the West. To a limited extent it has been used in the East in "Pennsylvania Dutch" homes for drying apple strips (apfel schnitz). Because sun drying is dependent upon weather conditions, which are rather unpredictable, this form of drying has been largely superseded by machine drying or dehydration except in those areas where the weather is particularly favorable.

Evaporation has been applied principally to the concentration of liquids. Examples of this type of drying are the concentration, with or without vacuum, of sugar solutions to syrups and of fluid milk to so called "evaporated" milk. In the strict sense, spray drying and roller drying of milk also are forms of evaporation. Spray-dried milk is made by forcing milk through a fine aperture under pressure to produce a spray that is directed into a heated chamber; roller-dried milk is made by causing milk to flow in films over hot rolls.

Dehydration methods utilizing heat and forced draft or vacuum were developed extensively during World War II. The necessity of sending large quantities of food to distant combat areas and holding the food in an edible condition for long periods of time under the peculiar climatic conditions of that area posed problems that partly were solved by dehydration. Since water constitutes 60–80% or more of such foods as vegetables, fruits, meat, and eggs, the reduction of the water content of these products by dehydration reduces the shipping tonnage and storage area tremendously. In addition, these products, when properly packaged, are not affected appreciably by various climatic conditions over considerable periods of time.

Various types of dehydrators are in use. In some, prepared food products are stocked on trays placed in humidity-controlled cabinets through which heated air is forced. A widely used type of equipment is the tunnel dehydrator. Food products in trays are racked on trucks, which are run through a heated tunnel in which humidity is also controlled. The time is so regulated that when the truck reaches the end of the tunnel the food has been properly dried.

Several methods for removing water do not involve the application of heat. For example, the use of a hydraulic press and centrifuge to remove water is common in sugar manufacture. The removal of water by freezing was used by the Indians and early settlers to concentrate maple sap solutions.

Dehydration or drying are effective as methods of preservation because bacteria must obtain their nutrients in soluble form and thus require the presence of water. The limiting moisture content for bacterial growth depends upon environmental factors, pH, and salt content. A moisture content of 15% or less in flour and cereal products prevents bacterial spoilage; a moisture content of 18–20% in dried meats inhibits bacterial growth. The limiting moisture content for bacterial inhibition in a foodstuff is greater if salts are present. Molds may grow on food products containing insufficient water for bacterial growth if the air surrounding the product contains sufficient moisture. Thus, humidity control is important in preventing mold growth on relatively dry products.

The quality of a dried product is determined largely by the quality of the fresh material and the drying technique. It should be emphasized that drying will not make an inferior food product any better, nor will it produce products indistinguishable from the fresh. In fact, dried foods should be considered as a distinct class of food products and be judged on their own merits, as canned foods are. The dried products derived from some foodstuffs have flavors and characteristics that are prized, e.g., dried corn and dried fruits, such as prunes, raisins, and peaches. In some cases, drying gives more convenient products (onion flakes, chili peppers, etc.). If drying produces a product not as acceptable as the corresponding fresh, frozen, or canned food, it should be used only in an emergency, such as war.

Foods to be dried should be of high quality. Fruits and vegetables should be harvested at the proper stage of maturity and properly processed. Fruits are frequently sulfured before drying to obtain a better color in the dried product and also to retain more of the vitamins. Vegetables are cleaned and prepared in the proper form (corn removed from the cob, peas shelled, beet or potato skins removed, etc.), blanched to inactivate the enzymes and "fix" or stabilize the chlorophyll in green products, and then dried. With all foods, drying should be done as rapidly as possible to eliminate deterioration by enzymes, microorganisms, and oxidation. Controlling the humidity increases the rate of drying and helps prevent the formation of a hard, dry layer around the product (case hardening), which makes the interior difficult to dry.

It should be emphasized that removal of water, by dehydration, drying, or evaporation is a tremendously important method of food preservation. All dry foods (flour, cereals, sugar, etc.) owe their keeping quality to a relatively low moisture content. It is conceivable that as more information is obtained on the production of dehydrated products acceptable to the public, more foodstuffs may be preserved by this method.

Refrigeration

The effectiveness of refrigeration as a preservation method requires maintaining food at a sufficiently low temperature (usually 0°–4°C) to retard microbial growth, enzymatic activity, and atmospheric oxidation. The length of time food can be preserved by refrigeration depends upon the nature and condition of the food.

Ice was used for many years in the home as well as commercially to accomplish refrigeration. In areas where ice was difficult to get, home refrigeration involved digging a shaft into the ground or covering a box with wet cloths, which were kept cool by evaporation of water. In more recent years electric refrigeration and dry ice (solid CO_2) have largely supplanted ice and other modes of refrigeration.

Freezing

Freezing employs a much lower temperature than refrigeration and, consequently, retards food deterioration to a greater extent. Slow freezing of products was practiced first and is still used for materials without cell structures (milk and other liquids); however, this freezing method is used only to a limited extent for fruits, vegetables, and meats. More common is quick freezing in which products are rapidly frozen preliminary to storage at freezing temperatures.

The quick-freezing technique has several advantages. Water is frozen rapidly in the cells, forming small ice crystals, which do not damage the cell structure. In contrast, during slow freezing, water is abstracted from the cells (dehydration of cells) and freezes into large crystals in the intercellular spaces, causing rupturing of the cell walls. On thawing a quick-frozen product, the water formed when the small ice crystals melt is quickly absorbed in the tissues, a condition quite different from that obtained with a slow-frozen product having large ice crystals and dehydrated, ruptured cells. Quick freezing also rapidly reduces the temperature of a product below that at which microorganisms can grow and enzymes can function. This is especially important in the freezing of fruits and vegetables.

Several types of freezing are commonly employed: (1) sharp freezing in which the product is placed in a space kept at −24°C to −34°C; (2) air-blast freezing in which warm air from the product is directed to cooling coils; (3) tunnel freezing (belt and tray freezing); (4) single-contact freezing; (5) double-contact freezing; (6) spray and fog freezing; (7) block-ice freezing; and (8) immersion freezing.

Freezing comes closer to preserving food in its natural state than other methods of preservation. To produce good frozen foods, however, it is essential to start with a high-quality product and to exercise great care in its preparation for freezing, as well as in the actual freezing, packaging, and storage. With fruits and vegetables, selection of the proper variety and harvesting it at the proper state of maturity are critical, as are prompt handling of perishable produce and proper preparation before rapid freezing, proper packaging, and storage. A preliminary scald or blanch is necessary for the successful freezing preservation of most vegetables. Blanching removes gases, saturates the tissues with water, inactivates enzymes, and minimizes discoloration. Packing fruits and vegetables in syrup or brine before freezing protects them from the air and enables them to retain their quality better.

Usually most microorganisms are killed by the temperatures used to freeze food products, but some may survive and grow. Microorganisms have been found in natural ices, hail, snow, frozen soil, frozen meat and fish, frozen vegetables and fruits, sherberts, ice cream, frozen milk, etc., in considerable numbers with their vitality apparently unaffected by the freezing treatment. Also some dangerous pathogens have been found to resist temperatures down to −252°C. For example, repeated freezings and thawings up to 40 times in rapid succession may fail to destroy completely the germs of thyphoid and cholera. Although *Clostridium botulinum* apparently does not grow below −0.6°C, living clostridium spores in artificially inoculated material can survive frozen storage.

Freezing does not inhibit fat oxidation completely and, if a pro-oxidant is present, deterioration in frozen products may be rapid. For example, the fat in pork sausage seasoned with salt deteriorates rapidly even when the product is frozen.

Heating

The application of heat to kill microorganisms and inactivate enzymes by denaturation is a very common form of food preservation. Such heat treatments are classified as heat sterilization or heat pasteurization.

Heat Sterilization

In strict terms, *sterilization* refers to the complete destruction of all living microorganisms. Since complete sterilization may be unattainable in some processed foods, the term *commercial sterilization* has been introduced in the canning industry. Commercially sterile canned foods may contain viable spores (e.g., those of thermophilic organisms) that do not develop under conditions of normal storage of the food. Some of these spores are so resistant that the heat required to destroy them would overcook the food and make it unsuitable for sale. In the dairy industry, the term sterilization has also been modified to mean the destruction of all milkborne pathogens and significant reduction in the numbers of all other bacteria.

Heat sterilization is of greatest importance in the canning industry. Effective application of heat sterilization necessitates a knowledge of the ease with which organisms are destroyed. Molds and yeasts are readily killed at 66°–82°C. Bacteria are more resistant, especially the spore formers. Bacteria present no particular problem in acid products, since acidity retards their development. In nonacid vegetables, however, it is necessary to heat above 100°C to achieve sterilization. This requires treatment with steam under pressure (5–15 psi at 108°–120°C) so that a temperature of 115°C is reached inside the can and maintained for several minutes. (A somewhat lower temperature with longer exposure period will give similar results.) The criterion applied to nonacid foods ($pH > 4.5$) is that the thermal processes shall be adequate to destroy *Clostridium botulinum*, which is the most resistant food-poisoning type.

Heat Pasteurization

The objective of heat pasteurization is not to achieve permanent preservation of foods, by destroying all life present, but rather to destroy certain species of organisms and also inactivate enzymes. Thus, heat pasteurization achieves partial, temporary preservation resulting from the destruction of pathogenic organisms.

Two types of pasteurization are commonly employed. Flash pasteurization involves heating the product to 71.7°C for a 15-sec period and then cooling it. In holding pasteurization, the product is held at a lower temperature (*ca.* 62°C) for a longer time (30 min) and then cooled.

Pasteurization is particularly important in the manufacture of dairy products, beer, and fruit juices. Pasteurization of milk destroys pathogenic organisms and reduces other microorganisms to a point where the milk will not deteriorate rapidly. Because pasteurized cream is used in making butter, a starter culture of lactic acid bacteria must be added. This procedure permits better control of the fermentation and results in a more uniform product than is possible by using unpasteurized cream and relying on bacteria naturally present. Beer and fruit juices are pasteurized to destroy organisms and enzymes that would alter the flavor, color, or physical characteristics of the product.

Exclusion of Air

Certain types of deterioration are prevented by excluding air from food products. This is achieved by various means depending upon the type of product involved. In the canning industry, products are sealed in a can and sterilized. Exclusion of air after sterilization prevents any further contamination by organisms. Frozen foods are blanched preparatory to freezing. The blanching drives air out of the tissues and thereby restricts oxidase systems from acting in the frozen product. Sausage and meat products may be packed in lard, which excludes air from the surfaces of the product and thereby prevents molding.

Exclusion of Light

Atmospheric oxidation of fats themselves and of products containing fats, esential oils, and pigments is retarded by the exclusion of light. This is true because blue and ultraviolet light materially accelerates the atmospheric oxidation of food products containing fats, although visible light, such as red and yellow, has very little effect. Light is most easily excluded by wrapping or packaging products in lightproof containers. However, this is not always feasible since it may be desirable to display a product for advertising purposes. In this event, a packaging material that excludes ultraviolet rays but permits the longer wavelengths of visible light to penetrate is advantageous. Wrappers or glass of any visible color except blue will serve satisfactorily. Yellow to orange transparent wrappers are preferable in that products wrapped in them appear to have their approximate natural colors.

Radiation

Both ultraviolet and ionizing radiations are used to a limited extent for food preservation. Although the idea of preserving foods by ionizing radiation is not new, the major research work in this field has been done since World War II. The increased interest in radiation as a preservation technique, and the accompanying research, was stimulated by the availability of radioactive isotopes and technological developments in ion-acceleration equipment. Despite more research in this area in recent years, there is still much to be learned about the use of ionizing irradiations as a means of food preservation.

Ultraviolet irradiation as a method of food preservation has found application in the meat, baking, beverage, dairy, fresh fruit and vegetable, and frozen food industries. Bakers, among the first to use this process, irradiate some of their ingredients (e.g., raisins) as well as their cooling, slicing, and wrapping rooms. Irradiation of a slicing machine, while it is operating, materially reduces mold contamination in the sliced bread. Another advantageous use of ultraviolet irradiation is in meat display cases to cut down spoilage on cut surfaces of meat. In the beverage industry ultraviolet irradiation is used to sterilize bottle caps and also the compressed air used in forcing beverages into bottles. Because ultraviolet light penetrates matter poorly, it is useful only for air and surface sterilization.

Ionizing radiations potentially useful for food preservation include X-rays, cathode rays, and beta and gamma rays from radioisotopes. Through the use of higher or lower dosages of radiation it is possible to totally sterilize a product or just "pasteurize" (i.e., partially decrease the microbial population). A distinct advantage in the use of ionizing radiations as a means of food preservation is that heat is not required (cold sterilization). For foods that either cannot be preserved by heat or are greatly altered by heat processing, a cold sterilization method is particularly beneficial.

All foods can be sterilized by radiation; however, most foods subjected to sterilizing doses are altered somewhat with respect to flavor, odor, color, or texture. In addition, a number of the vitamins are susceptible to irradiation, although their destruction in foods during irradiation is not too different from that experienced during heat processing. Many plastic materials used in packaging foods degrade under high dosages of ionizing radiation and the degradation products may impart an undesirable taste to the food. There is also the question of toxicity of irradiated foods. Extensive toxicity and feeding tests are being conducted to determine the wholesomeness of foods treated with ionizing radiations.

Considerable divergence of opinion exists about the future potential of ionizing radiations as a means for sterilizing foods. In addition to its undesirable effects on many foods and the question of possible toxicity, radiation sterilization is more expensive than more conventional preservation methods. Apparently, the most promising application of irradiation in the food industry is in the field of radiation "pasteurization," particularly for meats. At present there is no very effective method of preservation for fresh meats; thus radiation processing could well become the method of choice. Additional areas in which ionizing radiations would be useful are in controlling insect infestations and preventing sprouting of potatoes and onions in storage.

Sugaring

Sugar solutions of high concentration exert a high osmotic pressure and withdraw water from microorganisms, thereby preventing their growth. A concentration of 65% or more of sugar is required to inhibit the growth of microorganisms. However, even at this concentration, some strains of yeasts and molds can grow. Sugar syrups, honey, sweetened condensed milk, jam, and jellies are some of the food products preserved in this way. Obviously, this method of food preservation is limited to those products in which a large quantity of sugar can be used.

Aside from its preservative action against microorganisms, sugar may function as a preservative in other ways. Sugar is included in meat curing formulas to help protect the color of the meat. Likewise, fruits are frequently packed in a sugar syrup before freezing to aid in preventing discoloration on storage.

Salting

Salt functions as a preservative through its effect on osmotic pressure and the destructive effect of the chloride ion itself on microorganisms. Gram-negative bacteria are more sensitive to sodium chloride (concentrations of 8% or less inhibit growth) than gram-positive bacteria, which sometimes grow in a saturated solution. Salt can also retard enzymatic activity. The use of salt in high concentrations is again limited to a relatively few foods, mainly fish and meat products, in which its presence is not objectionable.

Salt is the main component of the formulas used for curing hams and bacon. It is used in the form of a brine (*pickle*) or applied in solid form to the outside of the meat. In meat curing, salt exerts a selective action in protecting the meat from the action of putrefactive organisms while permitting the nitrate-reducing organisms to functions.

Salt is also used in lower concentrations to control the growth of certain organisms, e.g., yeast in bread dough fermentation.

Smoking

Smoke formed by the incomplete combustion of hardwoods, sawdust, corncobs, or similar materials contains formaldehyde, acetaldehyde and other aldehydes, aliphatic acids from formic through caproic, primary and secondary alcohols, ketones, phenols, cresols, and resins. Although a number of these substances exert a preservative action against microorganisms, formaldehyde is believed to be the chief bacteriostatic and bactericidal agent.

Food products are usually smoked by subjecting them to the action of smoke in a closed room. In place of this, commercial smoked salt, containing the substances derived from smoke, may be applied directly to products. Meat, sausage, fish, poultry, and certain cheeses are the usual products that are smoked. The main reason, however, for smoking these products is to produce the desired smoked flavor. Few meat products are produced in which smoke constituents exert an important effect in protecting the product against microbial spoilage. The penetration of smoke constituents is limited to a few millimeters beneath the surface of the product, so that in sliced products the preservative effects of the smoke constituents becomes negligible. Since a smoked flavor is imparted to the products treated, smoking is limited to only those products in which the smoke flavor is desired, or at least not objectionable.

Spicing

Spices were once considered to have an appreciable preservative action due to the volatile oils they contained. This has been shown to be not entirely correct; in fact, untreated spices added to food have been found to contaminate it with large numbers of microorganisms. Workers at the Michigan Experiment Station found whole and ground untreated spices purchased on the open market to have a bacterial plate count ranging from 0 to 67,000,000 per gram. Ground cinnamon and cloves were the only spices that exhibited any inhibiting action on bacterial growth in low concentrations. Several spices showed inhibiting action against a majority of the bacteria tested on nutrient agar at relatively high concentrations, as follows: pepper and allspice (1%); mustard, mace, nutmeg, and ginger (5%); and celery (10–20%).

In general, then, spices do not act as preservatives in the concentrations used in foods. The historically popular belief that they functioned as preservatives stemmed, no doubt, from the ability of spices to mask undesirable food flavors and odors resulting from microbial spoilage or other causes. Again, the use of spices in foods is limited to those products in which the spice flavor is desired or not objectionable.

Pickling or Souring

The acidity associated with pickling or souring is unfavorable to bacterial growth and thereby acts as a preservative. Acidity may be permitted to develop naturally by fermentation in certain products (sauerkraut, naturally fermented pickles, sour milk and cream, etc.) or acid may be added directly to the product. Acetic, lactic, and citric acids are some of the more common acids in pickled foods.

Bacteria usually grow at pH values from 4.5 to 10, but not all bacteria can grow in such a wide pH range. Fermentative organisms are, in general, more acid-tolerant than proteolytic organisms. The interference of bacterial growth by acids is not entirely dependent upon pH, since organic acids (containing unionized acid molecules) interfere at higher pH values than inorganic acids, which are completely ionized. In this respect, citric, acetic, and propionic acids are more effective than lactic or pyruvic acid. The composition of the medium also influences the limiting pH value for bacterial growth in the presence of a particular acid, since salts of the acid, salts of other acids, and proteins exert some effect. In general, any influence that tends to lower the ionization of an organic acid without changing its concentration will make the acid more toxic. Molds and yeasts are normally more acid-tolerant than bacteria, and a number of them will grow at a pH as low as 1.5.

Nonacid vegetables can be sterilized safely at 100°C in acidified brines. This method has been used commercially for artichokes. Lemon juice is added to some nonacid canned baby foods ostensibly to increase the ascorbic acid content (vitamin C), but of perhaps greater interest to manufacturers is the easier sterilization possible in the presence of lemon juice. Although not common today, in the past vinegar was added to bread dough formulas to increase the acidity of the bread and thereby inhibit the bacterial growth that caused "ropy" bread.

Chemical Preservatives

The ideal chemical preservative should (1) materially extend the shelf life of a food product; (2) be safe for human consumption over extended periods of time when used in effective concentrations in food; (3) impart no undesirable odor, color, flavor, or texture to the food; (4) be capable of being detected and/or determined in foods by analysis;

(5) not deceive the consumer as to the quality of the food; and (6) provide an economical means of preservation. Most of the chemical preservatives that have been suggested or used at some time do not meet all these specifications.

Proponents for the use of chemical preservatives argue that they are harmless in small amounts, that some occur naturally in food products, and that some may be formed in the processing of food products. Opponents to the use of chemical preservatives state that such preservatives may interfere with enzymatic processes in the digestive track, irritate the digestive tract, interfere with the oxidizing power of the blood, be harmful to an ill person even if not to a healthy individual, and permit the use of inferior or spoiled products by unscrupulous individuals. The use of chemical preservatives in foods is regulated by the Food, Drug, and Cosmetic Act, as discussed in Chapter 1.

Benzoic Acid and Benzoates

Benzoic acid and benzoates are permitted in foods provided such products are labeled to contain the preservative and provided the quantity of preservative added does not exceed the amount specified by state or federal laws (most states require that sodium benzoate not exceed 0.1%). Sodium benzoate is a much more effective preservative in an acid medium than in a neutral one. For example, at neutrality, a concentration of approximately 4% sodium benzoate is required to prevent the growth of most fermentative microorganisms; at pH 2.3–2.4, only 0.02–0.03% is required, and at pH 3.5–4.0 (range of most fruit juices) only 0.06–0.1% is required.

The methyl and the propyl esters of *p*-hydroxy benzoic acid have been approved by the FDA for the preservation of foods. Their use has been approved in concentrations not to exceed 0.1% for effective activity against molds, yeast, and bacteria. They are said to be effective at acid, neutral, and alkaline pHs at which benzoate is ineffective.

Sulfur Dioxide, Sulfurous Acid, and Sulfoxide

Several sulfur-containing preservatives may be used in foods provided the foods are properly labeled. However, the addition of abnormal amounts of sulfur dioxide in order to market foods with an excessive moisture content or to conceal inferiority is considered an adulteration. Sulfur dioxide must not exceed 2000 mg/kg in dried fruits.

Sulfur fumes are used extensively in the treatment of fruits previous to drying. This treatment helps to prevent darkening of the fruit, hastens drying by plasmolyzing the fruit cells, aids in the preservation of vitamins A and C in the drying process, and prevents spoilage by inhibiting the growth of microorganisms. The sulfur dioxide content of dried fruits varies between a few hundred parts per million to 3000 ppm. The sulfur dioxide is gradually lost during storage of fruit (one-third to one-half the original amount is lost in 6 months). Similar losses of sulfur dioxide occur in soaking and cooking the fruit.

The addition of sulfites to meat is prohibited by law. When hemoglobin gives up its oxygen and takes on carbon dioxide, it turns bluish. The addition of sulfites to old meat causes the formation of a red SO-hemoglobin compound, which is comparable in appearance to the oxyhemoglobin of fresh meat.

Antibiotics

Antibiotices, substances produced by certain microorganisms that can inhibit others, have many of the qualities of an ideal food preservative. They do not impart undesirable flavors, odors, or colors to food, and they are considered relatively harmless to humans in the concentrations required for preservative action. Tetracycline antibiotic is permitted for fish preservation in the United States, with the permitted tolerance set at 5 ppm in raw fish. This antibiotic is frozen in block ice at concentrations of 5 ppm, and the treated ice, after flaking, is used to chill freshly caught fish aboard fishing vessels. The antibiotic may also be added to refrigerated brine in which fish are stored or used as a dip or spray for fresh fish. Chlortetracycline and oxytetracycline are permitted as preservatives for raw poultry. These antibiotics are added to ice water in which poultry carcasses are dipped. A tolerance of 7 ppm in any part of the raw bird has been established for these antibiotics. Cooking destroys these antibiotics, so that none is left when poultry is served.

Antibiotics have a number of limitations as food preservatives. Since they are *bacteriostatic* rather than *bactericidal* agents, they protect the food for a limited time only (a few days to a few weeks) and then the microflora break through the antibiotic cover. Many antibiotics are effective against only a limited spectrum of microorganisms, and even broad-spectrum antibiotics, which are effective against many bacteria, are ineffective against molds and yeasts. Furthermore, antibiotics are degraded and their effective concentration reduced in flesh products within a few weeks.

Organic Acids and Their Salts

Most of the fatty acids containing 1–14 carbon atoms possess bacteriostatic or fungistatic properties. The preservative effect of acetic and other acids produced in the natural fermentation of products has been discussed previously. The direct addition of acetic acid or vinegar to produce pickled products is also widely practiced.Propionic acid and either its calcium or sodium salts are effective preservatives in certain baked goods and dairy products. Salts of propionic acid are used commercially to prevent the rope organism from developing in bread and also as a preventative for mold growth in bread. Propionates are also used to protect cheese from surface mold growth.

Sorbic acid ($CH_3CH{=}CH{-}CH{=}CH{-}COOH$) is a good fungistatic agent. At proper concentrations, it inhibits the growth of molds, yeast, and some highly aerobic bacteria in foods. It is also used on wrappers for food products, such as cheese.

Antioxidants

As discussed in Chapter 5, Lipids, antioxidants are added to oils to delay the onset of oxidative rancidity. The addition of antioxidants is controlled by The Antioxidant in Food Regulations 1974, which permit the presence of antioxidants in various oils and fats up to certain limits.

FOOD CONTAMINANTS

Foods may be contaminated with a number of substances that are potentially toxic to humans or that otherwise make a food unfit or undesirable for consumption. Contaminants may occur naturally in foods or be added through the activities of man. Among the important contaminants are botulinum toxins, mycotoxins, aflatoxins, and unintentional food additives.

Botulinum Toxins

Botulism is a food-borne disease that afflicts man and several species of animals with a high fatality rate. The cause is the contamination of food materials by a specific group of spore-forming bacteria found in the soil and aquatic environments. The anaerobic growth of the organism results in the production of potent neurotoxins. Most warm-blooded animals are susceptible to the toxin.

Symptoms of botulism may be noted 2–4 hr after ingestion of food contaminated with the toxin. Typically the symptoms appear within 12–36 hr. Generally, the earlier the first symptoms appear, the greater likelihood of a fatal outcome. Nausea and vomiting usually begin at the onset of the disease. Vomiting can be an important way of reducing the toxin concentration available for absorption. The subsequent symptoms result from paralysis of various muscles related to the effects of the toxin at the neuromuscular junction. Double vision may occur early and is considered one of the most important pathological manifestations. Drooping eyelids, dilated pupils, and loss of the light reflex are usually observed. Difficulty in swallowing is an early symptom due to the paralyzing action of the toxin on the pharyngeal muscles. The tongue and throat become inflamed due to the patient's inability to swallow secretions. Neck muscles are weakened so that the patient cannot raise his head from a pillow. Often the mental faculties remain lucid until death, which may be attributed to respiratory paralysis. Death is likely to occur 3–6 days after eating the poisoned food. If recovery occurs, it is prolonged with considerable muscle weakness.

The rate of mortality in botulism varies according to the immunological type of toxin involved. Even though the disease is still likely to be fatal, early recognition and prompt administration of type-specific antiserum will increase the patient's survival chances.

The causative agent of botulism is the gram-positive, anaerobic bacillus *Clostridium botulinum.* These endospore-forming bacteria possess a heat-resistant form, which under suitable growth conditions germinate to form the rod-shaped vegetative form. The vegetative form is capable of active proliferation and exotoxin formation in substrates such as canned foods or other food products that have the required low oxygen tension. The endospore is usually not killed at 100°C (normal boiling point of water), whereas the rod-shaped vegetative bacillus is. Thus, in canning and similar food preservation methods, the enclosed food must be subject to temperatures above 100°C for several minutes to assure destruction of all botulinum spores present.

There are six types of botulinum bacilli, which are designated with the capital letters A through F. Almost all cases of human botulism in the United States have been caused by types A, B, and E. Each type produces a specific neurotoxin; these toxins differ in immunological properties but all cause the typical botulism symptoms. Thus, antisera must be prepared to each type if neutralization of the toxin is to occur. Therefore, effective disease treatment requires identification of the specific type of botulism. There is now available a polyvalent antiserum containing types A, B and E antitoxins.

Types A and B of *C. botulinum* are soil organisms. In the United States, type A has been found in states west of the Mississippi River and type B most frequently in the eastern states (as well as in Europe). Type E is found in lake, mud, or coastal sediment and therefore is likely to contaminate fish and other marine life. Type C and D are important disease agents in certain animals but appear not to infect humans. Type F is the most recently discovered, but only a few cases of human botulism caused by this type have been described.

Canned fruits and vegetables including beans, spinach, tomatoes, beets, and olives support botulinum toxin production. Cheese spreads, pot pies, smoked and canned fish, and meats also have been implicated. Since the 1930s, outbreaks of botulism have been caused by home-prepared foods. Complete inactivation of botulinum spores can be achieved if they are subjected to steam under pressure at temperatures up to 121°C for 10–15 min. Large volumes of food require relatively longer periods of heating. Acidic foods are rarely implicated in botulinum food poisoning since toxin formation usually is not possible at a pH below 4.5. However, once the toxin is formed, it is more stable at an acid pH.

Many times, the growth of *C. botulinum* in preserved foods and flavors results in objectionable odors or flavors. However, such unfavorable off-flavors may be slight and thus masked by natural or added food flavors. Any questionable foods should be heated uniformally at 100°C for at least 10 min before eating or even tasting. This procedure will destroy the toxin, but any spores that are present may still remain viable.

In addition to pH and temperature, the concentrations of salt and sugar in foods influence the growth of *C. botulinum*. For example, 10% sodium chloride and 50% sucrose prevent growth of *C. botulinum*.

The exotoxins of *C. botulinum* are simple proteins. Crystalline type A toxin has been shown to have a molecular weight of close to one million. This molecular weight value is based on both the toxic and hemolytic components. More recent measurements suggest that the toxin itself has a molecular weight of 150,000.

The botulinum toxins have been termed the most poisonous poisons. Only 2.0 μg of pure toxin would be lethal to humans. The principle toxic action of botulinum toxin is the blocking of neural transmission by inhibition of acetylcholine release. Paralysis of the muscles of respiration is generally the ultimate cause of death in cases of botulism.

Mycotoxins

The term *mycotoxin* refers to all toxic metabolites of the true fungi, Eumycetes. *Mycotoxicosis* is the general term used to describe the diseases caused by mycotoxins.

Ergotism, caused by the parasitic fungus *Claviceps purpurea*, is a well-known disease of this type. Historically, this disease was once called "St. Anthony's Fire." Ergot contamination of foods for human consumption seldom occurs in civilized countries today. The most recent notable outbreak in man was in 1951 in France. Considerable controversy exists today as to the cause of the 1951 outbreak. Although traced to a batch of commercial flour, there was also considerable evidence to indicate mercury contamination. It is possible that both were involved in the 1951 outbreak in France.

The term mycotoxin is usually restricted to the filamentous fungi called molds. In nature these molds are both parasitic and saprophytic and are widely distributed over the earth's surface. These organisms have sexual reproductive cells, but they also form more numerous asexual structures, which may find their way into everything that is contacted and ingested by man and animals.

The organoleptic changes that occur in food due to mold growth have been recognized for many centuries. Some of these changes are desirable, e.g., the pleasing flavor imparted to Roquefort cheese as a result of mold growth. On the other hand, molds may causes unpleasant odors and tastes and a change in texture of the foods. These changes are usually significant enough to prevent consumption by man and animals.

Moldy food should not be eaten. During World War II in Russia, grain that had been over-wintered in the fields because of lack of farm workers was shown to be responsible for a disease syndrome known as alimentary toxic alukia (ATA). This disease, called stachybotryotoxicosis, affected livestock, especially horses, that came in contact with hay containing the black fungus *Stachybotrys atra*. Farmers and veterinarians in the United States have recognized for several years that moldy feed can cause illness or death in livestock. Certain mold species can produce toxic substances in foodstuffs that cause conditions ranging from dermal necrosis to liver damage and body hemorrhages.

Aflatoxins

The *aflatoxins* include some 10 or more brightly fluorescing, furanocoumarin compounds of which aflatoxin B_1 is the prototype. They are hepatotoxic carcinogenic metabolites of *Aspergillus flavus*, which grows on many different foodstuffs when sufficient moisture is present.

The first aflatoxins were discovered in the early 1960s following a disease outbreak among poultry in England. In 1960, thousands of turkey poults died in that country with severe liver lesions. The disease was traced to substances contaminating the Brazilian groundnut meal used in formulating the poultry rations. A single feed manufacturing company was the common supplier of the peanut meal involved in the outbreak of the disease. Mycologists in England found dead fungal hyphae associated with the toxic meal, and in 1961 the common mold *Aspergillus flavus* was isolated from the toxic peanut meal. Using thin layer chromotography, scientists isolated at least four related compounds that caused acute toxicity and liver carcinogenicity in duckling feeding trials.

Foods naturally contaminated with aflatoxins include corn, barley, cassava, cottonseed meal, peanuts, peanut meal, peas, soybeans, rice, wheat, and sorghum seed. In the laboratory, aflatoxins have been produced on many foods by growing *Aspergillus flavus* on the food. Toxigenic isolates are components of the soil, air, seed, and forage microflora throughout the world. The contamination of peanuts occurs most often after the nuts are lifted from the ground during the drying period. Other factors favoring contamination and toxin production include damage by insects as well as improper drying and storage.

The aflatoxins form a group of highly oxygenated heterocyclic compounds with closely related structures. The structure of aflatoxin B_1 is represented as follows:

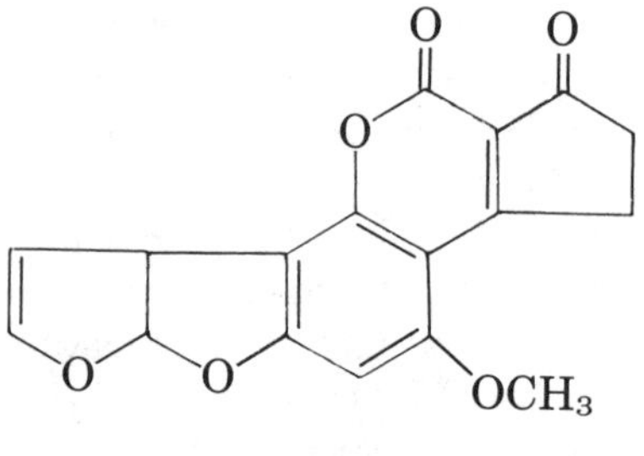

Aflatoxin B_1

Aflatoxin B_1 is the most potent natural hepatocarcinogen known. Studies have shown that approximately 10 μg aflatoxin B_1/day is sufficient to cause tumor induction in the liver of test animals. Low-level feeding of the toxin sensitizes the liver cells to additional toxic insults, resulting in a cancerous state. Other biological effects of aflatoxin have been investigated in plants and animals, tissue cultures, insects, and microorganisms. In a study in which crystalline preparations of aflatoxin B_1 were administered to rats, a rapid and marked inhibition of DNA and RNA synthesis in the liver by inhibition of the respective polymerases was observed.

Even though there is little direct evidence concerning the susceptibility of humans to aflatoxin, there is presumptive evidence that the high incidence of hepatitis in developing countries is associated with environmental exposure to the aflatoxins.

Unintentional Food Additives

As discussed in Chapter 1, unintentional food additives are materials that have no useful function in foods but are added due to some activities of man. Unintentional additives include materials added during the production, processing, or storage of food. Certain materials in the environment, such as radionuclides, dirty residues, and other substances found in water (especially heavy metals and industrial chemicals) may also be added to food unintentionally.

The unintentional food additives of greatest concern probably are the residues of insecticides, fumigants, and herbicides, used in the production of crops, that may be carried over into foods consumed by humans. The widespread use of these materials, the toxicity of some to humans and other organisms, and the persistence of some in the environment has stimulated considerable concern among both the general public and the scientific community. Antibiotics and other bacteriostatic and bacteriocidal compounds used in the production of livestock also may occur as residues in meat and poultry products.

Insecticide and Fumigant Residues

The use of insecticides for the control of insect pests frequently creates a residue problem in foods. A farmer normally applies an insecticide according to a definite spray or dusting schedule so as to obtain maximum protection of the crop with a minimum amount of residue left on the crop when it is harvested. Soon after harvesting certain crops may be washed to reduce the amount of insecticide residue. In some fruits (e.g., apples), the skin on which the residue would be retained is usually removed in processing operations. The FDA has the authority to establish residue tolerances of insecticides that may occur in particular food products. Fumigants may be absorbed by food products undergoing fumigation; however, they are normally dissipated by aeration or further processing of the product (boiling, heating, etc.) before consumption. If fumigant residues remain in a product, permitted tolerances are established as in the case of spray residues.

In addition to their use for insect control, chemical sprays, dusts, and fumigants are used to control impairment of foodstuffs by fungi, viruses, bacteria, mites, rodents, and weeds. Thus insecticides, fungicides, miticides, rodenticides, repellents, and herbicides are embraced in the general term *pesticide*, which has been defined as "a product, substance, or mixture of substances—gaseous, liquid, or solid—which may be used to destroy, prevent, control, repel, or mitigate any form of plant or animal life or viruses (except viruses, fungi, or bacteria on or in living man and other animals), and weeds." The use of any pesticide may result in a residue on foodstuffs. Consequently, the use of any pesticide in connection with the growing, processing, or marketing of foods is under governmental control. In general, the use of a given pesticide is permitted only on specified foodstuffs under specific conditions of application and with definite permitted tolerances (or no tolerances) for residues in the product (see Miller Pesticide Amendment in Chapter 1).

Carcinogens and Mutagens

Foods may contain a large number of potential carcinogenic and mutagenic agents. Precarcinogens such as aflatoxins have been shown to occur naturally in foods. Preparation of foods using heat treatment may produce both mutagenic and carcinogenic agents. Free radicals are formed at temperatures in excess of 475°C. Polycyclic aromatic hydrocarbons are formed when fat comes in contact with gas or electric heating elements. Smoke takes these compounds onto the food being cooked.

Many types of halogenated aromatic hydrocarbons have mutagenic and possibly carcinogenic effects on cells. Of greatest concern in the environment are occurrences of halogenated hydrocarbon pesticides and polyhalogenated biphenyls. Most notable neurotoxic pesticides are cyclodienes (aldrin, dieldrin, chlordane, and heptachlor), lindane, mirex, and DDT. These pesticides are noted for their environmental persistence, mainly in lipid tissues of animals, and concentrating toward the upper levels of natural food chains.

The industrial production of most of the halogenated hydrocarbons has ceased. Residues in food should decrease. However, imported meats, cheeses and other foods continue to contain halogenated hydrocarbon residues.

Polychlorinated biphenyls (PCBs) have been available commercially for 40 years. Since 1972, PCBs have been recognized to be of significant and potential toxicological concern. The largest use of PCBs have been in capacitors and transformers and in certain plasticizers. The possible routes into the atmosphere are vaporization and incineration of PCB-containing products.

Methods of transport of PCBs through the environment are complex: vaporized PCBs may be absorbed on particulates, transported by prevailing wind, and deposited in water or on land by sedimentations; PCBs introduced into water can be absorbed by particulates followed by diffusion into bottom sediment. The biota can assimilate, transport, and degrade the PCBs. Polychlorinated biphenyls have been shown to cause chromosomal aberrations and a high embryonic death in some organisms.

Bioassays for carcinogens attempt to estimate a MTD (maximum tolerated dose) which is the highest dose given during a chronic study. This dosage will not alter the longevity of the animal from effects other than cancer. From a practical standpoint, the MTD is the highest dose causing no more than 10% weight loss when compared to control animals. There are problems relating this data to humans. However, the Office of Technology Assessments has declared that all substances demonstrated to be carcinogenic in animals are regarded as potential carcinogens in humans. Short-term bioassays for mutagenesis and carcinogenesis include studies of chromosome aberrations in white blood cells, urine screening for mutagenic agents, exposure of cultured mammalian cells to mutagenic agents, and incubation of mutagenic agents with liver homogenates. Microbiological detection of mutagenesis involves the use of fungi, strains of bacteria, and bacteriophages. Using a specially developed strain of Salmonella the Ames Spot Test has been used to detect point mutations.

Detection of pesticides involves the use of gas–liquid chromatography which is presented in the analysis section of this chapter.

Trace Elements

Trace elements contaminating food include mercury and lead. Mercury is widely distributed in the biosphere and has been known as a toxic element associated with both ingestion and inhalation. Typical symptoms of subacute mercury poisoning may be salivation, stomatitis and diarrhea, or they may be primarily neurological, with Parkinsonian tremors, vertigo, irritability, moodiness, and depression. Oral intake of as little as 100 mg of mercuric chloride produces toxic symptoms, and a dose of 500 mg is almost always fatal unless immediate treatment is initiated. Levels of 0.005–0.035 ppm Hg in dairy products, fruits, cereal grains, meats, and vegetables have been reported with higher levels (0.020–0.18 ppm) in fish.

The increasing amounts of mercury injected into the environment has caused public health concern in some countries. This mercury comes from the burning of coal, the use of organic mercury compounds as pesticides and fungicides, and chemical industries. Methylated mercury compounds enter the food chain by way of microorganisms that methylate the mercury present in industrial wastes, and through the use of methylated mercury for protective treatment of grain.

Most human foods contain less than 1 ppm Pb when uncontaminated. Cow's milk normally contains 0.02–0.08 ppm Pb, and muscle meats about 0.1 ppm. Bones contain higher lead concentrations (5–20 ppm) because lead has a marked affinity for bone. Plant levels of lead are very low. Small amounts of lead are ingested with drinking water and are inhaled from the atmosphere and from cigarette smoke. Daily intake of lead from water supplies is 0.01 mg; from urban air, 0.026 mg; from rural air, 0.01 mg; and from cigarettes (30 per day), 0.024 mg. These levels of lead would probably not be toxic because lead is poorly absorbed in man and is excreted mainly in the feces. Symptoms of lead intoxication include abdominal colic, encephalopathy, myelopathy, and anemia.

☐ ANALYSIS

Mold in Tomato Juice

In the manufacture of tomato products, such as tomato juice, ketchup, tomato puree, etc., it is virtually impossible to prevent some contamination of the product with mold. The mold grows in cracks in the tomato and also in the internal portions of the tomato, so that it is not eliminated in the cleaning and sorting operations. Inasmuch as this fact is recognized by industry and the government, certain tolerances have been established for the amount of mold permitted in these products. Climatic conditions influence the prevalence of mold. Wet, humid weather when the tomatoes are ripening is conducive to mold growth and so aggravates processing problems.

Mold counts are routine tests in the industries making tomato products. Samples are taken at definite intervals to ensure that the product will meet government tolerances. These tolerances are subject to change from time to time but at present are tomato juice—20% of fields positive and tomato ketchup puree—40% of fields positive. The Howard mold counting chamber is most commonly used in this test. The technique employed when using this mold counting chamber on tomato products follows.

The diagram (Fig. 14.1) is a side view of the cell. The flat circular area A, the contact areas B, and the 33 × 33 mm cover glass are

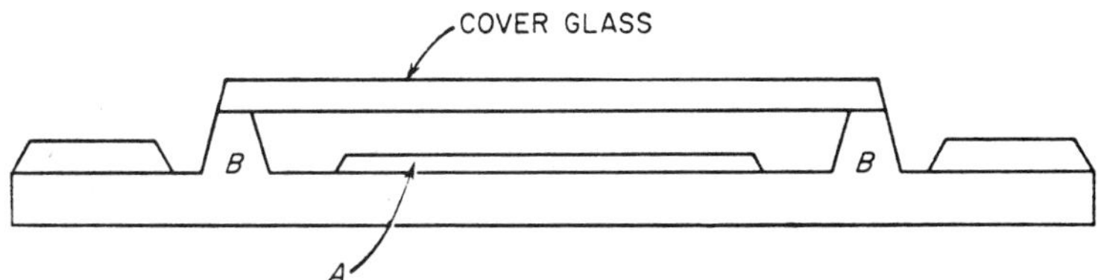

Figure 14.1. Side view of a Howard mold cell counting chamber.

optically worked. These areas should be thoroughly cleaned before using the cell. Cotton batting dipped in strong hydrochloric acid is used in cleaning the surfaces. Rinse thoroughly with distilled water, then with alcohol, and finally with acetone. All particles of lint or dust must be removed with a camel's hair brush.

Place a drop of the fluid to be tested on the circular disc A. Spread it evenly over the disc in order to avoid squeezing out the more liquid parts around the edge of the disc. Rest the cover glass against the edge of one of the contact areas B, and tilt it gently down onto the liquid. It is important to note that no liquid should be allowed to penetrate between the contact areas and the cover glass since this would interfere with the proper contact between these surfaces. The use of too much fluid is to be avoided for this reason.

The cell has been constructed to give a depth of 0.1 mm of fluid when the cover glass is in sufficiently close contact with the areas B so that Newton's rings may be observed. The rings appear between the surfaces in contact. These colored bands, or rings, are accurate criteria of the proper contact. Inability to obtain them is an indication of insufficient cleaning of the contact surfaces.

In making a mold count it is necessary to have a field of 1.5 mm^2 area; that is, the diameter of the field should be 1.382 mm. This may be obtained by using a 16-mm objective and a $10\times$ eyepiece, adjusting the draw-tube length to give the correct field size. For ready calibration a circle of 1.382 mm outside diameter is etched on the right-hand rail. Using the objective and eyepiece mentioned, the microscope is focused on the etched circle. The draw tube is then adjusted until the circumference of the microscope field coincides with the outside of the circle.

Dr. B. J. Howard gives the following instructions for mold counting:

> Observe each field as to the presence or absence of mold filaments and note the result as positive or negative. Examine at least 50 fields, prepared from two or more mounts. No field should be considered positive unless the aggregate length of the filaments present exceeds approximately one-sixth of the diameter of the field. Calculate the proportion of positive fields from the results of the examination

of all the observed fields and report as percentage of fields containing mold filaments. To assist in determining the aggregate length of the filaments a special eyepiece disc, the Howard Micrometer Disc, has been prepared. It is a glass disc engraved with a square of such size that the field is exactly inscribed in it. It is divided into sixths in both directions, forming 36 equal small squares. According to Dr. Howard's requisite for a positive field, then, the aggregate length of the filaments present should exceed the length of the side of a small square.

After use the cell should be rinsed with distilled water, then with alcohol, and finally with acetone.

Modified Mohler Test for Benzoic Acid

The use of benzoic acid as a preservative is permitted in foods provided such products are labeled to contain the preservative and provided the quantity of preservative added does not exceed the amount specified by state or federal laws. Sodium benzoate is a much more effective preservative in an acid medium than in a neutral. For example, at neutrality, approximately 4% of sodium benzoate was required to prevent the growth of most fermentative microorganisms, whereas at pH 2.3–2.4, only 0.02–0.03% was required, and at pH 3.5–4.0 (range of most fruit juices) only 0.06–0.1% was required.

Preparation of the Sample

Many nonalcoholic liquids may be used directly and extracted. Should such liquids develop troublesome emulsions during extraction, put 100 ml of the liquid in a 250-ml volumetric flask, add 5 g of sodium chloride, and shake until dissolved. Dilute to volume with ethanol, shake vigorously, and allow to stand 10 min with occasional shaking. Filter and treat filtrate in a manner similar to that for alcoholic liquids. For alcoholic liquids make 200 ml of the sample alkaline with 10% sodium hydroxide solution using litmus paper as indicator. Evaporate the solution on a steam bath to one-third its original volume, and dilute to 200 ml with water; filter if necessary. Grind solid or semisolid substances and mix thoroughly. Transfer 50–200 g of sample (depending on consistency) to a 500-ml volumetric flask and add water to make a volume of approximately 400 ml. Shake until the mixture is uniform, add 2–5 g of calcium chloride, and shake until dissolved. Make the solution distinctly alkaline to litmus paper with 10% NaOH solution. Dilute to the mark with water, mix thoroughly, and allow to stand for 2 hr or longer with occasional shaking, and filter.

Procedure

Dissolve the residue obtained by extraction in hot water, divide into two portions, and test the resulting solution for benzoic acid, as described in the next paragraph.

Add to the solution 1 or 2 drops of a 10% NaOH solution and evaporate to dryness. To this residue add 5–10 drops of concentrated sulfuric acid and a small crystal of potassium nitrate and heat for 10 min at 120°–130°C (do not exceed 130°C) in a glycol bath. Allow to cool, add 1 ml water, and make distinctly ammoniacal by addition of ammonia. Boil the solution to decompose any ammonium nitrite that might have been formed. Allow to cool and add a drop of fresh colorless ammonium sulfide solution without mixing. If benzoic acid is present, a red-brown ring is formed. When the solution is mixed, the color spreads throughout the liquid and, on heating, changes to greenish yellow. Salicylic and cinnamic acids form colored compounds which are not destroyed upon heating and so may be differentiated from benzoic.

Quantitative Determination of Benzoic Acid

To determine benzoic acid quantitatively, the bulk of the solid portion of the foodstuff is removed by filtration, leaving the benzoic acid in the remaining solution. The benzoic acid is then extracted from the solution with chloroform. This solvent, although not as good a solvent for benzoic acid as ether, is chosen for several reasons: (1) it dissolves only traces of mineral acids and other interfering substances; (2) it is nonflammable; and (3) it is heavier than water and so can readily be drawn off as the bottom layer in the extraction employing a separatory funnel. The quantitative extraction of benzoic acid is enhanced by saturating the solution with salt so that the benzoic acid is less soluble in the aqueous solution.

General Method for Preparation of the Sample

Thoroughly mix the sample, disintegrating it in a blender or food chopper if it is solid or semisolid. Transfer 150 g of the well-mixed sample into a 500-ml volumetric flask and add sufficient pulverized sodium chloride to saturate the water in the sample. Make the solution alkaline to litmus paper with 10% NaOH solution. Shake the mixture thoroughly, allow to stand for 2 hr with frequent shaking, and filter. If the sample contains large amounts of fat, add a few milliliters of 10% NaOH solution to the filtrate and extract it with ether to remove the fat. If the sample contains alcohol, use the sample preparation method described for cider and similar products containing alcohol in the next section. If a large amount of matter is precipitated by sodium chloride, prepare the sample as described for salted or dried fish.

Special Methods for Preparation of the Sample

Ketchup. Weigh 150 g of sample and add 15 g of pulverized sodium chloride. Transfer the mixture to a 500-ml volumetric flask, using approximately 150 ml of saturated sodium chloride solution for rinsing. Make slightly alkaline with 10% NaOH solution using litmus paper, and then dilute to volume with saturated sodium chloride solution. Mix thoroughly and allow to stand for 2 hr with frequent shaking. Remove the bulk of the solids by squeezing through a heavy muslin bag or centrifuging, and then filter.

Jellies, Jams, Preserves, and Marmalades. Digest 150 g of sample in about 300 ml of saturated sodium chloride solution and add 15 g of pulverized sodium chloride. Make alkaline to litmus paper with milk of lime solution. Transfer the mixture to a 500-ml volumetric flask and dilute to volume with saturated sodium chloride solution. Mix thoroughly and allow to stand 2 hr with frequent shaking. Centrifuge if necessary and filter.

Cider and Similar Products Containing Alcohol. Make 250 ml of the sample alkaline to litmus paper with 10% NaOH solution and then evaporate on a steam bath to a volume of approximately 100 ml. Transfer to a 250-ml volumetric flask, using a saturated sodium chloride solution for rinsing. Add 30 g of pulverized sodium chloride and shake to dissolve. Dilute to volume with saturated sodium chloride solution. Mix thoroughly and filter off the precipitated proteins and other materials.

Procedure

Pipette a suitable volume of the filtrate (100–200 ml) into a separatory funnel. Neutralize to litmus paper with dilute hydrochloric acid solution and then add 5 ml in excess. (With the filtrate from salted fish, protein matter usually precipitates on acidifying, but this does not interfere with the extraction.) Carefully extract with successive portions of 70, 50, 40, and 30 ml of chloroform, avoiding emulsion formation as much as possible by shaking cautiously each time with a rotary motion. Usually, a good separation occurs between the aqueous and the lower chloroform layer on standing for a few minutes. Should an emulsion form, try to break it by (1) stirring the chloroform layer with a glass rod; (2) drawing off the emulsified portion into a second separatory funnel and giving it one or two sharp shakes; or (3) centrifuging for a few minutes. Since the extraction, as practiced, is progressive, draw off as much of the clear chloroform layer as possible each time, but under no condition draw off any of the emulsion. If the chloroform extract does not contain any emulsion, it need not be washed.

Transfer the combined chloroform extracts to a porcelain evaporating dish, rinsing the container several times with small quantities of chloroform. Evaporate to dryness at room temperature in a current of dry air. If desired, the extracts may be transformed to a 300-ml Erlenmeyer flask and the chloroform removed by slow distillation at a low temperature until the volume is approximately one-fourth the original. The residue is then transferred to the evaporating dish, carefully rinsed with small portions of chloroform, and evaporated to dryness at room temperature in a current of dry air.

Dry the residue overnight (or until no odor of acetic acid can be detected in the case of ketchup samples) in a desiccator charged with sulfuric acid. Dissolve the residue of benzoic acid in 30–50 ml of neutral alcohol, add about one-fourth the volume of water, 1–2 drops of phenolphthalein indicator solution, and titrate with standard 0.05 *N* sodium hydroxide solution. One milliter of exactly 0.05 *N* sodium hydroxide solution is equivalent to 0.0072 g of anhydrous sodium benzoate.

Determination of Total Sulfurous Acid in Dried Fruits

The determination of total sulfurous acid is based on the Schiff reaction between *p*-rosaniline, formaldehyde, and sulfur dioxide forming a purple color.

Reagents

1. 0.015% Formaldehyde—Dilute 10 ml of 40% formaldehyde to 1 liter with water; then dilute 75 ml of this solution to 2 liters.
2. Acid-bleached *p*-rosaniline hydrochloride—Place 100 mg *p*-rosaniline hydrochloride in a 1-liter volumetric flask, add 200 ml water and 160 ml of 1:1 HCl and dissolve; dilute to the mark with water. Let solution stand for 12 hr before use.
3. Sodium tetrachloromercurate—Dissolve 23.4 g NaCl and 54.3 g $HgCl_2$ in 1900 ml water in a 2-liter volumetric flask, then dilute to the mark.
4. Sulfur dioxide standard—Dissolve 170 mg sodium bisulfite ($NaHSO_3$) in water and dilute to 1 liter. Standardize with 0.01 *N* iodine solution (1 ml is equivalent to about 100 μg SO_2).

Standard Curve

Add 5 ml sodium tetrachloromercurate reagent to a series of 100-ml volumetric flasks. Add 0, 1.0, 2.0, 4.0, and 8.0 ml of standard sulfur dioxide solution, dilute to volume, and mix. Add 5-ml portions of each diluted standard to a 200-mm test tube containing 5 ml of the rosaniline reagent. Add 10 ml of formaldehyde solution to each tube, mix, and keep at 22°C for 30 min. Read absorbance at 550 nm against zero standard. Plot standard curve.

Assay Procedure

Weigh 25±0.02 g of dried fruit (ground) onto an 11-cm Whatman No. 1 filter paper. Transfer the paper and sample to a 500-ml blender container, add 475 ml of 0.1 *N* NaOH, cover, and blend for 2 min. Let stand for 5 min, then transfer 10 ml of the clear lower layer to a 500-ml volumetric flask containing 15 ml of 0.1 *N* HCl. Mix (gently) and add 25 ml of the sodium tetrachloromercurate reagent and dilute to the mark. Run a blank on all reagents.

Transfer 5 ml of sample to a 200-mm test tube containing 5 ml of the rosaniline reagent, add 10 ml of formaldehyde solution, mix, and hold for 30 min at 22°C. Read absorbance at 550 nm against blank. Refer to standard curve and report results as ppm of SO_2. (If the same spectrophotometry tube is used, clean with 1:1 HCl and water.)

Scanning Electron Microscopy of Insect Fragments

Insect fragments recovered from food products are usually pieces of integument such as wing, leg, head capsule, antennae, and mandible. The insect integument is usually externally ornamented with hairs, scales, microtrichia, spines, ridges, and pores. Many fragments (e.g., mandibles), can be identified by their general structure. Other fragments may be identified by the uniqueness of their surface sculpturing.

Many living insects (beetles, fleas, ticks, caterpillars) can be viewed directly in the scanning electron microscope (SEM) without preparation. In most cases, if exposure to the vacuum during SEM examination does not exceed 20 min, the insect will revive after the examination. Apparently, the hemolymph of the living insect provides a conductive path for the electrons. Affixing a living insect to the stub with silver paint improves conductivity and reduces charging. Most insects become motionless (similar to chill coma) in the vacuum; mosquito larvae and ticks continue to move.

Insects that can survive SEM examination are apparently able to tightly close their spiracles, retaining sufficient air in their tracheae to endure the vacuum. Other insects, such as leafhoppers, rapidly die in the vacuum, presumably because air is not retained by a spiracular-closing mechanism. Such insects initially provide good images but they rapidly begin to charge as they dry out, becoming so bright that acceptable photographs cannot be obtained.

Extensive examination of living insects should be reserved for a time just before routine SEM column cleaning. The interaction of the electron beam with oils and waxes on the surface of living insects causes contamination problems. The condenser and objective apertures, in particular, become blackened with carbon deposits and "varnish."

Fixation

Living specimens are fixed in 2.5% glutaraldehyde solution (in 0.1 *M* phosphate buffer pH 7.2) for 1–2 hr at room temperature, and then washed twice in fresh buffer for 15 min.

Postfixation

Specimens are postfixed in a 2% osmium tetroxide solution for 2 hr in a hood. The fixation vial should be placed in an ice bath so that fixation occurs at 4°C. Osmic acid is toxic and should be handled with care. The osmium tetroxide solution is prepared by dissolving 0.5 g of OsO_4 in 25 ml of buffer.

Freeze-Drying

After the postfixation step, specimens should be thoroughly washed in three changes of fresh buffer for 15 min each. At this point specimens can be freeze-dried or dehydrated in ethanol for critical-point processing. Specimens are freeze-dried by rinsing them thoroughly in distilled water and placing in small drops of water in an aluminum weighing pan, which is then floated on liquid nitrogen on dry ice until the drops freeze. The frozen specimens are then placed in a lyophilizer until all water has been removed from the specimen. This process prevents severe shrinkage of the tissues. Specimens are then mounted on metal stub for examination in SEM.

Critical-Point Processing

Specimens are dehydrated by passing them sequentially through several concentrations (30, 50, 70, 95, 100, and 100%) of ethanol or acetone and keeping them for 15 min in each concentration. Specimens are then placed in a capsule with a 250-mesh screen. The capsules are placed in the critical-point apparatus. Liquid CO_2 (from a cylinder with an inverted syphon) is slowly admitted to the specimen chamber until a pressure of 850 psi is reached: then the CO_2 is allowed to diffuse into the specimen for 5 min. The first charge of CO_2 is then flushed from the chamber by slowly opening the exit valve. Gas and ethanol (or acetone) exit initially, but eventually flakes of dry ice sputter from the valve indicating that only liquid CO_2 is present. The second charge of liquid CO_2 is allowed to diffuse into the specimen for an additional 5 min. This procedure is repeated for a third time. Both valves are tightly closed and the specimen chamber of the apparatus is warmed slowly to 37°C. The specimen chamber is opened slowly to allow CO_2 to escape. The dried specimens are removed and mounted in preparation for SEM viewing.

Specimen Mounting

Silver-Paint Technique. The osmified, clean, dry specimen is mounted on a metal stub for viewing in the SEM. The conventional method is to apply a small drop of silver paint to the stub and to set the insect fragment in it.

Tape Technique. A small piece of double-stick tape is placed on the metal stub and the specimen affixed to the upper tape surface. Silver paint is applied to the edges of the tape until it covers a small rim of the upper surface. The surface of the tape becomes conductive after a film of gold is deposited on the stub.

Epoxy Technique. Various types of epoxy are used to mount specimens to the stubs. The epoxy is mixed with a hardener (usually 1:1) in a disposable weighing dish. Epoxy is transferred to the stub and the specimen placed in the epoxy.

Coating of Mounted Specimens

Metal Coating. Mounted specimens are coated with a gold or gold–palladium film in either a high-vacuum system (vacuum evaporator method) or a low-vacuum system (sputter-coater method). In the high-vacuum system, a 10-cm piece of folded gold or gold–palladium wire is placed in a tungsten spiral basket, which is heated slowly in a vacuum (10^{-4} torr) until the metal melts, persists for several minutes as a droplet, and then evaporates. The specimen is placed on a rotating tilting stage during the coating process. This method results in application of an even, thin coat (about 90 Å thick, depending on the distance from the surface) over the specimen. In the low-vacuum system, a gold or gold–palladium plate is ionized under a higher pressure (10^{-1} torr) in the presence of argon. This method also produces a thin, even coating, but in less time (about 3 min) and does not require tilting and rotation of the specimen.

Carbon Coating. An initial coating with carbon at 10^{-6} torr often aids in making refractory, complex structures conductive. Specifically shaped carbon electrodes are used to form the carbon layer. Care must be taken to prevent electrode sparks; these produce erratic, thick carbon deposits.

A preliminary examination in the SEM should be made after the first metal coating. Osmified specimens coated with a single 90 Å layer of gold–palladium frequently yield excellent results. Specimens that are unacceptable with a first coat of gold can be given a second coat, or a coat of carbon followed by a coat of gold. Exposure to OsO_4 vapors for several hours in a sealed chamber, followed by a coat of gold sometimes improves a refractory specimen.

After a specimen has received satisfactory preparation as viewed in the SEM, it should be stored in a cool, dry atmosphere (e.g., in a glass desiccator). Frequent changes in temperature and humidity can result in breakage of the metal or carbon film, thus causing the specimen to charge.

Viewing the Specimen

The specimen, which is mounted on a stub, is placed on the goniometric stage of the SEM. The goniometric stage is constructed in such a way that, by means of external controls, the specimen can be rotated and moved left, right, forward, backward, and, in some instruments, vertically. In addition, the specimen may be tilted from a −10 to + 70°. The controls are usually provided with position indicators that allow precise positioning and orientation of the specimen.

One of the first considerations in viewing the specimen is selection of the level of accelerating voltage to be applied to the filament. The greatest flux of secondary electrons occurs at lower voltages, in the vicinity of 4 kV. This peak flux produces the brightest image; however, resolution at 4 kV is around 200 Å or less in most instruments. Therefore, high-quality images with sharp focus can be obtained only for magnifications less than 3000×. This magnification range and level of resolution are adequate for most work with insects. The range of 50–2500× is most frequently used for anatomical examination of insect fragments for purposes of identification. For magnifications greater than 3000×, higher voltages must be used.

The majority of SEMs have a minimum of two cathode ray tube (CRT) display screens. One, for routine examination, has approximately 700 scan lines per inch. The second provides the photographic image and, in recent instruments, is referred to as a high-resolution CRT; this one has approximately 2000 lines per inch. The degree to which SEM negatives can be enlarged depends on the number of scan lines composing the CRT image and on the resolution of the CRT electron beam. Recent high-resolution CRTs allow enlargement up to 10× before scan lines are separated and the image quality is lost. Most recent SEMs electronically display a micron marker on the photo CRT as well as the magnification.

Gas Chromatographic Analysis of Organochlorine Pesticides

The procedure described in this section involves the extraction, purification, and gas chromatographic analysis of organochlorine pesticides, some of their degradation products, and some related compounds.

Such compounds are composed of carbon, hydrogen, and chlorine, but may also contain oxygen, sulfur, phosphorus, nitrogen, and other halogens.

The following compounds may be determined individually by this method with a sensitivity of 1 μg/liter: BHC, lindane, heptachlor, aldrin, heptachlor epoxide, dieldrin, endrin, Captan, DDE, DDD, DDT, methoxychlor, endosulfan, dichloran, mirex, pentachloronitrobenzene, and trifluralin. Under favorable circumstances, Strobane, toxaphene, chlordane (tech.), and others may also be determined. The usefulness of this method for other specific pesticides must be demonstrated by the analyst before any attempt is made to apply it to sample analysis.

When organochlorine pesticides exist as complex mixtures, the individual compounds may be difficult to distinguish in gas chromatograms. High, low, or otherwise unreliable results may be obtained because of misidentification and/or when one compound obscures another present in lesser concentration. Several steps in this method are intended to minimize the effect of such interferences.

Several analytical alternatives are described, which the analyst may select depending on the nature and extent of interferences and/or the complexity of the pesticide mixtures found. Specifically, the use of an effective co-solvent for efficient sample extraction and the elimination of nonpesticide interferences and the pre-separation of pesticide mixtures, by column chromatography and liquid–liquid partition, are included. Identification is achieved by selective gas chromatographic separations and may be corroborated through the use of two or more unlike columns. Separated components are detected and measured by electron capture, microcoulometric, or electrolytic conductivity detectors, Results are reported in micrograms per liter.

This method is recommended for use only by experienced pesticide analysts or under the close supervision of such qualified persons.

Interferences

Solvents, reagents, glassware, and other sample-processing hardware may yield discrete artifacts and/or elevated baselines causing misinterpretation of gas chromatograms. All of these materials must be demonstrated to be free from interferences under the conditions of the analysis. Specific selection of reagents and purification of solvents by distillation in all-glass systems may be required.

The interferences in industrial effluents are high and varied and often pose great difficulty in obtaining accurate and precise measurement of organochlorine pesticides. Sample clean-up procedures are generally required and may result in the loss of certain organochlorine pesticides. Therefore, great care should be exercised in the selection and use of methods for eliminating or minimizing interferences. It is not possible to describe procedures for overcoming all of the interferences that may be encountered in industrial effluents.

Industrial plasticizers and hydraulic fluids such as PCBs are a potential source of interference in pesticide analysis. The presence of PCBs is indicated by a large number of partially resolved or unresolved peaks, which may occur throughout the entire gas chromatogram. Special separation procedures are required if PCB interference is severe.

Phthalate esters, which are widely used as plasticizers, respond to the electron capture detector and are a source of intereference in the determination of organochlorine pesticides using this detector. Water leaches these materials from products containing these plastics, such as polyethylene bottles and tygon tubing. Samples containing phthalate esters respond to the electron capture detector but not to microcoulometric, electrolytic conductivity, or flame photometric detectors.

A number of organophosphorus pesticides, particularly those containing a nitro group (e.g., Parathion), also respond to the electron capture detector and may interfere with the determination of organochlorine pesticides. Such compounds can be identified by their response to the flame photometric detector.

Sample Preparation

Blend the sample if suspended matter is present and adjust pH to near neutral (pH 6.5–7.5) with 50% sulfuric acid or 10 *N* sodium hydroxide.

The sensitivity requirement of 1 μg/liter for microcoulometric or electrolytic conductivity detectors requires a sample of 100 ml or more when these detectors are used. If interferences pose no problem, the greater sensitivity of the electron capture detector should permit as little as 50 ml of sample to be used. Background information on the extent and nature of interferences will assist the analyst in choosing the required sample size and preferred detector.

Quantitatively transfer the proper aliquot of blended sample solution into a 2-liter separatory funnel and dilute to 2 liter.

Extraction

Add 60 ml of 15% methylene chloride in hexane (v:v) to the sample in the separatory funnel and shake vigorously for 2 min. Allow the mixed solvent to separate from the sample, then draw the water layer into a 1-liter Erlenmeyer flask. Pour the organic layer into a 100-ml beaker and then pass it through a column containing 3–4 in. of anhydrous sodium sulfate, and collect it in a 500-ml K-D flask equipped with a 10-ml ampul. Return the water phase to the separatory funnel and rinse the Erlenmeyer flask with a second 60 ml of methylene chloride solvent; add the solvent to the separatory funnel and complete the extraction procedure a second time. Perform a third extraction in the same manner.

Concentrate the extract in the K-D evaporator on a hot water bath, and then analyze it by gas chromatography unless cleanup is required to remove interfering substances.

Extract Cleanup

Use a glass chromatography column (22 mm inner diameter; 20–30 cm long) with a reservoir at the top. The outlet should have a coarse glass fril and a stopcock to regulate flow. Pack the column either with dry Florisil or with a slurry of Florisil in petroleum ether to about 4 in. of the column length settled. Add about $\frac{1}{2}$ in. of anhydrous Na_2SO_4 to the column top to take up traces of water that may be left over from the sample. Carefully add 40–50 ml of petroleum ether to the column; collect it in an Erlenmeyer flask and discard.

Transfer the concentrated sample extract, dissolved in about 5 ml of petroleum ether, to the column. Place a new 500-ml Erlenmeyer flask under the column to collect the eluate. Rinse the extract container twice with 5 ml of petroleum ether, transfer washings to the column, and then rinse the upper column walls and the solvent container. Let the extract and rinses of petroleum ether settle at a rate of 4 ml/min. When the solvent has settled down to the Na_2SO_4 layer, carefully add 200 ml of 6% peroxide-free ethyl ether in petroleum ether (30°–60°C) without disturbing the Na_2SO_4 layer. Elute these 200 ml into the flask at about 5 ml/min.

When the solvent layer has reached the Na_2SO_4, carefully add 200 ml of 15% ethyl ether, change receivers, and elute at 50 ml/min. Further eluates at 50% ethyl ether in petroleum ether and 100% ethyl ether can later be added depending upon the pesticides sought. Concentrate each eluate in a Kuderna–Danish concentrator to about 5 ml. If needed, continue with further cleanup on another Flourisil column or on a MgO–Celite column. Otherwise, the sample is ready for gas chromatographic (GC) analysis.

GC Calibration and Analysis

Gas chromatographic operating conditions are considered acceptable if the response to dicapthon is at least 50% of full scale when 0.06 ng or less is injected for electron capture detection and 100 ng or less is injected for microcoulometric or electrolytic conductivity detection. For all quantitative measurements, the detector must be operated within its linear response range and the detector noise level should be less than 2% of full scale.

The following column packings are recommended for pesticide analysis: 6-ft glass, packed with 1.5% OV-101 on High Performance Chromosorb W, 100/120 mesh (Fig. 14.2); 6-ft glass, packed with 1.5% OV-17 on High Performance Chromosorb G, 100/120 mesh; and 6-ft glass, packed with 2% OV-101 and 3% OV-210 on High Performance Chro-

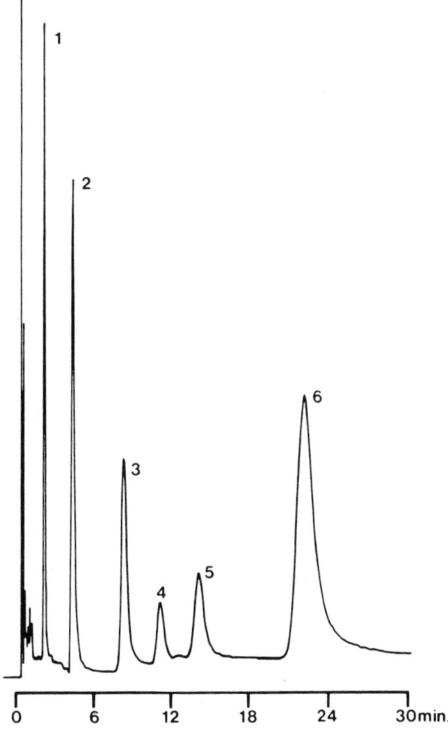

Figure 14.2. Column packing for pesticide analysis: 6-ft × $\frac{1}{4}$ in. glass column with 5% OV-101 on 100/120 mesh Chromosorb W; 185°C; N_2 at 30 ml/min. (1) Lindane. (2) Aldrin. (3) *p,p*′DDE. (4) *o,p*′DDT. (5) *p,p*′DDT. (6) Methoxychlor.

mosorb G, 100/120 mesh. Standards are injected frequently as a check on the stability of operating conditions.

Quality Control

Duplicate and spiked (known standard added to unknown sample) sample analyses are recommended as quality control checks. When the routine occurrence of a pesticide is being observed, the use of quality control charts is recommended.

Each time a set of samples is extracted, a method blank is determined on a volume of distilled water equivalent to that used to dilute the sample.

☐ SELECTED REFERENCES

AMES, B. N. 1972. A bacterial system for detecting mutagens and carcinogens. *In* Mutagenic Effects of Environmental Contaminants. H. E. Sutton and V. I. Harris (editors). Academic Press, New York.

AMES, B. N., DURSTON, W. E., YAMASAKI, E. and LEE, F. O. 1973. Carcinogens and mutagens: A simple test combining liver homogenates for activation and bacteria for detection. Proc. Natl. Acad. Sci. USA *70*(8), 2281–2285.

AMES, B. N., LEE, F. D. and DURSTON, W. E. 1973. An improved bacterial test system for the detection and classification of mutagens and carcinogens. Proc. Natl. Acad. Sci. USA *70*(3), 782–786.

ARENA, J. M. 1979. Poisoning: Toxicology, Symptoms. Thomas, Springfield, IL.

BAMBURG, J. R., STRONG, F. M., and SMALLEY, E. B. 1969. Toxins from moldy cereal. J. Agric. Food Chem. *17*, 443–449.

BLOOMER, A. W., NASH, S. I. PRICE, H. A. and WELCH, R. L. 1977. Pesticides in people. Pest. Monit. J. *11*, 111–113.

BOYD, ELDON M. 1973. Toxicity of Pure Foods. CRC Press, Cleveland.

CASCIANO, D. A. 1982. Mutagenesis assay methods. Food Technol. *36*(3), 48–52.

COMMONER, B. A., VITAYATHIL, A. J., DOLARA, P. NAIR, P., MADYASTHA, P. and CUCA, G. C. 1978. Formation of mutagens in beef extracts during cooking. Science *201*, 913–916.

DAS GUPTA, B. R., BERRY, L. J. and BOROFF, D. L. 1970. Purification of *Clostridium botulinum* Type A toxin. Biochem. Biophys. Acta *214*, 343–349.

FURIA, T. E. 1972. Handbook of Food Additives. 2nd ed. CRC Press Cleveland, OH.

GERWING, J., DOLEMAN, C. E., KASON, D. V. and TREMAINE, J. H. 1966. Purification and characterization of *Clostridium botulinum* Type B toxin. J. Bacteriol. *91*, 484–487.

GERWING, J. DOLMAN, C. E., REICHMANN, M. E. and VAINS, H. S. 1964. Purification and molecular weight determinations of *Clostridium botulinum* Type E toxin. J. Bacteriol. *88*, 216–219.

GORHAM, J. R. 1981. Principles of Food Analysis for Filth. Decomposition and Foreign Matter. FDA Technical Bulletin No. *1*.

GOSSELIN, R. W., SMITH, R. P. and HODGE, H. C. 1984. Clinical Toxicology of Commercial Products. Williams and Wilkins, Baltimore, MD.

HEID, J. L. nad JOSLYN, M. A. 1967. Fundamentals of Food Processing Operations. AVI Publishing Co., Westport, CT.

LAMANNA, C. 1959. The most poisonous poison. Science *130*, 763–772.

MATSUMURA, F., BOUSH, G. M. and MISATO, E. 1972. Environmental Toxicology of Pesticides. Academic Press, New York.

MATSUMURA, F. and KRISHNA, C. R. 1982. Biogradation of Pesticides. Plenum Press, New York.

OBRIEN, R. D. 1967. Insecticides, Action and Metabolism. Academic Press, New York.

ORGELL, W. H. 1963. Inhibition of human plasma cholinesterase *in vitro* by alkaloids, glycosides and other natural substances. Lloydia *26*, 36–40.

PARIZA, M. W. 1982. Mutagens in heated foods. Food Technol. *36*, 53–56.

PECKHAM, G. C. and FREELAND-GRAVES, J. H. 1979. Foundations of Food Preparation. 4th ed. Macmillan Publishing Co., New York.

POTTER, N. N. 1978. Food Science. 3rd ed. AVI Publishing Co. Westport, CT.

ROWE, W. D. 1983. Evaluation Methods for Environmental Standards. CRC Press, Boca Raton, FL.

SCHILLER, C. M. 1984. Intestinal Toxicology. Raven Press, New York.

SCHOENTAL, R. 1967. Aflatoxins. Ann. Rev. Pharmacol. *7*, 343–390.

SPERLING, F. 1984. Toxicology: Principles and Practice, Vol. 2. John Wiley & Sons, New York.

SUNSHINE, I. 1982. Methodology for Analytical Toxicology. CRC Press, Boca Raton, FL.

WARE, G. W. 1983. Pesticides, Theory and Applications. Freeman, San Francisco.

Index

C

D

J

K

Q

R

T